Studies in Systems, Decision and Control

Volume 462

Series Editor

Janusz Kacprzyk, Systems Research Institute, Polish Academy of Sciences, Warsaw, Poland

The series "Studies in Systems, Decision and Control" (SSDC) covers both new developments and advances, as well as the state of the art, in the various areas of broadly perceived systems, decision making and control–quickly, up to date and with a high quality. The intent is to cover the theory, applications, and perspectives on the state of the art and future developments relevant to systems, decision making, control, complex processes and related areas, as embedded in the fields of engineering, computer science, physics, economics, social and life sciences, as well as the paradigms and methodologies behind them. The series contains monographs, textbooks, lecture notes and edited volumes in systems, decision making and control spanning the areas of Cyber-Physical Systems, Autonomous Systems, Sensor Networks, Control Systems, Energy Systems, Automotive Systems, Biological Systems, Vehicular Networking and Connected Vehicles, Aerospace Systems, Automation, Manufacturing, Smart Grids, Nonlinear Systems, Power Systems, Robotics, Social Systems, Economic Systems and other. Of particular value to both the contributors and the readership are the short publication timeframe and the worldwide distribution and exposure which enable both a wide and rapid dissemination of research output.

Indexed by SCOPUS, DBLP, WTI Frankfurt eG, zbMATH, SCImago.

All books published in the series are submitted for consideration in Web of Science.

Natalia Kryvinska · Michal Greguš ·
Solomiia Fedushko

Editors

Developments in Information and Knowledge Management Systems for Business Applications

Volume 7

 Springer

Editors
Natalia Kryvinska 🆔
Department of Information Systems
Faculty of Management
Comenius University
Bratislava, Slovakia

Michal Greguš
Department of Information Systems
Comenius University
Bratislava, Slovakia

Solomiia Fedushko
Department of Social Communication
and Information Activities
Lviv Polytechnic National University
Lviv, Ukraine

ISSN 2198-4182   ISSN 2198-4190 (electronic)
Studies in Systems, Decision and Control
ISBN 978-3-031-25697-4   ISBN 978-3-031-25695-0 (eBook)
https://doi.org/10.1007/978-3-031-25695-0

This Springer imprint is published by the registered company Springer Nature Switzerland AG
The registered company address is: Gewerbestrasse 11, 6330 Cham, Switzerland

Preface

This book presents a combination of chapters assembled in different fields of expertise. This volume examines different aspects of business knowledge from both a philosophical and a practical standpoint. This data helps modern organizations by providing valuable insights and suggestions for future research and results. The increasing number of business disciplines studied necessitates the continued implementation of effective analytics practices within organizations. This volume explores disciplinary and multidisciplinary concepts and practical techniques to help analyze the evolving field.

We start out with the Chapter "Research on the Service Sector in the Modern Economy" analysis. This chapter examines the internal and external processes of the police as well as the perception of those processes by others. Wherever threats are averted or crimes are prosecuted, there is always the possibility of misconduct by individual police officers. If the police as an authority face public criticism, their reputation suffers and is at risk. It appears that the increase in assaults against operational police forces and the associated monetary expenses for their medical treatment represents an economic factor.

The next work titled Chapter "Management of Marketing Communication in State Administration" focuses on analyzing the starting points and opportunities for the public to participate in land consolidation processes. The current legislation defines the concept and meaning of land consolidation in Slovakia, examines the content of the concept of public interest under the land consolidation legislation, and analyzes the conditions for informing and subsequently involving the public in the protection of the environment. On this basis, it examines and explains one of the essential options by which the public influences environmental decision-making. This assesses the environmental impact of strategy papers and proposed actions. In dealing with the issue, we applied the primary analysis of the legislation given the nature of the topic under examination. However, we also use other scientific methods of investigation, scientific literature, case law, and analogy of law. Our research aims to provide qualified answers to the main pitfalls associated with land consolidation in the Slovak Republic. The added value of our scientific study is, especially in a broader context, to critically highlight selected application problems and suggest an improvement.

The chapter by Anton Lisnik and Milan Majerník Chapter "Aspects of Strategic Management and Online Marketing" attempts to answer a question about managing a company based on the company's most important documents, such as vision and mission. Communication with customers, whether in direct sales or sales and for further business, is mainly based on quality marketing. Marketing is thus part of strategic decisions in the field of business management. The present time uses virtual space as a primary means of communication for the presentation and sale of products. Especially as a result of the Covid period, virtual marketing gained a priority in marketing activities and other globalized trade and business. The presented article describes the place of online marketing in business management and mainly highlights its place in strategic decisions and management.

The subsequent Chapter "Intelligent Information System for Controlling International Innovation Activities of an Enterprise" authored by Mykola Odrekhivskyi, Oleh Kuzmin, Orysya Pshyk-Kovalska, and Volodymyr Zhezhukha, considers the features of designing and functioning of an intelligent information system for controlling international innovation activities of an enterprise, and also develops an appropriate mathematical apparatus in this area. It is proposed to consider an intelligent information controlling system as a logical and cognitive model of a social agent, and an integral system as a multi-agent system. A modular structure was chosen to build an intelligent information system for controlling the company's international innovation activities. Therefore, the corresponding modules of this system were identified and characterized. Since the dynamics of states of international innovation activity of an enterprise is stochastic, it is proposed to use the Kolmogorov system of differential equations to assess and predict such states and make optimal decisions on their management. The developed mathematical apparatus was tested in the study of the state of development of the Truskavets sociopolis, considering the volume of innovative services provided. During the calculations, 31 sanatoriums of sociopolis were studied. As a result of calculations, it is established that the proposed mathematical apparatus and the mathematical and program software developed based on the intelligent information system for controlling international innovation activity of the enterprise adequately describe the state of development of the sanatorium-resort complex of sociopolis Truskavets under study and can find wide application in the study of the state of international innovation activity of companies and their development in general. In addition, the structure of the management system for international innovation activities of the enterprise, as well as the structure of the controlling system for such activities, is given.

The research performed in the Chapter "Optimization Modeling of Urgent Investment Tools of Crisis Management at Enterprises" aims to implement optimization of economic and mathematical modeling of urgent investment tools for crisis management in enterprises. The research established prerequisites for building models of compulsory investment crisis management instruments and developed a model for optimizing the structure of these tools. The scientific work built a generalized model of optimization of the application process of urgent investment tools for crisis management at enterprises. The practical application of the developed models, in particular, showed that all the surveyed companies have the opportunity to overcome the financial crisis and avoid bankruptcy, using urgent investment tools for crisis management.

The Chapter "Assessment of the Impact of Biofuel Production on the Sustainable Development of Enterprises in the Agrarian Sector of Ukraine" defines the impact of biofuel production by agrarian enterprises on their sustainable development. The leading indicators of sustainable development of enterprises and the level of biofuel production are highlighted. The economic efficiency of production of different types of biofuel according to a sample of enterprises in the agrarian sector of the economy of Ukraine is estimated. The degree of the impact of biofuel production by Ukrainian agrarian enterprises on the indicators of their sustainable development has been established. For this purpose, both deterministic analysis and statistical methods were used. Applying these methods to a sample of Ukrainian agrarian enterprises showed that the impact of biofuel production on their sustainable development is quite significant. In particular, the dynamics of this production are positively correlated with the dynamics of operating income and profits of enterprises and with the change of most indicators that characterize the energy and environmental aspects of sustainable development of the surveyed companies. Among other things, there are statistically significant linear relationships between biofuel production growth rate and operating income growth rate, and product profitability. Accordingly, there is a quadratic relationship between the growth rate of biofuel production and changes in the operating profit of the surveyed agrarian enterprises. Thus, the growth of biofuel production is one of the essential drivers of the sustainable development of enterprises in the agrarian sector.

In the work titled Chapter "Simulation Models for Prioritizing the Implementation of Energy-Saving Investment Projects for the Enterprise", the structure of decision support tools for determining the priority of energy-saving projects has been developed. It is determined that for the effective operation of the decision support system in determining the priority of energy-saving projects, the system should include means of collecting and storing information about the energy efficiency of the enterprise; means of numerical evaluation of selection criteria; simulation model for determining the priority of energy-saving projects; means of forecasting energy efficiency of the enterprise taking into account the implementation of the energy-saving project; means of visualization of simulation results. The method of pairwise comparison is proposed, and the scale for estimating the weights of the importance of selection criteria is modified. A model for determining the priority of energy-saving projects has been developed, which is based on the method of hierarchy analysis and takes into account the interaction and interdependence of project selection criteria. A list of economic, technical, production, environmental, and organizational criteria for the selection of energy-saving investment projects has been formed. It is shown that the development of the model for determining the priority of investment energy-saving projects is performed in five stages: at the first stage—the task of determining the priority of investment energy-saving projects in the form of the three-level tree of hierarchies; at the second stage—expert numerical evaluation of economic, technical, production, environmental, and organizational selection criteria is carried out; at the third stage—a matrix of pairwise comparisons is formed and the eigenvector and the vector of priorities for the second level of the hierarchy are calculated; at the fourth stage—we form a matrix of pairwise comparisons and calculate the vector of priorities for each energy-saving project; in the fifth stage—the vector of global priorities is calculated. A block diagram of the operation algorithm has been developed, and a simulation model for determining the priority of investment energy-saving projects has been implemented using MATLAB. Approbation of the simulation model on the basis of the municipal enterprise "Sambir Vodokanal" was carried out.

The next chapter presents a study on Chapter "Designing an Information System to Create a Product in Terms of Adaptation". An adaptive approach to the design of control systems for functionally complex technical objects and technological processes in the conditions of uncontrolled changes in their properties and the properties of the environment has been developed. The scheme of development of the order for the release of new products is constructed. The structural connections that arise from the interaction between the client and the company that wants to master the release of a new product are analyzed. The scheme of business processes of the firm with an adaptive management system is constructed. The links on the approval and development of new products have been analyzed and developed. An adaptive scheme has been developed that can be applied to the firm's design.

The work titled Chapter "A Real-Time Room Booking Management Application" presents a model of a room booking management application. Booking and managing meeting rooms shouldn't be a hassle—it should just work. Our application allows logged-in users to search available rooms according to parameters, make reservations, and manage them. The practical part of the work was made in C# as a Windows Forms application. As an IDE, we used Microsoft Visual Studio Ultimate 2012. To create appropriate diagrams, we used dedicated tools. The application "Facility Management System" is designed and implemented. The application is the solution to make a workplace as smart and efficient as possible. It allows real-time availability and booking. Using the system helps users find the right solution, optimize how their office is used, save money, and make the workplace more flexible and accessible.

The next study on Chapter "Designing and Implementation of Online Judgment System" analyzes the creation process for an online judge solution. Design of architecture is inspired by methods used widely in professional applications and by experiences with existing systems of this kind. We present our vision of all system elements—web application, server application, and executors. To create a web application, we use Material Design technology. Scalability is given by the use of Google Kubernetes and Docker technologies. There is a wide range of probable security threats, and after extensive testing, eBPF solutions were chosen. In the first version of the system, C++, Java, and Python are supported. The system also offers incredible extensibility through plugins. In this way, even the support for new programming languages can be added.

The subsequent Chapter "E-commerce Drivers During the Pandemic and Global Digitalization: A Review Study" investigates the current problems of e-commerce drivers during the pandemic and global digitalization. Digital transformation has become a tool for business diversification and competitiveness, but it is the reason for the whole business paradigm to shift toward digital reality. The several fundamental groups for contributing factor types were systematically structured based on countries and business sectors. The structured research employs the pandemic impact issue, technological innovations' impact on the post-pandemic e-commerce economy, and the relation between infrastructural changes, social networks, and companies' competitiveness. In this research, the authors provided the literature review based on analysis and developed indicative categorization according to the initial research objective stated by the authors. In all databases the articles were taken from, we worked on publications from Business, Computer Science, and Economics. In selecting relevant articles, we used files with criteria to provide appropriate parameters for further analysis. We have paid significant attention to identifying new trends in the literature that are discussed during pandemics. The publications for review were identified from Business, Computer Science, and Economic databases. As the preferred reporting method, we chose meta-analyses for the systematic literature review. We start by explaining how the relevant journal papers were defined and then explore the limits of our review strategy, present combined infographics, and focus our attention on the 73 most relevant papers. An indicative categorization of the articles was developed due to the initial research objective stated by the authors. The analysis of the reports is then presented, highlighting current significant trends in the literature and research identified categories.

In the next work Chapter "Remanufacturing and Refurbishment of Electronic Devices—Their Future from a Business Perspective", Ann-Sophie Schweiger and Christine Strauss indicate future from a business perspective of the remanufacturing and refurbishment of electronic devices. Remanufacturing and refurbishing electronic devices can be profitable business strategies for companies and a way to counteract electronic scrap's rapidly growing waste stream. The industrial remanufacturing process provides a used device with a second life and a quality similar to a new device. However, companies and organizations that offer such devices still encounter difficulties. While earlier research has already partly covered the field of remanufacturing, it still lacks real-world examples and practical solutions to existing challenges. The identification and discussion of solutions for possible obstacles are thus a focus of this work. Furthermore, emerging trends in this industry, digitalization, and online trade with remanufactured devices are discussed. Attention is also paid to future developments in the field of remanufacturing, specifically to topics that could dominate this industry. The method applied to collect indicators for these various focal points was a Delphi study conducted with experts in the field of remanufacturing. This study included a total of three survey rounds. It is noteworthy that the experts emphasized the importance of networking and various forms of cooperation in the reuse sector. Furthermore, the significance of remanufacturers' transparency, especially concerning the remanufacturing process and the condition and assessment of the offered devices, was emphasized.

The Chapter "Nuclear Waste Potential and Circular Economy: Case of Selected European Country" focuses on the analysis of baselines for the use of spent fuel and radioactive waste in the circular economy within the broader framework of state energy security. Based on both European and national strategies and existing legislation, it defines the essence of the circular model of the economy and, within it, the concept and meaning of waste management. It analyzes the approach and regulation for managing a specific type of waste, nuclear waste, compared to other wastes. On this basis, it examines "de lege lata" possibilities for further use of spent nuclear fuel and radioactive waste in the increasing demands of the company for energy sources. It assesses the direction and impact of strategic documents at the international and national levels concerning the analyzed topic. When processing the study, the method of primary analysis of legislation and strategic documents was applied. As a result of our research, we provide answers to questions about the possibility of using nuclear waste and proposals de lege ferenda. The added value of our scientific study is especially in a broader context to critically highlight selected legislative problems and propose ways to address them.

The work called Chapter "Knowledge Management and Local Government" aims to study knowledge management in the context of ongoing and necessary changes in public administration institutions, which are constantly undergoing processes associated with modernization, the use of new technologies, and comprehensive streamlining of the provision of relevant public services for citizens. In this chapter, we focus on the level of quality in knowledge management at the local government level. The article aims to identify the level of preparedness of local governments for changes resulting from digital transformation, both in the context of the relevant strategy, its implementation in connection with the action plan or vision, but especially the real practice of local government in the use and management of knowledge for strategic management of the municipality. We analyze the situation in selected municipalities in Slovakia, and the research problem is the readiness of local government for challenges arising from the need for continuous improvement of public services, especially in terms of the ability of responsible municipal staff to use knowledge management to meet this goal. At the same time, whether employees increase their qualifications and knowledge through vocational training or education.

The work authored by Ondrej Čupka, Ester Federlova, and Peter Vesely Chapter "Comparison of Methodologies Used in Cybersecurity Reports" specifies that new legislation in the field of cybersecurity makes public and private institutions legally responsible for keeping the nation's IT infrastructure and data safe under the threat of legal consequences. This sudden shift saw many companies and public institutions scrambling to meet the new legal demands but often found a lack of professionals, institutions, systems, and information. The laws do oblige most organizations to report incidents and breaches. How to turn this data into information is still up for discussion. This has led us to our research, which consists of defining, comparing, and assessing the methodologies of four major cybersecurity reports, characterizing their interactions with cybersecurity legislation in the Slovak Republic to create a base for developing a methodology for Slovakia. The reports we used are Verizon's Data Breach Investigation Report, IBM's X-Force Threat Intelligence Report, Ponemon Institute's Separating Truth from Myth in Cybersecurity, and E-Governance Academy's National Cybersecurity Index. Despite the difference in their approaches, we were able to discover unifying best practices and issues that should not be repeated in developing a new methodology. Three key best practices are sound statistical methods, timeliness of data, and transparency. Three key issues were lack of transparency in methodology, limitations, bias and reasoning, inconsistency, and loss of relevance due to late publication. Additionally, we have identified that reports of this nature are in line with cybersecurity legislature and national strategies. Such reports may be used as a part of preventative measures by creating broad awareness, increasing interest in the topic, warning organizations of any dangers, and finding ways of creating and improving cybersecurity strategies to build capacities and resilience.

The next chapter presents a study on Chapter "Home Office—Benefit for Employer or Employee?". The main goal is to determine if home offices help businesses or workers. Other research topics include the following: What is a home office? What do companies gain from a home office? What are the advantages of working from home? Work/life balance is typically promoted via flexible work arrangements. The outcomes vary by worksite. Virtual office employees said telework had no impact on work/life balance. Virtual workers reported poorer work/life balance and personal/family success than conventional or home office workers. A multivariate analysis found that virtual workers damage work/life balance. Working from home affects work/life balance. One aspect may be the absence of external physical constraints. Virtual office workers may struggle to distinguish work from home. The benefits of working from home are many. A happy workforce may contribute to increased profitability and productivity. Letting workers work from home has numerous advantages, but this essay argues it helps employers and corporations more than employees. It is easy to communicate that Business communication includes email, meetings, phone calls, video conferencing, and casual conversations. Employers may use several communication alternatives to keep remote staff connected, resulting in a more efficient and collaborative work environment.

The research performed in Chapter "Knowledge Management of Private Banks as an Asset Improved by Artificial Intelligence Discipline—Applied to Strategic McKinsey Portfolio Concept as Part of the Portfolio Management" demonstrate a possibility to support the evolution of the knowledge management discipline for the Financial Services industry with the Artificial Intelligence (AI) technology, e.g. Machine Learning (ML). As one of the banking core capabilities, the part of strategic Portfolio Management will be focused as a use case along the results out of the step mentioned before. Knowledge Management (KM) is an essential core part of business development and prospering of each individual, group, and organization—this discipline is an asset for each organization to do the right things on a subject matter expert level. It is a major influencer in the setup of many IT systems with its functions. In addition, an existing well-organized Knowledge Management can help to flatten the shortfalls of professionals while the knowledge will be saved, improved, and up-skilled. KM in a private bank can divide into an external and an internal view—to have a look into the external direction, mainly the customer is in the focus in terms of its requests—the market knowledge. Opposite, the bank internal knowledge is more or less aligned with the processes and workflow chains to satisfy the mentioned customer demands and the bank's strategy. Moreover, as one approach, the knowledge can be separated into explicit and tacit knowledge categories based on the SECI model. Yet the exploit knowledge visible in documents, data records of knowledge databases, workflow descriptions, and content explanation is easier to retrieve, maintain, and use than the tacit knowledge with its thumb rules, heuristics, and intuition. The listed items are examples. Many expressions of these categories exist, depending on the industry sector. To improve/scale up to a higher level of the knowledge management of the private banks, the Artificial Intelligence technology can support to generate new knowledge entities, models, scoring points, measurements, and probabilities under the usage of various categories of the algorithm. One of the essential strategic parts of a bank there is to create its own portfolio concept to ensure the definition of the main indicators, the strategic business units (e.g. Fond business), the appropriate scorings, probabilities drill-downed and related to sub-indicators as well as deviations on the whole business units. The next part of the scope of this article is focused on using the Knowledge Management on the McKinsey "GE-Matrix" or "Nine-box-matrix" adapted for the generation of a Strategic Portfolio Management within Private Banking Financial Services. In addition, based on the previously mentioned aspects, this article reflects a short introduction to the possibility of a seamless AI-guided workflow circle of strategic portfolio creation based on explicit and tacit knowledge. It gives a preview of risk assessment, alternative portfolio model simulations, and finally the trade order management related to the regular portfolio revision (re-balancing and upgrading) based on the underlying technology described in this article.

The Chapter "Duties of National Civil Courts and European Commission in Their Cooperation" analyzes the degree to which national civil courts and the European Commission must collaborate under primary and secondary legislation. First, it is determined to what extent national civil courts have the authority to request statements and information from the European Commission, whether and to what extent the European Commission is required to respond, and to what degree the national courts are required to consider the response. The role of the European Commission is then considered insofar as it can appear before national courts as a so-called amicus curia with self-induced statements; specifically, the right of the European Commission to make such an appearance and the duty of national courts to take such statements into account is examined. Finally, the extent to which national courts are required to submit judgments to the European Commission is explored. The duties (and rights) outlined here are examined in more detail below and placed in the broader context of primary law.

The work authored by Nadja Pade and Rozália Sulíková Chapter "Significance of an Ethical Culture for Young Employees" considers ethical culture for young employees. Fast-changing businesses and developments such as the shortage of skilled workers, the Corona pandemic, and the consequences of political instability are putting the needs of employees once again in the spotlight. And here also— and especially—the needs of the youngest generation, which is already conquering the labor market. The Organizational Culture makes a major contribution to motivating and retaining employees, and the consideration of an Ethical Organizational Culture is becoming more and more interesting as the return to values and meaning is gaining relevance in these turbulent times. The importance of an Ethical Organizational Culture for young employees is the focus of attention as this cohort seems to have special requirements for their employers.

The Chapter "European Digital Strategy and Its Impact on the Conclusion of Selected Types of Business Contracts" analyzes digitalization processes. At the moment of the information age, the knowledge and digital economy have become part of the pan-European economy due to the fourth industrial revolution. Despite this fact, many unanswered questions about the correct use of digital technologies, particularly when concluding contracts, have long arisen in business practice. In our view, this is a serious problem which, due to the lack of interest of legal theorists, is an unexplored area. For this reason, answers to the ambiguities and problems that have arisen are relatively difficult to find. The aim of this study is, in particular, to examine the current legislative options and the associated problems of the electronic conclusion of selected types of contracts. The setting of this objective is based directly on current needs and emerging practical problems in business practice. In processing the issue, we applied the primary analysis of legislation given the nature of the topic examined. However, we also use scientific literature, case law, and analogy of law. This scientific study provides qualified answers to serious problems of business practice. The added value of our research is in a broader context to critically examine selected application problems and propose appropriate ways of improvement.

The work Chapter "AI in Customer Relationship Management" investigates artificial intelligence in customer relationship management. With the impact of globalization on market business conditions, companies must adapt their business more quickly to new situations. New strategic directions, especially Artificial Intelligence (AI), have become imperative for the successful business of today's companies. An intelligent company makes better decisions faster and outperforms its competitors. Consumer Relationship Management (CRM) is a business strategy for choosing customer management to optimize long-term value. CRM also requires a business philosophy focusing on consumers and a business culture that supports effective marketing, sales, and service processes. CRM integrates people, processes, and technology to maximize relationships with all consumers. Through the collaborative CRM, all communication toward clients is carried out, while their responses to the information system arrive via the operating part of the CRM. Analytical CRM, through a detailed analysis of a multitude of data based on expert knowledge, creates an image of each client, his needs, and desires to develop more vital interconnections.

The Chapter "Competitiveness of International IT Companies—Comparison of Strategies, Their Strengths, and Weaknesses" deals with the competitiveness of international companies. The article aims to guide managers and companies to properly determine the strategy for implementing cloud solutions by identifying the position of international IT companies in a competitive environment. The research subject is four selected IT companies, providers of cloud solutions. The relationships between individual providers of cloud solutions are examined by analysis. The starting point for the elaboration of the article is annual reports, professional articles, and publications of domestic and foreign authors. In the first chapter, we deal with theoretical knowledge in the field of competitiveness, strategies, and analysis of the external and internal environment of the company. The second chapter deals with the main and partial goals. In the third chapter, we deal with research methods and work procedures. We characterize selected IT companies. The last chapter summarizes the achieved results of the work and discussion.

The chapter authored by Ondrej Čupka, Ester Federlova, and Peter Vesely Chapter "Definition of Terminology in Security Sciences" investigates terminology in security sciences. Cybersecurity has only recently become a major topic in the Slovak Republic, mainly because of the introduction of new legislation, such as the General Data Protection Regulation and the Cybersecurity Act, as well as the impact of a pandemic that moved many human activities into the digital world. This legislation makes public and private institutions legally responsible for keeping the nation's IT infrastructure and data safe under the threat of legal consequences. This sudden shift saw many companies and public institutions scrambling to meet the new legal demands but often found a lack of professionals, institutions, systems, and information. Cybersecurity is a young science with its terminology still in debate. This paper will define critical terms necessary for discussions on this topic.

Succeeding Chapter "Code Smells: A Comprehensive Online Catalog and Taxonomy" provides a widely accessible Catalog that can perform useful functions both for researchers as a unified data system, allowing immediate information extraction, and for programmers as a knowledge base. Identifying all possible concepts is characterized as Code Smells and possible controversies. To characterize the Code Smells by assigning them appropriate characteristics. The authors performed a combined search of formally published literature and gray material strictly on Code Smell and related concepts where it might never have been mentioned, along with the term "Code Smell" as a keyword. The results were analyzed and interpreted using the knowledge gathered, classified, and verified for internal consistency. Results: We identified 56 Code Smells, of which 15 are original propositions, along with an online catalog. Each smell was classified according to taxonomy, synonyms, type of problem it causes, relations, etc. In addition, we have found and listed 22 different types of Bad Smells called hierarchies and drew attention to the vague distinction between the Bad Smell concepts and Antipatterns. This work has the potential to raise awareness of how widespread and valuable the concept of Code Smells within the industry is and fill the gaps in the existing scientific literature. It will allow further research to be carried out consciously because access to the accumulated information resource is no longer hidden or difficult. Unified data will allow for better reproducibility of the research, and the subsequent results may be more definitive.

In the work Chapter "Modeling of Information System to Determining the Degree of Coincidence in Text for Higher Education Institutions", the authors deal with the to model the activity of the information system to determine the degree of coincidence in the texts, which are the results of scientific achievements of consumers of educational services in higher education institutions. In this paper, modeling the information system to determine the degree of coincidence in the text is based on the organization of the process of detecting plagiarism. However, it should be noted that the scope of this system, in addition to the direct consumers of educational services, which targets this system, also includes other educational institutions and organizations that need such verification. The article models the processes of the information system for determining the degree of coincidence of texts based on CASE technologies. In particular, the class diagram shows a static representation of the structure of the information system to determine the degree of coincidence in the texts. The interaction between the actors and the entities of the information system is depicted using a diagram of use cases. Descriptions of the behavior of the information system at the level of individual objects that exchange messages are shown in the diagram of cooperation. The components of the information system and the relationships between them are presented in the component diagram, and computing nodes during the program operation, components, and objects running on these nodes in the proposed system are shown in the deployment diagram. Modeling information system processes to determine the degree of coincidence in the texts is the basis for developing the information system's architecture. Evaluation of software development, which is based on the indicators of the COCOMO evaluation system, is also described in this article. An important place in this study is the identification of risks in developing an appropriate information system based on the modeling. Using the proposed information system to determine the degree of coincidence of texts allows optimizing the process of plagiarism of scientific results of consumers of educational services to improve the quality of analysis of materials used and presentations of their scientific results.

The Chapter "Use of a Communication Robot—Chatbot in Order to Reduce the Administrative Burden and Support the Digitization of Services in the University Environment" focuses on a literature review of chatbot systems, methodology of the work, and the stated objectives and describes the current state of the ongoing processes at the Faculty of Management at Comenius University. On the selected process, a time and financial analysis were performed. The technology of chatbots has come a long way in the last decade, thanks to the growing popularity of artificial intelligence and machine learning. Chatbots are established as a beneficial tool in many scenarios of our daily lives. This paper aims to discuss the design, architecture, and potential application of a chatbot in a university environment. The output of this chapter consists of the identified critical path in the processes as an area for possible implementation of chatbots in the academic department to reduce the administrative burden. The result is the architecture and development of the proposed solution. The paper concludes with opportunities for improvement of the proposed solution and recommendations that we recommend to be applied in practice.

Bratislava, Slovakia   Natalia Kryvinska
Bratislava, Slovakia   Michal Greguš
Lviv, Ukraine   Solomiia Fedushko

Contents

Research on the Service Sector in the Modern Economy

Miloš Šajbidor and Marian Mikolasik ⓘ

Abstract The major objectives of this research paper are to give more comprehensive information about the service sector's role in the modern economy, to illustrate it with theoretical data and actual figures, and to explore how the service sector has changed and what changes it brought to the economy. The beginning of the current paper aims to explain the term "Service" given by various researchers, to identify the key unique characteristics of them that distinguish services from goods, and present common short classifications. Then the evolution of the service-based sector is described, namely the leadership of service industries, employment share in the service sector, and the increasing role of services worldwide. Finally, it provides information about the collaboration and interaction of E-business and Services, and how E-business and ICT have changed the service sector.

1 Service Definitions

Nowadays, services accompany our daily life. Services are substituting manufacturing. Examples of many developed countries show that a service-based economy prevails in a new millennium. Recognizing the importance of services and service jobs, fewer and fewer people consider that service sector jobs are low-skill and low-wage [1]. To give a deeper insight into the topic and provide a better understanding, it is necessary to give a definition of this term by various researchers.

Certainly, there are many definitions of services. One of the first was the definition of services as "activities, benefits, or satisfactions which are offered for sale, or provided in connection with the sale of goods" by the American Marketing Association (1960). This definition took a very limited view of services as it proposed that services are offered only in connection with the sale of goods.

M. Šajbidor · M. Mikolasik (✉)
Comenius University, Odbojarov 10, 831 04 Bratislava, Slovakia
e-mail: mikolasik3@uniba.sk

M. Šajbidor
e-mail: sajbidor1@uniba.sk

© The Author(s), under exclusive license to Springer Nature Switzerland AG 2023   1
N. Kryvinska et al. (eds.), *Developments in Information and Knowledge Management Systems for Business Applications*, Studies in Systems, Decision and Control 462,
https://doi.org/10.1007/978-3-031-25695-0_1

Many scholars define services as for example, Zeithaml and Bitner [2]: "deeds, processes, and performances".

Another definition given by Blois (1974) says that "a service is an activity offered for sale which yields benefits and satisfactions without leading to a physical change in the form of a good" [3].

Services are economic activities offered by one party to another, most commonly employing time-based performances to bring about desired results in recipients themselves or in objects or other assets for which purchasers have responsibility. In exchange for their money, time, and effort, service customers expect to obtain value from access to goods, labor, professional skills, facilities, networks, and systems; but they do not normally take ownership of any of the physical elements involved [4].

Or for example, more complex: According to Grönroos [5] "a service is an activity or series of activities of more or less intangible nature that normally, not necessarily, take place in interactions between the customer and service employees and/or physical resources or goods and/or systems of the service provider, which are provided as a solution to customer problems" [5].

The above-given definitions demonstrate the development of the notion of "Service" and the strengthening of its role in our life.

Service leadership culture is a strategy designed to carefully select and design service processes with the active participation of customers and employees in the design processes and strategically build on collective leadership efforts at every level of organisation to fully benefit from an organisations investment in its human resource [6].

The concept of service leadership is based on a multidisciplinary approach that draws on leadership theories and principal methods of strategic management, as shown in Fig. 1. It is defined as the culture that empowers the organization to strategize its promises, design its processes and engage its people in a proactive quest for a competitive advantage [6].

Fig. 1 Multidisciplinary approach to service leadership [6]

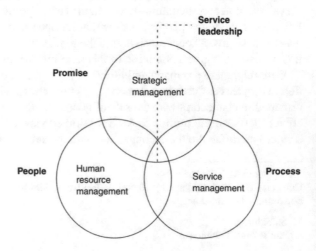

2 Characteristics of Services

To further understand the nature of services, the key unique characteristics of them that make them different from of goods should be identified. The most common characteristics of services consist of four features:

- Intangible
- Heterogeneous
- Inseparable
- Perishable

These features of what makes a service are referred to as 'IHIP' characteristics.

Intangibility—One of the most common features of services is intangibility. As services are performances or actions, they cannot be seen, touched or felt in the same way as tangible goods. Not all service product has similar intangibility. Some services are highly intangible, while others are low i.e. the goods (or the tangible component) in the service product may vary from low to high.

Inseparability—In most cases, a service cannot be separated from the person or firm providing it. Services are typically produced and consumed at the same time.

The inseparability of production and consumption increases the importance of the quality in services.

Perishability—refers to the fact that services cannot be saved, stored, resold, or returned. Services are deeds, performance or act whose consumption take place simultaneously; they tend to perish the absence of consumption. Hence, services cannot be stored.

Variability or Heterogeneous—By nature, services are highly variable, as they depend on the service provider and where and when they are provided. Heterogeneity also results because no two customers are precisely alike; each will have unique demands or experience the service in a unique way. Thus, the heterogeneity connected with services is largely the result of human interaction and all of the characteristics that accompany it.

2.1 Classification

Services are activities that are intangible in nature, therefore, standardization is one of the major issues in services.

In accordance with the classification of Bagad, services are classified according to [7, 8]:

1. The end-user of the services. It can be classified into the following categories:

    - Consumer: e.g. leisure, hairdressing, personal finance and package holidays

- Business to Business: e.g. advertising agencies, printing, accountancy, Consultancy
- Industrial: e.g. Plant Maintenance and repair, work-wear and hygiene, installation, and project management.

2. Service Tangibility. The degree of tangibility can be used to classify a service:

   - Highly tangible: car rental, vending machines, telecommunications
   - Service linked to tangible goods: domestic appliance repair, car service.
   - Highly intangible: psychotherapy, Consultancy, legal services.

3. Services can be broken down into people-based and equipment-based services:

   - People-based services: high contact: education, dental care, restaurants and medical services
   - Equipment based: low contact: automatic car wash, launderette, vending machine, cinema.

4. Expertise. The expertise and skills of the service provider can be broken down into the following categories:

   - Professional: medical services, legal services, accountancy, tutoring.
   - Non Professional: babysitting, caretaking, and casual labor.

5. Profit Orientation. The overall business orientation (profit) is a recognized means of classification:

   - Not for profit: The Scouts Association, charities, and public sector leisure facilities.
   - Commercial: banks, airlines, tour operators, hotel and catering services

One more classification should be mentioned, the classification by Lovelock (1983). He classified services by their placement along two dimensions [9]:

1. Who/what is the direct recipient of the service? (people or things)
2. What is the nature of service act? (tangible/intangible).

2.2 Service Management

Because of the unique characteristics of service, managing the service process can be a daunting task. Due to the intangibility of the service, it cannot be inventoried. Patented, or readily displayed or communicated. Pricing can be difficult due to the difficulty in determining the actual cost incurred by the provider or the value of the service to the customer. In spite of efforts to standardize services [10], it is obvious that it is impossible to plan or regulate every action taken in employee-customer contact. Most services are performed by people for the consumption or use of other people—a process that can never be fully foreseen ar standardized. The result is a heterogeneous set of characteristics of the interaction. Quality of service delivery and customer satisfaction depend heavily on employee actions. Service quality depends on many often uncontrollable factors, such as prior experiences, expectations and in some cases the particular moo or psychological motivation of the individual service provider or customer. Furthermore, the way customers react, behave or communicate can affect the employee. All of this adds to the management challenge. In fact, there is no guarantee that the service delivered will match what was planned or promoted by the organization.

The perishable nature of service makes it hard to manage both from the marketing and from an operational point of view. It is difficult to synchronize supply and demand for service. The fact that production and consumption of services are simultaneous events implies the fact that the customer is in effect a participant in the transaction. Service organization intends to provide certain st of services. The intent is usually based on the organization's corporate and service strategies and is tailored to leverage the organization's operational efficiency and fulfillment of expected customer needs. The intended services are then delivered through interactions with customers. The impact of that interaction, however, is a result of the match with customers expectations, the perception and satisfaction customers experience in the interaction with the organization and the value they perceive before, during and after the service has been delivered.

2.3 Service Product

In recent years, an increasing number of service companies have begun to refer to their offerings as "products," a term previously associated primarily with manufactured goods. What is the difference between manufactured goods and services in the contemporary business environment? What is an item? A product entails a consistent and well-defined "bundle of output" as well as the capacity to distinguish one bundle of output from another. In the context of manufacturing, the concept is simple to comprehend and visualize. Using the various "models" provided by manufacturers, service firms can also differentiate their products in a similar fashion. Fast food restaurants, for instance, display a menu of their products, which are, of course, very tangible. If you are a burger expert, you can easily tell the difference between Burger King's Whopper, its Whopper with Cheese, and McDonald's Big Mac. Providers of more intangible services also offer various "models" of products, which represent a collection of carefully prescribed value-added supplementary services constructed around a core product [11].

Credit card companies develop different cards, and each comes with a distinct bundle of benefits and fees; insurance companies offer different types of policies; and universities offer different degree programs, each composed of a mix of required and elective courses. The objective of service product development is to design bundles of output that are distinct and can be easily differentiated from another [11].

Designing service products transform intangible services into tangible, tradable objects. In the context of professional services, it was discovered that clients frequently lack a "clear understanding of what they need and what the company can offer." The common objective of developing service products] was to create simple, tangible, and easily understood offerings. Ideal service products have well-defined features, well-articulated descriptions, a clear value proposition, a brand, and a defined pricing structure and method of purchase. All of these can be utilized in a company's external communications, such as its marketing collateral, website, social media, and personal selling. Customers can better comprehend what the service entails and its distinctive characteristics, as well as appreciate it. A manager who sells digital marketing communications services describes these advantages as follows.

We gave the service a brand name and a clear description of the content, price, and even delivery time. Through the use of images and catchphrases, we've created promotional materials that convey the primary benefit of the service. It really breaks the ice in a sales situation!

With a well-designed service product (see Fig. 2), consumers can better comprehend what the service consists of (e.g., what are its key components), how it is created (e.g., what are the main phases of the process and experience, and when will they occur), and the value they will receive (e.g., what is delivered, what is the experience and outcome, why it works, and what is so special about it) [11].

To better comprehend the nature of service products, it is necessary to distinguish between the core product and the supplementary elements that facilitate its use and increase its value to customers. Next, experienced service marketers recognize the need to take a comprehensive view of the entire customer experience they wish to create. Specifically, it must be determined how core and supplementary services should be combined, sequenced, and delivered to create a value proposition that meets the needs of target segments. In conclusion, the creation of a service product requires the design and integration of the three components listed below:

(1) the core product,
(2) the supplementary services, and
(3) the delivery processes.

Core Product

The core product is "what" the customer is purchasing most fundamentally. When purchasing a one-night hotel stay, the essential services are lodging and safety. When paying to have a package delivered, the essential service is for the package to arrive on time and undamaged at the correct address. In brief, a core product is a central component that provides the most sought-after benefits and solutions to customers. The core product is the primary component that provides the desired customer experience (e.g., a relaxing spa treatment or an exhilarating roller coaster ride) or the problem-solving benefit that customers seek (e.g., a management consultant provides advice on how to develop a growth strategy, or a repair service restores a piece of equipment to proper working condition). Certain essential products are highly intangible. Consider the innovative design of credit card and travel insurance products in terms of their features, benefits, and pricing (note that many credit card fees are hidden transaction costs, especially if used abroad) [11].

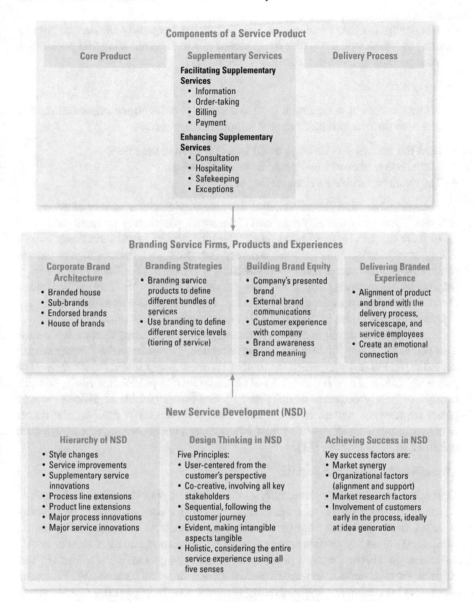

Fig. 2 Service products, branding and new service development [11]

Supplementary Services

Typically, the delivery of the core product is accompanied by a number of other service-related activities that we refer to as supplementary services. These enhance the core product, making it easier to use and increasing its value. As an industry matures and competition increases, core products typically become commoditized.

Consequently, the pursuit of competitive advantage frequently focuses on supplementary services, which can play a significant role in differentiating and positioning the core product relative to competing services [11].

Delivery Processes

The third element of designing a service concept pertains to the processes used to deliver both the core product and each supplementary service.

- How the various service components are delivered to the client.
- The nature of the customer's role in those processes.
- The required service level and style [11].

The Flower of Service is comprised of the core service and a variety of additional services. There are potentially dozens of different supplementary services, but almost all of them can be classified into one of the eight clusters identified as either facilitating or enhancing. Either service delivery (e.g., making a reservation) or the use of the core product necessitates supplementary services that facilitate their use (e.g., information). Improving supplementary services increases their value and allure for customers for example, consultation and hospitability can constitute important supplementary services in the healthcare environment.

Figure 3 depicts the eight clusters as flower petals surrounding the flower's center. The petals are arranged in a clockwise direction, beginning with "information," according to the order in which customers are most likely to encounter them. The petals and center of a well-designed and managed service product are fresh and well-formed. A poorly designed or delivered service is comparable to a flower whose petals are missing, wilted, or discolored. Even if the center is flawless, the flower is unappealing. Consider your own experiences as a consumer (or when buying on behalf of an organization) [11].

Managing the Flower of Service

The eight categories of supplementary services that comprise the "Flower of Service" collectively offer a variety of enhancement options for core products. The majority of supplementary services are (or should be) responses to customer demands. As mentioned previously, certain services, such as information and reservations, enable customers to utilize the core product more effectively. Others are "extras" that enhance or even reduce the core's nonfinancial costs (e.g., meals, Wi-Fi, magazines, and entertainment are hospitality elements that help pass the waiting time). The "information" and "consultation" petals represent this book's emphasis on the need for education and promotion of communication with service customers. In conclusion, the Flower of Service and its petals can be used as a checklist in the search for new ways to enhance existing core products and design new offerings. Regardless of which supplementary services a company decides to offer, each petal should receive the necessary care and attention to consistently meet predetermined service standards. Thus, the "flower" produced will always have a fresh and appealing appearance [11].

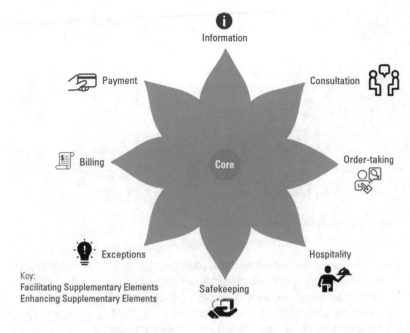

Fig. 3 The flower of service [11]

2.4 Branding Services

Almost any service business can implement branding at the corporate, product, service, and experience levels. In a well-managed company, the corporate brand is not only easily recognizable but also meaningful to customers, representing a particular business philosophy. Applying distinctive brand names to individual products enables the company to communicate to the target market the unique experiences and benefits associated with a particular service concept. In short, it aids in establishing a mental image of the service in the minds of customers and clarifies the nature of the value proposition [11]. Numerous large service providers possess a portfolio of businesses. Figure 4 depicts a spectrum containing four broad branding alternatives from which these corporations must choose. The branded house (i.e., using a single brand for all products) and the house of brands strategies represent the two ends of this spectrum (i.e., using separate standalone brands for each offering). Both subbrands and endorsed brands are a mixture of these two extremes [11].

Branded House

The term branded house refers to a business strategy in which a single brand (typically the corporate brand) is used to cover all of a company's products. These may be from unrelated fields. In addition to travel, entertainment, and lifestyle, Virgin Group also provides financial services, health care, media, and telecommunications services.

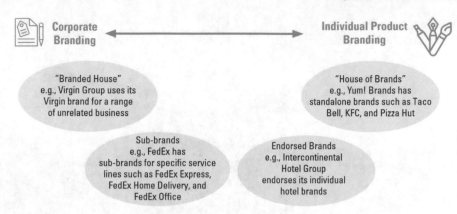

Fig. 4 The spectrum of branding alternatives [12]

This strategy has a number of advantages, including increased overall brand aware-ness, decreased marketing and communications costs, and the possibility of positive spillover effects between products. The risk of such a branding strategy is that the brand becomes overstretched and weakened and is exposed to a greater reputation risk because a problem with one product can affect the entire brand.

Sub-brands

Next on the spectrum are sub-brands, whose primary reference point is the corporate or master brand. The objective is to use the corporate brand's equity to support the sub-brands, such that the corporate brand gives the sub-brands significance.

Endorsed Brands

For endorsed brands, the product brand predominates while the corporate name remains prominent. Many hotel companies employ this strategy. Each of the endorsed brands must offer a unique value proposition aimed at a different customer segment for this strategy to be successful. It is important to note that in some cases, segmen-tation is situation-based: The same individual may have different needs (and will-ingness to pay) in different situations, such as when traveling with a family versus when traveling for business. A multi-brand strategy seeks to persuade customers to continue purchasing from the same brand family. Frequently, loyalty programs are used to incentivize this behavior, in which loyalty points are accumulated in one sub-brand during business travel and then redeemed in another brand during leisure travel [11].

House of Brands

At the extreme end of the spectrum is the house of brands strategy, which employs distinct standalone brands for each of its offerings. This strategy is employed when product brands differentiate their positioning and target distinct market segments. As a result, customers have a greater appreciation for the service because they comprehend what it entails and what makes it unique. Customers can be easily encouraged to mention these branded services in social media posts and other peer-to-peer communications, as evidenced by past performance [11].

Now that we understand branding in the context of services, we must learn how to build a strong brand and what factors contribute to brand equity. Brand equity is the value premium a brand possesses. It is the amount that customers are willing to pay for the service, in excess of what they are willing to pay for a comparable service that lacks a brand. Figure 5 illustrates the following six key brand equity components:

- Representation of the company's brand, primarily through advertising, service facilities, and personnel.
- External brand communications, including word-of-mouth and publicity These are beyond the company's control.
- Customer experience with the company—what the customer went through when using the company's services.
- Brand recognition is the ability to recognize and recall a brand when prompted.
- Brand meaning—what comes to mind when a customer hears a brand's name.
- Brand equity—the extent of a brand's marketing advantage over its rivals.

New Service Development

Nearly all service industries are being affected by intense competition and increasing consumer expectations. Thus, great brands not only provide existing services effectively but also innovate continually to create new service approaches.

Figure 6 demonstrates that for businesses to deliver branded service experiences, they must place the service product and its value proposition at the center and align it with the other six Ps of services marketing. The service product guides the development and delivery of the branded service experience, as all other Ps must support the customer's desired service experience.

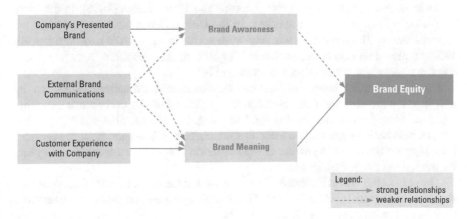

Fig. 5 A service-branding model [11]

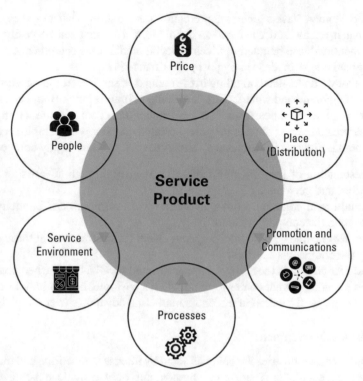

Fig. 6 All services marketing need to support the service product [13]

3 Global Importance of Services

Services are almost all around us. We are living in the time of service economy. It substituted the manufacturing economy.

As an example, according to Fitzsimmons and Fitzsimmons (2006), in the early 1900s, only three of every ten workers in the United States were employed in the services sector. The remaining workers were active in agriculture and industry. By 1950, employment in services accounted for 50% of the workforce. Today, services employ about eight out of every ten workers [14].

Nowadays, service industries lead in every industrialized nation. Thanks to the new jobs are created, and consequently, the quality of life of everyone is growing. Many of these jobs in the service field are aimed at high-skilled, knowledgeable workers and have the greatest projected growth in professional and business services [15]. Figure 7 illustrates the extent of the migration to services over the past 30 years for the top ten postindustrial nations.

The data of the World Factbook (2013) show that the service sector accounted for almost 64% of global GDP in 2012 and is expanding at a quicker rate than agriculture (5, 9%) and manufacturing (30, 2%) sectors.

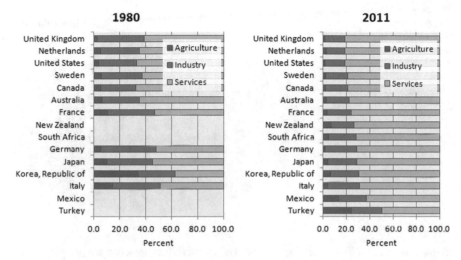

Fig. 7 Employment shares by sector, 1980 and 2011. *Source* http://www.bls.gov/fls/flscomparelf. htm

Moreover, trade in services is growing at a pace faster than the trade in goods since the 1980s.

According to WTO World Trade Report [13] and UNCTAD Handbook of Statistics [16], commercial services exports grew 11% to US$ 4.1 trillion—29, 82% coming from developing countries and 2.85% from transition economies in 2011.

Services include a broad spectrum of activities, beginning with financial, tourism and health and finishing with government services, most of which can be sold. In an industrialized economy, specialized firms can supply business services to manufacturing firms cheaper and more efficiently than manufacturing firms can supply these services for themselves. Thus, service firms often find advertising, consulting, and other business services being provided for the manufacturing sector. Except for basic subsistence living, where individual households are self-sufficient, service activities are necessary for the economy to function and enhance the quality of life. Consider, for example, the importance of the banking industry to transfer funds and the transportation industry to move food products to areas that cannot produce them. Moreover, a wide variety of personal services, such as restaurants, lodging, cleaning, and child-care, have been created to move former household functions into the economy.

Government services play a critical role in providing a stable environment for investment and economic growth. Services such as public education, health care, well maintained roads, safe drinking water, clean air, and public safety are necessary for any nation's economy to survive and people to prosper. Increasingly, the profitability of manufacturers depends on exploiting value-added services. For example, automobile manufacturers have discovered that financing and/or leasing automobiles can achieve significant profits. As personal computers become a commodity product with very low margins, firms turn to network and communication services

Fig. 8 Austria
GDP—composition by
sector, 2012 est. *Source* The
World Factbook, https://
www.cia.gov/library/public
ations/the-world-factbook/
geos/au.html [17]

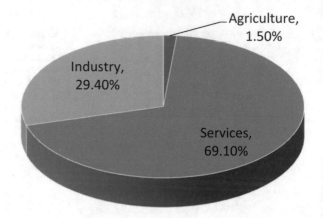

to improve profits. Thus, it is imperative to recognize that services are not peripheral activities but rather integral parts of society. They are central to a functioning and healthy economy and lie at the heart of that economy. Finally, the service sector not only facilitates but also makes possible the goods-producing activities of the manufacturing sectors. Services are the crucial force for today's change toward a global economy.

Service sector is very important for Austria too, because it generates the vast majority of Austria's GDP. Vienna has grown into a finance and consulting metropolis and has established itself as the door to the East within the last decades. Viennese law firms and banks are among the leading corporations in business with the new EU member states. Tourism is very important for Austria's economy. The Fig. 8 depicts the Austrian GDP composition by sector.

Austria stands out among its major trade partners as a country with the highest services exports and imports to GDP ratios.

According to the figures published by Statistics Austria, 26,172 new companies were founded in Austria in 2010. Overall 85% of business start-ups in Austria were in the service sector: the majority of these were in "trade" (5,032), in "freelance/technical services" (4,447) and in "hospitality and catering" (3,352).

4 Key Figures and Role of the Services in Economy

The service market grew incredibly in the XX century. Nowadays, services increasingly dominate in every economy worldwide. Today it is the largest and the fastest-growing sector, contributing to the global output and employing more people than any other sector; therefore it's termed as 'sunrise sector of the economy. The service sector is crucially important as for developed countries as for developing ones. Although developed countries accounted for three-fourths of world services output, the contribution of services in developing countries also increased significantly.

Today, service industries are the source of economic leadership for developed countries. For example, during the past 30 years, more than 44 million new jobs have been created in the service sector to absorb the influx of women into the workforce and to provide an alternative to the lack of job opportunities in manufacturing [14].

The service economy is also a key to growth for developing countries.

Consistent with OECD (2008), services constitute over 50% of GDP in low income countries and as their economies continue to develop, the importance of services in the economy continues to grow. The service economy in developing countries is most often made up of the following: Financial Services, Tourism, Distribution, Health, and Education.

Many developed countries are now termed as service economies. At the same time services are a new growth engine for poor countries. In accordance with the data of Ghani et al. [18], the share of developing countries in world service exports increased from 14% in 1990 to 21% in 2008 [18]. According to the World Bank their service exports are growing faster than goods exports.

The service sector promises to solve the problems of unemployment and poverty by means of creating and expanding new job opportunities. Directly, they provide the largest source of new job growth. Indirectly, they provide the income that, when spent, drives further demand for goods and services and jobs to produce these.

A greater role at a recent time plays Services exports and imports. Figure 9 shows the top 10 developing country exporters of commercial services accounted for almost 68% of developing country commercial service exports and 13% of world commercial service exports.

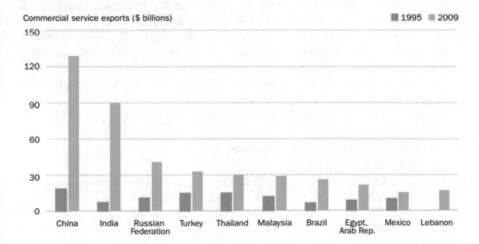

Fig. 9 The top 10 developing country exporters of commercial services. *Source* World Bank (April 15, 2011) World Development Indicators 2011 (http://siteresources.worldbank.org/DATASTATI STICS/Resources/wdi_ebook.pdf p. 217) [19]

Trade in services makes up 22% of world trade, up from 20% in 2000. In developing economies, the nominal value of trade in services grew 16% a year over 2000–09, doubling the rate of growth over 1990–2000 and surpassing that of high-income economies, which grew at 11% a year over 2000–09 [19].

Growth in the service economy also facilitates growth in the rest of the economy. Services such as energy, telecommunications, and transportation are important to all sectors of the economy; financial services facilitate transactions and investment; health and education services that contribute to a fit, well-trained workforce; and legal and accountancy services allow an institutional framework required to run a successful market economy.

In advanced economies, the growth in the primary and secondary sectors are directly dependent on the growth of services like banking, insurance, trade, commerce, entertainment etc.

For developing countries and least developed countries, service trade is the new frontier for enhancing their participation in international trade and, in turn, realizing development gains. One can say, it helps to integrate with the global economy.

5 Reasons for Growth of Services

The reasons for growth are the following:

- Increased globalization. It involves the fragmentation of value chains, the outsourcing of intermediate inputs and tasks and the increasing provision of integrated bundles of services and manufacturing activities;
- Growth of Information and Communications Technology (ICT). Due to the growth of E-business, use of mobile phones, Internet and as a result new innovative products, shorter product cycles, and reduction of distance, time and transaction cost many services have been reshaped. The application of ICT brought the shift towards more intensive online delivery;
- Rapid progress of technology and IT Revolution. The increasing role of IT in our life is impossible to deny;
- Urbanization and Migration to urban areas;
- Business growth;
- Increase in the Disposable Income (resulted in improved standards of living). Due to industrialization and globalization, the national income of many countries has increased. This has resulted in growth of per capita income of the people with more disposable income. This has effect in improvement of people spending on services.

They are attracted to new areas like travel and tourism, banking, investment, entertainment, retailing, insurance;

- Aging population;
- Low-cost labor in the developing countries & externalization (outsourcing) of service activities;
- Cultural Changes. Culture influences us, our lifestyle. e.g. working women.

Customer Feedback Evaluation, Reporting, and Distribution

Choosing the appropriate feedback tools and collecting customer feedback are ineffective if the company cannot disseminate the information to the appropriate parties for action. To promote continuous improvement and learning, a reporting system must provide feedback and its analysis to frontline employees, process owners, branch or department managers, and senior management. Figure 10 provides an overview of the information that should be distributed to the organization's key stakeholders. It also demonstrates how well the different tools complement one another: the higher-level tools provide benchmarking over time and against the competition, while the lower-level tool identifies why ratings go up or down and generates ideas for improving the service [11].

The feedback loop to the frontline should be immediate for complaints and compliments, as is the practice in a number of service industries where complaints, compliments, and suggestions are discussed during the daily morning briefing. In addition, we suggest three types of service performance reports to provide the necessary data for service management and team development:

1. A monthly Service Performance Update provides timely feedback on customer feedback and operational process performance to process owners. Here, the

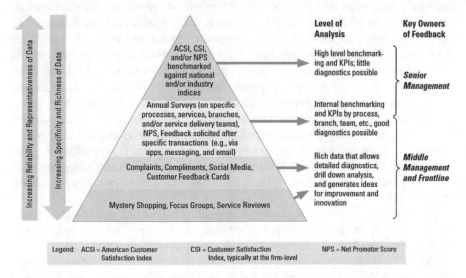

Fig. 10 Mapping reporting tools to levels of management [11]

verbatim feedback should be forwarded to the process managers, who can then share it with their service delivery teams.
2. A Service Performance Review conducted on a quarterly basis provides process owners and branch or department managers with process performance and service quality trends.
3. An annual Service Performance Report provides senior management with a representative assessment of the current status and long-term trends regarding customer satisfaction with the company's services [11].

6 E-business and Services

As the main structural changes in recent decades, one has to point out the shift in employment towards services in highly developed economies. As a result the competitive advantage of manufacturing firms is increasingly derived from the value contributed by service processes associated with their core activity, with competition shifting away from how companies build their products towards how well they serve customers before and after they produce and sell products. As it was mentioned above, one of the main reasons for the growth of service sector was growth of information and communications technology (ICT) and IT revolution. It resulted in acceleration and simplification of the information exchange, the emergence of new services and the development of existing, developing of new innovative offers, improving of interaction with customers and labor intensity, more conditions for customization and efficiency. Consequently, transaction costs and input prices can be reduced; marketing and administrative costs might become lower. New customers and markets can be accessed with less difficulty than before. Moreover, the introduction of IT technology makes the services more available for customers. Service has migrated from human interaction to substitution of machines for service employees or feasible to anywhere-anytime electronic service. Thus it abolishes the role of mediator and increases so called self-services. Services that can be digitalized and delivered via the Internet, such as entertainment, information, and training, represent new opportunities for Self-services such as tickets/hotel booking, selection of places, menus, registration, payment by credit cards etc. More and more services are available over the internet (from grocery shopping to online learning) and it liquidates obstacles of place and time. Service analytics provides companies with many advantages. On the other hand, companies face various challenges that arise from the use of service analytics [20].

Development of self-services affects at ousting low-skilled and low-paid work. At the same time, it increases the role of intellectual, creative and highly skilled jobs. Table 1 illustrates concrete examples development of self- services.

Electronic (virtual) services share many of the characteristics of traditional (physical) services. Electronic services are intangible because the service transactions and experiences are delivered via electronic channels. They are difficult to measure and describe fully. Table 2 shows the Comparison of Electronic and Traditional Services.

Table 1 Evolution of self-service

Service industry	Human contact	Machine assisted service	Electronic service
Banking	Teller	ATM	Online banking
Grocery	Checkout clerk	Self-checkout station	Online order/pickup
Airlines	Ticket agent	Check-in kiosk	Print boarding pass
Restaurants	Wait person	Vending machine	Online order/delivery
Movie theatre	Ticket sale	Kiosk ticketing	Pay-for- view
Book store	Information clerk	Stock-availability terminal	Online shopping
Education	Teacher	Computer tutorial	Distance learning
Gambling	Poker dealer	Computer poker	Online poker

Source Fitzsimmons, Service Management, p. 108 [21]

Table 2 Comparison of electronic and traditional services

Features	Electronic service	Traditional service
Service encounter	Screen-to-face	Face-to-face
Availability	Anytime	Standard working hours
Access	From home	Travel to location
Market area	Worldwide	Local
Ambiance	Electronic interface	Physical environment
Competitive differentiation	Convenience	Personalization
Privacy	Anonymity	Social interaction

Source Fitzsimmons, Service Management, p. 114 [21]

E-business made changes in service provision and service markets and lead to a new way of providing services. One of the most influential implications of e-business is that it changes the way services are generated and delivered to the customer. Thus services can be provided more efficiently, and according to Licht and Moch (1997) entirely new types of services are developed [22].

Information technology is a tool that enables companies to realize business models and production schemes that have been summarized as e-business. Firms can use e-business to access information as an essential input more efficiently than before.

Services can be delivered via electronic networks without the physical movement of either supplier or client. As a result, a global reorganization of service companies has been expected.

As one would expect, e-business and service sector play an outstanding importance in the world economy. The opportunity to provide services over the Internet facilitates the establishment of new businesses independent of size and origin, facilitates the entry of new players on the market and it can be said ICT and e-business have a strong impact on the competitive situation in the services sector. E-business technologies help to enhance sourcing activities globally and to manage a larger number of suppliers. In addition to that, the Sector Report of the European Commission (2005) states that E-business technologies could be used to support the search for the best-qualified suppliers and to improve the collaboration with them.

7   Conclusion

ICT and e-business have significantly changed the business and economics in general, and the service sector in particular. It can be confidently concluded that services play a major role in the modern economy.

As the data indicate, the size of the service sector dominates over the rest as in developed as in developing countries. Moreover, it is the main way of development for developing countries.

Development of ICT and E-business facilitated improvement and modernization of the existing service activities and created new types of services.

The borders between manufacturing and services sectors become more and more vogue. A lot of 'outputs' consist of combined services and manufactured products, which may be supplied by companies from either the "services" sector or "manufacturing" sector. Many companies use the "additional" service elements surrounding the core product as a major focus of competition in industrial and consumer markets. Manufacturers apply the competitive advantage from the value contributed by service processes associated with their core activities. Oftentimes, enhancing service quality and enhancing service productivity are two sides of the same coin, with powerful potential to increase customer and firm value. It is a significant challenge for any service business to provide quality service and customer satisfaction in a cost-effective manner. Strategies to improve service quality and productivity should complement one another rather than compete. In a world of constant innovation and competitive markets, only a handful of companies can afford to spend more money (i.e., permit lower productivity) for higher quality.

References

1. McCredie, A., Bubner, D.: Seven myths about services. Aust Serv Round Table (2010)
2. Zeithaml, V.A., Bitner, M.J., Dremler, D.: Services marketing, international edition. McGraw Hill, NY London (1996)
3. Blois, K.J.: The marketing of services: an approach. Eur J Mark (1974)
4. Lovelock, D., Mendel, M., Larry Wright, A : An Introduction to the Mathematics of Money: Saving and Investing. Springer (2007)
5. Grönroos, C.: Service Management and Marketing. Lexington books Lexington, MA (1990)
6. Gronfeldt, S., Strother, J.: Service Leadership: The Quest for Competitive Advantage. SAGE Publications (2005)
7. Bagad, V.S.: Total Quality Management. Technical Publications (2008)
8. Fedushko, S., Ustyianovych, T., Syerov, Y., Peracek, T.: User-engagement score and SLIs/SLOs/SLAs measurements correlation of e-business projects through big data analysis. Appl. Sci. **10**(24), 9112 (2020). https://doi.org/10.3390/app10249112
9. Lovelock, C.H.: Classifying services to gain strategic marketing insights. J. Mark. **47**, 9–20 (1983). https://doi.org/10.2307/1251193
10. Poniszewska-Maranda, D., Matusiak, R., Kryvinska, N., Yasar, A.-U. H.: A real-time service system in the cloud. J. Amb. Intell. Human Comput. **11**(3), 961–977 (Mar.2020). https://doi.org/10.1007/s12652-019-01203-7
11. Lovelock, C.H., Wirtz, J.: Services Marketing: People, Technology, Strategy, 9th edn. World Scientific, New Jersey (2022)
12. Devlin, J.: Brand architecture in services: the example of retail financial services. J. Mark. Manage. **19**, 1043–1065 (2003). https://doi.org/10.1080/0267257X.2003.9728250
13. Berry, L.L.: Cultivating service brand equity. J. Acad. Mark. Sci. **28**, 128–137 (2000). https://doi.org/10.1177/0092070300281012
14. Fitzsimmons, J.A., Bordoloi, S., Fitzsimmons, M.J.: Service management operations, strategy, and information technology: operations, strategy, and information technology. Mc Graw Hill Irwin, Nueva York (Estados Unidos) (2014)
15. World Trade Organization: Trade and Public Policies: A Closer Look at Non-tariff Measures in the 21st Century. WTO Publ, Geneva (2012)
16. United Nations Conference on Trade and Development: UNCTAD handbook of statistics. United Nations, New York (2012)
17. The World Factbook—The World Factbook. https://www.cia.gov/the-world-factbook/. Accessed 25 Jun 2022
18. Ghani, E., Goswami, A.G., Kharas, H.: Can services be the next growth escalator? In: VoxEU.org (2011). https://voxeu.org/article/can-services-be-next-growth-escalator?quickt abs_tabbed_recent_articles_block=1. Accessed 25 Jun 2022
19. World Development Indicators|DataBank. https://databank.worldbank.org/source/world-dev elopment-indicators. Accessed 25 Jun 2022
20. Halás Vančová, M., Mikolasik, M.: Use of E-service Analytics in Slovakia, pp 83–121 (2022)
21. Fitzsimmons: Service Management 5E W/Cd. McGraw-Hill Education (India) Pvt Limited (2006)
22. Licht, G., Moch, D., Licht, G., Moch, D.: Innovation and information technologies in services. ZEW Discuss (1997). https://doi.org/10.2307/136427

Management of Marketing Communication in State Administration

Ľubomíra Strážovská and Lucia Vilčeková

Abstract The scientific study focuses on the analysis of the starting points and opportunities for the public to participate in land consolidation processes. Under the current legislation, it defines the concept and meaning of land consolidation in Slovakia, examines the content of the concept of public interest under the land consolidation legislation and analyses the conditions for informing and subsequently involving the public in the protection of the environment. On this basis, it examines and explains one of the important options by which the public influences environmental decision-making processes. This is the assessment of the environmental impact of strategy papers and proposed actions. In dealing with the issue, we applied the primary analysis of the legislation given the nature of the topic under examination. However, we also use other scientific methods of investigation as well as scientific literature, case law, and analogy of law. With our research, we strive to provide qualified answers to the main pitfalls associated with land consolidation in the Slovak Republic. The added value of our scientific study is, especially in a broader context, to critically highlight selected application problems and suggest ways of improvement.

1 Introduction

Public administration has historically evolved in relation to the current requirements and functions it was supposed to fulfill, and therefore it is possible to find several different definitions in the literature. The inconsistency of the definitions of public administration is caused by the view that individual scientific disciplines, such as administrative law or public economics, look at this issue. Public administration

L. Strážovská (✉) · L. Vilčeková
Comenius University, Odbojárov 10, Bratislava, Slovak Republic
e-mail: lubomira.strazovska@uniba.sk

L. Vilčeková
e-mail: lucia.vilcekova@uniba.sk

represents a certain type of activity (administration of public activities) and institutions (office, organizations) that perform public administration. In the formal sense, public administration is defined as the activity of bodies designated as administrative authorities. In material terms, public administration represents the activity of state or other public institutions, the content of which is neither judicial nor legislative.

The concept of public administration is constantly the subject of scientific interpretation in the field of social, political but also legal theories, not only in Slovakia. In general, the view has become established that public administration consists of two components, namely state administration, and self-government, which are further divided according to individual hierarchical levels of organization. In examining the concept of public administration, we find many different opinions and attempts to express its content and its explanation. The large number of opinions and aspects used is caused by the historical differentiation of the very concept of "administration" under the influence of its development in the social, natural, and technical sciences. The report brings together traditional and historically constituted disciplines such as:

1. law,
2. economics,
3. social sciences,
4. and administrative theories from camera studies to cybernetics.

According to Hendrych [1], the concept of public administration is a central concept of administrative law. This concept has its roots in Roman law, from where it was translated into other languages. The current approach to the interpretation of such a complex, multi-layered and dynamic phenomenon as public administration can be criticized for its legal nature, as less attention is paid to the effectiveness of individual phenomena in public administration and society and the need to address the problems.

According to Prucha a Pomahač [2], public administration activities are directed either towards the public administration itself or also towards the citizen. They can therefore be divided into external and internal activities. The most important, from the point of view of the mission of public administration, are:

• external activities, which serve citizens (companies) in dealing with public administration bodies
• and internal activities are only here as supportive, which ensure the smooth running of public administration.

The main desirable areas of public administration are the reduction of undesirable administrative elements, the strengthening of the autonomy of entities or the reduction of the amount of management resources.

In the last two decades, attention has been increasingly focused on the need for change in public administration. The issue of administrative reforms is constantly evolving, and it is necessary to consider that the public sector is also influenced by political interests. Administrative reform focuses on the administrative aspects of public administration and how best to implement a political decision. Although shortcomings in public administration are often attributed to administrative officials, the main culprits are policy decisions that the reform can improve. It is therefore necessary for public administration reforms to go hand in hand with political reforms.

2 Overview of Public Administration and Its Definition

The poor functioning of public sector and its inefficiency are current topics that the company has been dealing with for several years. The current unsustainable system calls for reform and draws attention to the need for a greater focus on public administration. The literature dealing with the improvement of the functioning of public administration and its specifics in the field of management is referred to as public management, i.e., management in the public sector. Management in the public interest is characterized as a theoretical, practice-oriented discipline, art, and reform policy. Most of the literature dealing with this topic debates the differences between governance in the public interest and public administration. Management in the public interest is usually characterized as a way of managing an organization so that, under the guidance of a manager, it makes the best use of the resources at its disposal. The public administration is then presented as part of the management system, fulfilling the tasks assigned by the managers. Economic pressures to reduce public budgets, the crisis of economic structures and the internationalization of public affairs have also become central motives for public administration reform.

According to Balejová [3] we encounter several aspects of looking at the importance of public administration, especially in terms of law, economics, governance, but also the socio-political and social aspects. Their summary forms a certain sufficient basis for comparing and outgrowing the objects of public administration and economic policy.

According to Hamalová [4] from this integral point of view, public administration is characterized as a set of methods, measures, methods, and procedures that put in place the economic policy mechanism, ensure its objectives, regulation, and development. Most economists identify with this notion of public administration. Economic policy without public administration cannot exist and the political forces of the state cannot fulfill their mission. Conversely, public administration without economic policy loses its position and importance because it loses its object and direction for its action. Therefore, it is possible to see the role of public administration as a subject of public sector management, which results from the fact that the public sector is financed mainly from public finances. The taxpayers entrusted their taxes to the public administration, based on a public election through their elected representatives, to manage them based on a public election.

According to Kosorin [5], they define public administration in a very comprehensive and concise way, which "is represented by the performance of self-government and state administration,because the sector is financed from public finances and public finances are formed from sources from the population, there is no entity available for the management of the public sector other than the public administration." Public administration is primarily bound to the rule of law. It can administer and enforce only what is required by law, most often through a set of rules specifically designed for a particular administrative body. The regulation specifies the competencies of a public authority, which means the predominance of the element of normative behavior in public administration over the personal element. Many authors are more inclined to define public administration from a legal point of view, they do not directly mention its connection with economic policy, but they do not even exclude it.

According to [6], public administration is derived from the performance of the functions of state and self-governing bodies, which ensure the decisions of the legislature, the judiciary, and the executive in various structures and at many levels. Perhaps the simplest definition is that which describes public administration as an administration which aims at the administration of public affairs and its implementation is in the public interest.

According to Bogumil and Kuhlmann [7], in view of the above, it is possible to define public administration in a broader and perhaps more complex way, as "decision-making activity performed by authorized bodies on the basis of law, within the scope of mandated competence and authority and aimed at ensuring public administration", or public administration as public administration it is a public interest administration and the bodies that implement it implement it as a legally imposed obligation, by virtue of their status as public bodies.

2.1 Division of Public Administration

Based on the indicated understanding of public administration, it is necessary to be aware of a different nature of public administration carried out as a state administration and different from a public administration of the territorial self-government type. According to Hamalová [4] the structure and content of the concept of public administration can be understood by learning about its basic organizational components. State administration represents activities performed on behalf of the state through its bodies. The self-government carries out its activities through the bodies of municipalities, cities, and higher territorial administrative units. The basic differences between these two groups of public administration entities are shown in Fig. 1. The existence of two lines of public administration is already found in the objective existence of a conflict of interest between a citizen, a group of citizens living in a certain territory (municipality, micro-region, district, region) and all citizens. integration grouping.

2.2 Restructurin of State Organizations

Restructuring is the most common way out of crises in state organizations, as the cessation of activities is not possible in most cases. One of the most used forms of restructuring is rationalization. It primarily leads to a reasonable, justifiable, and cost-effective organization of activities performed by organizations on behalf of the

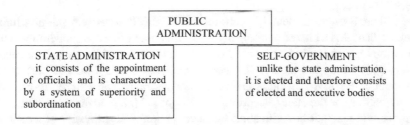

Fig. 1 Division of public administration. *Source* Papcunová, V., Gecíková, I. *Basics of public administration*. Bratislava: VŠEMVS, 2011. p. 60

state, their structuring, setting up internal and external processes and, to a significant extent, to increase the efficiency of working procedures, which results not only in more efficient use of financial resources but overall improvement of securing and providing activities within the competence of state organizations [8].

According to Kádarová (2011) the rationalization of organizations can take several forms and can lead to the demise of the organization as such. Although the complete cessation of a state-established organization is possible, the activities carried out by the organization do not disappear and the state property that has been invested in it remains state property. Although the state may dispose of unnecessary assets in accordance with the applicable legislation, it can never be assets directly related to the implementation of activities. Only surplus assets can be sold. Within the rationalization processes, such activities can also be identified that can be provided by forms of financing without an impact on public resources, and thus also on the state budget, without endangering the implementation of these activities. The state organization (as part of the rationalization process, even after its abolition) then goes through a process that redistributes activities and assets in a certain way. This process is called transformation [9, 10].

Throughout the process of restructuring public administration, the correct use of information technology to "overcome barriers to access to citizens and help create new services is more targeted and flexible than at present." Information technologies can help meet citizens' unrestricted and uninterrupted access information and government services, which is a prerequisite for ensuring an inclusive information society for all. "*In this* respect, *they play a key role in the efficient provision of public services and also improve cooperation between public administrations in different countries of the European Union to take over the competent authorities.*"

Strengths and weaknesses of e-government

According to Balejová [3] building an information society is a long way off. Negotiations on the electronic public administration in the Slovak Republic began for the first time in the 1990s. Since then, many steps have been taken, from changing the laws that bring electronic communications to the level of the ordinary, to the electronicization of paper documents. The information society includes many components.

One of them is e-government, ie electronic communication between public administration offices and a citizen or company. It is also about effective cooperation across the various offices and building a relationship between the public and private sectors.

Strengths—the introduction of e-government represents a benefit both for the consumer of the service, which can be a legal or natural person, and for the providers of this service, or the state. According to Wright [11], the advantages of the state administration include simplification and acceleration of the provision of the service, reduction of the number of employees (and thus cost savings), higher satisfaction of the consumers of the given service or reduction of the error rate of the provided data. According to Peráček [12] *"Advantages on the part of individuals or companies are an increase in the quality of the selected service, the provision of services outside the working hours of the institution or the provision of remote access. In general, the benefits of electronic government can be seen in the speed and quality of services to citizens and businesses, user-friendliness and simplicity, financial savings, transparency of processes and decisions and office hours 24 h a day, 7 days a week, and therefore time savings."*

Weaknesses—*"e-government e-government services are therefore universally beneficial for all stakeholders, but it should be noted that they also have certain limitations and cannot be applied globally. A certain limitation is the consumer's equipment with information and communication technologies and his literacy. There will always be certain individuals or entities that are not technologically and literate enough to be consumers of the electronic service, and for these groups of entities, the service will have to be kept in paper form. However, the variability of service provision is associated with a higher level of costs."*

According to Buchta [13], equipment with information and communication technologies and literacy in the given area does not only concern citizens or companies, but also the state administration itself. *"This involves costs in the form of an initial investment in appropriate equipment and training. The loss of personal contact between the official and the citizen is also related to the electronic services. Finally, it is also important to consider the profitability of the electronic services provided. For less used services, their electronicization can be significantly more expensive than providing the service in paper form."*

Priority goal: to build e-government in accordance with the needs and expectations of end users, the so-called tailor -made e-services

$$\text{population SR } + \text{ business segment } = \text{"customers" of}$$
$$\text{e-services public sector}$$

Survey $=$ a way to set the quality of e-services provided by the public sector.

The survey provides a realistic picture of the company in terms of awareness, use, quality assessment and satisfaction with the 20 basic e-government services in the residential and business sector, provides answers regarding the perception of e-government, willingness to use e-services, trust, expectations, and satisfaction of end users. VS e-service users with the current situation in this area.

The project is a long term plan of the Ministry of Finance of the Slovak Republic, its added value is the ability to compare results over time and evaluate the process of building a modern electronic public administration, offers new stimuli that respond to the interests and needs of existing and future users.

Digital literacy in Slovakia:

- Regular survey of the Institute for Public Affairs (IVO)—monitoring of digital literacy in the Slovak Republic, measurement every two years since 2005, current findings from 2009.
- Detection—opening the digital divide scissors.
- People who are already digitally literate (84%) are especially going to improve. People with low digital literacy only less (36%) and digital illiteracy almost not at all (9%).
- Regular Internet users in the population reach 60%, the Internet is becoming a common household equipment—92% access from home, the Internet is used more by people aged 18–39 and respondents with higher education (high school graduate and university), on the contrary, they use the Internet little older people (50+) and respondents with lower education.
- Even though only 12% of the population prefers an electronic way of communi-cating with VS, searching for information, including information from offices, is one of the most common activities on the Internet => **A good starting point for the development of the use of e-services**.
- The survey and their comparison of results with the previous year did not show any dramatic changes in the development of knowledge and use of e-services VS in the residential segment, at least one e -service of public administration was used by 33% of the population.
- The part of the population that dealt with the official matter via the Internet declares that currently the development of e-services corresponds to their ideas, perceives the benefits of e-services VS (saving time and money, simplicity, and convenience of processing over the Internet), but do not have enough information about e-services, resp. such a possibility of resolving an official matter.
- Job search, income tax filing, motor vehicle registration and social security are official issues that have the potential to be widely used and should therefore be a priority in the VS e-services development strategy.

2.3 Telecommunication and Cyber Security in the World

According to Rittinghouse, Fransome [14], the expected increase in the availability and affordability of mobile devices, especially those with e-mail capabilities, will significantly change the country of government services and the associated use of e-mail and RSS, but also investment in security. This change will be beneficial, espe-cially in developing countries, where the growth of these services in the social sector is expected to accelerate significantly with the growth of affordability and availability of mobile devices. This will not only bridge the digital divide between regions in

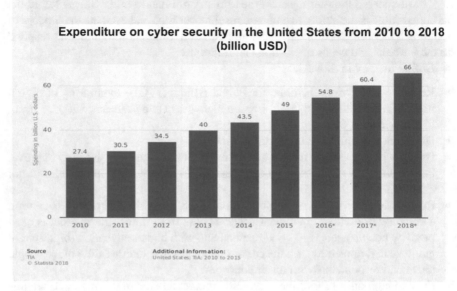

Fig. 2 Expenditure on cyber security in the United States from 2010 to 2018. *Source* Own processing according to https://publicadministration.un.org/egovkb/Portals/egovkb/Documents/un/2016Survey/

online services in the social sectors but will also contribute to sustainable development. In this area, however, the security of systems, which often must withstand global attacks, must be taken into account. An example is the investment in cyber security in the US in Fig. 2.

According to Peracek et al. [15], mobile applications and SMS services are showing a huge increase in almost all industries. The highest growth of mobile application services and SMS services was in the healthcare sector, where it increased from 11 to 34%, followed by the finance sector with an increase from 14 to 32%.

Other sectors also recorded a high increase, resp. 7–24% for the environment, 14 to 30% for education, 8–23% for well-being and 11–25% for work. Figure 3 provides an overview of mobile services by sector. Email or RSS updates were the highest in the sector compared to mobile apps or SMS. Mobile applications and SMS services have increased in the last two years and the gap between the two online services is narrowing. However, the gap is still high in education (30 countries), finance (23 countries), well-being (22 countries), environment (22 countries) and labor (13 countries).

Already in 2016, the survey pointed to two important phenomena. First, social sectors such as health and education have grown, representing a strong commitment by governments around the world to use technology for the benefit of all and to promote sustainable development. "According Milošovičová [16] *second, given the trends highlighted in the surveys, there is expected to be an increase in both services*

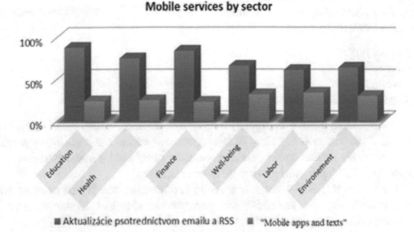

Fig. 3 Mobile applications and SMS services between 2014 and 2016. *Source* Own processing according to https://publicadministration.un.org/egovkb/Portals/egovkb/Documents/un/2016Survey/

in both services—updates via *email/RSS and mobile applications/SMS services. Such increases will be determined by the availability and availability of mobile devices."*

3 Digital Marketing Communication in Public Administration

According to Kita [17] communication is urgently needed in the public administration as well as in the efforts of the public institution to build a positive relationship with the public.

"In general, it is the exchange of information between people who want to communicate something, mediate something, share something. They can do this in different ways: in words, writing, images, gestures, head nods, body movements and silence. This knowledge from communication is also used in the field of marketing, with the difference that in their application the intention of communication is subordinated to the company's strategy."

3.1 Communication in Public Administration

According to Štarchoň [18] communication is one of the components of the marketing mix. The marketing mix is a set of tactical marketing tools that the company uses to adjust the offer according to the target markets. The marketing mix includes everything a company can its product. "The *possible ways are divided into four groups of variables, known as 4P:*

- *product policy,*
- *price policy,*
- *communication policy (promotion),*
- *distribution policy (place).*

Society is generally moving to digital forms of communication with the public administration. From this point of view, it is necessary to define the creation of positive links through the online world between the public and public administration. This implies the following digital marketing communication."

According to Kotler [19] digital marketing communication is part of other areas of digital marketing that are sometimes perceived as identical. However, there are significant differences between these concepts, which are defined as follows:

- Internet marketing is a set of marketing tools and techniques that uses a range of services and tools that are available on the Internet and used through devices connected to the Internet [20].
- Mobile marketing is a set of tools and techniques that use mobile technologies and devices that are easily portable and are always carried by the user. Most mobile phone applications use the Internet and there is an overlap between these groups, but there are also many technologies and tools that do not need the Internet, such as SMS/MMS marketing, Bluetooth geolocation via GPS or native mobile applications [21].

In relation to the cyber environment, it is necessary to define certain ethical barriers to communication, the so-called ethical hacking.

In today's globalized world, IS/IT hacking is an element that determines the IS/ICT architecture in companies. *"All companies are currently processing sensitive data such as personal data, therefore, according to the legislation, they have to create a security project and they have to deal with security management. Ethical hacking is a practical aspect of IS/ICT security testing in companies."* It is a matter of using commonly available, *"as well as less available techniques to get into or damage systems."* A manager cannot claim that his information system is secure unless he has developed a comprehensive set of security tests and evaluated them in terms of severity. There is currently an international standard OWASP—Open Web Application Security Project, which defines the steps to perform the tests. "However, *to perform these steps on behalf of the company, a specialist is needed, the so-called an "ethical hacker" who must have sufficient knowledge in this area. The training of these specialists requires not only theoretical knowledge, but especially practical experience."*

According to Labská [22] the results of technological development and its practical application become visible in production, trade, transport, education, increased productivity, etc. Flexibility represents the need for high adaptability to changing market conditions. It is created with fast customer requirements. Humanization manifests itself in overcoming existing methods of work organization, where operational specialization is preferred. Comprehensive qualifications are preferred—comprehensive development of employees with a focus on creativity, innovation, and effective multicultural existence. Intensification is an effort to effectively assess available resources: human resources, materials, machinery, equipment, buildings, and financial accounts.

3.2 Modern Marketing Communication Tools

According to Frey [23] narketing communication tools are often divided into two basic categories, namely above line (indicated by the abbreviation ATL, from the English above the line) and below the line (indicated by the abbreviation BTL, from the English below the line). The main difference between the two categories is their price. Line tools tend to be more costly and have less response. ATL tools include advertising, print, or radio. BTL tools include sales promotion, personal selling and direct marketing, these tools are significantly cheaper and usually have a greater effect.In recent years, the Internet, and social networks, which intersect the two mentioned categories (ATL/BTL) and carry the characteristics of both, have come to the forefront of communication. Their price and the possibility of targeting a selected target group, which increases their effectiveness, is like BTL and at the same time their nature of impersonal communication is reminiscent of ATL. The Internet is a place of endless possibilities, creating a new and completely different world of marketing communication, which is also suitable for tourism, as it is relatively cheap and effective.

Within marketing communication, it is necessary to use all its available tools and then combine them in a suitable way about the set goals and specific objectives of the market segment. By interconnecting individual marketing communication tools, it is possible to achieve maximum, mutually multiplying and strengthening synergy effects.

During the twenty-first century, there was a huge technological boom, new technical conveniences come to the fore, such as internet. There are changes in telecommunications, microelectronics, computer technology and others. Technological development at the turn of the century also brings changes to marketing communication, new trends began to appear in this field, which proved to be effective even for customers who are already immune to the classic tools of the communication mix. The Internet has undergone enormous development since its inception, and today life is unimaginable without it. The Internet provides each of us with virtually unlimited space in disseminating and obtaining information. Over time, entrepreneurs have realized that if they want to succeed in the market and gain awareness, then an online presence (understandable website, use of social media) is crucial for them. The gradual development of the Internet has brought changes in the field of distribution and distribution curves. With a well-chosen strategy, service providers can focus on entirely new markets and more specific segments, reach the right people at the right time in the right place, expand their distribution channels, build trust with customers, build good relationships with them, or analyze individual campaigns.

Viral marketing can be imagined as the ability to attract customers in such a way that they exchange information about the offered product. It is based on the spontaneous dissemination of commercial and non-commercial content (using images, video clips or games and other tools), which is so interesting and original for customers that they will need to share it with their surroundings. According to Palatková [24], social networks such as Facebook, Twitter, YouTube are mainly used in its dissemination. Its advantages include easy and fast implementation and high actionability at minimal cost. As a disadvantage, we can mention a small control over the running viral campaign, when the "virus" is placed in the hands of the recipient.

Search Engine Optimization (SEO) and Search Engine Marketing (SEM)— Search Engine Optimization (SEO) is a process that affects the visibility of a website in the unpaid part of the search engine result (Google, YouTube, List, Facebook). According to Kotler [25] in general, the more often and the higher a website appears in search engine results, the more visitors will be able to find the site from an Internet search engine. SEM also aims to increase the visibility of the website, but unlike SEO, the targeted effect is achieved with the help of paid ads. Compared to SEO optimization, the initial costs tend to be lower, but to maintain the optimization of the website (campaign), it is necessary to constantly spend money. As an example of paid advertising on the Internet used within SEM, we can mention PPC systems (AdSense, Google Adwords, Sklik), advertising banners, placement in the leading positions within the websites that compare products (Heureka). The disadvantage of SEM is the constant dependence on funds, as soon as their supply to this system stops, the ad serving campaigns will cease to be active and individual ads with a link to the website (banners) will stop running. On the other hand, the advantage of SEM is its quick response to the client's needs, as soon as the campaign is created and paid in the system, it is immediately activated and starts to appear immediately. For emerging sites, it is appropriate to combine these two approaches, as soon as the results of SEO optimization begin to show and the site strengthens its position in the search results, it is possible to turn off the SEM campaign, respectively. reduce the costs associated with its implementation [26].

Guerrilla marketing—The possibility of using creativity outside the Internet environment offers us one of the most creative marketing communication tools today, which is Guerrilla marketing. Guerilla marketing is a form of marketing communication that combines creativity, wit, idea, innovation, unconventionality or originality, all considering the low budget in its final implementation. We can observe the influence of this form in the most widespread forms of various slogans and slogans, which are spreading at a dizzying rate among people." Originality and at the same time uniqueness are key elements of guerrilla marketing. Firstly, there is no such thing as one proper marketing and secondly, it is not possible to use the same guerrilla action twice. If so, only with less effect on the audience than in the first event [27]. Its great advantage is low financial demands. It can go e.g., about placing the inscription in a busy place. Guerrilla marketing can be a very effective form of communication with regular and potential customers, but it must be done within the limits of legality and morality.

Guerilla marketing is often considered an aggressive communication tool, its strategy is to quickly strike an unexpected place on a precisely selected target and then its immediate retreat into the background.

Product placement and destination placement—This is an advertising technique used by companies to promote their products discreetly through non-traditional advertising techniques, usually through appearances on film, television, or other audiovisual media. Product placement is often started as an agreement between the manufacturer and the media company and is carried out for financial consideration or other consideration. Product placement can be an alternative in the presentation of a specific product or brand. In contrast to traditional advertising, product placement should be less aggressive [8]. If the product placement is implemented correctly, it should evoke in the viewer a desire for ownership of the object and the viewer should perceive it as part of the film and not as a disruptive element. Within audiovisual works, we can talk about three variants of product placement. The product is either talked about, used directly, or is directly captured in the scope of the work. Within tourism, product placement is a suitable tool for communicating the destination, then we are talking about the so-called destination placement. Destination placement can be defined as the target promotion of a destination in audiovisual works to increase destination traffic. In addition, placement destination is much more natural as product placement, because the film story must take place somewhere [26].

Mobile marketing "It *can be understood as any form of marketing, advertising or sales promotion activity targeted at the consumer and carried out* via *a mobile phone.*" Its application can be observed across various industries and has recently been applied in the field of mail order sales. In mobile marketing, mostly secondary functions of mobile phones are used, usually advertising text messages (SMS), with an informative character and they can also bring the customer an advantage, for example in the form of a discount coupon. One of the interesting forms in which it is possible to use mobile marketing are QR codes, which the advertiser adds to the printed advertisements and the reader can scan it with the help of a mobile phone and learn more about the offer [28, 29].

3.3   Crisis Management and Management Theory

The basic goals of crisis management are based on its tasks and are carried out based on the application of managerial functions. We can formulate the basic objectives of crisis management: "*1. Detection and analysis of possible sources of threat and its expected development 2. Development of crisis plans based on risk analysis and possible solutions 3. Development of a set of tools to manage crises and crisis situations and eliminate their consequences.*"

"*Crisis planning tasks are elaborated in the plans. We recognize two types of such plans.*" It's an emergency plan and a contingency plan. Contingency plans shall be *developed for the entity that may endanger the environment and for the territory that may be endangered by the entity."* Contingency plans shall be developed for the economic mobilization entity, rescue units or business plan. *"In order for contingency plans to fulfill their functions, they need to be drawn up for certain periods of time and updated and refined on an ongoing basis."* A crisis plan can be defined as *"… a specific planning document in which measures are developed to address specific crisis situations that may endanger the planning entity in each environment, or to support the solution of which it must implement certain measures."* Every employee of the Ministry of the Interior is obliged to report a security incident or identified security risk without undue delay and, if necessary, to cooperate and provide co-operation in its resolution and elimination.

Crisis prevention and crisis phenomena

According to Koprla [30] the process of prevention and the solution of crisis phenomena can be effective enough only if it is solved and secured comprehensively. *"If it is to operate effectively and effectively perform its extensive and important functions, it must be supported and ensured by:*

- *revised and effective legislation, which is consistently elaborated down to the level of implementing regulations, decrees, and regulations,*
- *functional staffing,*
- *a developed structure of governing bodies with the necessary authority,*
- *the relevant executive bodies, which are based on a professional but also voluntary basis,*
- *effective technical means and equipment, which must include a sophisticated monitoring, information, and communication system,*
- *allocating enough funds, finally, the understanding and personal commitment of mainly top management, but also of the ruling political and party officials"*

Crisis manager competences

According to Šimák [31] there are many opinions on what constitutes a competency manager. The term itself can have different content, e.g., competence as a set of qualities, the ability to do something, as a qualification, or as a set of established responsibilities and authority. *"In the context of management, we speak of managerial competencies as personality characteristics of executive managers."* However, these personality traits are only relevant if their bearer is endowed with a fixed range of powers. We can then define the competencies of the manager as a set of knowledge, skills, abilities, and experience as well as physical and mental readiness to use these qualities to effectively perform certain tasks (functions and roles) in managing crisis operations in various phases of crisis management in accordance with assigned powers. and general expectations.

According to Novák (2005) *"The crisis is at the height of the risk. In nature, the crisis occurs as an unforeseen or sometimes as a predicted disaster. The crisis in the company and society in general is the result of unresolved or unresolved risks. Risks are a necessary solution here, so the effort is to know them. Sometimes risks can be identified, albeit very difficult, other times it is almost impossible. It is similar with the crisis."* Crisis is a term that is very often used in many situations today: *"environmental crisis, political crisis and also corporate crisis are situations that represent permanent or long-term negative deviations from the normal state. Signs of the crisis:—serious—endanger the very existence of the company,—less serious— endanger the long-term goal of the company."* In both cases, the structure of the company is acutely endangered. The events that can trigger a crisis can be of a different nature. *"In this context, we distinguish:*

(a) *a slowly developing crisis, eg: the disappearance of the market, the growth of costs that cannot be transferred to the market price, the increased turnover of workers.*
(b) *a sudden crisis that arose without a previous signal, eg: o state bans, o massive supply disruptions. Businesses must be prepared for both types of crises. Different types of strategies need to be developed for both types of crises."*

Crisis management

Every action is connected to risk. *"Every human activity or every phenomenon is associated with the fact that, in addition to positive consequences, it can also have negative consequences, greater or lesser in nature."* In everyday life, risk is associated with loss or failure, especially health, occupational risk, safety risk, investment risk, with risk in economic or political development. We can therefore state that *"Every activity, event or phenomenon taking place in nature, in society, in technical or technological systems can, under certain conditions, become uncontrollable and thus cause damage and loss of health and life to people, property, the environment and the environment. intellectual and cultural heritage."*

According to Míka [32] although the concept of risk is one of the most frequent, it is characterized by ambiguity, complexity, and ambiguity. The fact that terms such as the source of risk, risk factors, and causes of risk are often confused by many authors contributes to this. *"Risk is a phenomenon that has accompanied humanity since its inception. It can be said that the risk is all around us."*

Following the above, we have identified such research questions, the statement of which is verifiable based on testing in the survey.

4 Research Question No. 1: The Establishment of the DKO Will Create an Effective Communication Tool for Resolving Crisis Situations in Transport?

4.1 Effective Communication in Transport

In the event of a crisis, we will assume that people tend to seek help or advice if they have become direct witnesses, victims, or perpetrators of a traffic crisis. We will express this fact separately.

The first group will be a group of civilians who were involved in the traffic incident.

The second group will be members of the police force. From the penetration of both groups, we find out their perception in a crisis situation, specifically how they perceive the need for communication of both groups in terms of intensity. We assume that their motives will be different. The average person's feeling of anxiety is mainly due to ignorance of how to proceed in such an event [33].

The same common component is the search as well as the availability of information, which in crisis situations is sought by both components of respondents with different preferences. The availability of information is in principle necessary throughout the period under review in terms of accident investigation as well as the situation that has arisen directly. However, when it comes to crisis, people realize that time is a particularly scarce commodity in this case. It is necessary to have a continuous source of information available on the spot with professionals who know how to reassure the called party or immediately provide him with remote support for the correct action.

We will examine both groups of respondents in terms of immediate access to professional information in the form of a specialized form of support and assistance. Subsequent analysis of the occupancy of the operations center as the current first contact point for the participant in the accident reporting the accident. Crisis in crisis situations describes the active communication of participants as well as police officers intervening in a traffic accident. The analysis of the utilization of the operations center also obtains objective data for our survey, which we can compare with subjective ones.

By an overall analysis of the current form of operation of the operations center as a point of first contact, we will compare the results more subjectively, obtained by the questionnaire and objectively by analyzing the number of calls and total minutes called by the operations center with respondents. By comparing the collected data, we obtain outputs that will indicate whether DKO is an effective tool for communication in crisis situations [34, 35].

5 Crisis Management Theory

According to Mikolaj (2001) the basic objectives of crisis management are based on its tasks and are implemented based on the application of managerial functions. We can formulate the basic objectives of crisis management: "*1. Detection and analysis of possible sources of threat and its expected development 2. Development of crisis plans based on risk analysis and possible solutions 3. Development of a set of tools to manage crises and crisis situations and eliminate their consequences.*"

«Crisis *planning tasks are elaborated in the plans. We recognize two types of such plans.* "*It's an emergency plan and contingency plan. An emergency plan shall be drawn up for an entity that may endanger the environment and for an area that may be endangered by the entity.*" Contingeny plan shall be drawn up for an economic mobilization entity, a rescue unit or a business plan. "In order for contingency plans to fulfill their functions, they need to be drawn up for certain periods of time and continuously updated and refined." A contingency plan can be defined as "…a specific planning document in which measures are developed to address specific." Every employee of the Ministry of the Interior is obliged to report a security incident or identified security risk without undue delay and, if necessary, to cooperate and cooperate in its implementation solutions and removal [29, 20].

5.1 Prevention of Crisis Phenomena

The process of prevention and the solution of crisis phenomena itself can be sufficiently effective only if it is solved and secured comprehensively. "If it is to operate effectively and effectively perform its extensive and important functions, it must be supported and ensured by:

- Revised and effective legislation, which is consistently elaborated down to the level of implementing regulations, decrees, and regulations,
- Functional staffing,
- Developed bodies with the necessary authority,
- The relevant executive bodies, which are based on a professional but also voluntary basis,
- Effective technical means and equipment, which must include a sophisticated monitoring, information, and communication system,
- Allocating enough funding, not least also by the understanding and personal commitment of mainly top management, but also of ruling political and party officials".

Competences of a crisis manager

There are many views on what constitutes a manager's competencies. The term itself can have different content, e.g., competence as a set of qualities, the ability to do something, as a qualification, or as a set of established responsibilities and authority. "In the context of management, we speak of managerial competencies as personality characteristics of executive managers." However, these personality traits are only relevant if their bearer is endowed with a fixed range of powers. We can then define the competencies of the manager as a set of knowledge, skills, abilities, and experience as well as physical and mental readiness to use these qualities to effectively perform certain tasks (functions and roles) in managing crisis operations in various phases of crisis management in accordance with assigned powers. and general expectations [36, 37].

"*The crisis is at the height of the risk. In nature, the crisis occurs as an unforeseen or sometimes as a predicted disaster. The crisis in the company and society in general is the result of unresolved or unresolved risks. Risks are a necessary solution here, so the effort is to know them. Sometimes risks can be identified, albeit very difficult, other times it is almost impossible. It is similar with the crisis*" Crisis is a term that is often used in many situations today:"*environmental crisis, political crisis and also corporate crisis are situations which represent, for a permanent or prolonged period, negative deviations from the normal situation. Signs of the crisis:—serious—jeopardize the very existence of the company,—less serious—threaten the long-term goal of the company.*" In both cases, the structure of the company is acutely endangered. The events that can trigger a crisis can be of a different nature. "In this context, we distinguish:

(a) *a slowly developing crisis, eg: the disappearance of the market, the growth of costs that cannot be transferred to the market price, the increased turnover of workers* ...
(b) *a sudden crisis that arose without a previous signal, eg: o state bans, o massive supply disruptions ... Businesses must be prepared for both types of crises. Different types of strategies need to be developed for both types of crises.*"

5.2 Building a Positive Public Relationships

Investing in public relations will help organizations to reach own goals efficiently and smoothly. Public relations do not create a "good image for a bad team". Because the fake image cannot be maintained for long. Although an organization's product or service is good, it needs an effective public campaign to attract, motivate the public to the product or service or to the purpose of the program. Not only does this encourage public involvement, but it also leads to a better image [22].

Effective public relations can create and build the image of an individual or entity. In times of adverse publicity or when an organization is in crisis, effective public relations can eliminate "misunderstandings" and create mutual understanding between the organization and the public [39].

Public relations are ongoing _ thoughtful action on the differentiated the public to create favorable image and positive relationships. *PR has three main one's function:*

1. *inform people*
2. *convince them*
3. *improve mutual relations*

The main objectives of PR include: building awareness of the organization and its products as a public service, credibility and crisis preparedness, stimulating public interest on public administration activities, strengthening internal communication and motivating employees. According to Kotler, PR activities are made up of a set of tools called *"PENCILS:*

- P—publications—corporate magazines, annual reports, etc.,
- E—events—sponsorship, lectures, fairs and exhibitions,
- N—news—news about the company, its employees, products,
- C—community involvement—meeting the needs of local society,
- I—identity tools—business cards, dress rules,
- L—lobbyin—influencing legislative and regulatory measures,
- S—social investments—building a good reputation in the field of corporate social responsibility."

Efficiency surveys help to measure the impact on public relations as well as activities affecting public opinion. Prior to the start of public relations activities, it is appropriate to carry out a survey to find out the attitudes of the people and to carry out a further survey after the end of the campaign to see whether there have been any changes and whether public opinion is in favor. public relations objectives. Public relations play an important role in improving the organization's image in the eyes of the public [40, 41]. Public relations can be defined as the process of communicating with a group with which it exists and with which the organization operates.

In Slovakia, there is a problem with the provision of information. They are based on principles from the time of the socialist establishment of the police force and have not fundamentally changed since then. Our preferred communication way is a new form of external communication between the police force and those affected by the traffic accident.

The concept we proposed was supposed to function externally as support for road accident participants. Inside as a communication department for police members who are at the scene of a traffic crisis or are investigating the fact. The Transport Communication Department (TCD) containing the databases of the audiovisual files would provide a kind of central repository of the Audiovisual Record for its further analysis and the creation of valuable knowledge from the accident.

In our work, we concluded that respondents in a crisis in transport perceive the need for communication intensely. Their motives are diverse—different. In the case of a police officer, professional procedure, correct documentation and omission of footprints or other circumstances are paramount. The accident participant's predominant feelings are mainly due to ignorance of how to proceed in such an event. However, both groups of people perceive the need for even remote assistance, e.g., over the phone very intensively.

We noted that respondents expect information to be available, especially if it is a crisis. At the scene of the accident, it is necessary to have a continuous source of information available to professionals who know how to reassure the called party or provide him with remote support for the correct action immediately.

Both groups of respondents in the questionnaires requested immediate access to professional information and support through specialized forms of support. The data from the legislative framework [42] can be achieved by the transport communication department. Overall, the possibility of an online/telephone source of professional assistance was welcomed by an absolute majority of both groups of respondents.

The analysis of the data of the operation center KRPZ NR district Nové Zámky revealed the fact that the operation center is a highly occupied component of the Police Force. Depending on the nature of the accidents, the number of calls to the operations center from accident participants increases in the event of serious and fatal injuries, as does the number of calls from intervening police officers. The frequency of communication is extremely tense in places.

The creation of communication channels will increase the capacity of professional staff able to provide professional full-fledged support through a telephone call or other electronic media. An essential fact is our proposed solution, which would be a department with a central data registry and system interconnection. It would speed up the distribution of data and outputs to senior police management as well as support for the population. Finally, the workplace would have data available to the media, which would no longer have to come to the scene immediately and thus interfere with the accident sites during the investigation [43].

5.3   The Reliable Functioning of the TCD

For the reliable functioning of the TCD it is necessary to adopt:

- Staffing (head or director, psychologist, experienced communication staff, to ensure continuous service),
- Spatial support within the Presidium of the Police Force, (location could be in the current tower of the operation center)
- Technical support high-speed connection of computers and connection to monitoring equipment as well as the operations center, it would also count on extended permissions to all relevant systems under the Ministry of Interior (Police, Firefighters), data server operating on Linux
- Information and communication security: approx. 4 desktop computers, 2 mobile phones, 2 shortwave radios, optical internet connection.
- Equipment: the uniform is not necessary in this case as it is a service located in a building without physical contact with the civilian population
- Operating costs (wages, hardware upgrades, regular training of an international nature) [44].

Advantages:

- The creation of a "TCD" could be an effective form of communication in crisis situations with disabled people, which would specialize mainly in issues related mainly to traffic and crisis situations in transport, which would use the rights and legitimacy of a police officer. Thanks to the remote access and professional composition, it would operate all-Slovak.
- Cooperation is of paramount importance, as the current way of communication has been trying to modernize in recent years, (Facebook, written or e-mail inquiry) it should be noted that this is always a general impersonal form of communication. TCD would act not only as a communication department for the civil sector but would be a basic element in creating the concept of accident sections and preventive measures within Slovakia and sharing information on-time.
- The media as such would no longer have to attend explicitly tragic road events. The audio-visual recording media would be available on a shared server equipment platform, in which data would be stored immediately from the event site.
- It is probable that the presence of remote cooperation with police officers in the field with the public gives the impression of some control by the Presidium. The police force has suffered from a lack of credibility in the eyes of the population in the last year, and the TCD traffic communication department proposed by us would be able to cope with this situation.
- During the investigation, due to inattention and time complexity, there are trampling during the investigation. Some tracks that are already unrepeatable are often overlooked. What would the TCD contribute to this case with specially trained personnel and would take care of the methodological approach in the field as well as the remote provision of the accident.
- From the point of view of a working police officer, this would be work that would not pose a direct danger to the profession.
- Concentration of work in one place. It can be considered as an advantage to a certain extent. However, many executive officers prefer to work in which they have some leeway. However, these are usually younger members.
- The operation of such a center is independent and not harmful to health. The possible exposure of a person to a possible chemical leakage of substances or hazardous biological material is impossible in this case.

Risks:

- Improper retraining and redistribution of competencies related to the performance of the service.
- Further bureaucratic interference in the process of implementing the TCD" project in order to centralize power in the hands of police officers.
- The risk may also lie in the case when persons close to the management enter the given department, while they would not meet the basic professional preferences.
- Non-trained police officers who have not served in the field of transport or have not dealt with traffic accidents will have a more difficult ability to practice.
- Lack of willing police officers with experience to work indoors.

Other supporting recommendations that we found during the work and consider important for the practice of both Police members and social workers.

More surveys must focus on the implementation of this goal, with emphasis on international training and exchange of information from police forces of other states. The police in Great Britain, but especially the police in Australia, have very good and elaborate principles.

It is recommended that a further needs assessment be carried out, with an emphasis on the impartiality of the implementation of the objective. It is necessary to realize that the way of functioning of state bodies in the Western world and in Anglo-American countries is different with the countries of the Eastern bloc. However, it is probable that with the growth of separate information systems, we expect a high success of the system with an emphasis on providing services to the citizen. This system, on the other hand, would prevent or reduce the level of stress on police officers in the performance.

We consider the aim of the dissertation elaboration of a draft justification of the justification of the establishment of the Transport Communication Department as a general tool as well as a specialized means of communication as fulfilled with the designated sub-tasks as fulfilled.

6   Conclusion

The aim of our scientific study is the analysis and synthesis and deduction of scientific and professional literature, the study of professional literature sources concluded that a change in communication with the public in public administration, the Police Force is a necessity. we can use the tools of the marketing communication mix to improve communication. Based on our extensive scientific research, we have found that the public expects sufficient communication value with the public. We think that the public should have sufficient communication, especially in traffic accidents. Marketing communication with your tools can also be used for this.

References

1. Hendrych, D.: Správní věda. Teorie veřejné správy. Praha: Wolters Kluwer ČR, a.s., 232 s (2009). ISBN 978-80-7357-458-1
2. Prucha, P., Pomahač, R.: Lexikon: Správní právo. 1. vyd. Sagit, Ostrava, s. 683 (2002). ISBN 80-7208-314-7. 246–247
3. Balejová, A.: Dôvody a podmienky pre profesionalizovanie výkonu verejnej správy na Slovensku. In: Kvalita samosprávneho manažmentu na miestnej úrovni, Zborník príspevkov, Trenčín: Inštitút aplikovaného manažmentu (2012). 65 s. ISBN 978-80-89600-09-0
4. Hamalová, M.: Teória, riadenie a organizácia verejnej správy. 1. časť: Teória verejnej správy. Bratislava: Merkury, 132 s (2007). ISBN 978-80-89143-58-0
5. Kosorín, F.: Verejná správa (koncepcia, reformy, organizácia). Ekonóm, Bratislava, 168 s (2003). ISBN 80-225-1696-1

6. Papcunová, V., Gecíková,: 1 Základy verejnej správy. Bratislava: VŠEMVS, pp. 160 (2011). ISBN 978-80-970802-0-4
7. Bogumil, J., Kuhlmann, S.: Zehn Jahre kommunale Verwaltungsmodernisierung Ansätze einer Wirkungsanalyse, edn. In: Status Report Verwaltungsreform-Eine Zwischenbilanz nach, vol. 10, pp. 63. Sigma, Jahren, Berlin (2004). ISSN 1640-3622
8. Sindleryová, I.B., Čajková, A., Sambronská, K.: Pilgrimage in Slovakia—a hidden opportunity for the management of secular objects? Religions **10**(10), 560 (2019). https://doi.org/10.3390/rel10100560
9. Săraru, C.S.: The European groupings of territorial cooperation developed by administrative structures in Romania and Hungary. Acta Juridica Hungarica **55**(2), 150–162 (2014). https://doi.org/10.1556/AJur.55.2014.2.5
10. Săraru, C.S.: Considera ţii cu privire la limitele libertăţii contractuale în dreptul public impuse de integrarea în uniunea europeană. Transylvanian Rev. Admin. Sci. **1**, 131–140 (2008)
11. Wright, G., Nemec, J.: Manažment verejne spravy. Teorie a praxe – zkušenosti z transformace verejné správy ze zemí stredné a východní Evropy. NISPAcee, Bratislava, 419 s (2003). ISBN 80-86119-70-X
12. Peráček, T.: E-commerce and its limits in the context of the consumer protection: the case of the Slovak Republic. Juridical Tribune - Tribuna Juridica, **12** (1), 35–50 (2022). https://doi.org/10.24818/TBJ/2022/12/1.03
13. Buchta, M.: Faktory a hodnotenie efektívnosti vo verejnom sektore v oblasti efektívnosti verejného sektora, Sbomik praci Asociace vetojné ekonomie, RSF MU, Brno, str. 105–106 (2007)
14. Rittinghouse, J. W., Fransome, J.: Grids, clouds and virtualization: implementation, manažment, and security. 1st. edn. pp. 301. Springer, New York (2010)
15. Peracek, T., Andrukhiv, A., Sokil, M., Fedushko, S., Sycrov, Y., & Kalambet, Y.: Methodology for increasing the efficiency of dynamic process calculations in elastic elements of complex engineering constructions. Electronics, **10**(1), pp. 1–20, 40 (2021). https://doi.org/10.3390/electronics10010040
16. Milošovičová, P. et al.: Medzinárodné ekonomické parvo, 1st edn. Wolters Kluver, Praha, 284 s (2017). ISBN 80-72369-456-1
17. Kita, J et al.: Marketing. Bratislava: Wolters Kluver. (2014) ISBN 8089047238
18. Štarchoň, P.: Priamy marketing alebo Priama cesta ako si získať a udržať zákazníka. Bratislava: Direct Marketing Beta, 340 s (2004). ISBN 80-969-0785-9
19. Kotler, P.: Marketing. Praha: Grada Publishing (2004). ISBN 80-82369-230-1
20. Lacuška, M., Peráček, T.: Trends in global telecommunication fraud and its impact on business. Stud. Syst. Decis. Control **330**, 459–485 (2021). https://doi.org/10.1007/978-3-030-62151-3_12
21. Peráček, T., Mucha, B., Brestovanská, P.: Selected legislative aspects of cybernetic security in the Slovak Republic. Lect. Notes Data Eng. Commun. Technol. **23**, 273–282 (2019). https://doi.org/10.1007/978-3-319-98557-2_25
22. Labská, H.: Marketingová komunikácia. EKONÓM, Bratislava, s. 324 (2004). ISBN 978-80-79153-86-0
23. Frey, P.: Marketingová komunikace: nové trendy 3.0. Vyd. 3. Praha: Manažment Press, s. 211 (2011)
24. Palatková, M.: Marketingový manažment. Grada Publishing, Praha, s. 207 (2011). ISBN 978-80-247-3749-2
25. Kotler, P.: Marketing manažment. Praha: Grada Publishing (2017) ISBN 80-73869-230-2
26. Peráček, T.: Flexibility of creating and changing employment in the options of the Slovak Labor Code. Probl. Perspect. Manag. **19**(3), 373–382 (2021). https://doi.org/10.21511/ppm.19(3).2021.30
27. Hajduova, Z., Peracek, T., Coronicova Hurajova, J., Bruothova, J.: Determinants of innovativeness of Slovak SMEs. Probl. Perspect. Manage. **19**(1), 198–208 (2021). https://doi.org/10.21511/ppm.19(1).2021.17

28. Plavčan, P., Funta, R.: Some economic characteristics of internet platforms. Danube **11**(2), 156–167 (2020). https://doi.org/10.2478/danb-2020-0009
29. Popa, M.F.: What the economic analysis of law can't do-pitfalls and practical implications. Juridical Tribune **11**(1), 81–94 (2021)
30. Koprla, M.: Nové trendy v marketingovej komunikácii: Zborník medzinárodnej vedeckej konferencie. Trnava: Fakulta masmediálnej komunikácie UCM, s. 238–264 (2010). ISBN 978-80-8105-167-8
31. Šimák, L.: Krízový manažment vo verejnej správe. Žilinská univerzita v Žiline. Fakulta Špeciálneho inžinierstva. Žilina. (2001). ISBN 80-88829-13-5
32. Míka, V.T:. Základy manažmentu.Virtuálna kniha. FŠI ŽU, Žilina, s. 125 (2006)
33. Mura, L., Ključnikov, A., Tvaronavičienė, M., Androniceanu, A.: Development trends in human resource management in small and medium enterprises in the Visegrad Group. Acta Polytechnica Hungarica **14**(7), 105–122 (2017). https://doi.org/10.12700/APH.14.7.2017.7.7
34. Mura, L., Mazák, M.: Innovative activities of family SMEs: case study of the Slovak regions. Online J. Model. New Europe **27**, 132–147 (2018)
35. Nováčková, D., Vnuková, J.: Competition issues including in the international agreements of the eropean union. Juridical Tribune **11**(2), 234–250 (2021). https://doi.org/10.24818/TBJ/2021/11/2.06
36. Čajka, P., Abrhám, J.: Regional aspects of V4 countries' economic development over a membership period of 15 years in the European union. Slovak J. Polit. Sci. **19**(1), 89–105 (2019). https://doi.org/10.34135/sjps.190105
37. Čajka, P., Olejárová, B., Čajková, A.: Migration as a factor of Germany's security and sustainability. J. Secur. Sustain. Issues **7**(3), 400–408 (2018). https://doi.org/10.9770/jssi.2018.7.3(2)
38. Cajková, A., Jankelová, N., Masár, D.: Knowledge management as a tool for increasing the efficiency of municipality management in Slovakia. Knowl. Manage. Res. Pract. (2021). https://doi.org/10.1080/14778238.2021.1895686
39. Čajková, A., Šindleryová, I.B., Garaj, M.: The covid-19 pandemic and budget shortfalls in the local governments in Slovakia. Sci. Papers Univ. Pardubice Ser. D Faculty Econ. Admin. **29**(1), 1243 (2021). https://doi.org/10.46585/sp29011243
40. Horváthová, Z., Čajková, A.: Social and economic aspects of the EU's education policy. Integr. Educ. **22**(3), 412–425 (2018)
41. Jančíková, E., Pásztorová, J.: Promoting eu values in international agreements. Juridical Tribune **11**(2), 203–218 (2021). https://doi.org/10.24818/TBJ/2021/11/2.04
42. Fedushko, S., Ustyianovych, T., Gregus, M.: Real-time high-load infrastructure transaction status output prediction using operational intelligence and big data technologies. Electronics **9**(4), 668 (2020).https://doi.org/10.3390/electronics9040668
43. Kryvinska, N., Kaczor, S., Strauss, C.: Enterprises' servitization in the first decade—retrospective analysis of back-end and front-end challenges. Appl. Sci. **10**(8), 2957 (2020). https://doi.org/10.3390/app10082957
44. Mucha, B., Peráček, T., Brestovanská, P.: Strategy of environmental policy of the Slovak republic until 2030—an effective tool for combating waste? Int. Multidisc. Sci. GeoConf. Survey. Geol. Mining Ecol. Manage. SGEM **19**(5.4), 541–548 (2019). https://doi.org/10.5593/sgem2019/5.4/S23.071
45. Equality monitoring 2019/20 DfT Group summary. In: House Analytical Consultancy (2021). [cit. 2021-01-01]. Dostupné na: https://assets.publishing.service.gov.uk/government/uploads/system/uploads/attachment_data/file/585055/equality-monitoring-dft-c.pdf
17. Štarchoň, P.: Vademecum reklamy. Vybrané teoretické aspekty. Bratislava: Univerzita Komenského, 122 s (2004). ISBN 80-223-2012-9
47. Adamisin, P., Kotulic, R., Mura, L., Kravcakova Vozarova, I., Vavrek, R.: Managerial approaches of environmental projects: an empirical study. Polish J. Manage. Stud. **17**(1), 27–38 (2018). https://doi.org/10.17512/pjms.2018.17.1.03
48. Jankurová, A., Ljudvigová, I., Gubová, K.: Research of the nature of leadership activities. Econ. Sociol. **10**(1), 135–151 (2017). https://doi.org/10.14254/2071-789X.2017/10-1/10

49. Kryvinska, N., Bickel, L.: Scenario-based analysis of IT enterprises servitization as a part of digital transformation of modern economy. Appl. Sci. **10**(3), 1076 (2020). https://doi.org/10.3390/app10031076

Aspects of Strategic Management and Online Marketing

Anton Lisnik and Milan Majerník

Abstract Strategic management is a way of managing a company based on the most important documents of the company such as vision and mission. Communication with customers, whether in the field of direct sales or sales and for further business are nowadays mainly based on quality marketing. Marketing is thus part of strategic decisions in the field of business management. Present time, which uses virtual space as a basic means of communication for the presentation and sale of products. Especially as a result of the Covid period, virtual marketing gained a priority in marketing activities and further globalized trade and business. The presented article describes the place of online marketing in business management, mainly highlights its place in strategic decisions and management.

Keywords Strategic management · Enterprise · Online marketing · Customer · Business management · Marketing activities · Management

1 Introduction

Planning, organizing and controlling are the basic functions and tasks of management. Today and the new circumstances in which we find ourselves today are a great challenge for us. Management is faced with many decisions to be as competitive as possible and to maintain its market position in the long run. To make decisions about future management, they must have up-to-date data, be able to recognize external influences and know their performance. Performance analysis is considered complete if it measures the external and internal environment of the company. The competitive environment puts pressure on effective business management conditioned by obtaining current and relevant information. Marketing has a place in the company's

A. Lisnik (✉) · M. Majerník
Institute of Management, Slovak University of Technology, Vazovova 5, 812 43 Bratislava, Slovak Republic
e-mail: anton.lisnik@stuba.sk

M. Majerník
e-mail: milan.majernik@stuba.sk

N. Kryvinska et al. (eds.), *Developments in Information and Knowledge Management Systems for Business Applications*, Studies in Systems, Decision and Control 462, https://doi.org/10.1007/978-3-031-25695-0_3

strategy and vision, which uses virtual space to sell products. Today, it is very popular given the circumstances we have experienced and are experiencing, such as the Covid pandemic and its consequences, but also the high inflation and the impact of the war in Europe.

In the first part of the article we deal with the principles and rules of strategic management. In the next part we describe the principles of online marketing and its possibilities and principles of use in marketing activities of entrepreneurs.

2 Strategic Management

Building a competitive advantage must become a part of an active strategic management. Active strategic management consists of creating effective mechanisms enabling the company management to anticipate future market developments and its impact on competitiveness, as well as influencing it for their own benefit. Strategic management must ensure synergy requirements and expectations of customers with internal and external business processes. Otherwise, businesses cannot sustain their competitive advantages in the long term.

Today, businesses need dynamic and credible indicators to measure their performance. This means discovering, tracking and exploiting key strategic and competitive factors for the company's future development. It is therefore necessary to find a link between the short-term mission and long-term vision rather than the control function of the annual budgets and the strategic vision. The goal of each company is to successfully implement the strategy. Transforming the strategy into a metering system allows for more effective realization of strategic goals. Business performance indicators thus provide a more dynamic and forward-looking picture of the company's competitive position and a much better idea to further improve the company's performance.

At the beginning of its establishment, each company set itself a vision and goals that it would like to achieve in a certain time. Transforming a vision into reality is about planning. The task of strategic management is also to choose the right planning methods. In practice, it is important to classify planning methods according to the degree of their objectivity into: qualitative methods: Delphi method, scenario method), benchmarketing, quantitative methods—applied economic statistics, mathematical statistics, methods of operational analysis and economic-financial analysis [1].

2.1 Basic Terms

Non-financial indicators are at the forefront of business performance assessment especially in the context of developing new theories of performance measurement and management. Attention is focused on the development of such measurement performance indicators that include not only financial inputs but also non-financial ones that would accept individual functional areas of the business and the business strategy and that would allow performance measurement at each management level.

Performance measurement is the process of quantifying the efficiency and effectiveness of business activities. The efficiency corresponds to the extent to which the customer's requirements are filled. Efficiency is a measure of how economically an organization's resources are used to provide a certain level of customer service.

Evaluation and measurement of enterprise performance are tied to the defined goals of the company formulated in the strategy. Through these objectives the owners formulate their capital appreciation requirements as well as other objectives, and the executive management manages the business to achieve these objectives. Therefore, evaluating a company's performance means comparing the results achieved with the objectives formulated in the business strategy (as well as comparing these results with other comparable results). These shortcomings of financial indicators eliminate non-financial indicators. The application of non-financial indicators ensures compliance with the requirement that other areas and aspects of performance aiming at long-term prosperity are incorporated into performance measurement and management. These indicators are considered as leading indicators. In order to determine non-financial indicators, it is necessary to start from the long-term goals and strategies of the company. In this case the measurability of these indicators is particularly demanding. These indicators are divided into measurable indicators such as an increasing number of customers, increasing market share, and shortening customer service times and other. The most difficult is to measure the second group of indicators - intangible indicators such as innovation, corporate culture, customer satisfaction, customer loyalty, etc. If one of these indicators cannot be expressed in physical units it is necessary to choose another indicator.

Generally, strategy can be defined as 'a way, method, mean or tool for achieving pre-set goals'. There are more ways how proposed goals can be achieved. Strategy is being created before the company starts to really implement it and it is a result of rational considerations. Strategy is an idea, conception and culture, which all the employees identify with and moves them into the future [2].

Henry Mintzberg characterizes the strategy using alternative "five P's"—the strategy is Plan, Pattern, Position, Perspective and Play.

Plan is a purposeful guidance and action, i.e. instructions on how to determine a particular situation.

Pattern is a certain regularity in behaviour and action. It determines the logic of strategic thinking.

Location means placing a company in an environment that is conditional on its competitors and market shares.

In this sense, perspective is an idea, a concept and a culture that goes inwards of the company, with which all employees of the company have identified and oriented them into the future.

Play is part of a plan and strategy in terms of the dodge by which we want to change our opponents.

2.2 Strategic Goals

Strategic goals of the company should come from a pre-set vision and a mission. They define the final state which the organisation is heading into. Their purpose is to define measurable result of the vision. Strategic goals are focused on strengthening the market position (segment position. higher production quality, lower costs compared to the competition, wide product range and other).

One of the simple and illustrative models of strategic management takes into consideration the continuity of basic characteristics of strategic management and deepens its internal structure with particular steps. Strategic management is mostly process in the new understanding. Process which basic attributes are continuity, compactness, and internal harmony. An iterative process, gradually going through individual phases and their steps. The strategic management process (Fig. 1) consists of examining the environment, formulating a strategy, implementing a strategy, evaluating, and controlling.

Part of the **environment research** is a detailed analysis of the internal and external environment of the organization. *The external environment analysis* focuses on identifying such opportunities and threats to the external environment that could affect (positively or negatively) the achievement of the set strategic goals. The *internal environment analysis* of the company aims to identify the strengths and weaknesses of the organization in relation to the set strategic goals, vision, and mission of the organization. The purpose is also to identify key competencies and key vulnerabilities of the organization.

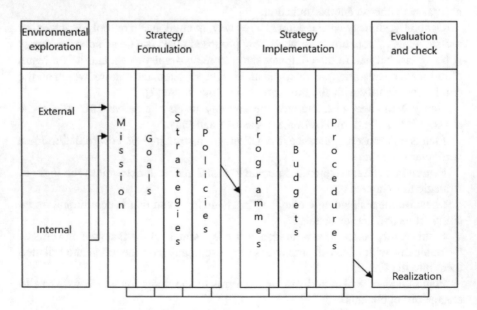

Fig. 1 Strategic management model. *Source* [3]

The formulation of the strategy conceals above all the determination of the business vision and mission of the organization. Based on them, the long-term strategic goals of the organization are set. The target orientation of the company is based on the company's perspective on the future, which is factually and timely structured. The idea of the future direction and position of the company in the farthest time horizons is expressed in the form of a vision, the more specific expression of which is in the mission of the company. The formulation of the strategy, following the results of the research of the environment, proceeds in steps:

- **definition of a mission**, clarifying the purpose and meaning of the company's existence,
- **formulation of strategic goals** that specify the direction of the company,
- **creation of strategy** as a way to fulfil strategic goals,
- **establishing an implementation policy** as a space for decision-making process.

The implementation of the strategy is a process of planned preparation for the implementation of the strategy. It includes the following formation steps:

- **Programme formation**, which mission is to concretize which procedures and steps of each functional fields will be inevitable to implement, thus programmes of organisational and informational character, implementation programmes and implementation plans,
- **Budget formation**, which task is to finalize the programs on the value side of the 2 perspectives of costs, revenues, and resources movement,
- **Procedures formation**, that elaborate programs into detailed sequential steps or techniques [3].

The last phase of strategic management is evaluation and control. This phase fulfils the monitoring, cognitive, control and evaluation function. It monitors and evaluates the achieved results, while giving impulses through the created feedbacks to make the necessary changes. It gives the whole process of strategic management character of a dynamic and continuous process [3].

Strategic management is a continuous process. Once the strategy is implemented, its functioning must be constantly monitored to make it clear to what extent the set goals are actually being met. Feedback serves to confirm existing goals and strategies or recommends a change.

The environment investigation is one of the essential phrases of the process of strategic management. Significance of this phase grows with the change of environmental character. The more unpredictable changes in the environment and the more the uncertainty of its development, the more grows the need to look at the company as a relatively closed system that responds to changes in the environment [4].

Analysis of the marketing/business environment of the company is one of the essential starting steps in the creation of the company strategy. Its purpose is to find out the effects of basic strategic factors—business opportunities and environmental threats, strengths and weaknesses of the company, respectively its specific virtues and weaknesses, what is used for determining the company's position [5].

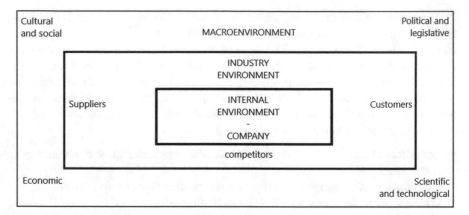

Fig. 2 Structure of the enterprise environment. *Source* [3]

The more unpredictable changes in the environment and the more the uncertainty of its development, the more grows the need to look at the company as a relatively closed system that responds to changes in the environment [4].

The external environment in which the company is located consists of two spheres. Higher sphere is called the microenvironment and consists of the economic, social, technical, demographic, legal, political, and environmental sectors. Lower sphere is called the industry environment and contains factors that directly affect the position of the company. These are companies' competition, customers, suppliers and other factors [6].

For a better orientation in the environment and the possibility of choosing appropriate procedures, corresponding to the needs of strategic management, we divide the company's environment into the following layers shown in Fig. 2.

The purpose of the macroenvironment examination is to reveal its internal structure and existing relationships, then on the basis of which potential opportunities and threats that arise in this environment are identified. We can only talk about opportunities when external needs allow a company to make higher profits. Threats arise when external trends call into question the existence and profitability of a business. However, a threat to one business is often an opportunity for another business.

2.3   Strategic Analysis

The industry environment is being referred in the professional literature as a competitive environment. An industry is defined as "a group of enterprises offering products or services that are interchangeable" [6, p. 66].

The aim of the sectoral and competitive analysis is to reveal strategic factors that affect the overall situation in the sector and are a source of opportunities and threats for individual companies. The analysis consists in examining the main economic parameters of the industry, the driving forces of the industry, the nature and strength of the competitive forces, the positions of key competitors and their likely behaviour, and in assessing the attractiveness of the industry.

The complex and most used analysis for examining the external and internal environment is the SWOT analysis (S-Strength, W-Weakness, O-Opportunity, T-Threat).

The SWOT analysis is based on the existence of internal strengths and weaknesses of the company and its external opportunities and threats. Strengths and weaknesses are signs of the organization's internal capabilities. Strengths are internal capabilities through which an organization is able to operate effectively in an external environment. Weaknesses indicate a lower capacity for the use of external opportunities, as well as weaker protective effect against possible threats of the external environment.

Opportunities and threats exist outside the organization in various fields. These include factors from local and global environment, the macro environment, the interaction environment and the industry environment. A typical feature of opportunities and threats is that they affect all actors in the defined environment, i.e. not only to our organization. The differences between organizations are precisely in the ability to adapt effectively, respectively to strategically prepare for the impacts of the environment.

The aim of the SWOT analysis is to:

- Identify the opportunities and threats in the external environment and strengths and weaknesses of the organisation,
- Evaluate their importance from the mission fulfilment point of view of and strategic aims of the organisation,
- Analyse and evaluate the relationship between external changes and internal abilities and qualifications in context in order to reveal:

  - how the organization, based on its internal capabilities, is prepared to respond to external influences (on what opportunities it is best able to respond, or to what threats it is least prepared to resist),
  - how it should change its abilities and capabilities if it is to respond effectively to anticipated externalities (which abilities should it change in order to be better prepared to take advantage of existing opportunities and face existing threats) [7].

The completion of an analytical work is a synthesis of the analysis results of the external and internal environment of the company. The synthesis consists of comparing external threats and opportunities with internal strengths and weaknesses of the company. Their combination and penetration creates a strategy as a balancing factor that brings the company into harmony with its environment. Such a strategy will allow the company to focus only on opportunities that match its capabilities and will allow it to avoid threats that it cannot defend itself against. The matrix of the result synthesis of the SWOT analysis (Fig. 3) shows different variants of the strategy due to the different significance and weight of the individual analysed items.

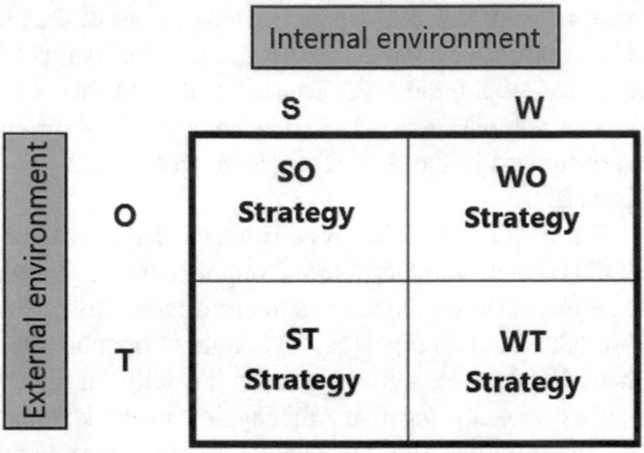

Fig. 3 Matrix Synthesis of
the SWOT Analysis Result.
Source [6, p. 116]

		S	W
		SO Strategy	**WO Strategy**
	O		
	T	**ST Strategy**	**WT Strategy**

Internal environment

External environment

The SO strategy is the most attractive strategic option. It can be chosen by a company in which strengths prevail over weaknesses and opportunities over threats. Due to its powerful strength, the company is able to take advantage of all the opportunities offered. An offensive strategy from a position of strength is recommended.

The ST strategy is the strategy of a strong company located in an unfavourable environment. A strong position should be used to block dangers, intimidate competition, or escape into a safer environment. A defensive strategy is recommended for the company protecting its already acquired position.

The WO strategy is chosen by a company in which weaknesses prevail over strengths but is located in an attractive environment. In order for the company to take the advantage of now-opening opportunities, for what it does not have enough internal skills to manage, it gradually tries to strengthen its position and eliminate shortcomings. An alliance strategy is recommended to increase inner strength and share opportunities with a reliable ally.

The WT strategy is suitable for a company that is weak and, in addition, located in an unattractive environment. Such a company must, at best, consider leaving this particular business and try to establish itself in a more favourable environment where its weaknesses would not be so significant, or, worse, will reduce and liquidate its business activities [6].

Porter's five-force model is used in the analysis of the external environment in strategic management.

Another option to analyse the external environment is the Porter 5F model, respectively Porter's Five-Force Model, which analyses the competitive environment from the point of view of market players [8]. The main five factors also called driving forces are [9]:

- Threat to new entrants to the market (competitors),
- The level of competition between entities on a given market,
- Negotiating customer force,
- Negotiating power suppliers,

- Threat of product substitution,
- GE matrix.

The GE matrix represents a more advanced version of the Boston Matrix (BCG matrix) with a finer division of 3X3, which makes it possible to classify the market position more accurately. The matrix has an internal factor on the horizontal axis—the competitive position and the external factor—the attractiveness of the market [10]. Mentioned factors are not understood as one-dimensional quantities but consist of elements that represent a complex of factors influencing a given factor [11]. The first is the identification of the partial elements and the assignment of the coefficient to the element—the coefficient representing the significance of the element for the subject. Sum of coefficients is equal 1. The assignment of coefficients to individual elements can also be understood as the degree of factor dependence on a given element. The resulting sums values transferred into the matrix represent the product/subject position in one of the matrix segments [12].

Aspects of strategic management significantly affect other parts of business management, such as marketing, costs or investing. Together, all of these form a complex of business management tasks.

3 Aspects of Marketing

American Marketing Association [13] defines marketing as "an activity, set of institutions, and processes for creating, communicating, delivering, and exchanging offerings that have value for customers, clients, partners, and society at large."

3.1 Basic Terms

As a result of the increasing need for quantitative parameters to characterize the marketing and the business itself the creation of countless indicators aimed at measuring almost everything. The concept of performance becomes the key since we consider performance to be the maximum achievable performance or the ideal performance that is required (Fig. 4).

Traditional marketing performance measurement has been focused on the evaluation of the effect of marketing activities on sales. Fundamental reason for it was, that the information about sales were continuously available, but the information about revenues wasn't, they require more summarised data processing. Measure of shares on the market as a main predictor of the revenues and profitability came into the fore in the 1980s.

Gradually, the growth of the market shares has started to be understood as a main function of the perceived quality. In 1996 Aaker and others pointed out on the fact, that the stock price movements could be explained by the combination of

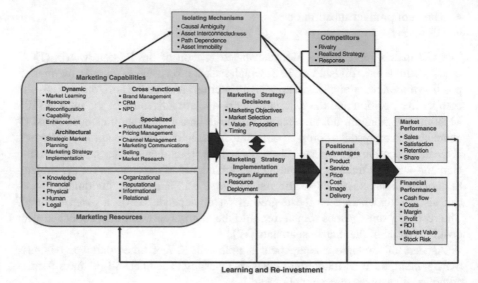

Fig. 4 An integrated conceptual framework linking marketing and business performance. *Source* [14]

the change in the return on investments and brand equity and not just through ROI itself. Quality as the competitive advantage could be used less and less because it affects the customers for a short time and also due to wide convergence of quality standards (such as the introduction of Total Quality Management—TQM systems). Such development also stimulated the beginning of the use of whole group of other indicators, which until then were on the 'lower ranks of importance' of measuring of the company's success.

3.2   New Parameters as the Response to Environment Changes

New parameters which came into the fore express the substance of the relationship between the customer and the product, service, the company itself and also the quality of the experience emerging from it (indicator called **user experience**).

The essential indicators in this context therefore include:

- Loyalty of the customer
- Capitalized value of the brand
- Customer satisfaction

Studies from the end of 1990s prove significant statistic relationship between the customer satisfaction and future financial performance of the company, which is positive [15]. Prevalent act was the publishing of the book Balanced Scorecard by

Kaplan and Norton (1996), which offered way of connecting the business strategy into four managerial perspectives: financial, processing, knowledge, and customer one.

Other authors pointed out that the financial indicators like revenue, sales and financial flow can be replaced by non-financial indicators like market share, quality, customer satisfaction, loyalty and capitalised value of the brand. Increase of the number of marketing measurement indicators created and used was the consequence of more factors. One of them is the expanded and improved database technologies, which enabled the companies to gather more information about their customers and to some extent, about their competitors and their customers.

Another impact is the establishment of new distribution channels for products and services such as the internet, which significantly increased the availability and complexity of marketing indicators, which have not been limited only to measuring customer value and return on the investments.

3.3 Modern Tools Enabling the Observation of the Users

Growth of the number of indicators has led to the considerations about the optimal number of monitored parameters, that the top management can effectively watch, but it also supported the development of techniques and tools that allow their observation. Modern tool used for displaying the values of individual indicators and process of their fulfilment is so-called **dashboard** or 'console'. It is a software allowing the observation in high graphic quality and well-structured form.

3.4 Marketing Factors

Factors influencing the choice of indicators.

The choice of marketing measurement indicators is influenced both by objective and subjective factors within the company. Authors Knowles and Ambler [15] present 4 main groups of factors influencing on the choice of marketing indicators, namely:

1. **Perception of the paradigm of modern marketing** in the last decade, which is characterized by areas such as building relationships with customers (not only in the perspective of business transactions), typical for relationship marketing. Then also the implementation of the aspects of companies' society responsibility (we can rather talk about the efforts and the search for appropriate approaches of the companies in this both socially and ethically sensitive area, but this is author's opinion); higher involvement of the marketing into affairs of the company also through phenomena like social networks. Current marketing significantly affects the way of finding profitability for the customer (through the value of the product

for the customer) and also for the company and making a profit from how the
company satisfies the customer's needs.

2. **Business model**. Company chooses those indicators which connect areas and
   impacts of the managerial decisions on the company's financial results the best.
   Then their task is to monitor the performance of the most-important strategies,
   i.e. their realization in the process of value creation.
3. **Aims**. Choice of the indicators should monitor the aims, which are so-called
   crucial for the creation of financial value and not only fulfilling their target, but
   also their fulfilment over time. They are not only financial, but also indicators
   from the customer field, from the company's internal processes and from learning
   and development of the company's skills.
4. **Time frame**. Companies with a long sales cycle or with a focus on building a
   corporate position and valuable references should quantify and monitor indicators
   with the benefit of profits in the future. Within this approach, there is an emphasis
   on the evaluation of so-called capitalized value of the brand or the valuation of
   the brand capital. We consider these terms to be equivalent and are based on the
   interpretation of the market capitalization, that means how the brand is evaluated
   by the market, what 'potential does it have to bring the capital to the owner by
   selling shares on the market'.

3.5 Quantitative Indicators of Measuring Performance of the Company

Classic quantitative indicators of measuring performance of the company include
great number of indicators, whose methodology is theoretically described and implemented in practice by experts in the field of market research. These include indicators
such as market share (value, volume, absolute, relative, share of the product, category share), brand penetration, spontaneous and promoted brand knowledge, advertisement knowledge, intention to buy, shopping habits, willingness to recommend,
indexes of customer satisfaction and other indicators.

Rating of the sale supporting activities include an important area of basic sale
modelling, (sale achieved by sale supporting activities); then it is the price of the sale
supporting activities (that usually means overall expenses with coupons and rebates
included), the share of sales within the sales promotion event on total sales, amount
of the share of sales time with activities to support sales in the total sales time (for a
certain period), the indicator of the so-called "Price waterfalls".

Management of distribution channels evaluates many parameters, beginning with
the performance of sales representatives e.g. through sales volume or by total sales.
Then achieved indicator of numerological distribution is examined, total supply level
of the customers, missing goods on the sale spot (so-called out of stock is examined).

Parameters of compensation or bonus part of the salary are also examined, number
of sales representatives and their performance by regions, their workload, individual

performance parameters like the number of visits, the success, visits in accordance with the schedule, unplanned visits etc.

The measurement of advertising, communication activities requires e.g. the measurement of so-called Gross Rating Points (share of the advertising 'interventions' on total number of people in the auditorium) and indicators such as: price calculated on 1000 'interventions', net 'intervention' = number of people reached by the advertisement, average frequency of it, share of total advertising etc. New types of advertising on the internet brought new measurement indicators like cost per one 'click', parameter of 'clicking through', number of webpage visits, costs of an order and for gaining a new customer, decrease in webpage visits etc.

3.6 Marketing Indicators

Permanently current is the field of measurement indicators, which is in focus of men of finance and includes margin evaluation (such as unit margin, margin based on distribution channels, percentage of margins), then monitoring of variable and fixed expenses, structured view on the marketing budget costs, development of average unit prices, evaluation of aim fulfilment in the volume of sales and goals in sales etc.

Another set of indicators, which are in recent years in highest demand are the area of connecting the financial and marketing indicators, although it is necessary to state that their usage is very selective, based on the ability of the company and managers to handle the implementation of these new indicators. Particularly the profitability is mentioned, which can be evaluated from several angles. The main indicators are net profit, (in terms of our methodology it is an economic result from the operating activity—the main activity, usually adjusted by tax or economy result from the common operations after tax), the profitability of the sales, return on investment— ROI, economic added-value—EVA, net present value—NPV, internal rate of return— IEE or return of marketing investments—ROMI.[1]

The internal rate of return is a metric used in financial analysis to estimate the profitability of potential investments. The internal rate of return is a discount rate that makes the net present value (NPV) of all cash flows equal to zero in a discounted cash flow analysis. IRR calculations rely on the same formula as NPV does.

Return on marketing investment or ROMI is a metric used in online marketing to measure the effectiveness of a marketing campaign. It examines results in relation to the specific marketing objective. ROMI is a subcategory of return on investment or ROI, because here the cost is incurred on marketing.[2]

In simple terms, it is measured by calculating total revenues against marketing investment. It should only reflect the direct impact of a marketing campaign. For ROMI to be effective, it is important for the campaign to have some measured metrics.

---

[1] https://corporatefinanceinstitute.com/resources/knowledge/valuation/economic-value-added-eva/, 28.5.2022.

[2] https://economictimes.indiatimes.com/definition/return-on-marketing-investment 28.05.2022.

4 Online Marketing

Following the previous sections in which we talked about the current needs of marketing today and the measurability of the effectiveness of marketing investments, we focus on online marketing. It is currently a modern form of marketing, which on the basis of the space in which marketing takes place and its activities take place, its effectiveness is very measurable.

Online marketing as part of digital marketing includes all ongoing, planned or past marketing activities of the company, which are implemented via the Internet or other connection with a particular management, usually electronic system. Companies use various digital channels for their marketing, such as search engines, social networks, media, e-mail, applications and their own or third-party websites.

The spectrum of digital channels is constantly expanding, and in line with this trend, new possibilities for their use for marketing purposes are growing. These channels are used for two-way connections between companies and current and future customers. The communication interaction conducted from the company towards the customer takes place, for example, through digital advertising or e-mail marketing, most often in the form of newsletters, online brochures or banners.

Due to the reciprocal nature of this communication interaction, it is necessary to monitor important indicators leading from customers back to companies. In this regard, customer communication, feedback and valuable information about customer behaviour are essential [16].

Online marketing broadly consists of initial analyzes performed before entering the online business. These analyzes include online advertising tracking, marketing content, website building, data analysis, search engine optimization or adaptability to various electronic devices, and viral marketing, as indicated in Fig. 5.

In online marketing, the decisive factor is how the client gets to the digital content. There are two ways: an organic visit and a so-called paid visit, or advertising. An organic visit is unpaid traffic from internet search engines. Online marketing, which deals with search engine optimization SEO (Search Engine Optimization), focuses on increasing the number of organic visits.

4.1 Basic Terms, History and Status in Year 2022

The origins of online marketing can be traced back to 1978, when Gary Thuerk first introduced e-mail advertising in the form of sending unsolicited advertisements for his company's computer products to 320 recipients. Since then, we can observe massive advances in technology in the European and global market, which offers widespread variability, especially today. Approximately $ 333.25 billion was spent on e-marketing in 2019, and the forecast for 2023 is more than $ 517 billion. Google alone generated $ 24.1 billion in revenue from Ad Sense and AdWords in the first quarter of 2018, where that revenue was generated largely from contextual digital

Fig. 5 Components of online (digital) marketing. *Source* [16]

ads. It is clear that the use of mobile devices and online marketing is an increasingly preferred tool for corporate communication. However, even on the basis of these data, it is possible to state an increasing intensity of work with the Internet not only in the professional sphere, but also among the general public [17].

4.2 Ethical Aspect in Advertising from the Point of View of Laws in Slovakia

Although in practice we encounter many reservations about advertising, from a moral point of view we consider advertising to be legitimate and permissible, as long as it is not based on lies, misleading and coercion. Advertising helps sellers of goods as well as buyers achieve their goal. It helps sellers target customers, while saving buyers time to find the goods they want.

However, marketing managers, as well as the creators of advertising themselves, must adhere to the ethical minimum, which is also supported by Act no. 634/1992 Coll. on Consumer Protection, as amended (hereinafter referred to as the "**Consumer Protection Act**"). Pursuant to § 20 par. 1 and par. 2 of the Consumer Protection Act, the following applies:

(1) Advertising, including advertising intended for consumers, may not contain:

    (a) Anything that would offend national or religious feelings,

    (b) Anything that would endanger morality, especially vulgarity,

    (c) Promotion of violence,

(d) Promotion of products demonstrably harmful to life or health, without the harmfulness being clearly stated in the advertisement,

(e) The promotion of products as beneficial to health, unless this is technically proven or generally accepted.

(2) Advertising of tobacco products is prohibited [18].

The requirement of the truthfulness of the published advertisement is one of the most important things that the public requires of the advertisement. False advertising leads to deception of the public, which is morally unacceptable. Another aspect that is needed to discuss from a moral point of view is the problem of conscious manipulation of human actions. There are reservations about advertising that abuses psychological knowledge, for example in the form of subliminal advertising, which is particularly dangerous for recipients from the group of children and adolescents. The primary responsibility lies with advertising producers, but advertising agencies, the media, the public and the state also have a large share [19].

4.3 Advantages and Disadvantages of Online Marketing

The main advantage of online marketing is that it is not absolutely necessary to set up an online store or website. Social networks themselves can serve as a means of connecting with customers for the purpose of selling goods and services, and at the same time serve as a two-way communication marketing channel. Such a start-up is a very cheap alternative to the solution, given the initial costs [20].

Without much initial investment, anyone has the opportunity to share their personal thoughts and ideas for profit with a global community of Internet users [21].

On the other hand, there is a large group of consumers who like to hold the goods in their hands before buying test them and check whether they sufficiently meet their requirements. As a result of these preferences, physical outlets must choose a different communication methodology when addressing customers than when communicating in an online environment that can engage potential consumers virtually 24 h a day, 7 days a week. However, in the digital world, competition has the same advantage, so it is necessary to constantly advance in the online environment, use the data obtained from the Internet and focus on the development of computer technology. Online retailers need to know much more about customer behaviour in order to apply the argument interface, and they need to monitor, for example, the dimensions of who buys, what buys, whether sales have taken place at all, and at what stage. Metrics such as number of conversions, time spent on a particular page, or email opening rate are also useful. Also, the creation of a website that promotes online sales and attracts new customers is a major challenge, especially when we perceive the fact that a start-up consumer competes for the consumer in his business sector with the whole world or at least with the country in which he does business [20].

When shopping online, it is necessary to arouse certain emotions and shopping experience in customers, which is very demanding in a limited space. An online retailer has only one screen area to use through a limited number of virtual channels.

In online marketing, it is important to be aware of the impact and impact on the success of a company, its reputation or recommendations from customers and business partners. This mechanism is also technically called the reputation system. It uses two-way communication between the customer and the seller, or mutual communication between the customers themselves. We understand the use of the Internet in this regard as the use of the communication capabilities of the online environment in order to artificially create large networks. Online reputation mechanisms allow community members to share their views on other members of the community.

The online configurator thus represents an important measure in the CRM process and will, in the near future—especially for the younger generation—become an essential prerequisite with a marketing portfolio. In addition to the marketing benefit, it is also a tool for the sales force that represents a validated possibility to create meaningful offers to end-users [22].

The feedback sent is analyzed and collected, and the feedback posted by other members is publicly available to the community in the form of individual member profiles. Online reputation mechanisms have proven to be a very beneficial mechanism for encouraging cooperation between strangers. This is the case, for example, if a member of an online community's behaviour becomes public knowledge, and that person can influence the behaviour of other members. We are talking about the influence of marketing.

Based on a summary of the facts, we can assess that the most significant benefits of online marketing include low budget demands, the possibility of global occurrence at all times, rapid information dispersion and the ability to collect large amounts of data through two-way communication with customers. On the other hand, the biggest disadvantage is the reluctance of some customer groups to pay for online marketing services.

4.4 Marketing and Marketing Mix

The marketing mix has long been one of the main elements for setting goals and budgetary measures in marketing. The importance of each element depends not only on the company itself or its activities, but also on the competition and the time when the marketing mix is re-evaluated. All marketing elements are interconnected and should be seen as a whole given their synthesis and impact. At the same time, these tools work most effectively when all elements are functionally combinable and work together. For specific companies, some components of the marketing mix may be more important than others, but this depends on the strategy and plan of the company. The elements used in classical theory are product, price, place and promotion [23].

The marketing mix is a set of elements and tools that a company can use in order to gain a foothold in the market and increase demand for its products or services, and thus affect sales. Marketing communication aims to improve the positive image of the company or product [19].

The marketing mix is not a scientific theory, just a conceptual framework that helps managers make decisions when configuring their offerings to meet consumer needs. All competing managers must allocate their available resources so that they are appropriately distributed among the components of the marketing mix.

The basic and most recognized elements of the marketing mix according to the 4 P model are thus the Product, the Price, the Place and the Promotion. We call the sum of these strategic tools the marketing mix.

However, 4Ps target the market only in terms of sellers and do not focus on buyers. Therefore, we can transform 4P from another point of view, from the point of view of customers. We refer to this type of marketing mix as 4C, namely Customer value, Cost to the customer, Convenience and Communication [24].

Every successful marketing strategy requires reassessment over time. If a company wants to have a 4P strategy in place in its business, it is important to understand that the elements of the first marketing mix should not be static. They are designed and devised to adapt and improve as a company's product grows and as potential buyers change with a focus on their desires. The marketing concept is therefore a constantly revolving circle.

The product

The term product refers to the goods or services that the company offers to customers. Ideally, the product should meet existing consumer demand. In another case, however, it may be so convincing in itself that consumers will believe that they need it and the product will then create its own demand. For marketing teams to be successful, they must understand the product life cycle, and sales managers must have a plan for handling products at every stage of their life cycle. The type of product also determines in part how many companies can charge for it, where they should place it and how they should promote it in the market.

Many of the most successful products were launched first in their category, without prior demand.

The price

Price is the amount consumers pay for a product. Traders must link price to the actual and perceived value of the product, taking into account delivery costs, seasonal discounts and competitors' prices. In some cases, sales managers may increase the price to give the product the impression of luxury goods. Alternatively, there may be a deliberate price reduction so that more consumers can try out the product and retailers expand their consumer audiences.

Traders must also determine when and whether it is appropriate to discount the product at all. The discount can sometimes attract more customers, but it can also give the impression that the product is less exclusive or less luxurious compared to the higher price. A very important and sometimes overlooked fact is that a product cannot be discounted unless it covers all costs, not just the direct ones, and unless it brings a sufficient profit margin. Market-stable companies usually have whole teams and studies on pricing research.

The place

When deciding on a location, the company seeks to determine where it should sell the product and how to bring the product to market with the greatest possible profit in line with the company's long-term strategy. The goal of sales managers is always to get their products in front of the consumers who are most likely to buy them.

Proper location identification applies to product placement in specific stores or chains, but also to the country in which the company wants to operate, the region and the overall geolocation. In today's world, product placement in the digital world is becoming increasingly important. Placing a product on TV shows, movies, or websites in order to draw attention to the product is almost essential for successful brand building.

The promotion

Promotion includes advertising, public relations and promotional strategy. The purpose of product promotion is to reveal to consumers why they need it and why they should pay a price.

Marketers tend to combine elements of promotion and placement in order to reach their main audience. For example, in the digital age, "place" and "promotion" factors are as online as they are offline. In particular, factors such as where the product appears on a company's website or social media, as well as what types of search features trigger relevant targeted ads for the product, are important. Companies can use an Instagram campaign, a PR campaign that presents a product, or an email campaign to reach their audience in the right place and at the right time.

4.5 Online Marketing in Marketing Mix Concept

Digital online business is the newer of the business contexts and one with a greater need to differentiate the marketing mix than the original 4 P. During this evolutionary process, researchers have always been divided among "conservatives" who think that the 4 P paradigm is able to adapt to environmental change by including new elements into each "P", and "revisionists" who claim that 4 P is outdated and proposes new paradigms. Researchers see the biggest problem in the fact that 4Ps are less customer-oriented and two-way communication, and claim that they are little inward-oriented. This parameter is needed throughout modern online marketing communications, where the relationship with the customer is the most important and the interaction with the customer is considered key [25].

Another parameter specific to online marketing is to increase the accuracy of the target segment selection process, especially thanks to database management systems, collection, storage and sorting of contacts and monitoring of the customer purchase process. However, all this is only possible thanks to advanced systems that one would never be able to process or calculate on one's own. It is not only important to reach a large number of people, but above all to reach the right people who are more likely to make a purchase [25].

In the past, online marketing has seen a trend of reaching as many people as possible on the smallest possible budget. Marketing organizations recommended flooding the website with advertising banners as much as possible, displaying so-called Pop-ups and mailboxes spam bulk emails. It was believed that the more a person was affected by the advertising campaign, the more likely they were to buy from the seller, and this was where the biggest profit would come [16].

The costs of creating such marketing campaigns were much lower than traditional forms of advertising (billboards, radio, television, leaflets …), which naturally enticed clients to try to reach the widest possible target group. In the beginning, it was also a very effective form of promotion, but as other companies added with the same intention, after a few years, the entire online space became flooded with all kinds of ads. However, such behaviour led to advertising hindering people and ceasing to be effective. For example, the browser-blocking AdBlocker feature was downloaded by more than 300 million users by 2015 alone. For this reason, a non-segmented campaign is no longer sufficient, and it is necessary to publish more thoughtful advertising campaigns only to people who have a greater potential to take the required action, whether it is a product purchase or a news subscription [20].

Personalization, in turn, refers to the ability to create a flexible interface that is able to adapt to the needs and will of customers. For many years, for example, Google has shown different results to different users, based on the most relevant content for them and their searches. One-to-one, for example, is an extreme form of personalization, where targeting is set to a precisely specified individual of the individual we want to find and address in the online space [26].

Remarketing is considered to be a trend today, which we refer to as intentional targeting of Internet users who have already visited a specific website, opened a specific application, seen a specific video, etc. It is also important to move from quantity to quality [26].

In the case of online sales, payment systems for customers must also be secure and easy to use. The purchasing process must be as simple as possible so that unnecessary complexity does not deter a customer who is already determined to buy [27].

Social networks

In the modern world, social media has exploded as a category of online discussions, where people create content, share it, bookmark it, and network with each other at tremendous speed. This provides an opportunity for companies to present their products to huge dynamic communities and individuals. Thanks to easy registration and how to use, social networks today have countless users and set trends in the topics discussed, which include the environment, politics, technology or shopping, and the entertainment industry. Examples of the world's best-known social networks include Facebook, Instagram, MySpace, Twitter, LinkedIn and Google+. Social media is basically promoted by their users distributing content to each other and voluntarily [27].

Promotion, marketing intelligence, segment research, public relations, marketing communication, product and customer management are sub-disciplines of marketing that social media can use. Every social media platform, such as blogs, online discussion forums and online communities, has an impact on marketing performance (eg sales), so it is essential to understand their relative importance and their interconnectedness. In addition, social media users are currently highly motivated web consumers. As many as 70% of social media users are interested in shopping online. Consumers can easily get what they want just by sitting in front of a computer screen and browsing online websites. This represents such great potential for businesses in the online world that it would not be wise to neglect it [27].

Website

The company's website, its clarity, visuals and customization have been identified as one of the key factors for e-commerce success. The most important factor influencing the success of a website is the attractiveness of content that is trustworthy for the customer, gives a sense of reliability and can also mention the shopping experience. The shopping experience includes the convenience of customers, the necessary effort, an unexpected experience or even added value. However, the success of the site also depends on the specific industry, the amount of money spent on the operation of the website, effort, timeliness, ease of use or scope. The site must be useful and useful in the first place and graphically and design-wise interesting in the second place [28].

A website domain is the name of a website on the Internet and usually matches the name of a company. A well-chosen domain alone can attract customers or serve to easily remember and navigate the site.

When you first create a website, it's important to determine who the page is created for and what we want your audience to be. It is necessary to consider for which age or social group the site is created, or it is necessary to evaluate the gender perspective of the customer or take into account the specific interests of customers. Many companies want to have their website exceptional, they want to be immediately attracted and different. However, the most important distinction should be the offer of services and quality descriptions of what the product or service includes. When creating a website, you need to think about the content of the website and search engine optimization. The site also needs to be regularly updated and adapted to current needs [29].

Copywriting

Copywriting is the writing of advertising and marketing texts. Copywriting is one of the basic elements of effective marketing. It is an act of promoting a product, business, person or idea, the aim of which is for the reader to take a specific action, for example to buy in an e-shop, subscribe to a newsletter and the like. Copywriting is used in different media, whether online or offline. Copywriters create texts on websites, newsletters, advertising brochures, catalogs, while different copywriters specialize in different types of copywriting, as, for example, creating a good newsletter of a company is different from creating text on a billboard.

In general, they focus mostly on the correct grammatical and stylistic side of the text. The copywriter must be a creative copywriter, an experienced salesperson, and to a lesser extent must be able to estimate the psychological aspects of customers. He needs to understand not only the idea or product he is promoting, but most importantly he must be able to empathize with the role of the target group to which he is speaking. The copywriter must understand the problems his reader is solving and be able to offer him solutions in the way he sells the promoted product [30].

The term copywriting is often confused with the term content marketing, which is content marketing, but there is a significant difference between these terms. Content marketing is the creation of valuable free content that attracts new customers and turns them into regular customers. These are blogs, podcasts, autoresponders, etc. Copywriting, on the other hand, is the creation of the texts they sell. It is up to the reader to take concrete action. These are texts on sales pages, advertisements and blogs with CTA—call to action, i.e., blogs with a call to a certain action, which the reader has to take [30].

SEO

SEO or other search engine optimization affects how commercial search engines display their search results. A search engine is generally a program that searches for results between documents, web pages, and web content that contain a search keyword or phrase, and searches the search results for those that contain that keyword or phrase and considers them optimal results for that search [31].

Today's most used search engines are Google, Bing and Yahoo, but the vast majority of SEO optimization focuses on Google. It was created in 1998 based on the PageRank algorithm, named after its founder Larry Page. This algorithm was based on the evaluation of backlinks referring to a given domain. Some links have been assigned more value than others. For example, websites that are often cited in other texts and link to other important high-ranking sites have been given priority in displaying results.

In general, these are, for example, various academic works, which are displayed according to the quality of the content and citations that lead to the work with the date of publication. Page Rank evaluates pages on a logarithmic scale, which means that a value of 8 is ten times stronger than a value of 7, but on the other hand it is easy to move from a value of 0 to a value, but getting a value of 8 from a value of 7 is incredibly difficult. Aull likens this algorithm to hives, where those looking for information are bees that have the ability to choose information from search results. The more bees move in one and the same place, the more information is exchanged between that place and the hive. Similarly, search engines evaluate and submit content based on how many people and where they have gathered, so they have visited the site [31].

We divide search engine optimization into two main activities, namely off-pafe and on-page optimization. On-page optimization takes place on the web domain or in the source code of the page and deals with the content of texts, headings, use of keywords, graphic design or page load speed. Off-page optimization takes place outside the page and falls under it, e.g. creating backlinks, thanks to which the site becomes more trustworthy in the eyes of search engines, we also include in the off-page optimization the building of reputation on social networks or the availability and visibility of the site. Key analysis is also important.

Keywords, so-called keywords, are really key for good site placement in search engines and correctly chosen keywords help to gain a large number of targeted customers [31].

Keyword optimization on the site helps blogs, news sites and e-shops the most. Finding keywords doesn't have to be easy at all, as the following example shows. For example, a website that restores historical sites should not choose "building renovation" or "renovation" as its keywords, because those keywords will attract customers who have nothing to do with historical sites and such people on the site will not lead to any conversions and increase profits. A small change in a keyword can mean targeting the right customer segment [31].

Link building, or building backlinks to a site, is like building a reference and awareness about the site. If a site is good, it links to many links, whether within the domain itself or, more importantly, from other sites on the Internet. Ideal backlinks are from thematically similar websites and are organically created, not paid [31].

Publishers should create links that lead from relevant sites with high quality of their own content and not from cheap sites that create a backlink to anyone who pays them for a few euros. First, there are practices against Google's Terms of Use, but the practical impact is that sites with good backlinks are ranked higher in search. Another rule regarding SEO optimization for Google is to acquire backlinks naturally and after a certain time and not suddenly in one day and then to pause for a long time, which would most likely cause Google to mark the page as spam.

Google Page Rank not only evaluates sites and scores for them, but also penalizes them if unauthorized practices are found. Among the so-called Black hat techniques in SEO include, for example, spamming, and thus spamming (flooding) with keywords in your own text, as well as link farms, URL redirection, cloaking, doorway pages, wallpapering catalogs or stealing content [31].

In general, people don't bother to click on the other side of search engines, which makes it necessary to appear on the first page of Google. Up to 60% of all clicks will take the top 5 positions in this search engine, with the first three results displayed being the paid results [31].

SEM

Search engine marketing (SEM) is a strategy that uses paid advertising to be visible in search engines. It is also known as PPC (pay-per-click), which means that the displayed ad itself is not paid for, the advertiser only pays for the customer's actual click on their own website. Unlike the use of SEO tactics that help websites organically bring customers to the web, SEM uses PPC advertising platforms such as Google Ads and Bing Ads to use available ad formats to reach their target audience. SEM includes everything from setting up and optimizing paid ads to managing your account to increase conversions and returns. SEM usually begins with a comprehensive keyword survey and competition survey, which can be done with tools such as the PPC Ads toolkit, and then appropriately set up and enter targeted campaigns that place their products and services for the target audience (Mike, 2015).

SEM has proven to be an effective audience acquisition strategy, and when properly set up and implemented, SEM generates a stable level of site traffic and a high Return On Investment (ROI). SEM is even considered one of the most effective forms of online marketing channels for many companies. Another advantage of SEM over SEO is that in effective SEO, it takes months to years for real results to emerge in the form of purchases and growing clients, while the effect of SEM is visible and measurable immediately. An advertiser can take full control of when their ads appear and to whom they appear, making the channel ideal for testing new strategies, boosting website traffic, or increasing off-season sales [32].

Email marketing

Email marketing is another online way to promote your business by sending ready-made emails to current or potential customers. Every company needs to maintain a close relationship with customers, and this is exactly what an email marketing campaign does. Any online store "lives" thanks to a database of customers to whom it sends notifications about new offers, promotions or liquidation of stocks with discounts. Email marketing brings tremendous benefits to any company, with minimal promotional costs and easier content preparation.

As with all marketing activities, e-mail marketing requires clear goal-setting, such as strengthening the relationship between customers and society; business support; revenue growth; highlighting strengths over competitors; increase online sales; branding; attracting new customers; presentation of the offer to existing or potential customers; promotion of new products; or information about changes. Before sending an e-mail campaign, a database of people who have already expressed interest in the company's products must be created, because in many countries the company must obtain permission from consumers to send promotional materials in advance, otherwise they will be categorized as spam [33].

Influence marketing

Influence marketing is the most important newly developed approach to marketing in the last ten years. The word "influence", or Slovak influence, is understood in marketing terms as the ability and power to influence another person, thing or course of events into a desired effect. Brown and Hayes define influenza as a third party that significantly influences a customer's purchasing decision and can sometimes be fully responsible for it. We call influencers individuals who have the power to influence the purchasing decisions of others because of their authority, knowledge, status, or relationship.

Ordinary people are considered to be social influencers who influence consumers when deciding to buy a particular product. Anyone who influences the affinity for the brand and the decision to buy another person can be a social influencer.

However, when the term influencer is mentioned, it generally means a person better known in a particular community. They can be, for example, athletes, artists, scientists, etc., but the essential element is that they have their own community on social media and are willing to produce sponsored and professionally created content for their followers. Influence marketing can take the form of blog posts, videos or photos on influencer's social media channels, which mean content collaboration, and can be content for a company's marketing campaign. The marketing campaign is then associated with the name or photo of the influencer, which results in greater sales of the product or service produced. The Influencer can also act as a brand ambassador, organizer of a competition for products sold or other activities. The collaboration most often takes place on various social media channels, such as the company's Instagram, Snapchate or Twitter, known as Fisherman's model [34].

Fisherman's model shows the location of the influencer at the centre of this event, around which are his followers. Therefore, companies interested in and promoting their products and services must first and foremost identify people who have a wide reach within communities focused on specific interests or keywords. Fisherman's model of influence can help companies identify potential influencers who were affected by the original influencer, and their communities can later be used as a basis for further research and analysis of these relationships. Fisherman's model refers to the application of the concept of "throwing a wide net to catch as many fish as possible". Therefore, most of the people sought are the most viewed and reach large communities of people, potentially raising brand awareness. However, this form of marketing is a relatively more expensive option, especially if the company is interested in long-term and extensive cooperation with a well-known influencer.

For influence marketing, it is very important to retrospectively measure the effectiveness of such product promotion, so that the company knows whether the company will pay to invest in such cooperation in the future, or to identify problems and avoid them [34].

Viral marketing

Viral marketing is part of an online marketing strategy that can generate interest and potential brand or product sales through messages that spread "like a virus." Viral literally means anything that spreads rapidly among users, and in marketing there is an effort to subtly incorporate the products or services being sold into such virally spreading content. The goal of viral marketing is for users to share content themselves. The basis of viral marketing is the dissemination of information by word of mouth, but modern technologies allow viral marketing to include many internet platforms [35].

Thanks to their speed and easy sharing, social networks are a natural environment for marketing of this kind. The most common example today is the creation of moving, interesting or engaging videos on YouTube, which are then shared on Facebook, Twitter and other social media. This is especially attractive for smaller businesses or companies because viral marketing can be a cheaper alternative to a traditional marketing strategy.

Google online marketing management tools

Digital promotion is a new and very innovative idea of the twenty-first century. Promoting products through digital promotion is a costly activity, especially if you don't make effective use of tools that speed up and simplify the setup of individual promotions [25].

Google My Business

My Business on Google is a service that Google provides to businesses for free. By simply creating a company profile in this service, the company can excel and gain customers. This service is more intended for local targeting of customers, because the created profile will also be displayed on the Google map. This service from Google is especially recommended for online stores that also have a brick-and-mortar operation or offer services in a certain geographical area.

With a business profile on Google, you can manage your local business views across Google, maps, and search, and help customers find your business better. Google-verified businesses are twice as likely to be trusted by their users. In order to increase your chances of developing a company, you need to optimize your Google My Business account on a regular basis. For example, by systematically adding contributions or adding individual products. This step will ensure that the company profile also displays these products in the organic search [36].

Google Search Console

The Google Search Console, originally called the Google Webmaster tools, is one of Google's basic search engine optimization tools that show how a search engine sees the content of a particular webpage. This tool provides bug reports on the pages you crawl when it encounters indexing pages. Google Search Console is a tool for business owners, SEO specialists and marketers, website administrators and web developers.

Search Console provides tools and reports for various actions:

- Confirms that Google knows or sees and indexes the newly added site,
- Fixes and reports indexing issues, can index updated content on request,
- Shows how many pages Google indexes
- Shows how often a domain appears in search results.
- Shows the number and rate of clicks to the site
- sends notifications when spam or other errors are detected on the website,
- draws attention to erroneous and duplicate tags,
- Shows what images and keywords visitors clicked on to the site's submitters.
- indicates whether the client has SEO penalties
- Alerts you to problems with AMP and mobile devices [36].

Ads Keywords planner

Keywords planner is part of Google AdWords, which allows merchants to plan paid search campaigns. Keyword Planner is a great tool to help you identify keywords that are relevant to the complexity of your ad campaign or to the relevant blog audience and website content. The Keyword Planner helps you find relevant keywords by viewing their search volume. It's easy to see a comparison of different keywords, search volumes, and how many people are likely to use those keywords in your campaign.

The more popular a keyword is, the more targeted it will cost for that keyword. Accordingly, it is possible to anticipate a budget for PPC campaigns. Keywords planner is suitable for searching so-called long-term keywords that are relevant and probably not as widely used in competing campaigns as short-term keywords [36].

Google Trends

Google Trends is a tool that allows you to view keyword search volumes to make it easier to search for smart keywords. Google Trends helps you research the latest news, stats, stories, searches, and more. This Google Merchant Tool allows you to understand how search trends change across regions over time and how they relate to the world. Google Trends helps you compare the differences between multiple search keywords over time [36].

Google AdWords

Google AdWords is a marketing tool for merchants to run paid search campaigns. Google AdWords helps you reach the right people at the right time. With this service, you can target potential customers in different geographic areas, through image content ads, YouTube video ads, and more [36].

Google Analytics

Google Analytics is one of the most powerful, effective and used Google tools for merchants. Analytics collects data and performs in-depth analysis of all types of website data that the company collects in order to improve its performance [37].

In addition to the above, it also helps to monitor website traffic. With Google Analytics, you can track a variety of parameters, such as the people who visit the website, how much time they spend on it, what search queries they use, how they got to the page, what their purchase path was, and much more. The information obtained can be used to further optimize the website, which provides an opportunity to improve the user experience and increase site traffic. Google Analytics helps optimize test variations of websites and applications, and ensures that relevant feedback is obtained from real people. Google Analytics helps you work seamlessly with other Google platforms and tools, which saves time and increases efficiency [36].

Facebook Business Suit

The Facebook Business Suit application, part of Meta's Facebook platform tools, is used to manage multiple accounts and combine all the basic functions of Facebook and Instagram. Specifically, the most used functions are post scheduling, communication with users from comments, Messenger and Instagram chat, access to analyze and statistics, or work with advertisements and campaigns.

Metrics

Metrics are a basic indicator of marketing effectiveness. Each of the virtual environment clients can use these tools to determine their effectiveness. They are an objective informant and a basic source for further decision-making for strategy creation. Virtual environment tools such as Meta business even offer scenarios for strategy creation based on long-term and short-term data analysis, client behaviour (communication, sharing, communication with virtual space by type, method and time of adding contributions) communication time, area traffic from which are visitors, population segmentation of clients. They take into account the demographic breakdown of clients. Based on this information, it offers at least two scenarios that the client chooses and helps the owner to choose a scenario. Metrics are used as a source of information in this scenario creation, and learning machines define the strategy. This is most often used in direct marketing or B2B.

Reach

Reach, translated as Reach, translates to the total number of unique social network users who have viewed certain content. We can measure impact on organic content or content from a paid sponsored promotion. Reach metrics alone give a good idea of how many people are creating content and how large a potential audience is. In general, the greater the impact, the better the marketing strategy, but it is also important to monitor the level of involvement of the so-called "Engagement" and how many people interact with the post.

Impressions

Impressions are the display of a post on a social media platform, regardless of whether the user clicked on the ad or the content or not. A more accurate term instead of displaying would be loading. The difference from reach is that one content can appear to a single user more than once, and it is the impressions that indicate the total number of views of the content. Impressions are calculated when the user's post is loaded into the feed.

Link Clicks

Link clicks is a metric that indicates the number of clicks on links in an ad that have led to destinations on or outside of Facebook, such as an advertiser's website.

Results

The result, or Slovak result, is what we require to happen and what the campaign optimizes for. The desired result can be of many types, such as vision parameter, reach for unique profiles, audience engagement (comment and liking), brand awareness, application download, video viewing, or conversion, by which is meant any programmed element.

Cost per Result

The cost per result is calculated by dividing the total amount used for the ad by the number of results (e.g. reach, clicks, conversions, video views) [37].

It's a good way to measure the overall performance of each individual campaign at a high level, which can also help you draw attention to a specific campaign whose cost per result is starting to fall. In this case, we see that the campaign is successful and appropriate.

CPI

Most studies on the effectiveness of promotional marketing methods work with CPI (Click Per Impression) metrics, which express the number of clicks to the number of content impressions. Therefore, CPI is the number of clicks your ad receives divided by the number of impressions [37].

STDC model

An important tool in marketing is the so-called See-Think-Do-Care model. STDC represents a comprehensive marketing model of the customer's shopping path. The customer goes through uniform phases, during which he gets acquainted with the product he has expressed interest in. The result of the whole model is an action— conversion, which can be, for example, a purchase, subscription to a newsletter or the transfer of contact details.

SEE phase

- Potential customers do not know the services. The communication is aimed at the widest possible audience, who is trying to be inspired by the need for professional photography—as a memory, a gift for loved ones, a portfolio, building a personal relationship with the brand, etc. The effort to use the Fear-of-missing-out (FOMO) model, in Slovak "fear of losing an opportunity" and creating the impression that quality photography is not a current trend, but a habit and necessity.
- Trying to communicate in the natural environment of the occurrence of customers, where they get even without thinking about it yet (they are not looking for a service yet). We turn foreigners into visitors.
- Suitable marketing channels: PPC, print media, online magazines, PR in the media, SEO optimization of the website, social media, influence marketing, Youtube advertising.

THINK phase

- The initial interest has already been created in the audience and a closer acquaintance with the services is taking place. The audience is narrowing and showing the first signs of shopping interest. The audience is thinking about the offer.
- Communication is beginning to focus on the specific benefits of the service— professional technology and equipment available, photographer's know-how, proximity to the studio for geolocation of the target group, a wide range of photo genres, comfortable environment.
- Encouraging audience involvement to ask questions, share content, educate through published materials, seek inspiration in the portfolio.
- Visitors turn into leads, which is a sign for a potential customer.
- Suitable marketing channels: Web, blog, SEO, social media, own Youtube channel, emailing (newsletter), forums and FB groups for photographers, modeling actors, wedding agencies, events.

DO phase

- The audience is determined for the event. It is ready to purchase / order the service and visit the photo studio. This phase supports conversion and the effort to mediate a seamless purchase path (optimizing the shopping path on the web—order form/e-shop, adding the option of payment by card, setting up automatic order confirmation and delivery method).
- Leaves turn into customers.
- Appropriate marketing channels: Web, PPC retargeting, automated email marketing (order confirmation, abandoned cart).

CARE phase

- The customer has successfully purchased or used the service.
- At this stage, customer loyalty is built. It is necessary to thank for the purchase or use of services, to offer a discount or benefit for customer friends, or for a repeat purchase. It's a good idea to collect reviews on Google Maps or social networks and motivate the customer to share their experience with Company A on their own social networks.
- The customer changes into a brand promoter.
- Appropriate marketing channels: remarketing, e-mail, retention (customer retention), social networks [38].

Web analysis in terms of UX and On-page SEO

The website and its functionality is one of the most important communication channels. It is usually right after social media, given the type of business and the possibility of customer engagement in the online space.

Up to two specialized website quality assessment tools are used for website analysis, namely:

- www.seobility.net
- www.seositecheckup.com

At the same time, these indicators correspond to the user reality when browsing the website. Web auditors consider UX optimization, on-page SEO and even off-page SEO.

5 Conclusion

The role of marketing and especially online marketing is an important tool for communicating with the company's external environment today. in addition to helping to sell goods and services, at the same time, it is a source of information about a particular business. It replaces many obsolete sources of information and allows easy and simple access to information for everyone. On the one hand, it creates a number of opportunities for sharing, but on the other hand, it is this information that becomes a source of competition or can be misused.

That is why, especially today, ethical standards that protect values and help to dispose of them right are at the forefront. Online marketing has its future only if, in addition to professional foundations, it respects moral standards. Only then can this tool be used for the good of man.

References

1. Veselovská, L., Janičková, J.: Vybrané metódy, techniky a prípadové štúdie z manažmentu. Belanium, Banská Bystrica. ISBN: 978-80-557-1955-9 (2022)
2. Antošová, M.: Strategický manažment a rozhodovanie. Wolters Kluwer (Iura Edition). ISBN: 9788080785307 (2012)
3. Sakál, P., Podsklan, A.: Strategický manažment. STU vo vydavateľstve STU. Bratislava. ISBN: 80-227-2153-0 (2004)
4. Sakál, P., Baran, D., Božek, P., Cibulka, V., Cook, T. M., Černá, Ľ., Dolinková, Z., Drahňovský, J., Gilejeva, T. A., Hajnik, B., Hatiar, K., Hatiarová, K., Horňák, F., Ismagilova, L. A., Jacinto, D., Jahnátek, Ľ., Jedlička, M., Kubovič, M., Lenort, R., Liberko, I., Liberková, L., Lozenko, V. K., Machyniak, M., Mamojka, M., Mihok, J., Mrvová, Ľ., Paulová, I., Rybanský, R., Sablik, J., Sakál, M., Sokolová, M., Sergejeva, I. G., Šujanová, J., Uhrovčíková, P. a Závadský,J.,: Strategický manažment v praxi manažéra. SP Synergia, Tripsoft Trnava, 699 s. ISBN 978-80-89291-04-5 (2007)
5. Štefánik, J., Melíšek, M., Naňák, M.: Strategický manažment. TnUAD, Trenčín. ISBN 978-80-8075-249-1 (2007)
6. Slávik, Š.: Strategický manažment. Sprint 2. S.r.o. Bratislava. ISBN: 978-80-89393-96-1 (2013)
7. Papula, J.: Podnikanie a podnikateľské myslenie I. Wolters Kluwer, Bratislava. ISBN: 9788074789502 (2015)
8. Porter, M.E.: Competitive Advantage: Creating and Sustaining Superior Performance. Free press, New York. ISBN: 9780743260879 (2004)
9. Smith,S.M., Albaum, G.S.: Basic Marketing Research: Official Training Guide from Qualtrics. Qualtrics Labs, Incorporated. ISBN: 9780984932818 (2012)
10. Boučková, J.: Marketing., C. H. Beck, Praha. ISBN: 80-7179-577-1 (2003)
11. Kajanová, J.: Economical and managerial challenges of business environment. Comenius university Bratislava. ISBN: 978-0-9860419-7-6 (2017)
12. Pelsmacker, P.D., Geuens M., Van den Bergh J.: Marketingová komunikace. Grada, Paraha. ISBN: 8024702541 (2003)
13. Dann, S.: Definition for social marketing. Adaptation and adoption of the American marketing association. Soc. Market. Q. **14**(2) (2008)
14. Morgan, N.A.: Marketing and business performance. J. Acad. Mark. Sci. **40**(1), 102–119 (2011)
15. Knowles J., Ambler T.: Orientation and marketing metrics. In: Maclaran, P. et al. (eds.): The Sage Handbook of Marketing Theory. Los Angeles, Sage. ISBN 978-1-84787-505-1 (2009)
16. Hanlon, A.: Digital Marketing: Strategic Planning & Integratio. Sage Pubn. ISBN: 9781529742800 (2022)
17. Hajarian, M., Camirelli, M. A., Díaz, P., Aedo, I.: A Taxonomy of Online Marketing Methods. Bingley: Strategic Corporate Communication in the Digital Age. 235–250 pp. Emerald (2021). https://doi.org/10.1108/978-1-80071-264-520211014
18. Zákon č. 634/1992 Zb. o ochrane spotrebiteľa v znení neskorších predpisov („**Consumer Protection Act**")
19. Jaderná, E., Volfová, H.: Moderní retail marketing. Grada, Praha. ISBN: 978-80-271-1384-2 (2021)

20. Santos, K.E.S.: Online marketing: benefits and difficulties to online business sellers. Int. J. Adv. Eng. Res. Sci. (IJAERS) 7(3), 159–163. ISSN: 2349-6495(P) (2020)
21. Torabi, O., Mirakhor, A.: Crowdfunding with Enhanced Reputation Monitoring Mechanism (Fame). De Gruyter Oldenbourg. ISBN: 9783110582925 (2020)
22. Niedermayer, M.H., Zatrochová, M.: Online Configuration as an Accompanying Measure in the CRM-Process. In: PEFnet 2019—23rd European Scientific Conference of Doctoral Students, pp. 99–100. Mendel University in Brno. ISBN 978-80-7509-692-0 (2019)
23. Jun-Jie, Z., Li, Y.: A simple analysis of revolution and innovation of marketing mix theory from big data perspective. In: IEEE 2nd International Conference on Big Data Analysis (ICBDA) Beijing, China. ISBN: 978-1-5090-3619-6 (2017)
24. Lahtinen, V., Dietrich, T., Rundle-Thiele, S.: Long live the marketing mix. Testing the effectiveness of the commercial marketing mix in a social marketing context. J. Soc. Market. 10(3), 357–375. ISSN: 2042-6763 (2020). https://doi.org/10.1108/JSOCM-10-2018-0122
25. Kingsnorth, S.: Digital Marketing Strategy: An Integrated Approach to Online Marketin. Kogan page. ISBN: 1398605999 (2022)
26. Marcus, A., Wang, W.: Design, User Experience, and Usability: Users, Contexts and Case Studies. Springer. ISBN: 978-3319918051 (2018)
27. Mutuku, C.: Advantages and Disadvantages of Using Social Networks in Business. Green Verlag, Munich. ISBN: 978 3 66863 042 (2018)
28. I-Sung , L., Yung-Fu , H., Jie-Hua , S., Ming-Wei, W.: Evaluation of key success factors for web design in Taiwan's bike case study. J. Asian Fin. Econ. Bus. 7(11), 927–937 (2020). ISSN: 2288-4637
29. Wood, K.: Confident Web Design: How to Design and Create Websites and Futureproof Your Career. Kogan Page, London. ISBN: 978 1 978966 364 4(2020)
30. Luh, C.-J., Yang, S.-A., Huang, T.-L.D.: Estimating Google's search engine ranking function from a search engine optimization perspective. Online Inf. Rev. 40(2), 239–255 (2016). https://doi.org/10.1108/OIR-04-2015-0112
31. Duong, V.: SEO Management: Methods and Techniques to Achieve Success (Information System, Web and Pervasive Computing), Wiley-ISTE. ISBN: 978-1786304599 (2019)
32. Moran, M., Hunt, B.: Search Engine Marketing, Inc.: Driving Search Traffic to Your Company's Website. IBM, Pearson pls. ISBN: 978-0133039177 (2015)
33. Tarruella, R.: Email Marketing Success: How to Build an Email List and Create Successful Email Marketing Campaigs. Amazon Digital Services LLC—Kdp Print Us. ISBN: 9781094710471 (2019)
34. Yesiloglu, S., Costello J.: Influencer Marketing: Building Brand Communities and Engagement. New Your: Routledge. ISBN: 978 0 429 32250 1 (2021)
35. Guo, J., Wu, W.: Viral Marketing for Complementary Products. Nonlinear Combinatorial Optimization, pp. 309–315. Springer. ISBN: 978-3-030-16194-1_16 (2019)
36. Google Help Center. https://support.google.com/
37. Meta Business Help Center. https://www.facebook.com/business/help/
38. Alhlou, F., Shiraz, A. S., Ferrman, E.: Google Analytics Breakthrough: From Zero to Business Impact. Willay. New Jearsey. ISBN: 978-1119144014 (2016)

Intelligent Information System for Controlling International Innovation Activities of an Enterprise

Mykola Odrekhivskyi, Oleh Kuzmin, Orysya Pshyk-Kovalska, and Volodymyr Zhezhukha

Abstract The article considers the features of designing and functioning of an intelligent information system for controlling international innovation activities of an enterprise, and also develops an appropriate mathematical apparatus in this area. It is proposed to consider an intelligent information controlling system as a logical and cognitive model of a social agent, and such an integral system as a multi-agent system. To build an intelligent information system for controlling the company's international innovation activities, a modular structure was chosen, therefore, the corresponding modules of this system were identified and characterized. Since the dynamics of states of international innovation activity of an enterprise is stochastic, it is proposed to use the Kolmogorov system of differential equations to assess and predict such states, as well as to make optimal decisions on their management. The developed mathematical apparatus was tested in the study of the state of development of the Truskavets sociopolis (Ukraine), taking into account the index of the volume of innovative services provided (percentage compared to the previous year). During the calculations, 31 sanatoriums of sociopolis were studied. As a result of calculations, it is established that the proposed mathematical apparatus and the mathematical and program software developed on the basis of the intelligent information system for controlling international innovation activity of the enterprise adequately describe the state of development of the sanatorium-resort complex of sociopolis Truskavets under study and can find wide application in the study of the state of international innovation activity of companies and their development in general. In addition, the structure of the management system for international innovation activities of the

M. Odrekhivskyi · O. Kuzmin · O. Pshyk-Kovalska · V. Zhezhukha (✉)
Lviv Polytechnic National University, Bandera Street, 12, Lviv, Ukraine
e-mail: volodymyr.y.zhezhukha@lpnu.ua

M. Odrekhivskyi
e-mail: mykola.v.odrekhivskyi@lpnu.ua

O. Kuzmin
e-mail: oleh.y.kuzmin@lpnu.ua

O. Pshyk-Kovalska
e-mail: orysia.o.pshyk-kovalska@lpnu.ua

© The Author(s), under exclusive license to Springer Nature Switzerland AG 2023
N. Kryvinska et al. (eds.), *Developments in Information and Knowledge Management Systems for Business Applications*, Studies in Systems, Decision and Control 462,
https://doi.org/10.1007/978-3-031-25695-0_4

enterprise, as well as the structure of the controlling system for such activities, is given.

Keywords Innovation · Innovative development · Controlling · Kolmogorov equation · System approach · Stochastic analysis · Sociopolis

1　Introduction

At the present stage of international business development, manufacturing companies in economically developed countries are trying to intensify their international inter-firm relations as much as possible to ensure innovative development. In this context, it is worth talking about various well-known forms of inter-firm cooperation, namely: cooperation with research, design, experimental and production activities; the creation of joint innovative enterprises or associations (consortia); the absorption of innovative small enterprises by large companies to obtain the latest technologies; the division between companies of costs for the implementation of research and development projects, which are implemented in the form of scientific and technical cooperation, the exchange and transfer of technologies, know-how (non-proprietary technological solutions and design developments), experimental samples, production and technological experience, training of personnel, installation and configuration of equipment, etc.; creation of scientific and technical alliances and various international innovation systems, etc. Especially important among all this list today should be considered the formation and use of effective international innovation systems of organizations. The international innovation system of an enterprise should be understood here as a set of organizational, structural and functional components involved in the international process of creating and applying scientific knowledge and technologies that determine the legal, economic, organizational and social conditions of the international innovation process of a business entity.

The effectiveness of its innovation activities significantly depends on the innovation system formed at the enterprise. It provides the company with innovative development through the development and manufacture of high-tech products, as well as providing services to ensure the effectiveness of the entire chain of the innovation process (education, finance, legal support, information, etc.). The innovation system, respectively, includes subsystems that form an innovation policy and strategy aimed at providing subjects of innovation activity with the necessary resources, and take a direct part in the process of creating, commercializing and practical use of new knowledge (in the innovation process), as well as strengthen integration processes between systems for creating, transforming and practical use of new knowledge, etc.

Effective functioning and development of international innovation activities of the enterprise are impossible without building an effective intellectual and information system for controlling such activities. We are talking about a set of interrelated elements that would control all components of international innovation activities at the strategic and tactical levels, as well as contribute to information and analytical support for managerial decision-making in this area.

It is advisable to form an intelligent information system for controlling the international innovation activity of an enterprise by combining organizational, structural and functional components of the enterprise that are involved in the international process of creating and applying scientific knowledge and technologies. These components determine the legal, economic, organizational and social conditions of the company's international innovation process. Building a system of controlling international innovation activity allows for solving the problem of effective management of international innovation activities of an enterprise.

2   Methodology

Taking into account the nature of the work and the content of the tasks set, a wide range of research methods was used. Thus, it is assumed that the basis of the tools necessary for the design and functioning of an intelligent information system controlling international innovation activities of enterprises can be object-oriented, integrated and distributed databases and knowledge bases, expert systems, decision support systems, integrated neural networks and fuzzy logic tools. Taking into account that distributed artificial intelligence and integrated intelligent information systems as multi-agent systems are the most suitable class of models for the implementation of integrated intelligent information systems controlling international innovation activities of enterprises, therefore, the intelligent information system controlling is proposed to be considered as a logical and cognitive model of a social agent, and an integrated such system as a multi-agent system.

To build an intelligent information system for controlling international innovation activities of enterprises, a modular structure was chosen, which will provide it with flexibility and adaptability to environmental conditions.

Since the dynamics of states of international innovation activity of an enterprise is stochastic, it is proposed to use the Kolmogorov system of differential equations to assess and predict such states, as well as to make optimal decisions on their management.

3   Paper Preparation

3.1   Structure of the Enterprise's International Innovation Management System

Innovative systems of modern enterprises should be international, that is, integrated into international innovation processes. Therefore, in the formation of international innovation systems of enterprises, the role of the managed subsystem should be played by the international innovation activity of the organization, the components of which are the innovation process and subsystems that form the strategy and policy of innovative development, organize international cooperation, provide and serve the innovation process, as well as interact with consumers and other stakeholders (Fig. 1). The leading subsystem in the innovation management system is the company's managers, who make final management decisions in this area.

The effectiveness of international innovation systems of enterprises largely depends on the effectiveness of their management organization. In turn, the development of these systems depends on their integration into horizontal and vertical innovation cycles, as well as on the interaction between levels, since innovation systems at all levels are formed like networks. To ensure the effectiveness of international innovation activities of the enterprise as a whole, such forms of its organization are a priority, where the result of each stage is the basis for further movement to the next

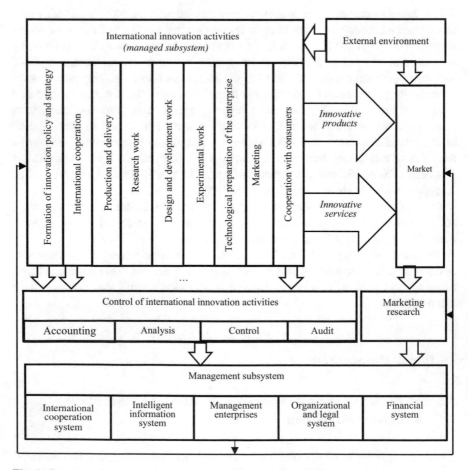

Fig. 1 Structure of the enterprises' international innovation management system

stage. Especially important is the connection of stages, which ensures continuity, flexibility and dynamism of the entire process. All this is provided by the means of effective organization of innovation activities.

3.2 Formation and Functioning of the Controlling System of International Innovation Activity of the Enterprise

The development and functioning of innovative processes of an enterprise are impossible without creating an effective system of controlling innovation activities, which would control all components at the strategic and tactical levels, as well as contribute to information and analytical support for managerial decision-making. The proposed structure of formation and functioning of the controlling system for the international innovation activity of the enterprise is presented in Table 1.

The main functions of managing the international innovation activity of an enterprise include:

- Forecasting of international innovation activity: development of forecasts of scientific and technological development for the long term;
- Formation of the policy and strategy of international innovation activity: formation of the policy and strategy of international cooperation, justification and selection of the main directions of innovation activity following the accepted forecasts, goals, policies, strategies and tactics of development, as well as following potential opportunities, market demand, etc.;
- Organization of international innovation activities: ensuring international cooperation in research, design and experimental work; organizing the transfer of knowledge, innovations and technologies; ensuring the implementation of planned tasks and bringing together employees who jointly implement innovative plans, programs, projects based on relevant rules and procedures;
- Motivation of personnel to carry out innovative activities: encouraging employees to be interested in the results of their work on creating and selling innovative products.

A special role in these conditions is played by controlling the international innovation activity of the enterprise using economic accounting, analysis, control and audit, that is, controlling incoming resource flows, implementing the innovation process, selling innovative goods and services, meeting planned indicators, etc. (Fig. 2).

Economic accounting is one of the main sources of information that characterizes the actual state of the results of activities and development of enterprises, so it should be objective and accurate, accessible and understandable, timely and economical. Objectivity and accuracy of accounting mean that all accounting data must be reliable, reflect the actual state of economic activity, and contain both achievements and shortcomings in the work of enterprise divisions. Clarity and accessibility of accounting are necessary to become a means of public control, as well as

Table 1 Formation and functioning structure of the controlling system of international innovation activity of the enterprise

Elements of the controlling system	Innovative enterprise
Organizational elements of the system	Choosing the form of doing business with state-established rights and obligations
	Choosing an organizational and legal form
Legal elements of the system	Choosing the form of ownership
	Defining resource sources
Functional elements of the system	Forming an organizational structure
	Formation of innovation policy, strategy and tactics, communication strategy with the state, partners, suppliers, customers, consumers, competitors, investors, etc
	Formation of the composition and structure of divisions for the implementation of policies, strategies and tactics of international innovation activities
	Formation of a controlling system for international innovation activity
International cooperation elements	Formation and implementation of international cooperation policies and strategies
	Joint research, design, experimental and production activities
	Transfer of knowledge, innovation and technology
Technical and technological elements of the system	Forming a diversification strategy
	Formation of a system of machines and mechanisms for the necessary production facilities and controlling system
Financial elements of the system	Formation of fixed and current funds
	Formation of the level of capital investment
	Enterprise value regulation
Elements of HR policy	Formation of corporate culture
	Formation of personnel management culture
	Formation of personnel and social policy
	Formation of measures for training and retraining of personnel

to help attract the general public to actively participate in economic management. Accounting indicators, therefore, should be simple and clear and properly reflect all aspects of economic activity. Timely accounting should be aimed at providing the company's managers with timely information necessary for prompt decision-making and management of enterprises' activities, rational use of labour, information, and natural and other types of resources. In other words, accounting should

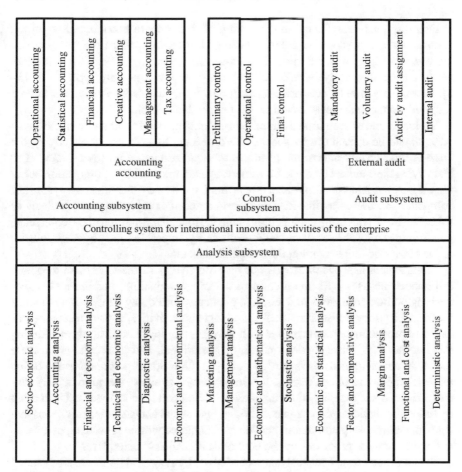

Fig. 2 Controlling system for international innovation activities of the enterprise

allow employees of enterprises to quickly develop and implement specific measures aimed at improving the quality of management. With this in mind, the organization should organize an accounting system that includes interrelated and internally coordinated types of accounting, in particular operational, statistical and accounting, where the accounting one is divided into financial, managerial, tax and creative accounting. Economic accounting should provide all levels of management of international innovation activities of the enterprise with the necessary information that is used for planning, stimulating, monitoring and analyzing such activities. That is, accounting in the system "planning—accounting—control—audit—analysis—decision—making—implementation of decisions" performs information, control, analytical and evaluation functions [1, 2].

The economic analysis examines the economic, technological and social spheres of international innovation activity of an enterprise, the multiplicity of its forms in their interrelation, reflected in the system of indicators of economic accounting, control, audit, reporting and other sources of information. Since the international innovation activity of an enterprise is a managed system, so to ensure the completeness of economic analysis, it is advisable to use the following types of it [3–6]: accounting (audit), economic-environmental, economic-statistical, marketing, socio-economic, technical-economic and financial-economic.

Technical and economic analysis of international innovation activities of the enterprise should be carried out by specialized divisions. Its tasks include studying the interdependence and interaction of innovation and economic processes, as well as identifying their impact on the achievement of economic results of innovation activities. Financial and economic analysis is implemented by the financial system of companies, as well as by financial and credit authorities regarding the efficiency of using borrowed and equity capital, improving the financial condition of the enterprise, identifying profit reserves, increasing profitability, improving the payment abilities of the enterprise, implementing the financial plan, etc.

Management analysis is carried out by all divisions of the enterprise to obtain information necessary for monitoring, processing, evaluating, making and implementing optimal management decisions, planning, developing strategies and tactics for financial policy, logistics, marketing and logistics activities, improving the organization of production, equipment and technologies, employee motivation, etc. Taking this into account, management analysis is operational, and its results are a trade secret. Accounting analysis is carried out by auditors for expert diagnostics of the financial conditions of enterprises. It is implemented to assess and predict the financial condition and stability of companies. A socio-economic analysis is carried out by economic services of management bodies, and social and statistical divisions that study the relationship and interdependence of socio-economic processes, as well as the impact of these processes on the economic activities of enterprises.

Economic and statistical analysis is carried out by statistical bodies to study many social events at different organizational levels of enterprise management. Environmental and economic analysis is implemented by national, regional and local management bodies, relevant services of enterprises, studying the interaction and interdependence of environmental and economic processes, as well as their impact on the activities of enterprises. Marketing analysis is carried out by the marketing department to study the state of the external environment of the functioning of an innovative enterprise, in particular: the resource market, sales markets, the competitiveness of the enterprise, supply and demand, and pricing policy commercial risks, etc.

Factor, diagnostic, comparative, margin, economic-mathematical, economic-statistical, deterministic, stochastic, functional-cost and other methods can be used to analyze economic activity within the framework of the system of international innovation activity of enterprises. Factor analysis (deterministic and stochastic, deductive and inductive, single-stage and multi-stage, static and dynamic, retrospective and perspective) is aimed at a comprehensive and systematic study of the impact of certain factors on the growth and level of effective indicators of international innovation activity, which contributes to the identification of key success factors of companies [7–11]. The algorithm of factor analysis of international innovation activity of the enterprise can be reflected as follows: selection of factors for the analysis of the studied indicators of international innovation activity; their classification and systematization to provide a systematic approach; modelling of relationships between the studied one and factor indicators; calculation of the influence of factors and assessment of the role of each of them in changing the value of the studied indicator; use of the factor model in management decision-making. An important task of factor analysis is to model the relationships between the studied indicators of international innovation activity of the enterprise and the factors that determine their value. In general, these relationships can be represented as the following system of equations:

$$
\begin{cases}
y = f(x_1, x_2, \ldots, x_n), \\
y' = f_1(y, x_1, x_2 \ldots, x_n), \\
y'' = f_2(y', x_1, x_2, \ldots, x_n),
\end{cases}
\tag{1}
$$

where y—the studied performance indicator; $x_1, x_2, \ldots, x_n$—factor indicators; f, f_1, f_2—functions that indicate the types of factor models; y', y''—the first and second derivatives of the studied indicator, used in the study of their dynamics, in particular during the study of the effective indicator for an extremum.

When using comparative analysis, they are limited to comparing the indicators of reporting documents on the results of international innovation activities of the enterprise with indicators of strategic and current plans, as well as with indicators of leading enterprises. Based on the results of such a comparison, management decisions are made. Let y_0 be the target and planned value of the studied indicator of international innovation activity of the enterprise, and y—actual value. Then decisions are made based on deviations of y from y_0, that is, based on the absolute value of the difference $/y_0 - y/$.

The diagnostic analysis is a mean and method of studying the nature of deviations from the planned course of the innovation process based on typical features inherent in a particular deviation. For example, if the quantitative growth of gross output of enterprises outstrips the growth of marketable products, then this indicates a quantitative increase in work-in-progress balances. If the quantitative growth of the company's gross output is higher than the growth of labour productivity, then this indicates non-fulfilment of the action plan for computerization of technological processes, improvement of labour organization and, accordingly, reduction in the number of employees. Awareness of the signs of deviations from the plan allows you to quickly and accurately determine their nature, without additional effort, which, accordingly, requires additional resources, time and money.

Margin analysis [12–15], which is based on margin income, allows us to evaluate and justify the effectiveness of management decisions based on cause-and-effect relationships, sales volumes, cost and profit, and division of costs into variables and constants. The obtained values of specific margin income for each specific type of enterprise product are important for decision-making. If this indicator is negative, it indicates that revenue from product sales does not even cover variable costs. Each subsequent unit of a certain type of product produced will increase the total loss of the enterprise.

The economic and mathematical analysis makes it possible to optimize the solution of the economic problem, as well as to identify reserves for improving the efficiency of international innovation activities of the enterprise due to the optimal use of the necessary resources. Deterministic analysis can be used to study functional interdependencies and relationships between factor and performance indicators of the company's international innovation activity.

Stochastic analysis [16–20] is used to study probabilistic relationships between the states of the studied processes of various organizational levels of enterprises and the states of international innovation activity in general. Taking into account the fact that the dynamics of enterprise states are stochastic [21], models based on mathematical methods of Markov process theory using Kolmogorov differential equation systems may be most suitable for controlling the states of enterprises, evaluating them and predicting them to further make optimal decisions [21–28]. Kolmogorov's system of differential equations for describing states of an element at the enterprise hierarchy level will look like:

$$\frac{dP_{i,j,l}}{dt} = \lambda_{i,j,l-1,l} \cdot P_{i,j,l-1} - \left(\lambda_{i,j,l,l-1} + \lambda_{i,j,l,l+1}\right) \cdot P_{i,j,l} + \lambda_{i,j,l+1,l} \cdot P_{i,j,l+1},$$

(2)

where $i = 1, 2, \ldots, N$ is a sequence number of the hierarchy level; N—number of hierarchy levels; $j = 1, 2, \ldots, M_i$—sequence number of the element at the i hierarchy level; M_i—number of elements at the i hierarchy level; $l = 1, 2, \ldots, L_j$—status sequence number of the j element at the i hierarchy level; L_j—number of states of the j element; $P_{i,j,l}$—probability of the l states of the j element at the i hierarchy level; $\lambda_{i,j,l,l+1}$—the intensity of the transition of the studied system from the state l to the l+ state of l of the j element i at the hierarchy level.

Functional and cost analysis [29–31] is used as a method of identifying the reserves of an enterprise due to the study of all possible functions implemented in the innovation process. It is aimed at methods of optimizing the implementation of these functions at all stages of the innovation cycle (research, design, experimental work, production, operation, modernization, renovation, etc.). The main purpose of functional cost analysis is to identify unnecessary costs and prevent them.

3.3   Tools for Designing and Functioning an Intelligent Information System for Controlling International Innovation Activities of Enterprises

The basis of the tools necessary for the design and functioning of an intelligent information system controlling international innovation activities of enterprises can be object-oriented, integrated and distributed databases and knowledge bases, expert systems, decision support systems, integrated neural networks and fuzzy logic tools. Decision support systems allow you to model and automate these processes, as well as the model and automate the management processes of enterprises in general. Distributed artificial intelligence and integrated intelligent information systems as multi-agent systems [32–35] are the most suitable class of models for implementing integrated intelligent information systems controlling the international innovation activities of enterprises. It is advisable to consider an intelligent information controlling system as a logical and cognitive model of a social agent, and such an integral system as a multi-agent system.

The creation of a unified system of economic analysis at enterprises, integrated into the corporate information system, occurs through the use of intelligent information systems controlling the international innovation activity of the enterprise, the use of software and hardware tools for automatic and automated collection, processing, analysis and transmission of information, support for management decisions on the implementation of strategic and operational tasks, effective use of innovation potential, working time, resources, equipment, etc.

To build an intelligent information system for controlling international innovation activities of enterprises, a modular structure was chosen (Fig. 3), which will provide it with flexibility and adaptability to environmental conditions. The structure of decision support modules (M3 and M4 modules) that can support decision-making and justification processes is proposed to include subsystems for accumulating knowledge of the first and second kind, a knowledge base, a user interface, and subsystems for making and explaining decisions.

When using an intelligent information system controlling the international innovation activity of enterprises, decisions will be made based on the knowledge of experts, which, accordingly, can be highly qualified specialists from specialized branches of knowledge (knowledge of the first kind), as well as knowledge obtained taking into account a priori information and the results of research on the results of international innovation activity of the enterprise (knowledge of the second kind). This knowledge can be formalized and entered the knowledge base as knowledge based on which decision-making on the state of international innovation activity of the enterprise is supported by the M3 module, and the guiding influences on them (which are implemented using the M6 module) are supported by the M4 module.

Management decisions on the state of international innovation activity of the enterprise within the intellectual information system controlling such activities with the proposed structure can be supported by using Monte Carlo modelling, discrete modelling, modelling the dynamics and statics of systems, digital business models and visual modelling, operations research (simulation modelling, business games, stochastic programming), decision trees, impact diagrams, fuzzy logic tools, modelling based on agents and multiagents, etc.

The study of dynamic and static characteristics of real states of international innovation activity of the enterprise with subsequent management decision-making is carried out using an intelligent information system controlling such activities. Data collection and their primary processing to clarify management problems is carried out using the M1 module controlling international innovation activities of the enterprise.

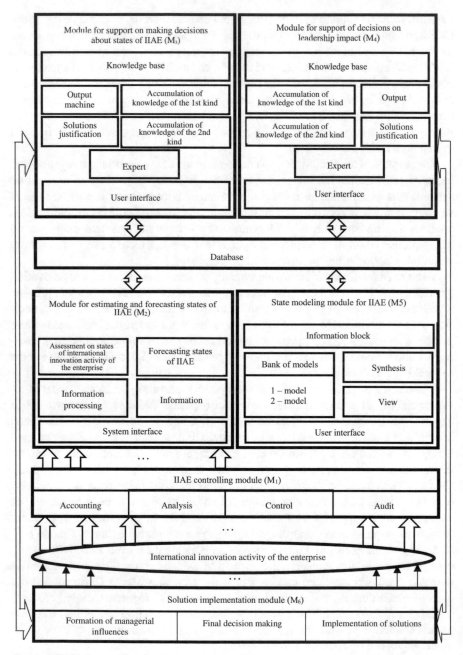

Legend: IIAE - international innovation activity of the enterprise

Fig. 3 Structure of the intelligent information system for controlling international innovation activities of the enterprise

Storage and processing of the collected data, further assessment and forecasting of the state of the specified activity are carried out through the M2 module. If real research cannot be implemented, then it is proposed to use virtual information from the information block and modulate the virtual States of international innovation activity of the enterprise using the M5 module. This will contribute to further assessment and forecasting of possible conditions, making situational decisions and their implementation through the M6 module. If information about the state of international innovation activity of an enterprise is well structured, then mathematical methods should be used to process this information and then choose management decisions. If the problem is poorly structured or unstructured, then it is suggested to use expert judgments and assessments to prepare solutions. Since the dynamics of states of international innovation activity of an enterprise is stochastic, it is proposed to use the Kolmogorov system of differential equations to assess and predict such states, as well as to make optimal decisions on their management.

3.4 Testing of the Mathematical Apparatus Within the Framework of the Intelligent Information System Controlling the International Innovation Activity of the Enterprise

The developed mathematical apparatus was tested in the study of the state of development of the Truskavets sociopolis (Ukraine) [21, 35] by the index of the volume of innovative services provided (percentage compared to the previous year). Truskavets is a well-known balneological resort in Ukraine, founded in 1827. Nowadays, it is considered one of the largest European centres of the health industry and annually receives about 350 thousand vacationers. During the study of 31 sanatoriums, it was found that 2 sanatoriums in the corresponding year were in the S1 state (*BH*), 18 sanatoriums—in the S2 state (*H*), and 11 sanatoriums—in the S3 state (*HH*). In this case: *BH*—the number of sanatoriums where the index of the volume of innovative services provided was higher than the average value; *H*—the number of sanatoriums where the index of the volume of innovative services provided was within the average value; *HH*—the number of sanatoriums where the index of the volume of innovative services provided was less than the average value. The initial conditions of the process under study in this case will be the following state probability values: P0 $(S1) = 2/31 = 0{,}065$; P0 $(S2) = 18/31 = 0{,}58$; P0 $(S3) = 11/31 = 0{,}355$. During the study period, according to the index of the volume of innovative services provided, the state of development of health-improving complexes in Truskavets has changed. The intensities of transitions of complexes from state to state are indicated by the corresponding values above the transition arcs on the graph (Fig. 4).

The study on the dynamics of the development state of the sanatorium-resort complex in Truskavets sociopolis according to the index of the volume of services provided was carried out with the Kolmogorov system of differential equations using computer technology:

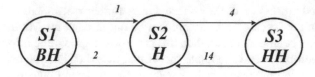

Fig. 4 Graph of the state of development of the Truskavets sanatorium and resort complex (according to the index of the volume of innovative services provided, % compared to the previous year)

$$\frac{dP_1}{dt} = -\lambda_{12} \cdot P_1(t) + \lambda_{21} \cdot P_2(t),$$

$$\frac{dP_2}{dt} = \lambda_{12} \cdot P_1(t) - (\lambda_{23} + \lambda_{21}) \cdot P_2(t) + \lambda_{32} \cdot P_3(t), \tag{3}$$

$$dP_3/dt = -\lambda_{32} \cdot P_3(t) + \lambda_{23} \cdot P_2(t).$$

The system of equations for $t \rightarrow \infty$ and $\frac{dP}{dt} = 0$ is transformed into a system of algebraic equations. This makes it possible to study the state of development of the Truskavets sociopolis in a stationary mode and make appropriate forecasts.

When analyzing the obtained dynamic and static characteristics of the probabilities of states for the studied complex according to the index of the volume of innovative services provided (Fig. 5), it is concluded that the most likely for the sanatorium complex of Truskavets sociopolis is an increase in the volume of innovative services sold, since the dynamics of characteristics revealed that the probability of the first state, in which the index of the volume of innovative services provided is greater than the average value, becomes the largest and reaches a static value equal to 0.61. That is, the development of the health complex of the Truskavets sociopolis was stable until 2020 (before Covid-19).

Thus the proposed mathematical apparatus and the mathematical and program software developed on its basis of the intelligent information system for controlling international innovation activity of the enterprise adequately describe the state of development of the sanatorium-resort complex of sociopolis Truskavets under study and can find wide application in the study of the state of international innovation activity of companies and their development in general.

4 Conclusion

The proposed mathematical apparatus and mathematical models developed on its basis can be widely used as mathematical support for an intelligent information system for controlling the international innovation activities of an enterprise. This, in particular, is confirmed by the example of the sanatorium-resort complex under study in Truskavets (Ukraine). These proposals contribute to the expansion of tools

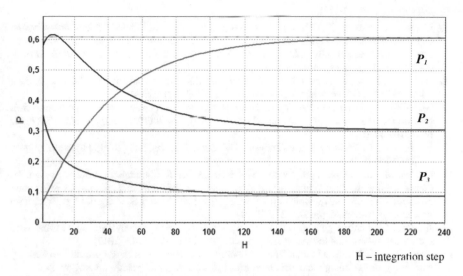

Fig. 5 Probabilities characteristics of states of development stability of the Truskavets sociopolis sanatorium-resort complex

for studying the state of international innovation activity of companies and their development in general. Understanding the structure of the controlling system of international innovation activity of the enterprise as a whole and the structure of such an intelligent information system will allow for their reasonable construction and effective management.

References

1. Poniszewska-Maranda, D., Matusiak, R., Kryvinska, N., Yasar, A.-U.-H.: A real-time service system in the cloud. J. Ambient Intell. Human Comput. **11**(3), 961–977 (2020). https://doi.org/10.1007/s12652-019-01203-7
2. Vančová, M.H., Ivanochko, I.: Factors behind the long-term success in innovation—in focus multinational IT companies. In: Developments in Information & Knowledge Management for Business Applications. Studies in Systems, Decision and Control, vol. 376 (2021). Springer, Cham. https://doi.org/10.1007/978-3-030-76632-0_16
3. Wohlfarth, M.: Data portability on the internet: an economic analysis. Bus. Inf. Syst. Eng. **61**(5), 551–574 (2019). https://doi.org/10.1007/s12599-019-00580-9
4. Christensen, H.B., Hail, L., Leuz, C.: Mandatory CSR and sustainability reporting: economic analysis and literature review. Rev. Acc. Stud. **26**(3), 1176–1248 (2021). https://doi.org/10.1007/s11142-021-09609-5
5. Jace, C.: An economic theory of economic analysis: the case of the school of salamanca. Public Choice **181**(3–4), 375–397 (2019). https://doi.org/10.1007/s11127-019-00662-y
6. Odrekhivskyi, M.; Pshyk-Kovalska, O.: Problems of building an analytical support system for innovative enterprises. Efektyvna ekonomika **1** (2019). https://doi.org/10.32702/2307-2105-2019.1.11

7. Gomez, R., Brown, T., Watson, S., Stavropoulos, V.: Confirmatory factor analysis and exploratory structural equation modeling of the factor structure of the questionnaire of cognitive and affective empathy (QCAE). PLoS ONE 17(2), e0261914–e0261914 (2022). https://doi.org/10.1371/journal.pone.0261914

8. Burian, S., Brčák, J.: Economic convergence criteria—factor analysis. Int. Adv. Econ. Res. 22(4), 475–476 (2016). https://doi.org/10.1007/s11294-016-9610-z

9. Bilandžić, A., Marina, J., Šarlija, N.: Dealing with interpretability issues in predicting firm growth: factor analysis approach. Bus. Syst. Res. 7(2), 23–34 (2016). https://doi.org/10.1515/bsrj-2016-0010

10. Finch, W.H.: Exploratory Factor Analysis. SAGE Publications, Inc. (2020). https://doi.org/10.4135/9781544339900

11. Gregus, M. ml., Fedushko, S., Syerov, Yu., Shurmelova, T., Steininger, E.: Service system of doctor's office. In: Proceedings of the Symposium on Information Technologies & Applied Sciences (IT&AS 2021), Bratislava, Slovak Republic, March 5, 2021. CEUR Workshop Proceedings, vol. 2824, pp. 209–215 (2021)

12. Juetten, M.: Identifying your ideal clients: a contribution margin analysis using KPIs can help measure profitable and less profitable projects. Law Pract. 43(5), 43 (2017)

13. Cross, P.L., Mulford, M.: Realizing collaborative systems design for missile seekers by combining design margin analysis with multi-disciplinary optimization. Concurr. Eng. Res. Appl. 23(3), 226–235 (2015). https://doi.org/10.1177/1063293X15586837

14. Vernimmen, P., Le Fur, Y., Dallochio, M., Salvi, A., Quiry, P.: Margin Analysis: Structure. In: Vernimmen, P., Le Fur, Y., Dallochio, M., Salvi, A., Quiry, P. (Eds.) Corporate Finance (2017). https://doi.org/10.1002/9781119424444.ch9

15. Poniszewska-Maranda, Kaczmarek, D., Kryvinska, N., Xhafa, F.: Studying usability of AI in the IoT systems/paradigm through embedding NN techniques into mobile smart service *system*. Computing 101(11), 1661–1685 (2019). https://doi.org/10.1007/s00607-018-0680-z

16. Pinar, M., Milla, J., Stengos, T.: Sensitivity of university rankings: implications of stochastic dominance efficiency analysis. Educ. Econ. 27(1), 75–92 (2019). https://doi.org/10.1080/09645292.2018.1512560

17. Stead, A.D., Wheat, P.: The case for the use of multiple imputation missing data methods in stochastic frontier analysis with illustration using English local highway data. Eur. J. Oper. Res. 280(1), 59–77 (2020). https://doi.org/10.1016/j.ejor.2019.06.042

18. Lin, Y., Huang, T.: Creative destruction over the business cycle: a stochastic frontier analysis. J. Prod. Anal. 38(3), 285–302 (2012). https://doi.org/10.1007/s11123-012-0273-3

19. Dimelis, S.P., Papaioannou, S.K.: Human capital effects on technical inefficiency: a stochastic frontier analysis across industries of the greek economy. Int. Rev. Appl. Econ. 28(6), 797–812 (2014). https://doi.org/10.1080/02692171.2014.907246

20. Anaya, K.L., Pollitt, M.G.: Using stochastic frontier analysis to measure the impact of weather on the efficiency of electricity distribution businesses in developing economies. Eur. J. Oper. Res. 263(3), 1078–1094 (2017). https://doi.org/10.1016/j.ejor.2017.05.054

21. Odrekhivskyy, M., Kunanets, N., Pasichnyk, V., Rzheuskyi, A., Tabachyshyn, D.: Information-analytical support for the processes of formation of "Smart Sociopolis" of Truskavets. In: ICTERI Workshops (2019)

22. Tan, L., Jiang, J.: Digital Signal Processing: Fundamentals and Applications, 2nd edn. Elsevier (2013)

23. Shpak, N., Odrekhivskyi, M., Doroshkevych, K., Sroka, W.: Simulation of innovative systems under industry 4.0 conditions. Soc. Sci. 8, 202 (2019). https://doi.org/10.3390/socsci8070202

24. Kuzmin, O., Zhezhukha, V., Gorodyska, N., Benova, E.: Benefits from engineering projects implementation. In: Barolli, L., Nishino, H., Miwa, H. (Eds.) Advances in Intelligent Networking and Collaborative Systems. INCoS 2019. Advances in Intelligent Systems and Computing, vol. 1035, pp. 431–441. Springer, Cham (2020). https://doi.org/10.1007/978-3-030-29035-1_42

25. Markoska, K., Ivanochko, I., Greguš ml, M.: Mobile banking services—business information management with mobile payments. In: Kryvinska, N., Gregus, M. (Eds.) Agile Information

Business: Exploring Managerial Implications, Sushil (Ed.) Flexible Systems Management, pp. 125–175. Springer (2018). https://doi.org/10.1007/978-981-10-3358-2

26. Garg, H.: An approach for analyzing the reliability of industrial system using fuzzy Kolmogorov's differential equations. Arab. J. Sci. Eng. **40**(3), 975–987 (2015). https://doi.org/10.1007/s13369-015-1584-2

27. Addona, D., Angiuli, L., Lorenzi, L., Tessitore, G.: On coupled systems of Kolmogorov equations with applications to stochastic differential games. ESAIM. Control Optim. Calcul Variat. **23**(3), 937–976 (2017). https://doi.org/10.1051/cocv/2016019

28. Vančová, M.H., Ivanochko, I.: Factors behind the long-term success in innovation—in focus multinational IT companies. In: Developments in Information & Knowledge Management for Business Applications. Studies in Systems, Decision and Control, vol. 376. Springer, Cham (2021). https://doi.org/10.1007/978-3-030-76632-0_16

29. Guerrieri, M.: Catenary-free tramway systems: functional and Cost-Benefit analysis for a metropolitan area. Urban Rail Transit **5**(4), 289–309 (2019). https://doi.org/10.1007/s40864-019-00118-y

30. Yoshikawa, T., Innes, J., Mitchell, F.: A Japanese case study of functional cost analysis. Manage. Account. Res. **6**(4), 415–432 (1995). https://doi.org/10.1006/mare.1995.1029

31. Shvets, V., Baranets, H., Tryfonova, O.: Evaluation of the conditions of effective logistic strategy implementation of an enterprise on the basis of functional and cost analysis. Baltic J. Econ. Stud. **4**(5), 405–411 (2018). https://doi.org/10.30525/2256-0742/2018-4-5-405-411

32. Fedushko, S., Ustyianovych, T., Gregus, M.: Real-time high-load infrastructure transaction status output prediction using operational intelligence and big data technologies. Electronics **9**(4), 668 (2020). https://doi.org/10.3390/electronics9040668

33. Dennis, A., Wixom, B.H., Tegarden, D.P.: System Analysis & Design: An Object-Oriented Approach with UML. Wiley (2021)

34. Rashidi, M., Ghodrat, M., Samali, B., Masoud Mohammadi, M.: Decision Support Systems. In: Management of Information Systems. IntechOpen (2018). https://doi.org/10.5772/intechopen.79390

35. Verma, D.: Study and analysis of various decision making models in an organization. IOSR J. Bus. Manage. **16**, 171–175 (2014). https://doi.org/10.9790/487X-1621171175

Optimization Modeling of Urgent Investment Tools of Crisis Management at Enterprises

Oleh Kuzmin, Oksana Yurynets, Olexandr Yemelyanov, Tetyana Yasinska, and Iryna Prokopenko

Abstract This work aims to implement optimization of economic and mathematical modeling of urgent investment tools for crisis management in enterprises. The research established prerequisites for building models of compulsory investment crisis management instruments and developed a model for optimizing the structure of these tools. The scientific work built a generalized model of optimization of the application process of urgent investment tools for crisis management at enterprises. The practical application of the developed models, in particular, showed that all the surveyed companies have the opportunity to overcome the financial crisis and avoid bankruptcy, using urgent investment tools for crisis management.

Keywords Optimization modeling · Urgent management tools · Emergency management · Crisis management · Enterprise

1 Introduction

In the context of the COVID-19 pandemic, the problem of ensuring the proper level of the financial condition of companies has become significantly more acute [1]. At the same time, many enterprises found themselves in a financial crisis, the way out of which requires the development and implementation of various measures

O. Kuzmin (✉) · O. Yurynets · O. Yemelyanov · T. Yasinska · I. Prokopenko
Lviv Polytechnic National University, Bandera Street, 12, Lviv, Ukraine
e-mail: oleh.y.kuzmin@lpnu.ua

O. Yurynets
e-mail: oksana.v.yurynets@lpnu.ua

O. Yemelyanov
e-mail: oleksandr.y.yemelianov@lpnu.ua

T. Yasinska
e-mail: tetiana.v.yasinska@lpnu.ua

I. Prokopenko
e-mail: iryna.v.prokopenko@lpnu.ua

for crisis management [2–7]. One of the possible directions of such management is to invest in a company in a state of the financial crisis with specific amounts of investment resources [8–10]. In general, as is well known, investments are a powerful source of economic development of enterprises [11–14]. In particular, investment is necessary for the development of innovations [15] and the implementation of progressive technological change [16–20]. At the same time, investment activities are usually associated with significant risk [21, 22], especially if the object of this activity is a financially unstable enterprise, which shows a negative effect on financial leverage [23–25]. However, under certain conditions, investing can be a lifeline for such an enterprise if the financial crisis is primarily due to reasons that require investment costs. This situation may arise if the company is in dire need of technological upgrades [26–28], including the introduction of energy-saving technologies [29–31].

Thus, one of the possible means of crisis management in enterprises is the use of urgent investment instruments. At the same time, their use requires prior careful justification. It applies both to investments at the expense of business owners, employees, outside private investors, and financial support for enterprises from the state [32–35]. An essential basis for establishing the feasibility of implementing urgent investment instruments in enterprises is the construction of these instruments' optimization economic and mathematical models. Given this, this work aims to implement optimization of economic and mathematical modeling of urgent investment tools for crisis management in enterprises.

2   Paper Preparation

2.1   *Prerequisites for Modeling Urgent Investment Tools for Crisis Management in Enterprises*

Determining the need for enterprises to use urgent investment tools for crisis management should involve a consistent solution to several tasks, namely:

(1) identification of the causes of the crisis in enterprises and its type, which requires a detailed analysis of the financial condition of these enterprises;
(2) preliminary assessment of the amount of investment required to eliminate the most acute manifestations of the crisis of enterprises (especially investments in the replacement of debt capital by their own);
(3) assessing the potential for increasing the financial results of enterprises because the crisis of companies, as a rule, is significantly caused by the problem of profitability;
(4) allocation of that part of the potential to increase the financial results of economic entities, for the implementation of which it is necessary to invest, and assess the need for these investments;
(5) preparation of forecasts of the expected increase in the creditworthiness of enterprises subject to the implementation of measures aimed at increasing the profits

of business entities and determining the final need for investment resources needed to implement these measures;

(6) establishing the total need for investment needed to overcome the crisis in enterprises and identifying sources of such requirements;

(7) calculation of the expected level of economic efficiency of the planned investments.

Given the above, it is possible to model the process of planning urgent investment tools for crisis management in enterprises. This model, shown in Fig. 1, is based on assessing the expected results of implementing appropriate instruments in terms of participants in the investment process. At the same time, such participants can be both enterprises in general and their current owners and future investors, who will later become co-owners. In particular, from the point of view of enterprises, the criterion for the success of the use of urgent investment instruments of crisis management will be overcoming the financial crisis in these enterprises.

We should note that the efficiency and economic feasibility of implementing urgent investment tools in crisis management for current business owners will depend on two main factors:

(1) from obtaining an excellent economic effect due to such tools. At the same time, this effect can be estimated, in particular, by the increase in the capitalization of a specific part of the net profit of enterprises, which will correspond to the share of contributions of their current owners in the total share capital. The difference between the estimated value of the specified capitalization and the amount of funds that business owners would receive from their sale of less debt can also determine this effect;

(2) continued control by the current owners of their enterprises after the new co-owners contribute to their share capital. The share of such contributions should not be too large.

As for the new co-owners of enterprises, from their point of view, the introduction of urgent investment instruments of crisis management will be appropriate provided that the proper level of profitability of the investments made by these co-owners and moderate investment risk.

Among the urgent investment instruments of anti-crisis management of enterprises, those that aim to fully or partially cover the debt obligations of enterprises and implement profitable investment projects deserve special attention. We should also note that at the expense of profits from such projects, we can repay the debt obligations of enterprises. At the same time, if the repayment terms of loans taken by enterprises are short, businesses may not have time to accumulate the required amount of funds for such repayment, using only the profits from expected projects. Taking into account these considerations, we model the use of urgent investment tools for crisis management of an enterprise in a state of a financial crisis.

We first introduce several assumptions and limitations:

• first, we believe that the company must repay the total amount of the loan taken after a specific predetermined time interval;

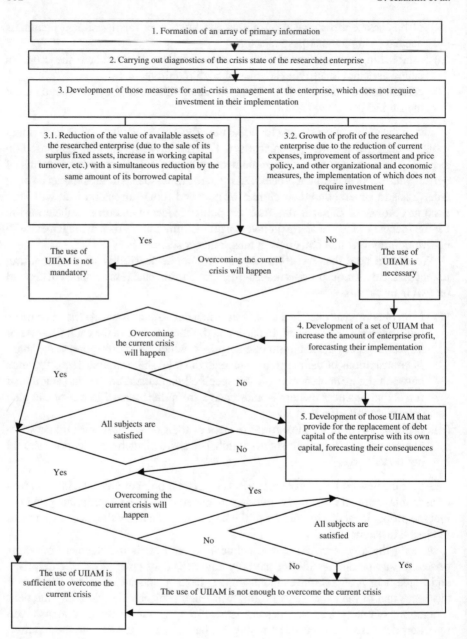

Fig. 1 Model of planning at the enterprise of urgent investment instruments of anti-crisis management (UIIAM). *Note* Developed by the authors

- secondly, we assume that the current owners of the company will agree to invest some of the capital that they can withdraw from other projects or take from their existing own savings (to get the surveyed company out of crisis);
- thirdly, we will assume that, apart from the contributions to the enterprise by its current owners, there are no other effective ways to avoid the financial crisis in which the investigated entity found itself;
- fourthly, we will also consider that the amount of profit from the investment of investment resources in the enterprise is proportional to the volume of this investment. This assumption, in general, is not fundamental; its elimination requires only a particular complication of the process of solving the above tasks, but these difficulties are primarily technical.

2.2 Model of Optimization of the Structure of Urgent Investment Instruments of Crisis Management at Enterprises

Let the enterprise owners have additional funds that they agree to invest in the financial recovery of the researched enterprise. Under such conditions, the enterprise will determine the amount of these additional funds accumulated at the discount rate at the time of payment by the company of the principal amount of the loan taken by it using the following formula:

$$C_d = C_0 \cdot (1 + d)^{T_{c1}}, \tag{1}$$

where C_d—the amount of additional cash of the current owners of the enterprise, accumulated at a discount rate at the time of payment by the enterprise of the principal amount of the loan taken by him, monetary units; C_0—currently available to the current owners of the enterprise the number of additional funds that they agree to invest in the financial recovery of the studied enterprise, monetary units; d—annual discount rate, unit shares; T_{c1}—the period during which the company must repay the principal amount of the loan taken by him, years.

The current owners of the enterprise may invest the available capital in the amount C_0 in implementing a profitable project (projects) at the same company. Then the amount of this capital, accumulated at a discount rate at the time of full repayment of the principal amount of the loan previously taken by the company, will be determined as follows:

$$C_{pd} = \frac{I_{p0} + \Delta I_p}{d} \cdot \left((1 + d)^{T_{c1}} - 1\right) \tag{2}$$

where C_{pd}—the amount of capital available to the current owners of the enterprise, accumulated at the discount rate at the time of full repayment of the principal amount of the loan previously taken by this enterprise, provided that these owners invest capital in the amount of C_0 in a profitable project (projects); I_{p0}—initial (i.e., available before investment) annual profit of the enterprise before tax, monetary units; ΔI_p—expected growth of the annual profit of the business entity before taxation after the implementation of the investment project (projects), monetary units.

We can present expression (2) in a somewhat modified form:

$$C_{pd} = \frac{I_{p0} + C_0 \cdot P}{d} \cdot \left((1+d)^{T_{c1}} - 1\right) \tag{3}$$

where P—the level of annual profitability of the project (projects), which the owners of the enterprise can implement it, the share of the unit:

$$P = \frac{\Delta I_p}{C_0} \tag{4}$$

It is also possible to divide the additional capital available to business owners into two parts. In this case, one of these parts we will accumulate at the rate of d, and the other we will direct to the implementation of some project (projects):

$$C_0 = C_{d0} + C_{i0} = C_0 \cdot \gamma + C_0 \cdot (1 - \gamma) \tag{5}$$

where C_{d0}—part of the capital of the current owners of the enterprise, which we will accumulate at a discount rate at the time of full repayment of the principal amount of the loan previously taken by the enterprise, monetary units; C_{i0}—part of the capital of the current owners of the enterprise, which we will invest in the implementation of the project (projects) at the same enterprise; γ—the share of the capital of the current owners of the enterprise, which will accumulate at the time of full repayment of the loan previously taken by the enterprise at the rate of d, in the total amount of this capital, the share of the unit.

We will assume in the future that the profitability of investing in the project at the enterprise exceeds the discount rate. Then such an investment will be more profitable in advance than reinvestment at a rate of d. Therefore, the choice of the second method of capital investment (i.e., reinvestment at a discount rate) will be due only to the need to fully and timely repay the principal amount of the loan previously taken by the company. Therefore, by meeting at least one of these two conditions, the company will be able to solve this problem:

$$C_0 \cdot (1+d)^{T_{c1}} \geq C_{r1} \tag{6}$$

$$\frac{I_{p0} + C_0 \cdot P}{d} \cdot \left((1+d)^{T_{c1}} - 1\right) \geq C_{r1} \tag{7}$$

where C_{r1}—the amount of loan previously taken by the company to finance its activities, monetary units.

However, it is necessary to fulfill this condition at the same time:

$$\frac{I_{p0}}{d} \cdot \left((1+d)^{T_{c1}} - 1\right) < C_{r1} \tag{8}$$

If we didn't meet an inequality (8), the company could accumulate the necessary funds to repay the loan at a profit, which would mean that the company is not in crisis.

Given the above, it is possible to model the process of accumulation of funds both at the expense of the initial profit of the entity and by investing its owners in the implementation of a particular project (projects) at this company and (or) in reinvesting capital at a discount rate:

$$C_{ac}(\gamma) = C_0 \cdot \gamma \cdot (1+d)^{T_{c1}} + \frac{I_{p0} + C_0 \cdot (1-\gamma) \cdot P}{d} \cdot \left((1+d)^{T_{c1}} - 1\right) \tag{9}$$

where $C_{ac}(\gamma)$—the amount of accumulation of funds both due to the initial profit of the entity and by investing its owners in the implementation of a project (projects) at the enterprise and (or) in the reinvestment of capital at a discount rate as a function of γ, monetary units.

It is also possible to simulate the total discounted profit of the company after repaying the loan, capitalizing on the relevant cash flows:

$$C_{acs}(\gamma) = C_{ac}(\gamma) - C_{r1} + \frac{I_{p0} + C_0 \cdot (1-\gamma) \cdot P + \Delta I_{p1}}{d} \tag{10}$$

or:

$$C_{acs}(\gamma) = C_0 \cdot \gamma \cdot (1+d)^{T_{c1}} + \frac{I_{p0} + C_0 \cdot (1-\gamma) \cdot P}{d} \cdot \left((1+d)^{T_{c1}} - 1\right)$$
$$- C_{r1} + \frac{I_{p0} + C_0 \cdot (1-\gamma) \cdot P + \Delta I_{p1}}{d} \tag{11}$$

where $C_{acs}(\gamma)$—total discounted profit of the enterprise after repayment of the loan taken as a function of the parameter γ, monetary units; ΔI_{p1}—expected increase in the company's annual profit after repayment of the loan (due to the lack of need to pay interest on the loan after such repayment), monetary units.

Given the above, the economic and mathematical model for optimizing the use of urgent investment tools in the enterprise crisis management will include:

(1) target function, which will be the total discounted profit of the enterprise after the return of the loan taken by him:

$$C_{acs}(\gamma) = C_0 \cdot \gamma \cdot (1+d)^{T_1} + \frac{I_{p0} + C_0 \cdot (1-\gamma) \cdot P}{d} \cdot \left((1+d)^{T_{c1}} - 1\right)$$
$$- C_{r1} + \frac{I_{p0} + C_0 \cdot (1-\gamma) \cdot P + \Delta I_{p1}}{d} \to max \tag{12}$$

(2) restrictions on the number of funds accumulated by the enterprise both due
to its initial profit and due to the investment of capital by the owners of the
business entity in the implementation of the project (projects) on it and (or) in
reinvestment at the rate d:

$$C_0 \cdot \gamma \cdot (1+d)^{T_1} + \frac{I_{p0} + C_0 \cdot (1-\gamma) \cdot P}{d} \cdot \left((1+d)^{T_{c1}} - 1\right) \geq C_{r1} \quad (13)$$

(3) restrictions on the share of the total capital of the current owners of the enterprise,
which we will accumulate at the rate d:

$$0 \leq \gamma \leq 1 \tag{14}$$

Regarding the required value in the model (12)–(14), it will be the share of the
capital of the current owners of the surveyed enterprise, which we will accumulate at a rate of d at the time of repayment of the loan. Taking into account the
previously assumed assumptions (in particular, that the return on investment in the
project is higher than the discount rate) to maximize (12), the value of γ should
be the smallest possible value (preferably zero). Accordingly, the solution of model
(12)–(14) requires, first of all, the transformation of inequality (13) into equality and
finding γ from it. Having made several mathematical transformations, we finally get:

$$\gamma_c = \frac{C_{r1} - \left(\frac{I_{p0}+C_0 \cdot P}{d}\right) \cdot \left((1+d)^{T_{c1}} - 1\right)}{C_0 \cdot (1+d)^{T_{c1}} - \frac{C_0 \cdot P}{d} \cdot \left((1+d)^{T_{c1}} - 1\right)} \tag{15}$$

where γ_c—the estimated share of the total capital of the current owners of the enterprise, which we will accumulate at a discount rate at the time of full repayment of
the loan, the share of the unit.

With the information on the indicator (15), it is possible to determine the optimal
value of the share of the total capital of current owners of the enterprise according
to model (12)–(14), which we will accumulate at a discount rate at the time of full
loan repayment. To do this, use the following expression:

$$\gamma_{opt} = \begin{cases} 0 \ ''\text{provided that}'' \ \gamma < 0; \\ \gamma_c \ \text{provided that} \ 0 < \gamma_c \leq 1; \\ \text{the problem has no solutions} \ if \ \gamma_c > 1, \end{cases} \tag{16}$$

де γ_{opt}—optimal according to model (12)–(14) the value of the share of the total
capital of the current owners of the enterprise, which we will accumulate at a discount
rate at the time of full repayment of the loan, the share of the unit.

2.3 Generalized Model of Optimization of the Process of Application of Urgent Investment Tools of Crisis Management at Enterprises

Above we considered the case in which the company only needs to repay the principal amount of the loan taken out once. However, it is also possible to repay the loan in several installments. If the company has previously taken several loans, the repayment of the principal amounts will occur in a particular order at specific intervals. Then the presented model of optimization of the process of application of urgent investment tools of crisis management at the enterprise is formally somewhat complicated, although its economic essence remains unchanged. In particular, under such conditions, the input data for the calculation of the model will feature many vectors of this kind:

$$V_k = (T_k, C_k, \Delta I_{pk}) \tag{17}$$

де T_k—the period in which the company must repay the k part of the principal amount of loans taken by him, years; C_k—the value of the k-part of loans previously taken by a particular enterprise to finance its business activities, monetary units; ΔI_{pk}— increase in the annual amount of profit of the entity after repayment of the k part of the principal amount of loans taken by him, since after such refund, the company should not pay interest on this part of loans, monetary units.

Then you can construct an objective function similar to function (12), which will characterize the total discounted value of the company's profit after repayment of the loan, but, unlike expression (12), take into account that the process of refund of the principal amount of loans is not one, but several times. According to the new objective function, the mathematical form of the constraint is also modified in some way (13). More precisely, instead of one such restriction, several correspond to each procedure of repayment of the principal amount of loans received by the company. However, the required value and the essential interpretation of such a complex model will not differ fundamentally from the previously described basic model, which the expressions (12)–(14) represent.

We should also note that in the study of the processes of accumulation of profits of the enterprise, you can consider the possibility of adding to the flow of profits depreciation deductions for the renovation of the entity's fixed assets. We can assume that to repay the principal amount of loans taken, and the company will use its operating profit and the flow of depreciation deductions for the renovation of labor. However, this assumption can be correct only when accumulating the necessary funds to repay the loan is not so long that the use of depreciation deductions for this purpose significantly affected the implementation of the required scale of simple reproduction of fixed assets of the investigated enterprise.

It is also necessary to consider the fact that some types of urgent investment instruments of crisis management may have several alternatives for their implementation. Given this, it is possible to perform generalized modeling of the process of optimizing the use of urgent investment tools in crisis management [36]. The corresponding model will involve setting the values of the variables χ_{ij}. In this case, each of these variables can take one of two values, namely—zero or one. Then the optimization model will contain:

(1) target function, which involves maximizing the expected reduction in the probability of bankruptcy:

$$W = \sum_{i=1}^{m} \sum_{j=1}^{l_i} (\Delta n_{ij} \cdot \chi_{ij}) \to max \qquad (18)$$

(2) restrictions on the number of values of the required variables:

$$\sum_{j=1}^{l_i} \chi_{ij} \leq 1 \qquad (19)$$

(3) restrictions on the level of the minimum allowable efficiency of investment in the enterprise:

$$\frac{\left(\sum_{i=1}^{m} \sum_{j=1}^{l_i} (\Delta I_{pij} \cdot \chi_{ij} + I_{p0})\right)\left(\sum_{i=1}^{m} \sum_{j=1}^{l_i} (\varpi_{ij} \cdot \chi_{ij})\right)}{\sum_{i=1}^{m} \sum_{j=1}^{l_i} (s_{ij} \cdot \chi_{ij})} \geq G \qquad (20)$$

(4) restrictions on the share of share capital to be owned by the new co-owners of the enterprise (investors):

$$\sum_{i=1}^{m} \sum_{j=1}^{l_i} (\varpi_{ij} \cdot \chi_{ij}) \leq \varpi \qquad (21)$$

(5) restrictions on the minimum allowable amount of economic profit that the current owners of the enterprise will receive:

$$\frac{\left(\sum_{i=1}^{m} \sum_{j=1}^{l_i} (\Delta I_{pij} \cdot \chi_{ij} + I_{p0})\right)\left(\sum_{i=1}^{m} \sum_{j=1}^{l_i} (1 - \varpi_{ij} \cdot x_{ij})\right)}{K_r} \geq N_e \qquad (22)$$

де m—the total number of urgent investment instruments of crisis management under consideration; l_i—the number of possible options for the implementation of the i-tool; Δn_{ij}—the expected reduction of the probability of bankruptcy of the enterprise due to the application of the j-variant of the i-urgent investment instrument; ΔI_{pij}—the expected increase in net profit of the enterprise due to the use of the j-variant of the i -urgent investment instrument, monetary units; I_{p0}—the initial value of the financial result of the research enterprise, monetary units; ϖ_{ij}—the expected share of new owners (investors) in the share capital of the enterprise due to the use of the

j-variant of the i urgent investment instrument; s_{ij}—investment costs to be incurred when using the j-variant of the i urgent investment instrument; G—the minimum allowable return for new business owners is their investment; ϖ—the maximum allowable for the current owners of the enterprise is the share of its share capital that will belong to the new owners; K_r—the capitalization rate of the net profit of the enterprise; I_{nk}—the capitalized income of the current owners of the enterprise in the abandonment of urgent investment instruments of crisis management, monetary units.

We can use the model (18)–(22) in the practice of enterprises in a state of the financial crisis. With the help of this model, it is possible to form a scientifically sound program of urgent investment measures aimed at overcoming the problem and, accordingly, reducing the likelihood of bankruptcy of these enterprises.

2.4 Empirical Analysis of the Use of Urgent Investment Tools for Crisis Management in the Enterprise

To assess the scale and effectiveness of enterprises' use of urgent investment tools for crisis management, we formed a sample of 200 enterprises from the western region of Ukraine, belonging to three types of economic activity: food industry, construction, and the woodworking industry. According to the data presented in Table 1, more than half of these enterprises, as of January 1, 2018, were in a state of a financial crisis. In particular, the unprofitability of the vast majority of these enterprises and the low share of their own capital in the structure of sources of economic resources reflected this situation. Thus, among the 36 enterprises in the food industry, which were in a state of the financial crisis on January 1, 2018, 33 had a negative net financial result, and 19 of them had a share of the equity of less than 30%. We should also note that the loss of the enterprises under consideration, according to their gross financial result, is much less common than the loss of net financial result. In addition, the number of enterprises that were in a state of the financial crisis on January 1, 2018, and in which the value of the coefficient of autonomy was more than 50%, is minimal: among the food industry, there were 2, among construction companies—3, and among woodworking enterprises —only one. Thus, the majority of surveyed enterprises, which on January 1, 2018, were in a state of the financial crisis, are characterized by a combination of profitability crisis with a low level of financial stability.

It is evident that enterprises in a state of the financial crisis have three main directions for further action: overcoming this crisis, continuing to be in a state of emergency, and terminating their activities. As evidenced by the data presented in the Table 2, the share of enterprises that were in a state of the financial crisis on January 1, 2018, and which came out of this state on January 1, 2020, in the total number of surveyed business entities is quite large and is: for food industry enterprises—61.11%, for construction companies—65.85%, for woodworking enterprises—64.44%. At the

Table 1 Data on the number of enterprises surveyed for their use of urgent investment instruments of crisis management by groups of these enterprises

Names of features of researched enterprises grouping	Number of enterprises by type of economic activity		
	Food industry	Construction	Woodworking industry
1. The total number of enterprises as of January 1, 2018	60	70	70
Among them —enterprises that were in a state of the financial crisis at that date	36	41	45
2. Number of enterprises that were in a state of the financial crisis as of January 1, 2018, and in which we observed a loss in 2017:			
2. 1. Gross	14	19	18
2. 2. From operating activities	24	27	32
2. 3. Net	33	37	41
3. The number of enterprises that were in a state of the financial crisis on January 1, 2018, and in which the value of the coefficient of autonomy was			
3. 1. Less than 0.1	3	7	6
3. 2. From 0.1 to 0.2	10	9	8
3. 3. From 0.2 to 0.3	6	12	15
3. 4. From 0.3 to 0.4	14	8	7
3. 5. From 0.4 to 0.5	1	2	8
3. 6. More than 0.5	2	3	1

Note Calculated by the authors

same time, many companies in crisis use urgent investment tools for crisis management. The share of such enterprises in the total number of companies in a state of the financial crisis on January 1, 2018, is: for food industry enterprises—63.89%, construction companies—65.85%, and woodworking enterprises—66.67%.

At the same time, the following trend is quite clear: among those enterprises that have emerged from the financial crisis, the share of economic entities that have used urgent investment instruments of anti-crisis management is much higher than among those enterprises that have not been able to overcome the crisis.

The developed models of optimization of the process of application of urgent investment tools of anti-crisis management at the enterprise on the example of three enterprises, which as of the beginning of 2020, were in a state of the financial crisis, were also tested. The input data and the results of the calculation of the proposed model (12)−(14) for the case of these enterprises we presented in the Table 3.

Table 2 Information on the structure of enterprises investigated for the use of urgent investment instruments of crisis management, which on January 1, 2018, were in a state of the financial crisis by groups of these enterprises

Names of features of researched enterprises grouping	The share of enterprises in the total number of enterprises that were in a state of the financial crisis on January 1, 2018, by type of economic activity		
	Food Industry	Construction	Woodworking industry
1. Number of enterprises that were in a state of the financial crisis on January 1, 2018, and that came out of this state on January 1, 2020	61.11	65.85	64.44
Among them—those who used urgent investment tools of anti-crisis management	55.56	58.54	55.56
2. Number of enterprises that were in a state of the financial crisis on January 1, 2018, and which remained in this state on January 1, 2020	22.22	19.51	24.44
Among them—those who used urgent investment tools of anti-crisis management	5.56	7.32	6.67
3. Number of enterprises that were in a state of the financial crisis on January 1, 2018, and which ceased to operate on January 1, 2020	16.67	14.63	11.11
Among them—those who used urgent investment tools of anti-crisis management	2.78	0.00	2.22
4. The total number of enterprises that were in a state of the financial crisis on January 1, 2018, and which used urgent investment tools for crisis management	63.89	65.85	64.44

Note Calculated by the authors

As can be seen from the data given in the Table 3, for all surveyed enterprises, the value of the optimal share of capital available to the current owners of the enterprise, which we will accumulate at the time of repayment of the principal amount of the loan at a discount rate, in the total amount of this capital is between zero and one. Thus, all companies have the opportunity to overcome the financial crisis and avoid bankruptcy by using urgent investment tools for crisis management. However, we should accumulate the bulk of the owners' capital at a discount rate for some

Table 3 Input data and results of calculation of the offered model of optimization of the process of application of urgent investment tools of anti-crisis management at the enterprises

Names of indicators, units of measurement	Values of indicators by enterprises		
	Vamir Gal LLC	Zahid-Bud-Service LLC	PE Avtotekhnobudservis
1. The initial amount of free cash held by the current owners of the enterprise and which they agree to invest in financial recovery, monetary units	650	792	485
2. The period in which the company must repay the principal amount of the loan taken by him, years	1. 0	0. 5	2. 0
3. The initial annual value of the company's profit before tax, monetary units	242	320	194
4. Annual profitability of the investment project (projects), which the current owners can implement at the enterprise, shares per unit	0. 22	0. 25	0. 20
5. Annual discount rate, unit share	0. 10	0. 10	0. 10
6. The amount of credit previously taken by the company to finance its business activities, monetary units	734	860	519
7. The optimal share of capital at the disposal of the current owners of the enterprise, which we will accumulate at the time of payment of the principal amount of the loan previously taken by the enterprise at a discount rate, in the total amount of this capital, shares per unit	0. 610	0. 826	0. 391

Note calculated by the authors

companies, and a smaller portion of this capital should be invested directly in investment projects. At the same time, at one of the surveyed enterprises, it is expedient to invest most of the owners' capital in implementing investment projects at the same enterprise and accumulate a minor part of this capital at a discount rate.

3   Conclusion

The use of urgent investment tools for crisis management in enterprises is characterized by an increased level of risk, as the investment may not lead to the exit of economic entities from the crisis. Therefore, assessing the needs of enterprises in the use of urgent investment instruments of crisis management requires careful calculations and detailed justification. With this in mind, the authors built economic and mathematical models for optimizing the process of applying urgent investment tools for crisis management in the enterprise.

In particular, the article developed a model for optimizing the structure of urgent investment tools for crisis management in enterprises. This model includes target function—the total discounted value of the company's profit after repayment of the loan; restrictions on the amount accumulated amount of funds due to the initial profit of the enterprise and due to the investment of its owners of certain capital in the implementation of the investment project (projects) at this enterprise and (or) in reinvestment at a discount rate; restrictions on the value of the share of capital at the disposal of the current owners of the surveyed entity and which will be accumulated at a discount rate at the time of payment of the principal amount of the loan previously taken by the company, in the total amount of this capital. The proposed model allows for increasing the degree of validity of the use of urgent investment tools for crisis management in enterprises.

References

1. Steshenko, O., Masalugina, V.: Anti-crisis management under a pandemic condition. Bull. Transp. Econ. Ind. 70–71, 75–78 (2020). https://doi.org/10.18664/338.47:338.45.v0i70-71. 222131
2. Demchuk, N., Tkalich, O., Tkachenko, H.: Anti-crisis management of the enterprise with the use of reingineering instruments. Econ. State 4, 29–32 (2020). https://doi.org/10.32702/2306-6806.2020.4.29
3. Kovbas, G.: Strategic aspects of crisis management in the context of eliminating threats to the motivation of personnel of enterprises. Scient. Notes Lviv Univ. Bus. Law 21, 45–50 (2019). https://doi.org/10.5281/zenodo.4661808
4. Poniszewska-Maranda, D., Matusiak, R., Kryvinska, N., Yasar, A.-U.-H.: A real-time service system in the cloud. J. Ambient. Intell. Human Comput. 11(3), 961–977 (2020). https://doi.org/10.1007/s12652-019-01203-7

5. Ugolkov, I., Karyy, O., Skybinskyi, O., Ugolkova, O., Zhezhukha, V.: The evaluation of content effectiveness within online and offline marketing communications of an enterprise. Innov. Mark. **16**(3), 26–36 (2020). https://doi.org/10.21511/im.16(3).2020.03

6. Poniszewska-Maranda, Kaczmarek, D., Kryvinska, N., Xhafa, F.: Studying usability of AI in the IoT systems/paradigm through embedding NN techniques into mobile smart service system. Computing **101**(11), 1661–1685 (2019). https://doi.org/10.1007/s00607-018-0680-z

7. Tulchinskiy, R., Kyrychenko, S., Ruzhytskyi, A., Saloid, S.: Strategic aspects of formation of the system of anti-crisis management of the enterprise. Invest.: Pract. Exp. **9**, 15–19 (2021). https://doi.org/10.32702/2306-6814.2021.9.15

8. Kuzmin, O., Stanasiuk, N., Yastrubskyi, M., Mohylevska, O., Artiushok, V.: Industrial potential: assessment, modeling and administration under the condition. Naukovyi Visnyk Natsionalnoho Hirnychoho Universytetu **6**, 128–135 (2021). https://doi.org/10.33271/nvngu/2020-6/128

9. Markoska, K., Ivanochko, I., Greguš ml, M.: Mobile banking services—Business information management with mobile payments. In: Agile Information Business: Exploring Managerial Implications. Flexible Systems Management. Springer, pp. 125–175 (2018). https://doi.org/10.1007/978-981-10-3358-2.

10. Odrekhivskyi, M., Pshyk-Kovalska, O., Zhezhukha, V.: Optimization of management decisions of recreational innovative companies. In: Developments in Information & Knowledge Management for Business Applications. Studies in Systems, Decision and Control, vol. 420. Springer, Cham (2022). https://doi.org/10.1007/978-3-030-95813-8_18

11. Valitov, S., Khakimov, A.: Innovative potential as a framework of innovative strategy for enterprise development. Procedia Econ. Financ. **24**, 716–721 (2015).https://doi.org/10.1016/S2212-5671(15)00682-6

12. Ivanochko, I., Greguš, M., Urikova, O., Alieksieiev, I.: Synergy of services within SOA. Procedia Comput. Sci. **98**, 182–186 (2016). https://doi.org/10.1016/j.procs.2016.09.029

13. Gradzewicz, M.: What happens after an investment spike-investment events and firm performance. J. Bus. Econ. Stat. **39**(3), 636–651 (2021). https://doi.org/10.1080/07350015.2019.1708369

14. Barber, B.M., Morse, A., Yasuda, A.: Impact investing. J. Financ. Econ. **139**(1), 162–185 (2021). https://doi.org/10.1016/j.jfineco.2020.07.008

15. Ortega-Argilés, R., Piva, M., Potters, L., Vivarelli, M.: Is corporate R&D investment in high-tech sectors more effective? Contemp. Econ. Policy **28**(3), 353–365 (2010). https://doi.org/10.2139/ssrn.1332586

16. Zhezhukha V., Kuzmin O., Gorodyska N., Benova, E.: Benefits from engineering projects implementation. In: Barolli, L., Nishino, H., Miwa, H. (eds) Advances in Intelligent Networking and Collaborative Systems. INCoS 2019. Advances in Intelligent Systems and Computing, vol. 1035. Springer, Cham, pp. 431–441 (2020). https://doi.org/10.1007/978-3-030-29035-1_42

17. Vančová, M.H., Ivanochko, I.: Factors behind the long-term success in innovation—In focus multinational IT companies. In: Developments in Information & Knowledge Management for Business Applications. Studies in Systems, Decision and Control, vol. 376. Springer, Cham (2021). https://doi.org/10.1007/978-3-030-76632-0_16

18. Sahai, A., Sankat, C., Koffka Khan, K.: Decision-making using efficient confidence-intervals with meta-analysis of spatial panel data for socioeconomic development project-managers. Int. J. Intell. Syst. Appl. (IJISA) **4**(9), 92–103 (2012). https://doi.org/10.5815/ijisa.2012.09.12

19. Caliscan, H.K.: Technological change and economic growth. Procedia. Soc. Behav. Sci. **195**, 649–654 (2015). https://doi.org/10.1016/j.sbspro.2015.06.174

20. Piva, M., Vivarelli, M.: Technological change and employment: is Europe ready for the challenge? Eurasian Bus. Rev. **8**, 13–32 (2018). https://doi.org/10.1007/s40821-017-0100-x

21. Detemple, J., Kitapbayev, Y.: Optimal investment under cost uncertainty. Risks **6**(1), 5 (2018). https://doi.org/10.3390/risks6010005

22. Kot, S., Dragon, P.: Business risk management in international corporations. Procedia Econ. Financ. **27**, 102–108 (2015). https://doi.org/10.1016/S2212-5671(15)00978-8

23. Akinleye, G.T., Olarewaju, O.: Credit management and profitability growth in Nigerian manufacturing firms. Acta Universitatis Danubius. OEconomica **15**, 445–456 (2019)

24. Pedrosa, Í.: Firms' leverage ratio and the financial instability hypothesis: An empirical investigation for the US economy (1970–2014). Camb. J. Econ. **43**(6), 1499–1523 (2019). https://doi.org/10.1093/cje/bez004
25. Haynes, R., McPhail, L.: When leverage ratio meets derivatives: running out of options? Financ. Mark. Inst. Instrum. **30**(5), 201–224 (2021). https://doi.org/10.1111/fmii.12154
26. Obradovic, D., Ebersold, Z., Obradovic, D.: Role of technology strategy in competitiveness increasing. Econ. Annals-XXI **1–2**(1), 32–35 (2015)
27. Honchar, M., Kuzmin, O., Zhezhukha, V., Ovcharuk, V.: Simulating and reengineering stress management system—Analysis of undesirable deviations. In: Data-Centric Business and Applications. Lecture Notes on Data Engineering and Communications Technologies, vol. 20. Springer, Cham, pp. 311–330 (2019). https://doi.org/10.1007/978-3-319-94117-2_13
28. Fajri, A., Sinaga, A.: Implementation of business intelligence to determine evaluation of activities (Case Study Indonesia Stock Exchange). Int. J. Inf. Eng. Electron. Bus. (IJIEEB) **12**(6), 51–67 (2020). https://doi.org/10.5815/ijieeb.2020.06.05
29. Bakhodir, A.N.: Research on the use of alternative energy sources in Uzbekistan: problems and prospects. Int. Multidiscip. Res. J. **10**(11), 763–768 (2020). https://doi.org/10.5958/2249-7137.2020.01429.9
30. Lesinskyi, V., Yemelyanov, O., Zarytska, O., Symak, A., Petrushka, T.: Development of a toolkit for assessing and overcoming barriers to the implementation of energy saving projects. Eastern-Eur. J. Enterpr. Technol. **5** (107), 24–38 (2020). https://doi.org/10.15587/1729-4061.2020.214997
31. Tsurkan, M., Andreeva, S., Lyubarskaya, M., Chekalin, V., Lapushinskaya G.: Organizational and financial mechanisms for implementation of the projects in the field of increasing the energy efficiency of the regional economy. Prob. Perspect. Manag. **15**(3), 453–466 (2017). https://doi.org/10.21511/ppm.15(3-2).2017.13
32. Yemelyanov, O., Petrushka, T., Lesyk, L., Symak, A., Vovk O.: Modelling and information support for the development of government programs to increase the accessibility of small business lending. In: 2020 IEEE 15th International Conference on Computer Sciences and Information Technologies (CSIT), pp. 229–232
33. Fedushko, S., Mastykash, O., Syerov, Y., Peracek, T.: Model of user data analysis complex for the management of diverse web projects during crises. Appl. Sci. **10**(24), 9122 (2020). https://doi.org/10.3390/app10249122
34. Kuzmin, O., Ovcharuk, V., Zhezhukha, V.: Administration systems in enterprises management: assessment and development. In: Developments in Information & Knowledge Management for Business Applications. Studies in Systems, Decision and Control , vol. 330. Springer, Cham, pp. 201–229 (2021) https://doi.org/10.1007/978-3-030-62151-3_4
35. Aleksandrzak, T., Mikolášik, M., Ivanochko, I.: Project patterns and antipatterns expert system. In: Developments in Information & Knowledge Management for Business Applications. Studies in Systems, Decision and Control, vol, 420. Springer, Cham. (2022). https://doi.org/10.1007/978-3-030-95813-8_15
36. Kuzmin, O., Yurynets, O.: Modeling the use of urgent crisis management tools in enterprises. In: Modernization of Today's Science: Experience and Trends: Collection of Scientific Papers «SCIENTIA» with Proceedings of the I International Scientific and Theoretical Conference, vol. 1 , pp. 20–22 (2021).

Assessment of the Impact of Biofuel Production on the Sustainable Development of Enterprises in the Agrarian Sector of Ukraine

Ihor Petrushka, Olexandr Yemelyanov, Olena Zagozetska, Oksana Musiiovska, and Kateryna Petrushka

Abstract The purpose of the study is to assess the impact of biofuel production by agrarian enterprises on their sustainable development. The main indicators of sustainable development of enterprises and the level of biofuel production are highlighted. The economic efficiency of production of different types of biofuel according to a sample of enterprises of the agrarian sector of the economy of Ukraine is estimated. The degree of the impact of biofuel production by Ukrainian agrarian enterprises on the indicators of their sustainable development has been established. For this purpose, both methods of deterministic analysis and statistical methods were used. The results of the application of these methods for a sample of Ukrainian agrarian enterprises showed that the impact of biofuel production on their sustainable development is quite significant. In particular, the dynamics of this production is positively correlated with the dynamics of operating income and profits of enterprises and with the change of most indicators that characterize the energy and environmental aspects of sustainable development of the surveyed companies. Among other things, there are statistically significant linear relationships between the growth rate of biofuel production and the growth rate of operating income and product profitability. Accordingly, there is a quadratic relationship between the growth rate of biofuel production and changes in the operating profit of the surveyed agrarian enterprises. Thus, the growth of biofuel production is one of the important drivers of the sustainable development of enterprises in the agrarian sector.

I. Petrushka · O. Yemelyanov · O. Zagozetska (✉) · O. Musiiovska · K. Petrushka
Lviv Polytechnic National University, Bandera Street, 12, Lviv, Ukraine
e-mail: zagoreckao@gmail.com

I. Petrushka
e-mail: ihor.m.petrushka@lpnu.ua

O. Yemelyanov
e-mail: oleksandr.y.yemelianov@lpnu.ua

O. Musiiovska
e-mail: Oksana.B.Musiiovska@lpnu.ua

K. Petrushka
e-mail: kateryna.i.petrushka@lpnu.ua

N. Kryvinska et al. (eds.), *Developments in Information and Knowledge Management Systems for Business Applications*, Studies in Systems, Decision and Control 462,
https://doi.org/10.1007/978-3-031-25695-0_6

Keywords Biofuel · Enterprise · Agrarian sector · Sustainable development · Efficiency

1 Introduction

The need to ensure the proper level of well-being of present and future generations in conditions of limited natural resources requires the sustainable development of countries and regions [8]. This problem should be primarily solved at the enterprise level as the activities of enterprises have a significant impact on the pace and proportions of socio-economic development of countries [6].

The complexity of the concept of sustainable development and a significant number of its goals should be taken into account [21]. In particular, at the national level, sustainable development should be interpreted as "the development that meets the needs of the present without compromising the ability of future generations to meet their own needs" [4]. If we consider individual companies, their sustainable development can be equated with the long-term stable improvement of economic, social, environmental and other results of their activities [29].

Because sustainable development covers almost all aspects of the activities of governments, enterprises, households and individuals, the goals of this development are extremely diverse [7]. However, there are relationships between these goals, and in some cases, some contradictions. In particular, among the important goals of sustainable development are economic growth and an increase in the availability of clean energy. However, economic growth may lead to an increase in the need for energy resources, which will lead to an increase in the consumption of fossil fuels. To prevent this, there are two main ways of action. First, it is necessary to increase the level of efficiency in the use of non-renewable energy sources with the transition to the energy-saving economic development of enterprises, regions and countries as a whole [30]. Second, it is necessary to increase the share of green energy in national energy balances [23]. Under such conditions, a gradual reduction in the consumption of non-renewable energy sources will be ensured. This, in turn, will help to preserve their reserves for future generations [2], reduce production costs [13] and the energy dependence of countries [5] and improve the external environment [17].

In recent years, along with solar, wind and other types of green energy, the role of bioenergy has significantly increased in some countries around the world [28]. Particular attention is paid to the development of this type of alternative energy in the United States [22] and some countries of the European Union [16]. However, at present, bioenergy is also developing quite intensively in some Asian countries, in particular, in China [9]. The potential of bioenergy in Ukraine is also significant [32].

The growing influence of bioenergy in the modern world is due to the advantages that are inherent in it compared to traditional energy based on fossil energy sources [1]. In particular, these advantages include the possibility of making good use of significant amounts of waste of biological origin [15, 24]. The relative cheapness of certain types of biofuels should also be noted. In addition, biofuels can take a variety of forms, leading to a wide range of use of this fuel by both businesses and households [24]. Due to these reasons, the world has seen a growing demand for all main types of biofuel in recent years [19].

At the same time, various barriers often arise in the implementation of projects that increase energy efficiency and the transition to the use of alternative energy sources [11]. In particular, such barriers include insufficient financial resources available to enterprises and households [31], low level of economic efficiency of certain energy-saving projects [12], lack of competence in energy-saving issues [3], etc. A separate group of barriers to energy efficiency are information barriers [10]. The lack of information on the expected results of the implementation of energy projects may be one of the main reasons why businesses and households do not always agree to invest in such projects [20].

Concerning projects for the production of various types of biofuel, the main subjects in these projects are enterprises of the agrarian sector of the economy. At the same time, it is obvious that these companies, as potential investors, must have suffi-ciently complete and comprehensive information about the results of these projects. In particular, it is necessary to forecast the impact that biofuel production projects will have on the indicators of sustainable development of agrarian enterprises. The results of the retrospective analysis can be used for this purpose. These results are data obtained through the assessment of economic, social, environmental and other consequences of the implementation of biofuel production projects by those agrarian enterprises that have already implemented these projects. If these results are positive enough, it will encourage other agrarian enterprises to switch to the production of certain types of biofuel.

Given the above, the purpose of this study is to assess the influence of biofuel production by agrarian enterprises on their sustainable development. To achieve this goal, the following main tasks were solved: the main indicators of sustainable development of enterprises and the level of biofuel production were identified; the economic efficiency of production of different types of biofuels according to the sample of enterprises of the agrarian sector of the economy of Ukraine was estimated; the degree of influence of biofuel production by Ukrainian agrarian enterprises on the indicators of their sustainable development was established.

2   Research Methodology

Assessment of sustainable development of enterprises requires a preliminary selec-tion of the indicators by which such an assessment can be performed. Thus, it seems expedient to carry out the division of these indicators into three such groups:

(1)   indicators that characterize the economic aspects of sustainable development of enterprises. These indicators include income and profit of enterprises, as well as relative indicators of economic efficiency, including product profitability;

(2) indicators that characterize the social aspects of sustainable development of
     enterprises. These indicators include the number of staff and the average salary
     per employee;
(3) indicators that characterize the energy and environmental aspects of sustainable
     development of enterprises. These indicators include the energy intensity of
     products, the share of reused waste, the share of clean energy in the structure of
     energy consumption, etc.

The assessment of the level of sustainable development of enterprises should,
among other things, include measuring the dynamics of these indicators.

The issue of assessing the level of production of biofuel by a certain enterprise
deserves special attention. Such an assessment can be based on the calculation of
absolute and relative indicators. Absolute indicators can be the volume of biofuel
production by the enterprise for a certain period of time in physical or cost units.
Regarding the relative indicators of estimating the level of biofuel production by
a certain enterprise, for their calculation, it is expedient to compare the absolute
indicators with the total value of the enterprise's income from the manufacture and
sale of all its products.

It should be noted that biofuel produced by agrarian enterprises can be divided
into groups according to the way of use and sources of production.

In particular, by the way of use, we can distinguish biofuel produced by agrarian
enterprises for resale and biofuel intended for internal consumption by these
enterprises.

Regarding the division of biofuel by sources of production, it is necessary to
distinguish between biofuel obtained from primary sources and biofuel obtained from
secondary sources. If the secondary sources are mainly different types of waste, the
production of biofuel from its primary sources often requires the creation of special
"energy plantations". As a result, the sown area of traditional crop products (i.e. those
used to meet food needs) may be reduced. Herefore, the use of microalgae [14, 18]
осадів and sewage sludge from sewage treatment plants [25, 26] as a raw material
for biogas production is highly viable.

The division of biofuels described above in the context of this study is important because the production of different groups of biofuels can affect the pace and
proportions of sustainable development of agrarian enterprises in different ways. As
can be seen from Fig. 1, the increase in the production of biofuel by enterprises for
their internal consumption does not directly affect sales volume. However, such an
increase can positively affect the profitability of enterprises (if the cost of producing
their own biofuel is less than the cost of purchasing equivalent amounts of traditional
energy resources). In addition, from Fig. 1, it follows that the mechanism of impact
of biofuel production on the indicators of sustainable development of agrarian enterprises is quite complex, which is reflected in the relevant relationships between these
indicators. In particular, the growth of biofuel production may lead to an increase in
the average wage of agrarian workers. This will happen if such growth causes both
an increase in revenue and an increase in profits. Accordingly, under such conditions,

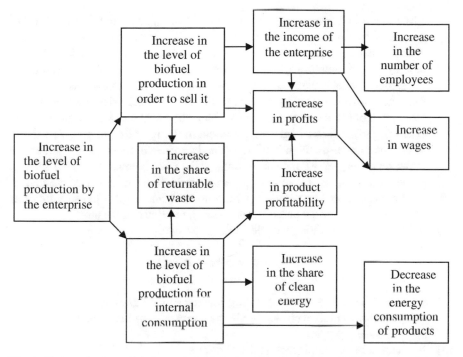

Fig. 1 The mechanism of influence of biofuel production on the indicators of sustainable development of agrarian enterprises

the owners of the enterprise will be able to increase the payroll of employees without reducing the amount of profit due to them.

It can also be assumed that the growth of biofuel production from secondary sources will more positively affect the sustainable development of agrarian enterprises than the increase in biofuel production from its primary sources. This can be explained by the fact that the first option will increase the level of use of industrial waste and will not reduce the area under which traditional crop products will be grown.

In general, assessing the impact of biofuel production on the sustainable development of agrarian enterprises requires the use of both methods of deterministic analysis and statistical methods. In particular, due to the deterministic analysis, it is possible to establish the share of the growth of certain indicators of sustainable development of enterprises due to their biofuel production. For this purpose, first, it is necessary to determine the change in the value of the relevant indicator in the reporting period compared to the base period due to the growth of biofuel production.

This change can be calculated using the following expressions:

(1) for the value of operating profit of the enterprise:

$$\Delta P = \Delta R_b \cdot P_{ab} - \Delta R_t \cdot P_{at} + R_t \cdot \Delta P_{at}, \tag{1}$$

where ΔP is an increase in operating profit of the agrarian enterprise due to the growth of its biofuel production in the reporting period compared to the base period, monetary units; ΔR_b is an increase in income from the sale of biofuel produced by the enterprise, monetary units; P_{ab} is the profitability of production and sale of biofuel in the reporting period, unit share; ΔR_t is a decrease in income from the sale of traditional agrarian products by the enterprise due to increased production of biofuel, monetary units; Pat is the profitability of production and marketing of traditional agrarian products in the reporting period, unit share; R_t is an income from the sale of traditional agrarian products by the enterprise in the reporting period, monetary units; ΔP_{at} is an increase in profitability of production and sales of traditional agrarian products due to increased production of biofuel for own consumption (which, in turn, may reduce the specific costs of the enterprise for the purchase of energy), the share of the unit;

(2) for the value of operating income of the enterprise:

$$\Delta R = \Delta R_b - \Delta R_t, \qquad (2)$$

where ΔR is an increase in operating income of the agrarian enterprise due to the growth of its biofuel production in the reporting period compared to the base period, monetary units;

(3) for the number of employees of the enterprise:

$$\Delta L = \frac{\Delta R_b}{V_b} - \frac{\Delta R_t}{V_t}, \qquad (3)$$

where ΔL is an increase in the number of employees of the agrarian enterprise due to the growth of biofuel production in the reporting period compared to the base period, monetary units; V_b is the amount of income of the agrarian enterprise from the sale of biofuel per employee in the reporting period, monetary units per person; V_t is the amount of income of the agrarian enterprise from the sale of traditional agrarian products per employee in the reporting period, monetary units per person;

(4) for the average salary per employee of the enterprise:

$$\Delta P_l = \frac{\Delta S_{pl}}{L}, \qquad (4)$$

where ΔP_l is an increase in average wages per employee of an agrarian enterprise due to the growth of biofuel production in the reporting period compared to the base period, monetary units per person; ΔS_{pl} is an increase in the wage fund of the enterprise due to the growth of its biofuel production, monetary units; L is the number of employees of the agrarian enterprise in the reporting period, persons;

(5)   for the volumes of non-renewable energy sources used by the agrarian enterprise:

$$\Delta E = \Delta R_b \cdot e_b - \Delta R_t \cdot e_t - \Delta E_b, \tag{5}$$

where ΔP is an increase (decrease) in the volume of non-renewable energy sources used by the agrarian enterprise due to the growth of biofuel production in the reporting period compared to the base period, physical units of energy consumption; e_b is physical costs of non-renewable energy sources to obtain a monetary unit of income from the sale of biofuel produced by the enterprise in the reporting period, physical units of energy consumption per monetary unit; e_t is physical costs of non-renewable energy sources to obtain a monetary unit of income from the sale of traditional agricultural products by the enterprise in the reporting period, physical units of energy consumption per monetary unit; ΔE_b is the value of the decrease in consumption of non-renewable energy sources by the agrarian enterprise due to the increase in its biofuel production for domestic consumption in the reporting period compared to the base period, physical units. Using the above expressions (1)–(5), we can assess the impact of the increase in biofuel production by agrarian enterprises on the relevant indicators of sustainable development. For this purpose, it is necessary to establish the share of the growth of the relevant indicator by increasing the volume of biofuel production in the total amount of this growth. It is also possible to assess the effectiveness of increasing the production of biofuels by a particular agrarian enterprise or a set of such enterprises. To do this, it is necessary to compare the values of indicators (1)–(5) with the amount of investment in biofuel production.

(1)   formation of a sample of agrarian enterprises that belong to a certain subsector of agriculture;
(2)   determination of the current level of biofuel production for each of the enterprises in the base and reporting periods;
(3)   assessment of the dynamics of the level of biofuel production by enterprises and the division of the whole set of enterprises into groups according to the value of the dynamics;
(4)   selection and calculation of the main indicators of sustainable development of the surveyed agrarian enterprises;
(5)   assessment of the statistical significance of the impact of biofuel production by the enterprise on sustainable development.

Also, to assess the impact of biofuel production on the sustainable development of agrarian enterprises, statistical relationships should be established between the indicators of the dynamics of the level of biofuel production and certain indicators of sustainable development. Then the presence of such an effect will be characterized by the appropriate degree of approximation of empirical data by regression equations and the statistical significance of the parameters of these equations.

3   Results and Discussion

To assess the impact of biofuel production on the sustainable development of agrarian enterprises, a sample of fifty such enterprises located in the western region of Ukraine was formed. The main activity of these enterprises is crop production. In particular, these enterprises grow corn, sunflower, soybeans, rapeseed and other types of crop products.

The survayed enterprises were divided into three groups depending on the level of dynamics of biofuel production in 2020 compared to 2016, namely:

(1)   enterprises with a low level of dynamics of biofuel production, in which the growth rate of such production for the study period did not exceed 10%. There were 17 such enterprises;

(2)   enterprises with an average level of dynamics of biofuel production, in which the growth rate of such production for the study period exceeded 10% but was less than 30%. There were 21 such enterprises;

(3)   enterprises with a high level of dynamics of biofuel production, in which the growth rate of such production for the study period was not less than 30%. There were 12 such enterprises.

The volumes of biofuel were measured in tons of conventional fuel.

For each of the groups of surveyed agrarian enterprises, the indicators of the impact of biofuel production on their sustainable development were determined and the efficiency of investing in such production was calculated (Table 1).

As follows from the data presented in Table 1, for all indicators of assessing the impact of biofuel production on the sustainable development of agrarian enterprises, there is a significant increase in their value with increasing levels of dynamics of this production. Thus, according to these indicators, biofuel production has a significant impact on the sustainable development of the surveyed enterprises. Also, according to Table 1, the economic efficiency of biofuel production projects is on average higher at the enterprises with a higher growth rate of its production. This trend may be due to the effect of scale.

The level of economic efficiency of investing in biofuel production may depend on the structure of such production. Therefore, according to the surveyed enterprises, the return on investment in the production of certain types of biofuel was calculated. As evidenced by the results of such calculations, which are shown in Fig. 2, the most profitable are investments in the production of solid biofuel. Also, for each type of biofuel, the return on investment increases with the increase in the level of dynamics of their production.

It is expedient to pay special attention to the assessment of the influence of the level of dynamics of biofuel production by the surveyed agrarian enterprises on the change of the main indicators of their sustainable development. For this purpose, a method of analysis of variance can be used. According to the data presented in Table 2, for most indicators of sustainable development (except for employment and average wages), the actual value of the F-criterion exceeds its critical value with

Table 1 Indicators of the impact of biofuel production on the sustainable development of agrarian enterprises and the results of evaluating the efficiency of investing in such production

Indicators	Values of indicators averaged by groups of enterprises		
	Enterprises with a low level of dynamics of biofuel production	Enterprises with an average level of dynamics of biofuel production	Enterprises with a high level of dynamics of biofuel production
1. The share of the growth of operating profit of enterprises due to the increase in biofuel production in the total growth of their operating profit,%	14.32	25.76	39.19
2. The share of the growth in operating income of enterprises due to the increase in biofuel production in the total growth in their operating income,%	12.48	23.43	37.62
3. The share of the growth of the number of employees of enterprises due to the increase in biofuel production in the total growth of the number of their employees,%	9.13	10.56	14.87
4. The share of the growth of the average salary of employees of enterprises due to the increase in biofuel production in the total growth of the average salary,%	6.02	7.48	9.70
5. The share of the decrease in the use of non-renewable energy sources due to the increase in biofuel production in the overall decrease in the use of these sources,%	15.87	22.93	40.75

(continued)

Table 1 (continued)

Indicators	Values of indicators averaged by groups of enterprises		
	Enterprises with a low level of dynamics of biofuel production	Enterprises with an average level of dynamics of biofuel production	Enterprises with a high level of dynamics of biofuel production
6. Profitability of investments in the implementation of biofuel production projects, shares per unit	0.26	0.31	0.34
7. Increase in operating income per USD 1 of investments in the implementation of biofuel production projects, shares per unit	1.13	1.21	1.33
8. Decrease in the volumes of used non-renewable energy sources by USD 1 of investments in the implementation of biofuel production projects, kg of conventional fuel per USD 1 investment	0.07	0.12	0.19

Fig. 2 Return on investment in production (P_{ai}): **a** pellets; **b** biodiesel; **c** bioethanol, where 1, 2 and 3—enterprise groups according to low, medium and high levels of dynamics of biofuel production

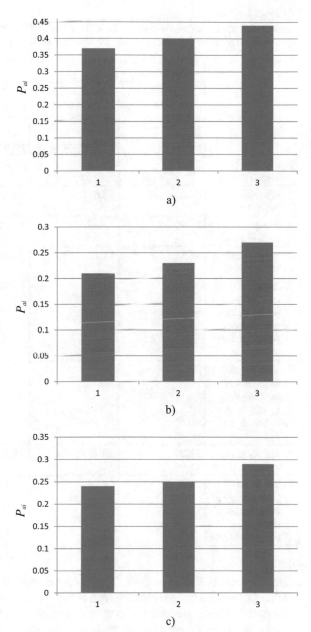

Table 2 Initial data and evaluation results of the impact of the dynamics of biofuel production by agrarian enterprises on the change of the main indicators of their sustainable development

Indicators	Values of the growth rates of indicators, averaged by groups of enterprises, %			The value of the F-criterion
	Enterprises with a low level of dynamics of biofuel production	Enterprises with an average level of dynamics of biofuel production	Enterprises with a high level of dynamics of biofuel production	
1. Operating income, USD	33.09	49.18	71.84	5.32
2. Operating profit, USD	35.43	63.61	88.22	5.78
3. Number of staff, persons	29.06	31.65	32.14	1.97
4. Average salary, USD/person	56.27	61.12	59.43	1.65
5. Energy consumption of products, kg of conventional fuel per 1 USD of income	16.58	27.36	38.29	4.51
6. Volumes of clean energy consumption, kg of conventional fuel	22.72	30.66	37.81	4.28
7. The share of reused waste,%	12.24	19.47	25.52	4.70
8. The share of clean energy in the structure of energy consumption,%	21.78	29.24	40.97	4.95

a significance level of $\alpha = 0.05$. Thus, we can state that the level of dynamics of biofuel production by agrarian enterprises affects the change of most indicators of their sustainable development. The same conclusion can be made for the indicators of the growth rate of incomes and profitability of production of the surveyed enterprises based on the results of the regression analysis (Fig. 3).

As follows from the data shown in Fig. 3, for the surveyed enterprises, there are statistically significant linear relationships between the growth rate of biofuel production and the growth rate of operating income and product profitability. The profitability of products was defined as the ratio of operating profit to operating income. Since the operating profit of enterprises can be presented as the product of

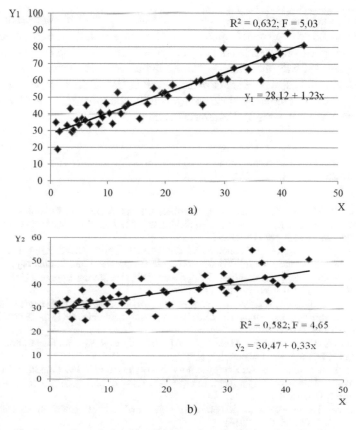

Fig. 3 Graphs of regression relationships between the growth rate of biofuel production by the surveyed enterprises (x) and: **a** operating income (y_1); **b** product profitability (y_2)

the value of income on product profitability, there is a quadratic relationship between the growth rate of biofuel production and changes in operating profit of the surveyed agrarian enterprises.

4　Conclusion

Assessing the impact of biofuel production on the sustainable development of agrarian enterprises requires a preliminary selection of indicators that characterize this development. These indicators should sufficiently characterize the economic, social, energy and environmental aspects of sustainable development. At the same time, assessing the impact of biofuel production on the sustainable development of agrarian enterprises requires the use of both methods of deterministic analysis and statistical methods. The results of the application of these methods for a sample of Ukrainian agrarian enterprises showed that the impact of biofuel production on their sustainable development is quite significant. In particular, the dynamics of this production is positively correlated with the dynamics of operating income and profits of enterprises and with the change of most indicators that characterize the energy and environmental aspects of sustainable development of the surveyed companies. Thus, the growth of biofuel production is one of the important drivers of sustainable development of enterprises in the agrarian sector.

References

1. Ahorsu, R., Medina, F., Constantí, M.: Significance and challenges of biomass as a suitable feedstock for bioenergy and biochemical production: a review. Energies **11**(12), 3366 (2018). https://doi.org/10.3390/en11123366
2. Alvarado, R., Deng, Q., Tillaguango, B., Méndez, P., Bravo, D., Chamba, J. , Alvarado-Lopez, M., & Ahmad, M. (2020). Do economic development and human capital decrease non-renewable energy consumption? Evidence for OECD countries. Energy, 215(B), 119–147. https://doi.org/10.1016/j.energy.2020.119147
3. Backman, F.: Barriers to energy efficiency in swedish non-energy-intensive micro- and small-sized enterprises—A case study of a local energy program. Energies **10**(1), 1–13 (2017). https://doi.org/10.3390/en10010100
4. Brundtland, G.H., Khalid, S., Agnelli, M., Al-Athel, S., Chidzero, B.: Our Common Future. Oxford University Press, London, UK (1987)
5. Chalvatzis, K.J., Ioannidis, A.: Energy supply security in the EU: Benchmarking diversity and dependence of primary energy. Appl. Energy **207**, 465–476 (2017). https://doi.org/10.1016/j.apenergy.2017.07.010
6. Dvořáková, L., Zborková, J.: Integration of sustainable development at enterprise level. Procedia Eng. **69**, 686–695 (2014)
7. Fonseca, L.M., Domingues, J.P., Dima, A.M.: Mapping the sustainable development goals relationships. Sustainability **12**, 3359 (2020). https://doi.org/10.3390/su12083359
8. Gaspa, G.S., Marques, A.S., Fuinhas, J.A.: The traditional energy-growth nexus: A comparison between sustainable development and economic growth approaches. Ecol. Ind. **75**, 286–296 (2017). https://doi.org/10.1016/j.ecolind.2016.12.048
9. Jiang, D., Zhuang, D., Fu, J., Huang, Y., Wen, K.: Bioenergy potential from crop residues in China: Availability and distribution. Renew. Sustain. Energy Rev **16**, 1377–1382 (2012)
10. Kangas, H.L., Lazarevic, D., Kivimaa, P.: Technical skills, disinterest and non-functional regulation: barriers to building energy efficiency in Finland viewed by energy service companies. Energy Policy **114**, 63–76 (2018). https://doi.org/10.1016/j.enpol.2017.11.060
11. Lesinskyi, V., Yemelyanov, O., Zarytska, O., Symak, A., Petrushka, T.: Development of a toolkit for assessing and overcoming barriers to the implementation of energy saving projects. East-Eur. J. Enterpr. Technol. **5**(107), 24–38 (2020). https://doi.org/10.15587/1729-4061.2020.214997
12. Lesinskyi, V., Yemelyanov, O., Zarytska, O., Symak, A., Koleshchuk, O.: Substantiation of projects that account for risk in the resource-saving technological changes at enterprises. East-Eur. J. Enterpr. Technol. **6**(1), 6–16 (2018). https://doi.org/10.15587/1729-4061.2018.149942
13. Maio, F.D., Rem, P.C., Baldé, K., Polder, M.: Measuring resource efficiency and circular economy: a market value approach. Resour. Conserv. Recycl. **122**, 163–171 (2017). https://doi.org/10.1016/j.resconrec.2017.02.009
14. Malovanyy, M., Nikiforov, V., Kharlamova, O., Synelnikov, O.: Production of renewable energy resources via complex treatment of cyanobacteria biomass. Ch&ChT **10**(2), 251–254 (2016). https://doi.org/10.23939/chcht10.02.251

15. Malovanyy, M., Masikevych, A., Kolotylo, M. & Yaremchuk, V. (2019). Analysis of environmental safety of recreational territories of mountain ecosystems and development of technical measures for its stabilization. East-Eur. J. Enterpr. Technol. **6/10**(102): 15–24. https://doi.org/10.15587/1729-4061.2019.185850

16. Mandley, S., Daioglou, V., Junginger, H.M., van Vuuren, D.P., Wicke, B.: EU bioenergy development to 2050. Renew. Sustain. Energy Rev. **127**, 109858 (2020). https://doi.org/10.1016/j.rser.2020.109858

17. Nabavi-Pelesaraei, A., Bayat, R., Hosseinzadeh-Bandbafha, H., Afrasyabi, H., Chau, K.-W.: Modelling of energy consumption and environmental life cycle assessment for incineration and landfill systems of municipal solid waste management—A case study in Tehran Metropolis of Iran. J. Clean. Prod. **148**, 427–440 (2017). https://doi.org/10.1016/j.jclepro.2017.01.172

18. Nykyforov, V., Malovanyy, M., Kozlovska, T., Novokhatko, O., Digtiar, S.: The biotechnological ways of blue green algae complex processing. East-Eur. J. Enterpr. Technol. **5**(10), 11–18 (2016). https://doi.org/10.15587/1729-4061.2016.79789

19. Oosterveer, P., Mol, A.: Biofuels, trade and sustainability: a review of perspectives for developing countries. Biofuels, Bioprod. Biorefin. **4**(1), 66–76 (2010). https://doi.org/10.1002/bbb.194

20. Palm, J., Backman, F.: Energy efficiency in SMEs: overcoming the communication barrier. Energ. Effi. **13**, 809–821 (2020). https://doi.org/10.1007/s12053-020-09839-7

21. Salvia, A.L., Leal Filho, W., Brandli, L.L., Griebeler, J.S.: Assessing research trends related to sustainable development goals: local and global issues. J. Clean. Prod. **208**, 841–849 (2019)

22. Song, Y., Jain, A.K., Landuyt, W., Kheshgi, H.S., Khanna, M.: Estimates of biomass yield for perennial bioenergy grasses in the USA. Bioenerg. Res. **8**, 688–715 (2015). https://doi.org/10.1007/s12155-014-9546-1

23. Steeves, B.B., Ouriques, H.R.: Energy security: China and the United States and the divergence in renewable energy. Contexto int. **38**(2), 643–662 (2016). https://doi.org/10.1590/S0102-8529.2016380200006

24. Szarka, N., Scholwin, F., Trommler, M., Fabian, J.H., Eichhorn, M., Ortwein, A., Thränac, D.: A novel role for bioenergy: a flexible, demand-oriented power supply. Energy **61**, 18–26 (2013). https://doi.org/10.1016/j.energy.2012.12.053

25. Tymchuk, I., Malovanyy, M., Shkvirko, O., Vankovych, D., Odusha, M., Bota O.: Monitoring of the condition of the accumulated sludge on the territory of Lviv wastewater treatment plants. In: Conference Proceedings, International Conference of Young Professionals «GeoTerrace-2020», pp. 1–5 (2020). https://doi.org/10.3997/2214-4609.20205714.

26. Tymchuk, I., Malovanyy, M., Shkvirko, O., Chornomaz, N., Popovych, O., Grechanik, R., Symak, D.: Review of the global experience in reclamation of disturbed lands. Inzynieria Ekologiczna **22**(1), 24–30 (2021). https://doi.org/10.12912/27197050/132097

27. Voytovych, I., Malovanyy, M., Zhuk, V., Mukha O.: Facilities and problems of processing organic wastes by family-type biogas plants in Ukraine. J. Water Land Dev. **45**(IV–VI): 185–189 (2020). https://doi.org/10.24425/jwld.2020.133493.

28. Welfle, A., Thornley, P., Röder, M.: A review of the role of bioenergy modelling in renewable energy research & policy development. Biomass Bioenergy **136**, 105542 (2020)

29. Yemelyanov, O., Petrushka, T., Symak, A., Trevoho, O., Turylo, A., Kurylo, O., Danchak, L., Symak, D., Lesyk, L.: Microcredits for sustainable development of small Ukrainian enterprises: efficiency, accessibility, and government contribution. Sustainability **12**, 6184 (2020). https://doi.org/10.3390/su12156184

30. Yemelyanov, O., Symak, A., Petrushka, T., Zahoretska, O., Kusiy, M., Lesyk, R., Lesyk, L.: Changes in energy consumption, economic growth and aspirations for energy independence: sectoral analysis of uses of natural gas in Ukrainian economy. Energies **12**, 4724 (2019). https://doi.org/10.3390/en12244724

31. Yemelyanov, O., Symak, A., Petrushka, T., Lesyk, R., Lesyk, L.: Evaluation of the adaptability of the Ukrainian economy to changes in prices for energy carriers and to energy market risks. Energies **11**, 3529 (2018). https://doi.org/10.3390/en11123529
32. Zulauf, C., Prutska, O., Kirieieva, E., Pryshliak, N.: Assessment of the potential for a biofuels industry in Ukraine. Probl. Perspect. Manag. **16**, 83–90 (2018)

Simulation Models for Prioritizing the Implementation of Energy-Saving Investment Projects for the Enterprise

Vasyl Teslyuk, Ivan Tsmots, Roman Sydorenko, and Serhii Stryamets

Abstract The structure of decision support tools for determining the priority of energy-saving projects has been developed. It is determined that for the effective operation of the decision support system in determining the priority of energy-saving projects, the system should include: means of collecting and storing information about the energy efficiency of the enterprise; means of numerical evaluation of selection criteria; simulation model for determining the priority of energy-saving projects; means of forecasting energy efficiency of the enterprise taking into account the implementation of the energy-saving project; means of visualization of simulation results. The method of pairwise comparison is proposed and the scale for estimating the weights of the importance of selection criteria is modified. A model for determining the priority of energy-saving projects has been developed, which is based on the method of hierarchy analysis and takes into account the interaction and interdependence of project selection criteria. A list of economic, technical, production, environmental and organizational criteria for the selection of energy-saving investment projects has been formed. It is shown that the development of the model for determining the priority of investment energy-saving projects is performed in five stages: at the first stage—the task of determining the priority of investment energy-saving projects in the form of the three-level tree of hierarchies; at the second stage—expert numerical evaluation of economic, technical, production, environmental and organizational selection criteria is carried out; at the third stage—a matrix of pairwise comparisons is formed and the eigenvector and the vector of priorities for the second level of the hierarchy are calculated; At the fourth stage we form a matrix of pairwise

V. Teslyuk (✉) · I. Tsmots · R. Sydorenko · S. Stryamets
University Lviv Polytechnic National University, 28a Bandera Street, Lviv 79013, Ukraine
e-mail: vasyl.m.teslyuk@lpnu.ua

I. Tsmots
e-mail: ivan.h.tsmots@lpnu.ua

R. Sydorenko
e-mail: roman.v.sydorenko@lpnu.ua

S. Stryamets
e-mail: serhii.p.striamets@lpnu.ua

comparisons and calculate the vector of priorities for each energy-saving project; in the fifth stage—the vector of global priorities is calculated. A block diagram of the operation algorithm has been developed and a simulation model for determining the priority of investment energy-saving projects has been implemented using MATLAB. Approbation of the simulation model on the basis of the municipal enterprise "Sambir Vodokanal" was carried out.

Keywords Decision support tools · Imitation model · Energy efficiency · Method of analysis of hierarchies · Energy-saving project

1 Introduction

Given the growing cost of energy and significant energy intensity, the current challenges for Ukrainian enterprises are to reduce the energy intensity of production, reduce non-production losses, and costs of use and optimize the structure of fuel and energy resources by choosing the best investment. Solving such problems requires extensive use of modern information technology and decision support systems that use the results of information processing, forecasting and modelling. One way to increase energy efficiency is to use investments to upgrade fixed assets using energy-efficient materials, energy-saving technologies and modern equipment. However, as a rule, decisions are made on the available funds that can create an enterprise to improve energy efficiency. Accordingly, the choice of object and type of work for the first investment is complex, which covers all the tasks of objects and types of work and creates their relationships and interactions to improve energy efficiency. Due to this special urgency, there is the problem of choosing the object and type of work for investment among the many possible ones that provide the company with the greatest increase in energy efficiency.

One of the main indicators that ensure the level of competitiveness of the enterprise in the domestic and foreign markets is the energy intensity of gross domestic product, which characterizes the level of consumption of fuel and energy resources per unit of gross domestic product. The main ways to reduce the energy intensity of the domestic product and increase the efficiency of the enterprise are: improving the organizational and economic mechanisms of energy-saving; conducting an energy enterprise; development of means to support decision-making on the implementation of energy-saving measures; reduction of specific costs of energy resources per unit of output; implementation of measures for the rehabilitation of buildings, including the introduction of software control of room temperature, replacement of windows and front doors, as well as insulation of building structures; replacement of worn-out equipment; reduction of energy losses during their transformation and transportation; reducing the share of natural gas in the structure of organic fuels and reducing its contribution to the energy intensity of the gross product; introduction and use of alternative and renewable energy sources; reduction of the level of non-production losses of fuel and energy resources; modernization of energy supply, equipment and implementation of modern energy-efficient technologies.

Each of these ways to improve the energy efficiency of the enterprise for the implementation of energy-saving investment projects, which can be divided into alternative, interconnected and independent. Alternative investment projects are projects that contain proposals for the implementation of the general direction of energy saving. The peculiarity of alternative investment projects is that only one of the many projects is selected for implementation. Interrelated investment projects are characterized by the fact that the implementation of one requires the implementation of others. Independent investment projects are projects whose implementation is not related to the necessary implementation of other projects, i.e. from many independent projects can be implemented in only one or all of them.

Determining the priority of the project for the first investment is a complex task that benefits all economic types of workers and insignificant, technical, industrial, environmental and organizational factors influence on improving the energy efficiency of the enterprise. Due to this particular urgency, there is the problem of choosing a project to invest in among the many possible ones that provide the greatest increase in the energy efficiency of the enterprise.

The object of research—the processes of collecting, evaluating, processing energy, economic, technical, production, and environmental information and choosing the best option for energy-saving projects.

The subject of research—models, methods, and algorithms for evaluating and processing information to determine the priority of energy-saving investment projects for the company.

The purpose of the work is to develop a simulation model to determine the priority of energy-saving investment projects at the enterprise.

To achieve this goal, the following main objectives of the study are identified:

- analysis of recent research and publications;
- development of the structure of decision support tools;
- assessing the importance of selection criteria;
- development of a model for determining the priority of energy-saving investment projects;
- implementation of a simulation model for prioritizing the implementation of energy-saving investment projects.

1.1 Analysis of Recent Research and Publications

Assessing the quality of data is one of the biggest problems encountered in developing models for prioritizing the implementation of energy-saving projects. All estimates must be quantified, which requires the conversion of qualitative expert assessments into numerical values. A recognized method of such transformations is fuzzy logic [1, 2], according to which the truth or falsity of a statement is not well defined. Using fuzzy logic to move from qualitative to numerical data and vice versa uses the membership function, which is not always good because it requires additional actions and data that are not in the project.

In the works [3, 4] considerable attention is paid to the study of the energy management system as part of enterprise management, which ensures the rational use of energy resources in the operation of the enterprise. However, the existing energy management systems of enterprises are not aimed at implementing an innovative strategy for enterprise infrastructure development and addressing the use of energy resources and the principles of sustainable development.

From the analysis of publications [5, 6], it follows that to increase the energy efficiency of enterprises it is necessary to widely use smart technologies. The energy efficiency of a modern enterprise largely depends on the ability to constantly adapt to new requirements through the introduction of smart technologies that use artificial intelligence, machine learning and big data analysis, sensors, information and communication and computing tools. The literature [7, 8] shows that the use of smart technologies in the development of energy-saving projects is aimed at preserving the natural habitat and more rational and efficient use of existing resources by regulating and automating energy consumption. Also in the publication [9] it is noted that improving the energy efficiency of modern enterprises is based on monitoring, analysis and management of energy consumption processes in real-time.

In the works [10, 11] the decision support systems for choosing the best variant of the investment project in order to increase the energy efficiency of the enterprise are considered. Such decision support systems are based on models, methods and algorithms for processing large data sets obtained during the economic and energy survey of the enterprise. The result of such a survey is the information used to assess energy consumption, economic, technical, production, environmental and organizational factors influencing the enterprise during project implementation [12, 13].

The publication [14, 15] analyzes the models and methods of decision support to choose the best option for the energy-saving projects and shows that existing models and methods do not use systematic procedures and do not take into account the interaction and interdependence of factors influencing enterprise operation.

From the analysis of publications, it follows that for the formation of effective management decisions in determining the priority of energy-saving projects it is necessary to develop a simulation model for determining the priority of investment energy-saving projects for the company.

2 Research Results and Their Discussion

2.1 Development of the Structure of Decision Support Tools

Improving the energy efficiency of the enterprise can be achieved by performing the following tasks:

1. Reduction of technological and non-production losses of energy resources through modernization of the energy supply scheme, modernization of equipment, and introduction of modern energy-efficient technologies, namely:

    reducing the energy intensity of production per unit of output;

    optimization of energy supply schemes and reduction of losses of fuel and energy resources;

    modernization of the heat supply system and autonomous heat supply.
2. Optimization of the structure of consumption of fuel and energy resources, in particular, the replacement of traditional types of energy resources with other types, which include renewable energy sources and alternative fuels, namely:

    optimization of the structure of consumption of fuel and energy resources;

    identification of technical potential and replacement of traditional types of energy resources through the use of renewable energy sources and alternative fuels;

    implementation of complex energy-efficient projects.
3. Improving the management system of energy efficiency and energy consumption, namely:

    Development of information and analytical system for evaluation, forecasting and management of energy efficiency of the enterprise;

    conducting an energy audit and implementation of energy management;
        training of personnel (training courses, seminars, conferences) and implementation of energy-saving measures.

To determine the effectiveness of energy-saving projects are selected depending on the models that take into account the interaction and evaluation criteria of the project and their impact on energy-saving projects of enterprises and organizations. When developing such models, other lists of evaluation criteria and their numerical values are needed. Developed models for evaluating energy-saving projects can have different complexity, depending on the methods of construction and complexity of projects. Models for evaluating and determining the continuity of energy-saving projects can be classified:

- dynamic or static;
- stochastic or deterministic;
- continuous or discrete;
- linear or nonlinear;
- statistical, expert, developed on the methods of Data Mining;
- predictive, classification and descriptive.

For the effective functioning of the decision support system for certain parameters of energy-saving projects offer a structure of tools, which is shown in Fig. 1.

The main components of the decision support system for certain parameters of energy-saving projects are the following tools:

Fig. 1 Structure of decision support tools for certain indicators of energy-saving projects

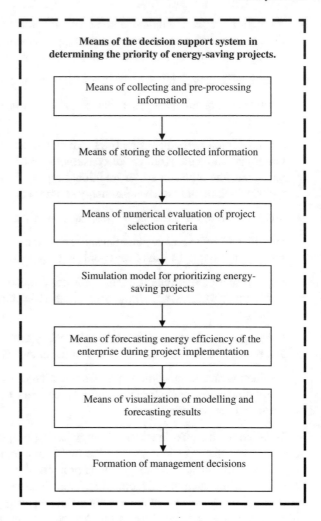

Means of the decision support system in determining the priority of energy-saving projects.

Means of collecting and pre-processing information

Means of storing the collected information

Means of numerical evaluation of project selection criteria

Simulation model for prioritizing energy-saving projects

Means of forecasting energy efficiency of the enterprise during project implementation

Means of visualization of modelling and forecasting results

Formation of management decisions

- collection and preliminary work of information;
- means of storing the collected information;
- numerical evaluation of election criteria;
- simulation model for determining the quality of energy-saving projects;
- forecasting the energy efficiency of the enterprise taking into account the implementation of the energy-saving project;
- visualization of modelling and forecasting results;
- formation of management decisions.

Means of collecting and preliminary work of information on the following functions:

- collection of raw data;
- data transfer from one source to another;
- conversion of data from one form to another;
- data search;
- formation of data or information in a user-friendly form.

The data proposed for the evaluation of criteria can be classified in two ways. The first way is to place sources of information, and the second—is by their purpose. The first way of information is divided into internal and external.

Internal information in its system can be divided into three groups: the first is information about the production (production dynamics, the actual level of capacity utilization, the rate of renewal of fixed assets, the rhythm of production, structure and technological level of equipment, etc.); second—information on labour resources (number of employees, qualifications, staff turnover, length of service, age of employees, payroll and its structure, training costs, morale, level of injuries, the general level of culture in the enterprise, etc.); third—internal financial information (data from the balance sheet of income, expenses, non-current and current assets, cost of goods sold, financial results from operating activities.).

External information can also be divided into the following groups: market information (analysis of trends in domestic and global markets); information about competitors; macroeconomic and geopolitical information; information about suppliers; external financial information (exchange rates, dynamics of stock prices, movement in the capital market, etc.); regulation information and data.

External and internal information can be primary, obtained from research or analysis, or secondary, obtained from existing sources.

In order to effectively formulate and evaluate the selection criteria, it is necessary to have access to information from both the external environment and internal sources.

To collect information offer integrated collection tools that provide [1, 5]:

- automation of input of documents from paper carriers in the electronic form;
- registration, accounting of the entire volume of incoming, outgoing and internal documents;
- primary processing and registration of documents;
- the prompt search of documents and search of documents in accordance with the request for attributes of the document (registration number, data, authors, performers, etc.), keywords and descriptions of fragments of documents;
- integration and interaction with e-mail, file systems and Web-technologies;
- support for various sources of information;
- protection against unauthorized access.

Preliminary processing of the collected information is reduced to the following operations: filtering of primary input data, threshold processing; recovery of partially lost data at certain time intervals; intellectualized pre-processing. The reasons for the inaccuracy are the order of physical damage to the sensor, failures in power supply systems, interference from the power equipment of production facilities and failures in the communication channels. The tasks of filtering primary data and threshold processing are solved by ETL using primary data collection systems. Intelligent pre-processing is designed to possibly reproduce lost measurements from the sensor and analyze the reliability and consistency of a particular parameter compared to others. Implementation of the tasks of recovery of partially lost data at certain time intervals requires the use of artificial neural networks.

Information storage tools are focused on the reliable storage of large amounts of data, which allows you to save data, which consists of two main parts: the database and the database. Data in the subsystem can be stored in different databases and their analysis can cause problems with support for different data formats, as well as their encoding. This problem is solved by creating data warehouses that are subject-oriented, integrated, and immutable and that support the history of the dataset. At the heart of the data warehouse concept is the idea of sharing data that is needed for operational analysis and for solving problems of intellectual analysis. In addition, the conservation subsystem is part of the knowledge base, which stores the accumulated experience in energy efficiency management.

Evaluation of important selection criteria. The list of criteria for selecting energy-saving investment projects is formed by economic, technical, production, environmental and organizational impact groups. One of the main tasks of input data quality is the development of project appraisal evaluation models. Information for MVPP can come both quantitatively and qualitatively. It is clear that in order to perform any computational or comparative action, it is necessary to lead to a single numerical form. The main tasks of the established MVPP are the development of tools for the optimal conversion of qualitative information into numerical form. The biggest problem you face when working with a specific feature is the accuracy of its translation. This quality is the appropriate assessment of the data that will be evaluated in the future.

For correct data processing in the information MVPP, which must be unified and numbered. Assessing the quality of the MVPP requires, first of all, assessing the quality of the data that ensure its effective functioning.

As data quality indicators are not physical quantities that can be measured, expert evaluation methods are used. All assessments must be quantified, requiring the conversion of qualified expert assessments into numerical values. The use of quantitative methods of expert evaluation is also expedient for the quantitative presentation of data. Quantitative methodological expert technologies are based on the use of logical-mathematical and statistical methods of generalization of experts' opinions, verification of statistical significance of examination results, and confirmation or refutation of examination in general.

To assess the weight of the importance of the selection criteria, we use the method of pairwise comparison, which is a separate method of examination. Qualitative comparison of two objects is considered simpler and more reliable than expressing the advantage in the reference or rating scale Analysis of methods of transition from qualitative to a quantitative form of input shows that qualitative comparison of two objects is considered simpler and more reliable than expressing the advantage in point reference or rating scale. For MVPP it is advisable to use the method of pairwise comparison. In many cases, detailed numerical information about the objects of study is not required, and comparisons can be made on a more or less basis without specifying how many times or how much more or less.

The method of pairwise determination of the definition of groups and criteria that have the greatest impact on the level of implementation of energy-saving investment projects The most effective method of identifying groups and criteria of influence is the method of expert assessments. The essence of this method is in the conducted intuitive-logical analysis in combination with quantitative evaluation. Since we obtain subjective results in the evaluation process, it is advisable to use the opinion of a separate group of experts to increase the reliability of the results. The process of expert assessment of factors influencing the company is divided into the following stage:

- formation of goals of expert survey;
- preparation of the questionnaire;
- determining the number of experts;
- survey of experts;
- elaboration of the results of the expert survey.

Based on the results of expert experience, the groups and criteria that have the greatest impact on determining the effectiveness of energy-saving investment projects have been identified. Expert assessment of the criteria for selection of energy saving investment projects uses the results of energy information of enterprises, accumulated by the cost of fuel and energy resources, organizational and technical mechanisms of energy-saving, existing equipment and use of modern energy technologies. For the scales of selection criteria for energy-saving investment projects, a modified scale of importance for the formation of selection criteria was used (Table 1).

Based on the answers to the question about the importance of the selection criteria, a number of evaluations are performed.

Using the method of hierarchy analysis, a simulation model for determining the efficiency of energy-saving projects is developed and implemented. This simulation model provides the choice of the best project. To make a management decision, it is necessary to perform energy efficiency forecasting of the enterprise, taking into account the implementation of energy-saving projects.

The next step is to visualize the results of modelling and forecasting. Visualization of modelling and forecasting results in a broad sense is a graphical representation that is displayed in the form of graphs, charts, infographics, charts, dashboards, maps and cartograms. Graphs and charts are the most common means of visualizing data used for both presentation and analysis. There are dozens of well-known types of charts, and their number is increasing—new types are offered to visualize complex and unusual data. Infographics visualize certain parameters, interactions of elements, etc. Data presentation and analysis is one of the common ways to use data visualization in the form of charts or infographics. When analyzing data with the help of visualization, the method of visual presentation of data is used in order to find hidden relationships and dependencies, as well as the initial evaluation of data sets in order to further

Table 1 The modified scale of weights of importance of selection criteria

The importance of weight	Definition	Explanation
1	Equal importance	Equal impact of two selection criteria on energy efficiency
3	Moderate advantage of one criterion over another	The analysis shows a slight advantage of one information selection criterion over another
5	Significant or strong advantage of one criterion over another	Analytical information indicates a significant advantage of one selection criterion over others
7	A significant advantage	One selection criterion has a great advantage, which becomes practically significant
9	A very big advantage	The obvious advantage of one selection criterion over others
2, 4, 6, 8	Intermediate values between adjacent levels	Recorded in compromise cases

use more complex analysis tools. This approach is called Exploratory data analysis (EDA), which is one of the tools of Data Mining. The main difference between EDA and data presentation is that visualization is performed using a variety of software tools. Business analytics and dashboards are now widely used when it is necessary to monitor certain indicators on an ongoing basis.

The problem of forming management decisions when choosing to implement energy-saving goals is as follows:

- the uniqueness of the situation of choice;
- difficult to assess the nature of the considered alternatives;
- uncertainty of aftereffects;
- many different factors that are necessary when making decisions;
- the presence of the person or persons responsible for the decision.

Selection decision support tools for the implementation of an energy-saving project using the following functions:

- the manager evaluates the evaluation, selects the criteria and evaluates their relevance;
- generate decision opportunities and scenarios;
- scenarios for evaluation and selection of solutions and arias;
- provides constant exchange and coordination of information on the progress of the decision-making process;
- modelling and analysis of possible consequences of decision-making.

Decision support tools have the right to have the potential to automate decision-making, but the decisions made by the system must be clear to the individual. For the necessary level of validity of management decisions, the adoption procedure itself should be transparent and open for discussion and analysis.

2.2 Development of a Model for Prioritizing the Implementation of Energy-Saving Investment Projects

The development of WFP for energy-saving investment projects is carried out in three stages.

At the first stage, the decomposition of the task of determining the priority of investment energy-saving projects and its reflection in the form of a tree of hierarchies, which structures the task and identifies the factors (criteria) by which the project will be selected. Based on the results of the expert survey, we determine the criteria for selecting projects using which to develop a tree of hierarchies. To determine the priority of energy-saving investment projects, the tree of hierarchies will consist of three levels.

At the first hierarchical level there is a common goal—an energy-saving investment project.

The second level of the hierarchical tree consists of criteria used to determine the priority of energy-saving investment projects. The number and types of criteria are determined by experts by processing the information obtained during the economic and energy survey of the enterprise. The result of such a survey is the criteria that assess the economic, technical, production, and environmental and organizational impacts on the enterprise in the implementation of projects.

At the third level of the hierarchical tree there are m possible investment energy saving projects, one of which must be selected.

The result of the first stage of MVPP development is the development of a tree of hierarchies, which is shown in Fig. 2.

At the second level, the following criteria are proposed for the evaluation of energy-saving investment projects:

- *economic*—net present income (e_1); payback period of investments (e_2); a total economic result of the project (e_3); the amount of investment (e_4); investment efficiency ratio (e_5);
- *technical*—technical feasibility (t_1); availability of appropriate equipment and tools (t_2); annual energy consumption (t_3); energy utilization factor (t_4); quality of equipment and materials (t_5);
- *production*—the depth of processing of raw materials (b_1); energy intensity of production (b_2); energy intensity of fixed assets (b_3); labour energy equipment (b_4); change in the quality of the final product (b_5);
- *environmental*—reduction of environmental pollution by industrial waste (p_1); reduction of emissions due to reduction of fuel and electricity consumption (p_2);

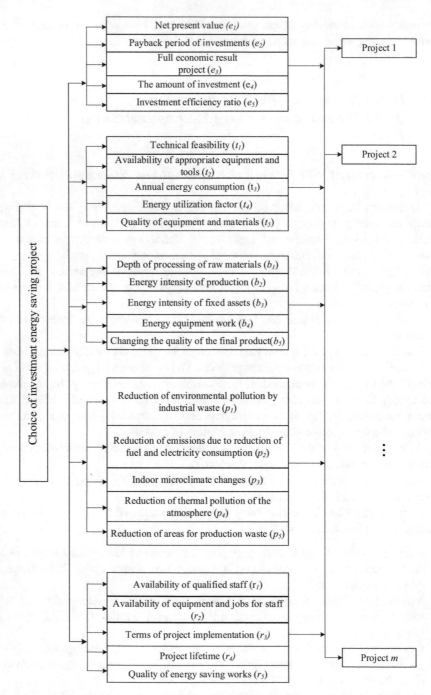

Fig. 2 Tree of hierarchies for the model of prioritization of energy saving investment projects

changes in the microclimate of the premises (p_3); reduction of thermal pollution
of the atmosphere (p_4); reduction of areas for production waste (p_5);

- *organizational*—the availability of personnel of appropriate qualifications (r_1);
availability of equipment and jobs for staff (r_2); terms of project implementation
(r_3); project lifetime (r_4); quality of energy-saving works (r_5).

To develop a model for determining the priority of investment energy-saving
projects, the tree of hierarchies, which is shown in Fig. 1, is considered as a baseline,
where the number of criteria depends on the type of investment energy-saving project
and its detail.

At the second stage of model development the expert assessment of economic (e_1,
e_2, e_3, e_4, e_5), technical (t_1, t_2, t_3, t_4, t_5), production (b_1, b_2, b_3, b_4, b_5), environmental
(p_1, p_2, p_3, p_4, p_5) and organizational (r_1, r_2, r_3, r_4, r_5) criteria for each investment
energy saving project. For expert evaluation of the project j–o criteria ($j = 1,\ldots,$ m
where m—number of energy-saving investment projects) we will use the results of
economic and energy survey of the enterprise, features j–o investment energy saving
project and a modified scale of weights of importance criteria (Table 5.1). The results
of the expert evaluation of each criterion for the j - o project determine the weight of
its importance. For economic (e_1, e_2, e_3, e_4, e_5), technical (t_1, t_2, t_3, t_4, t_5), production
(b_1, b_2, b_3, b_4, b_5), environmental (p_1, p_2, p_3, p_4, p_5) and organizational ($r_1, r_2, r_3,$
r_4, r_5) weight importance criteria are written as: $w_{e1} = w_1, w_{e2} = w_2, w_{e3} = w_3, w_{e4}$
$= w_4,$ and $w_{t5} = w_5; w_{t1} = w_6, w_{t2} = w_7, w_{t3} = w_8, w_{t4} = w_9,$ and $w_{b5} = w_{10}; w_{b1}$
$= w_{11}, w_{b2} = w_{12}, w_{b3} = w_{13}, w_{b4} = w_{14},$ and $w_{b5} = w_{15}; w_{p1} = w_{16}, w_{p2} = w_{17},$
$w_{p3} = w_{18}, w_{p4} = w_{19},$ and $w_{p5} = w_{20}; w_{r1} = w_{21}, w_{r2} = w_{22}, w_{r3} = w_{23}, w_{r4} = w_{24},$
and $w_{r5} = w_{25}.$

The number of criteria n depends on the detail of the evaluation of the investment
energy-saving project.

At the third stage of model development at the second level of the hierarchy tree for
each j–o investment energy-saving project of each matrix of pairwise comparisons
A_j is formed and priority vectors A_{jp} are calculated.

When forming a matrix of pairwise comparisons A_j we use the weights of the
criteria that it is desirable to achieve in the implementation process j - o investment
energy-saving project. The formation of a matrix of pairwise comparisons is carried
out by pairwise comparison of the weight of each criterion with the weights of other
criteria by the expression:

$$A_j = \begin{matrix} a_{j11} = \frac{w_{j1}}{w_{j1}} & a_{j12} = \frac{w_{j1}}{w_{j2}} & \cdots & a_{j1n} = \frac{w_{j1}}{w_{jn}} \\ a_{j21} = \frac{w_{j2}}{w_{j1}} & a_{22} = \frac{w_{j2}}{w_{j2}} & \cdots & a_{j2n} = \frac{w_{j2}}{w_{jn}} \\ \cdot & \cdot & \cdots & \cdot \\ \cdot & \cdot & \cdots & \cdot \\ \cdot & \cdot & \cdots & \cdot \\ a_{jn1} = \frac{w_{jn}}{w_{j1}} & a_{jn2} = \frac{w_{jn}}{w_{j2}} & \cdots & a_{jnn} = \frac{w_{jn}}{w_{jn}} \end{matrix} \qquad (1)$$

The result of such a pairwise comparison for everyone j - o investment energy-saving project we obtain a matrix of pairwise comparisons A_j of size $n \times n$. The next step is the calculation for the matrix A_j priority vector A_{jp}. The procedure for calculating the priority vector A_{jp} is carried out in accordance with the following expression:

$$
A_j = \begin{vmatrix}
a_{j11} = \frac{w_{j1}}{w_{j1}} & a_{j12} = \frac{w_{j1}}{w_{j2}} & \cdots & a_{j1n} = \frac{w_{j1}}{w_{jn}} \\
a_{j21} = \frac{w_{j2}}{w_{j1}} & a_{22} = \frac{w_{j2}}{w_{j2}} & \cdots & a_{j2n} = \frac{w_{j2}}{w_{jn}} \\
\cdot & \cdot & \cdots & \cdot \\
\cdot & \cdot & \cdots & \cdot \\
\cdot & \cdot & \cdots & \cdot \\
a_{jn1} = \frac{w_{jn}}{w_{j1}} & a_{jn2} = \frac{w_{jn}}{w_{j2}} & \cdots & a_{jnn} = \frac{w_{jn}}{w_{jn}}
\end{vmatrix} \Rightarrow \begin{vmatrix} a_{j1B} \\ a_{j2B} \\ \cdot \\ \cdot \\ \cdot \\ a_{jnB} \end{vmatrix} \Rightarrow \begin{vmatrix} a_{j1} \\ a_2 \\ \cdot \\ \cdot \\ \cdot \\ a \end{vmatrix} \quad (2)
$$

In accordance with expression (2), the calculation of the priority vector A_{jp} involves the formation of a pairwise comparison matrix A_{jp} and the definition of the eigenvector $A_{j\theta}$. The calculation of the elements $a_{ji\theta}$ ($i = 1,\ldots, n$, where n is the number of selection criteria) of the eigenvector A_j in is performed by the formula:

$$
a_{jiB} = \sqrt[n]{\prod_{i=1}^{n} a_{jii}} \quad (3)
$$

Having the values of the elements $a_{ji\theta}$ of the eigenvector $A_{j\theta}$ in we can calculate the values of the elements a_{jip} priority vector A_{jp} by the formula:

$$
a_{ji} = \frac{ji}{\sum_{i=1}^{n} ijB} \quad (4)
$$

At the fourth stage, for each j-th project of the third level of the tree of hierarchies, we form a matrix of pairwise comparisons A_j and calculate the priority vector A_{jp}. Informing the matrix of pairwise comparisons A_j, the criteria obtained as a result of the j-th project are used.

Based on the results of the fourth stage, we obtain m matrices of pairwise comparisons A_j and m priority vectors A_{jp}.

In the fifth stage for the selection of energy-saving investment projects, the vector of global priorities G is calculated.

To calculate the vector of global priorities G it is necessary to form a matrix of priorities size $m \times n$, where m is the number of projects, and n is the number of influencing factors. Priority matrix is formed from m calculated priority vectors A_{jp} for each project. The calculation of the global priority vector G is performed by multiplying the priority matrix on vector priorities A_p in accordance with the formula:

$$
G = \begin{vmatrix} a_{11} & a_{12} & \ldots & a_{1n} \\ a_{21} & a_{22} & \ldots & a_{2n} \\ \cdot & \cdot & \cdots & \cdot \\ \cdot & \cdot & \cdots & \cdot \\ \cdot & \cdot & \cdots & \cdot \\ a_{m1} & a_{m2p} & \ldots & a_{mn} \end{vmatrix} \times \begin{vmatrix} a_1 \\ a_2 \\ \cdot \\ \cdot \\ \cdot \\ a \end{vmatrix} = \begin{vmatrix} a_{11p} \times a_{1p} + a_{12p} \times a_{2p} + \ldots a_{1np} \times a_{np} \\ a_{21p} \times a_{1p} + a_{22p} \times a_{2p} + \ldots a_{2np} \times a_{np} \\ \cdot & & \cdot & \cdots & \cdot \\ \cdot & & \cdot & \cdots & \cdot \\ \cdot & & \cdot & \cdots & \cdot \\ a_{m1p} \times a_{1p} + a_{m2p} \times a_{2p} + \ldots a_{mnp} \times a_{np} \end{vmatrix}
$$

$$
= \begin{matrix} G_1 \\ G_2 \\ \cdot \\ \cdot \\ \cdot \\ G \end{matrix} \tag{5}
$$

The selection of a project for implementation is carried out by analyzing the values of the elements of the vector of global priorities G, each of which corresponds to a specific project. Before implementation, the project is recommended in the first place, the number of which coincides with the number of the element of the vector of global priorities G, which has the maximum value.

3 Implementation of a Simulation Model for Determining the Optimal Implementation of Energy-Saving Investment Projects

The development and model for determining the optimal implementation of energy-saving investment projects are based on the developed tree of hierarchies and the formula 1–5 used. In addition, in the process of a simulation model as input data is used database, which stores the results of expert evaluation of economic (e_1, e_2, e_3, e_4, e_5), technical (t_1, t_2, t_3, t_4, t_5), production (b_1, b_2, b_3, b_4, b_5), environmental (p_1, p_2, p_3, p_4, p_5) and organizational (r_1, r_2, r_3, r_4, r_5) criteria for each j-th investment energy-saving project. The algorithm for implementing the simulation model for determining the priority of investment energy-saving projects is shown in Fig. 3.

The simulation model for prioritizing the implementation of energy-saving investment projects has been developed using MATLAB. The main window of the user interface of the simulation model is shown in Fig. 4.

Consider in more detail the process of forming a matrix of pairwise comparisons A_j, the calculation of the eigenvector $A_{j\theta}$ in and priority vector A_{jp}. The formation of a matrix of pairwise comparisons begins with the introduction of the size of the matrix in the field n, where n is equal to the size of the matrix. Formation of a matrix of pairwise comparisons A_j is carried out by comparing the weight of the impact criteria with the weights of other impact criteria. The results of such a pairwise comparison are entered sequentially column by column.

Fig. 3 Block diagram of the algorithm for the implementation of the simulation model for determining the priority of investment energy-saving projects

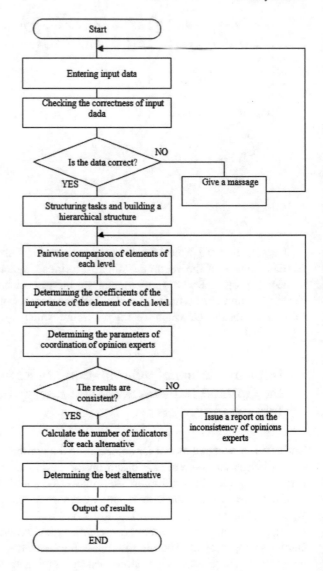

After entering the values of all columns, we obtain a matrix of pairwise comparisons A_j (Fig. 5).

Based on the pairwise comparison matrix A_j, we calculate the eigenvector A_{je} in and priority vector A_{jp}. The results of such a calculation are displayed in the field of eigenvector values A_{je} in and the vector of priorities A_{jp} (Fig. 6).

All calculations are saved in a text file and a Microsoft file format Office Excel, using the "Save to File" button, to a specific position in the file (if the calculations are performed more than once). An example of the result of exporting data to Microsoft Office Excel is shown in Fig. 7.

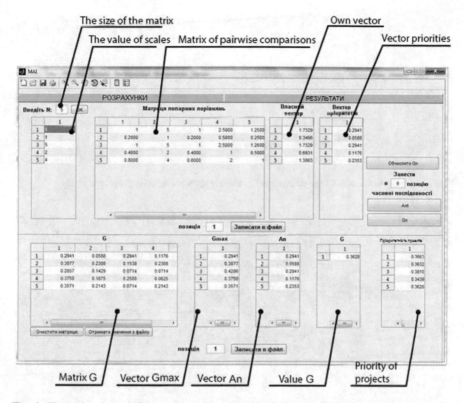

Fig. 4 The main window of the user interface of the simulation model for determining the priority of investment energy-saving projects

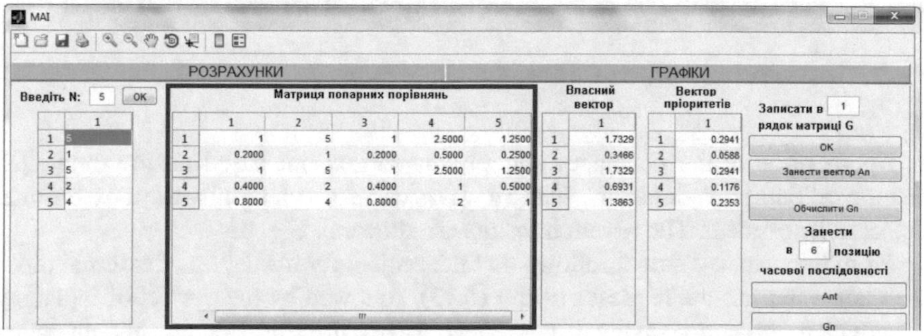

Fig. 5 Display the results of calculating the matrix of pairwise comparisons

The next step in the program implementation process is to calculate the global priority vector G.

Fig. 6 Display the results of the calculation of the eigenvector and the priority vector

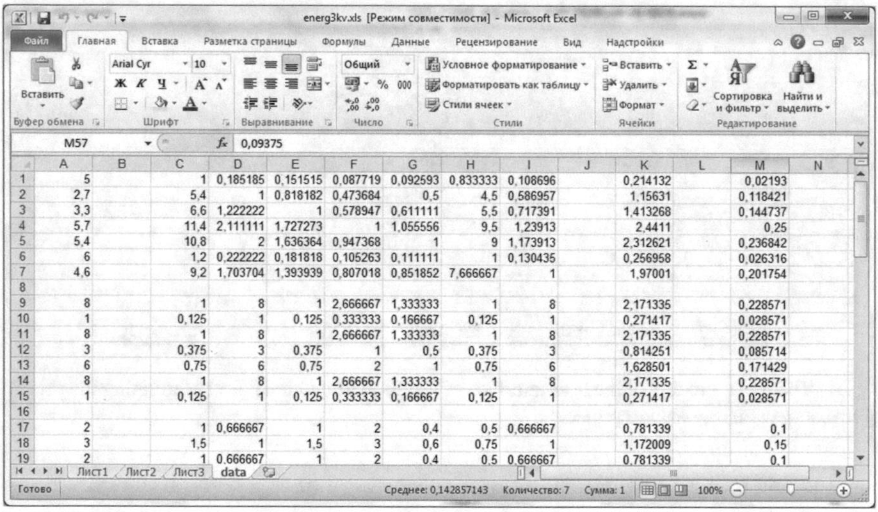

Fig. 7 The result of exporting data to Microsoft Office Excel

Approbation of the developed model was carried out based on the municipal enterprise "Sambir Vodokanal", where it was necessary to determine the most optimal object of investment. The research results are shown in Fig. 8.

From the obtained data it follows that the project for insulation of external pipes with hot water has the highest priority (0.13), followed by the project of replacing pipes with more energy-saving (0.124). Accordingly, these projects are recommended to be implemented primarily to increase the energy efficiency of the municipal enterprise "Sambir Vodokanal".

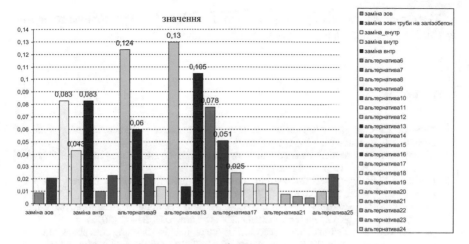

Fig. 8 The results of the simulation model

4 Conclusion

It is shown that the development of a model for prioritizing the implementation of energy-saving investment projects at the enterprise should be carried out using the method of hierarchy analysis, which takes into account the interaction and inter-dependence of factors influencing energy efficiency and determines the dominant factors.

A three-level tree of hierarchies has been developed, which structures the task of selecting energy-saving investment projects and identifies economic, technical, production, environmental and organizational criteria according to which the project will be selected.

It is proposed to assess the economic, technical, production, environmental and organizational factors influencing the choice of projects based on the enterprise's economic and energy surveys and using the scale of importance.

It is shown that the main stages of developing a model for determining the priority of investment energy-saving projects at the enterprise are:

(1) decomposition of the problem of determining the priority of investment energy-saving projects and reflecting it in the form of a tree of hierarchies;

(2) expert assessment of economic, technical, production, environmental and organizational criteria;

(3) the matrix of pairwise comparisons is formed for the second level of the tree of the hierarchy and the vector of priorities is calculated;

(4) formation for m projects of matrices of pairwise comparisons and calculation of priority vectors;

(5) the vector of global priorities is calculated and the investment energy-saving project is selected.

The developed model of priority of implementation of energy-stirring projects on the basis of the municipal enterprise "Sambir Vodokanal" is tested.

References

1. Zade, L.: The concept of a linguistic variable and its application to the adoption of approximate decisions. L.V. Rear. - M.: Mir, 167 p (1976)
2. Matviychuk, A.V.: Modeling of economic processes with the use of fuzzy logic methods. A.V. Matviychuk. - К.: КНЕУ, 264 с (2007)
3. Denisyuk, S.P.: Energy transition—requirements for qualitative changes in energy development. Energy: Econ. Technol. Ecol. **1**, 7–28 (2019)
4. Denisyuk, S.P., Sokolovsky, P.V.: Analysis of flexible generation operation at the stage of transition to smart grid intelligent networks. Electr. Transp. **15**, 31–42 (2018)
5. Veremeenko, O.O.: Estimation of energy efficiency of machine-building enterprises and development of projects with its increase. Scientific Bulletin of Uzhgorod National University: Series: International Economic Relations and the World Economy, Vip. 19. Ch. 1. S. 43–46 (2018)
6. Markevich, K.: Smart-infrastructure in Sustainable Urban Development: World Experience and Prospects of Ukraine. Razumkov Center, Zapovit Publishing House, Kyiv. 400 s (2021). https://razumkov.org.ua/uploads/other/2021-SMART-%D0%A1YTI-SITE.pdf. Accessed data 28 July 2021.
7. Vyshnevsky, V.P., Вієцька, O.V., Garkushenko, O.M., Knyazev, S.I., Leah, O.V., Chekina, W.D., Cherevatsky, D.Y.: Smart industry in the era of the digital economy: prospects, directions and mechanisms of development: a monograph. In: Acad. NAS of Ukraine VP Vyshnevsky; NAS of Ukraine, Institute of Industrial Economics. Kyiv, 192 p (2018)
8. Vyshnevsky, V.P., Вієцька, O.V., Вієцький, O.A., Vorgach, O.A., Garkushenko, O.M., Dasiv, A.F., Zanizdra, M.Y., Zbarazska, L.O., Knyazev, S.I., Kravchenko, S.I., Lipnytsky, D.V., Madih, A.A., Mazur, Y.O., Nikiforova, V.A., Ohten, O.O., Sokolovskaya, O.V., Turlakova, S.S., Chekina, V.D., Shevtsova, G.Z., Shchetilova, T.V.: Smart-industry: directions of formation, problems and solutions: monograph. In: Vyshnevsky, V.P. (ed.), NAS of Ukraine, Institute of Industrial Economics. Kyiv, 464 p (2019). ISBN 978-966-02-8994-9 (electronic edition)
9. Palagin, A.V., Petrenko, N.G., Malakhov, K.S.: Information technology and tools to support the processes of research design of smart-systems. Contr. Syst. Mach. **2**, 19–30 (2018)
10. Saati, T., Cairns, K.: Analytical planning. Organization of systems: Per. with English. M.: Radio and Communication, 224 p (1991)
11. Dzhedzhula, V.V. Modeling of the decision-making process to increase energy efficiency/VM Dzhedzhula. V. Dzhedzhula. Bulletin of the National University "Lviv Polytechnic". Management and entrepreneurship in Ukraine: stages of formation and problems of development **722**, 337–343 (2012)
12. Sichko, T.V.: Popadynets NP Estimation of efficiency of the energy-saving system of the enterprise. Agrosvit. **9**, 52–59 (2018)
13. Dzhedzhula, V.V.: Energy Saving of Industrial Enterprises: Methodology of Formation, Management Mechanism: Monograph. VNTU, Vinnytsia, 346 p (2014)
14. Teslyuk, V., Tsmots, I., Gregus ml., M., Teslyuk, T., Kazymyra, I.: Methods of effective energy management in a smart mini-greenhouse. CMC-Comput. Mater Continua **70**(2), 3169–3187 (2022)
15. Teslyuk, V., Kazarian, A., Kryvinska, N., Tsmots, I.: The optimal method of choosing the type of artificial neural network for use in smart home systems. Sensors (Switzerland) **21**(1), 1–14, 47 (2021)

Designing an Information System to Create a Product in Terms of Adaptation

Hanna Nazarkevych, Mariia Nazarkevych, Maryna Kostiak, and Anastasiia Pavlysko

Abstract An adaptive approach to the design of control systems for functionally complex technical objects and technological processes in the conditions of uncontrolled changes in their own properties and the properties of the environment has been developed. The scheme of development of the order for release of new production is constructed. The structural connections that arise from the interaction of the client and the company that wants to master the release of a new product are analyzed. The scheme of business processes of the firm with adaptive management system is constructed. The links on the approval and development of new products have been analyzed and developed. An adaptive scheme has been developed that can be applied to the design of the firm.

1 Introduction

The approach to the problem of adaptation in nonlinear dynamical systems is studied. Adaptability as a property of the device is considered in relation to information processing problems in nonlinear dynamical systems, the mathematical model of which is not fully known [1]. First of all, the theory and methods of adaptation are focused on control problems in open dynamical systems. Adaptability as a property of the device is considered in relation to information processing problems in nonlinear dynamical systems, the mathematical model of which is not fully known. The theory of adaptive control and identification systems, as a set of system-wide provisions and methods, is used to solve standard problems of object control [2].

An adaptive approach to the design of control systems [3] for functionally complex technical objects and technological processes in the conditions of uncontrolled changes in their own properties and the properties of the environment is relevant. At

H. Nazarkevych · M. Nazarkevych (✉) · M. Kostiak · A. Pavlysko
Lviv Polytechnic National University, Stepan Bandera Street, 12, Lviv 79013, Ukraine
e-mail: mariia.a.nazarkevych@lpnu.ua

H. Nazarkevych
e-mail: hanna.ya.nazarkevych@lpnu.ua

© The Author(s), under exclusive license to Springer Nature Switzerland AG 2023
N. Kryvinska et al. (eds.), *Developments in Information and Knowledge Management Systems for Business Applications*, Studies in Systems, Decision and Control 462,
https://doi.org/10.1007/978-3-031-25695-0_8

the same time, the application of classical methods is largely difficult or ineffective [4].

With the development of any project, the complexity of its management increases. This, in turn, leads to a loss of flexibility and a reduction in the pace of project work. The more people working on one task, the more difficult it is to control them and ensure the proper level of communication. Therefore, an important part of project management is the use of various project management tools [5].

The specific tasks to be solved by the information system depend on the application area for which the system is intended. Areas of application of information applications are diverse: banking, insurance, medicine, transport, education and more. It is difficult to point out the area of business activity in which today it would be possible to do without the use of information systems [6].

Having determined the quantity and range of ordered goods, the manager of the purchasing department begins to analyze the proposals of suppliers. This process is carried out monthly or as needed. The most favorable terms of delivery are selected. To do this, prices of suppliers are compared. This information is taken from the price list for purchases. When choosing a supplier, it is important to take into account the deferred payment provided. This information is taken from contracts marked as priority (active). As a result, a list of suppliers is formed, each position is assigned a sign of the main and reserve suppliers in descending order of priority.

2   Objectives and Methods of Designing Control Systems

To create a new product, you need to be organized with an idea based on information resources, technology and information. It is necessary to develop a strategy and study the environment [7], see Fig. 1. If the new product has protective properties, which are described in detail in. Infrastructure release new production show Fig. 1.

Fig. 1 Infrastructure release new production

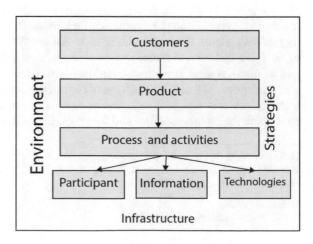

At the beginning of the project it is necessary to find out: for whom the product will be created; what is the goal of the project; what problems the product solves; who is the competitor of the product and how the final product will compete with them in the market [8]. It is necessary to set clear deadlines, the budget that the customer can afford to start the project, and what he expects after these deadlines. These preparatory work is usually performed during the so-called zero sprint.

Managers and other stakeholders come to the conclusion that the existing system of working with clients is not simple and therefore it is necessary to allocate the necessary resources. During this stage, some variables in variable-oriented research can be mentioned, such as perceived usefulness, perceived ease of use, attitude to use, expected productivity, expected duration of effort, and behavioral intentions. These variables and actual use cannot be observed at this stage, as a new or improved system has not been implemented in the organization.

- Development phase. This phase involves creating or acquiring the resources needed to implement improvements in the organization. This phase is part of the adoption, as the adoption cannot be completed without the necessary resources.
- Implementation phase. Implementation in the organization includes implementation planning, training, any necessary reconfiguration and transition to a new work system. Implementing in an organization requires implementation efforts, which can range from providing new opportunities for enthusiasts and publishing their voluntary use models all the way to a fully planned corporate implementation with mandatory schedules and expectations for use.
- Operation and maintenance. This is a temporary final state of acceptance of a new or improved version of the operating system. If it were necessary to determine the exact point of transition from the not yet accepted state to the accepted state, this point would come at the moment when the new operating system was declared fully operational. There is a high probability that different parts of the newly adopted work system will change organically, as participants in the work system will find more efficient methods of work and respond to exceptions, contingencies and changes in the environment.

    System theory of IT innovations, implementation and adaptation [9].

    With regard to the implementation of IT in business settings [10], the question is whether the adoption of IT means the acceptance of a hardware/software artifact, or whether it means the acceptance of a working system [11], the operation of which depends on this artifact.

- Technology. Implementation involves installing, configuring, updating, or modifying software provided by a vendor.

    In the design of management systems, in addition to expert methods, modeling is widely used, which is used in the design of management decisions.

    The number of management decisions used in an organization depends on the number of structural units of the organization. A group of decisions is assigned to each subdivision (for example, to accounting - decisions on accounting and analysis of economic activity, to the planning and economic department - decisions on technical and economic planning, etc.).

The number and composition of decisions made by each unit is compared to the ideal model, which is selected by designers at the preliminary stage. As such a model, you can use (with partial refinement) a standard information model.

The release of new products is appropriate [12] if:

Decreased profits, inefficient use of capital,

Low rate of development of the organization,

Insufficient product quality, dissatisfied customers, loss of market,

Low salaries, dissatisfaction of employees with their work.

Modeling management decisions requires improving the process of preparing and making a management decision [13] at all levels of management and at all stages of its development up to the final decision. Modeling is based on the designed information model.

The integrated design of the management system involves its improvement based on the selected qualitative goals of functioning and includes the design of the decision-making process and the management structure at the level of management decisions.

3   Designing a Control System Includes 6 Stages

Stage 1. A complete list of management decisions necessary for the functioning of the organization's management system is determined.

Stage 2. A list of documents used in the preparation and adoption of management decisions is compiled. Some of the documents are common in the preparation of various management decisions.

Stage 3. A list of groups of management decisions by function is compiled in such a way that each of the groups uses the minimum number of documents for preparing decisions.

Stage 4. The purpose of distributing decisions by management levels is to determine a group of decisions, for the preparation of which the head of the corresponding management level is responsible. This takes into account the load and throughput of that control layer. The manager's qualifications and competence should be sufficient to make decisions for this group.

The distribution of decisions covers all levels of management in the organization [14]. Especially if the system uses protective technologies that take into account biometric systems [15–17].

Stage 5. The formation of a structural scheme for managing an organization, as a rule, is based on typical management structures. The matrix-staff control scheme is used as a typical one. The formation of a management scheme involves the distribution of management decisions by management levels, the calculation of the load on the management level for the coordinating, problematic or functional level, which as a result will provide justification for choosing the type of management structure. The final choice of the variant of the structure scheme and all further calculations are carried out within the framework of the selected control structure [18].

Stage 6. Documents are being developed that regulate the activities of the management system: regulations on divisions, job descriptions, rules for the work of performers.

The concept of improving the organizational management mechanism, which is embodied in the management system design scheme. The advantage of the proposed concept is that a number of stages are solved using computer technology, which simplifies system design [19].

The necessary components of such a theory of adaptive control, as follows from the logic of its development, are:

(1) an apparatus for analyzing the properties of nonlinear systems, which does not require accurate knowledge of the mathematical models of the objects under study and does not depend on whether the object is stable.

(2) principles and methods of adaptation to uncontrolled [20], unmeasurable perturbations and uncertainties of the environment and object model, using only their system-wide, fundamental properties 3) search, analysis and synthesis of structures for the implementation of nonlinear control algorithms, adaptation and identification.

Stage 1. Choice of a typical management structure [21].

The design of a management system in real conditions is based on typical management structures, in which the number of management levels, the name and number of functional units, etc. are always fixed. For a long time, such a structure was a linear-functional management structure.

With the development and formation of market relations, new requirements are imposed on the object of management, the range of tasks is expanding, due to the constantly changing goals of functioning [22]. The matrix-staff structure, which combines all possible variants of hierarchical subordination: linear, thematic, functional, quite successfully copes with such tasks [23].

At the first stage of designing, the matrix-staff model of the structure is selected.

Stage 2. Distribution of solutions by levels.

Distribution of decisions by management levels within the framework of the matrix-staff structure.

Stage 3. Calculation of the control level load.

This is actually the process of designing one or another version of the management structure for the organization [24]. The question of the expediency of the presence in the structure of the functional, thematic or coordinating levels is being decided.

The main attention of the authors is focused on the theory and methods of adaptive control of nonlinear dynamic objects:

- with potentially unstable according and non-equilibrium target dynamics;
- subject to the potential impossibility of specifying the target sets in an explicit form;
- using minimal, high-quality information about the object, as well as in the conditions of unavailability of information about the exact mathematical models of the object;
- using models of uncertainties that are maximally adequate to the physical essence of processes and phenomena in the object itself;
- with the possibility of implementing control mechanisms in typical and homogeneous structures such as artificial neural networks of direct action.

A vision is an idea of a new product, a sketch of a future product. This short document describes what a product is, what customers will have, and so on. It replaces the whole package of documents that are commonly used in heavyweight methodologies. You can use innovative games, brainstorming and frameworks to create a vision. This stage is carried out jointly by the product owner, the customer or the person who decides on the project by the customer, the team. The following product questions need to be answered: 1. Who will buy this product? Who is his target customer? Who will use the product? Who are its target users? 2. What needs will the product meet? What value does it create? 3. What properties of the product are vital to meet the selected needs and, consequently, for the success of the product? What will the product look like and how will it work? In what aspects will it be especially good? 4. What will it look like against other products released by competitors or the company itself? What are the unique commercial arguments of the product? What will be its price? 5. Which company will make a profit from selling the product? What are the sources of revenue for it and what will the business model look like? 6. Has the product been manufactured? Will the company be able to develop and sell it? In fact, this document allows you to succinctly describe the target audience, and features of the project, and the current market, and the planned boundaries of the project. A good vision should briefly communicate the essence of the future product, describe a common goal that gives direction to work, but vaguely enough to promote creativity.

Quality of vision: short and concise; common and that unites the developer and the customer; motivational and one that gives space for creativity.

The client is applying for a business from an application for the preparation of certain products. Він submits an application to the manager of the sale, which is reviewed, the application is filed with the full contract (Fig. 2), in order to clarify the details, the application is transferred to the process engineer. The client may submit inaccurate information before the application, he may not have the competence to form an exact application, and even the application is sent to the client for additional processing.

If the engineer-technologist is blamed for the possibility of such a variability, he sends the application to the economist, which fakhovo may be able to supply food.

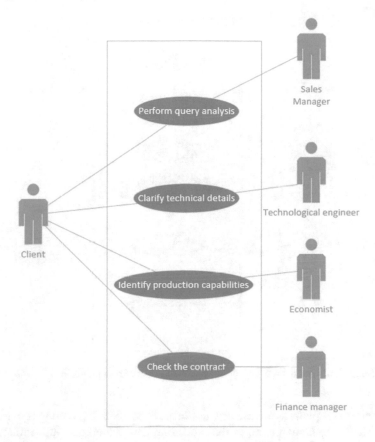

Fig. 2 Diagram of precedent propositions about the release of new products

As a matter of fact, commendably higher, we are preparing an agreement on the release of new products.

After the signing of the agreement, the marketing firmed up the company's strategy (Fig. 3). Expand the concept of a new product and use it with the help of sales. If you blame food, then the concept of a new product should be transferred from the head designer. To blame the inconsistencies, constructive shortcomings, incompetence could not develop the technology of development of old terms. If there are no food deficiencies, then the stench is in full swing with the head designer's head, the head technologist's head, and they blame the vikonators for manufacturing. If there is a need, then an application is submitted to the procurement office [20] for the supply of syrovin for a new product. Then we expand the technology of manufacturing and release the final batch of products.

Then the assessment of the trial batch was carried out by the head of marketing, the head of sales, the head of the head designer, the head of the head technologist and without intermediary technologists at the manufacturing plant. Then a plan is developed for the promotion of a new product.

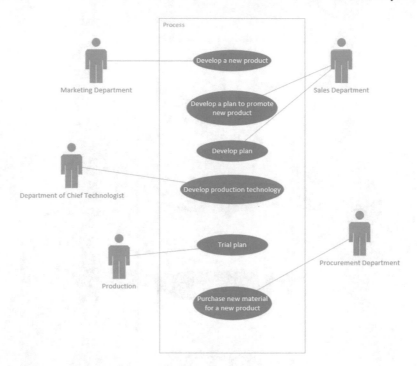

Fig. 3 Diagram of precedents for adapting the development of new products

The company plans to release new products. Procurement planning is carried out in the Marketing Department, in the marketing and planning group. Procurement planning is carried out as follows:

The manager of the planning and marketing group daily receives external and internal sales statistics data from counterparties in the form of sales reports (Fig. 5).

To plan the purchases of products, the manager of the planning and marketing group calculates the need for goods on a weekly basis based on sales statistics. As a result of the calculation, a table of product needs is formed.

The manager of the purchasing department on a monthly basis, based on the Table of needs for goods and the list of selected suppliers, generates delivery schedules indicating the timing and frequency, but without the quantity of delivery.

Every month, after determining the need for the product, the logistics group manager calculates the required number of purchases. The required number of purchases is calculated based on the actual stock in the warehouse, the required minimum and maximum stock levels. The norms of the minimum and maximum quantity of stocks are set in days. When calculating the required purchase quantity, the time of the goods in transit is also taken into account. Thus, this calculation should ensure the possibility of uninterrupted release of goods from the warehouse. Based on the results of the calculations, a plan of applications for the month is formed.

4 Considering the Risks of Creating a New Product

Risk is inherent in any area of human activity, which is associated with many conditions and factors that lead to positive or negative results of decisions made by individuals.

As a result of risk analysis, a picture of possible risk events, the probability of their occurrence and consequences. After comparing the obtained values of risks with the maximum allowable, a risk management strategy is developed, and on this basis—measures to prevent and reduce risk.

To be successful in business, people need to be enterprising. But there is no economic activity without risk. Therefore, the question is not to move away from the risk, but to find out its causes, to be able to calculate it, to plan in advance measures to minimize unforeseen losses, i.e. to reduce the consequences of risk.

Risk began to be studied mainly in connection with commercial and managerial activities, stock market games, profits, etc. In economic life, it is advisable to consider three main types of risk:

- risk of the entrepreneur (borrower);
- creditor risk (deviation of the debtor from debt payment);
- reduction in the value of the currency.

A classic example in the field of turnover forecasting, which requires the use of data mining methods, is the problem of changing customer preferences in a competitive market. The key to solving this problem is a database of customer profiles and purchases. Patients' behavior patterns can be analyzed to determine their distinctive characteristics. If such characteristics can be found, they can be used to identify non-regular customers or those who do not currently use certain products, but may be interested in them.

5 The Concept of Uncertainty and Risk

A distinction must be made between "risk" and "uncertainty". Entrepreneurship is always associated with uncertainty that arises, for example, in the following cases:

- uncertainty of the external environment;
- uncertainty of the economic situation, resulting from the uncertainty of supply and demand for goods, money, factors of production, etc.;
- uncertainty in the diversity of areas of capital raising;
- uncertainty in the diversity of preference criteria;
- investing funds;
- uncertainty associated with limited knowledge of the subject area of business, etc. Thus, we found that risk is an objectively inevitable element in making any business decision due to the fact that uncertainty is an inevitable characteristic of business conditions in modern conditions. But uncertainty should not be confused

with unpredictability, because uncertain situations and the risks that come with it characterize a situation where the occurrence of unknown events is quite probable and can be quantified. Unpredictability means the inability to accurately determine the time and sometimes the place of occurrence of the event.

Risk categories. The number of identified risks by category.

1. Risks of commercial viability
2. Risks of competitors
3. Consumer acceptance and marketing risks
4. Risks of public perception
5. Risks of intellectual property
6. Risks of production technology
7. Organization and risks of project management
8. Product family and brand positioning risks
9. Technological risks of the product
10. Risks of screening and evaluation
11. Supply and source chain risks
12. Risks of trading clients

1. Commercial risk is the risk that arises in the process of selling goods and services made or sold at the enterprise. Risks of commercial viability are those risks that are formed during the life cycle of the product.

Risks of competitors. The modern market cannot exist without competition. It takes different forms and methods of competition depending on the industries and areas that are constantly improving. The task of business is to be aware of changes in time and to adapt to them. Competition significantly improves and develops the employer's brand. Companies should provide services that will interest potential customers. Competition provides an incentive to improve the quality of services, introduce innovations and new approaches to doing business, as well as harden before inspections and crises. Business, ready for equal rules of the game, is less vulnerable to crises and inspections of regulatory authorities. Methods of competition in this case—only a consequence of external changes, as well as the impact of major trends. In addition, competitive businesses are better prepared to enter foreign markets. The business fully adheres to the sports rules in its segment. This is the same competition for the championship. To become a world champion and win the Olympics, you need to be the best in your category at the national level and overcome many of the stages that precede it.

Consumer acceptance and marketing risks. In modern market environment, the need to choose areas of innovation of the enterprise. For these reasons, managerial marketing decisions are made on the basis of incomplete, inaccurate or inconsistent information, i.e. in conditions of uncertainty and conflict. The consequence of decision-making in these conditions is the emergence of risk, so the relevance of their identification, evaluation and development of risk management system is growing [25].

Risks of public perception

Researchers emphasize that for every company, reputation should be the subject of strategic management. A good reputation in the long run can provide significant competitive advantages in the market, as well as, if necessary, facilitate the passage and recovery after crises [26].

Risks of intellectual property. The lower the level of risk in assessing market conditions, the greater the accuracy of calculations of the commercial potential of intellectual property, the more efficient the processes of transfer of intellectual technologies. domestic research and development (reduction of market capacity) [27].

Risks of production technology. Innovations play an important role in economic development, namely: they contribute to increasing the efficiency of factors of production, development of promising areas, intensive type of economic growth, increase the competitiveness of the national economy, the country's place in the global economic environment [28].

Organization and risks of project management. Decision-making processes for project management take place in conditions of uncertainty, ie under the influence of factors: incomplete knowledge of the situation, the presence of chance, the presence of force majeure. Thus, the project is implemented in conditions of uncertainty and risk. These two categories are interrelated [29].

Product family and brand positioning risks. One of the most effective means of product differentiation is branding. The introduction of branding as a modern marketing technology today is one of the key conditions for effective enterprise development. Today, there are many studies that show the significant impact of brands on the activities of companies in various fields. Recent research shows that brands have a positive effect on a company's financial performance and value, and having a strong brand reduces investment risks. Therefore, there is a need to create a full-fledged system of strategic brand management as a key asset of the organization. Under such conditions, the formation of sustainable competitive advantages due to well-thought-out product positioning and active ways of promoting it becomes important. The need to develop and implement branding technology in order to form consumer loyalty of a particular company and ensure the capitalization of brands as intangible assets of the enterprise, emphasizes the relevance of the research topic [30].

Technological risks of the product. Technologies with a medium level of innovation are progressive, but have relative advantages (modern low-waste, waste-free, resource-saving, safe, environmentally friendly technologies). That is, the level of their usefulness and uniqueness will be different for different recipients, which should also be taken into account in the pricing mechanism for such intellectual property. Technologies with a low level of innovation power, traditional technologies, reflect the average level of production achieved by most manufacturers in this industry, which leads to a low level of adjustment factor for this criterion. The next step in adjusting the price of technology is to determine the level of risk of its successful development. Yes, if there is a high probability that the intellectual property will not be successfully commercialized, the adjustment factor will be lower. Moreover, if the risk of implementation and efficient use of the technological product is high enough, then in this case we recommend setting its value at less than one [31].

6 The Process of Creating a New Product at the Enterprise

Then, in the logistics group, orders to suppliers are formed daily according to the application plan, delivery schedule, price lists of suppliers.

If an order is to be placed with an import supplier, then the logistics team manager calculates the certification costs, a certification cost report is generated. Certification costs are checked against in-house regulations. This operation is performed as needed.

If the costs of certification exceed the intracompany norms, then the logistics group manager repeats the process of forming orders to suppliers. New orders are being formed.

The daily prepared order to the supplier is accepted, the order must be signed by the logistics manager and the director of the Marketing and Inventory Management Department.

Every day, the logistics group manager forwards the order to the purchasing department. The purchasing manager sends the order to the supplier.

The good design of the interface is the responsibility of allowing coroners to experiment and work pardons, showing tolerance for pardons. This stimulates the subsequent activity of the user, the shards allow him to pardon the succession of children with the possibility to turn on the cob at some point.

Tolerance until pardons transfers the multilevel system of cancel the operation. This principle is important for the implementation, especially for re-insurance on a rich users and applications data bases.

The stage of operation and maintenance from the successful transfer to the customer of each last development option, ultimately the entire software product. The agreement is only an integral part of the software life cycle when it comes to the time and effort that goes into maintenance.

Operate refers to a changeover from an old software- or non-software based business arrangement to a new switchover, usually a gradual process [7]. Whenever possible, old and new systems should work in parallel to ensure that the old solution fails if the new solution fails. Maintenance consists of three different stages.

Project planning. Planning covers the entire life cycle of a software project [23]. It begins after the organization's business strategy has been defined as a result of system planning work and a software project has been identified. Project planning is the activity of evaluating the scope of supply, costs, risks, milestones, and required resources. It also includes the choice of development methods, processes, tools, standards, team organization, and so on.

Project plans are like an elusive goal. They are least of all reminiscent of something once and for all set and unchanging. Project plans are subject to change throughout their life cycle [32]. At the same time, these changes do not go beyond the limits set by several fixed restrictions.

Time and money are typical constraints—each project has a clear deadline and a strictly limited budget. One of the first tasks of project planning is to assess the feasibility of a project under time, budget, and other constraints [33]. If the project is feasible, then the restrictions are documented and can only be changed as part of a formal approval procedure, the project is evaluated taking into account several factors.

Metrics are used to measure quality characteristics such as correctness, reliability, productivity, integrity, usability, maintainability, flexibility, and testability [34]. For example, software reliability can be assessed by measuring the frequency and severity of failures, mean time between failures, accuracy of output results, failure recovery, and so on.

Another important application of metrics is the evaluation of development models (development products) at various stages of the software life cycle [35]. In addition, indicators are used to evaluate the effectiveness of the process and improve the quality of work at various stages of the life cycle.

Typical indicators that apply to the software development process and can be adopted for use at different stages of the life cycle, the variability of requirements.

To create a new type of product it is necessary to conduct surveys using IM: low cost; automation of the process and analysis of its resources; the ability to accurately focus surveys on the target audience (Fig. 4). Observation is carried out by recording events and studying the behavior of an object [4]. In IM observations are carried out by collecting and subsequent analysis of log-files of the Web-server; data obtained on the basis of analysis of user requests; user behavior; web server navigation; use of cookies (statistics of visits to the Web server) [36]. The experiment involves studying the behavior of the object (certain output parameters) based on changes in input parameters (change in the structure of buyers when changing the means of advertising/price). Advertising—commercial promotion of consumer qualities of goods/services in order to persuade potential buyers to buy them [37]. Types of advertising: image, stimulating. Possibilities of IM compared to traditional types of advertising: constant relevance of information; modifications of advertising depending on the reaction of the buyer; acceptance of information from the buyer; instant dissemination of information; providing different users with different information; purchase/sale transaction. Basic methods of advertising in IM: domain name registration; registration of the server in search engines (usual placement (banners), taking into account the subject (contextual display), by keyword); placement of links to the server in Web-directories; placing links in the "yellow pages"; link exchange with other servers; use of banner networks; use of ratings; placement of paid advertising on sites; use of newsgroups and discussion groups; use of affiliate and sponsorship programs (Fig. 5).

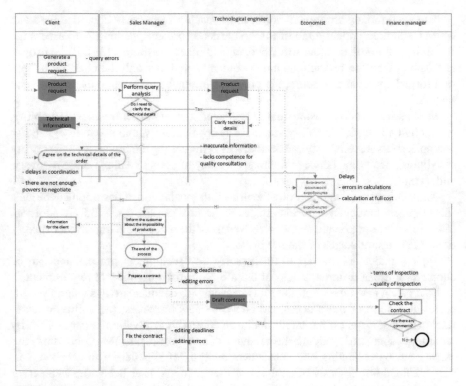

Fig. 4 Generate a product request

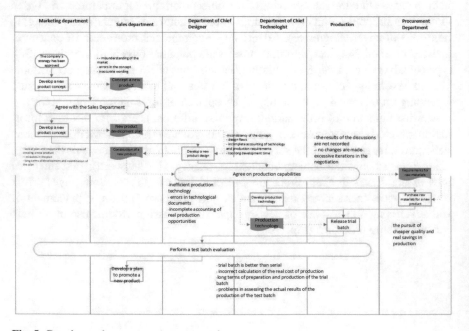

Fig. 5 Develop a plan to promote a new product

7 Conclusions

An attempt to solve the development of a new product using the requirements implemented during the implementation of IT projects to create information systems for various purposes. A new concept of submitting requirements to the information system is proposed and the definition of the pattern of requirements design is formulated. Models of patterns of designing requirements to the information system at the levels of information, data and knowledge are developed.

The most important decision criterion is the assessment of the impact of the introduction of a new product on sales in the enterprise and sales of other products, as well as an assessment of the payback period for development, production, sales, advertising usually for a period not exceeding five years, including the three-year period from the beginning of mass production to the moment of break-even.

The success of the product is largely guaranteed by setting the price for it. The main criterion for determining the price of a product is the price of an old or similar product on the market. In addition, the cost and profit ratios, sales point and break-even point of the product are used to set and calculate the price of a new product. In the presence of a number of projects, large enterprises invest in several at once, and in the subsequent verification of products make their selection. Sometimes a project that has failed in the past becomes successful in the future. To establish the correspondence between the price and the cost of the product, special forms of reports are provided, which supplement the accounting reports.

After the release of the product, the sales process is inspected. Information on the rate of sales of a new product and the factors influencing the value of sales is recorded in the financial report.

The main features of the architectural framework and intelligent information technology of accelerated development of information systems based on the above results are considered. The evaluation of the efficiency of using the considered technology within the framework of the IT project of creation of various information systems is carried out.

Acknowledgements This work was performed within the research work "Experimental mobile robotic platform with intelligent control system and data protection", performed by the National University "Lviv Polytechnic" and funded by the state budget of the Ministry of Education and Science of Ukraine for 2022–2023.

References

1. Syahputra, F., Dalimunthe, A., Sidabutar, U.B.: Industrial field work practice information system design. J. Phys.: Conf. Series **2193**(1), 012089 (2022, February). IOP Publishing
2. Lytvyn, V., Dosyn, D., Emmerich, M., Yevseyeva, I.: Content formation method in the web systems. In COLINS, pp. 42–61(2018)
3. Atta, A., Abbas, S., Khan, M.A., Ahmed, G., Farooq, U.: An adaptive approach: smart traffic congestion control system. J. King Saud Univ.-Comput. Inf. Sci. **32**(9), 1012–1019 (2020)

4. Nazarkevych, M., Lotoshynska, N., Brytkovskyi, V., Dmytruk, S., Dordiak, V., Pikh, I.: Biometric identification system with ateb-gabor filtering. In: 2019 XIth International Scientific and Practical Conference on Electronics and Information Technologies (ELIT), pp. 15–18. IEEE (2019, September)

5. Logoyda, M., Nazarkevych, M., Voznyi, Y., Dmytruk, S., Smotr, O.: Identification of biometric images using latent elements. In: CEUR Workshop Proceedings (2019). EID: 2-s2.0-85074659529

6. Hrytsyk, V., Grondzal, A., Bilenkyj, A.: Augmented reality for people with disabilities. In: CSIT'2015. IEEE. Lviv, 14–17 Sept, pp. 188–191

7. Dronyuk, I., Nazarkevych, M., Poplavska, Z.: Gabor filters generalization based on ateb-functions for information security. In: International Conference on Man–Machine Interactions, Oct 3, pp. 195–206. Springer, Cham (2017)

8. Qu, K., Liu, Z.: Green innovations, supply chain integration and green information system: a model of moderation. J. Clean. Prod. 130557 (2022)

9. Singh, N., Krishnaswamy, V., Zhang, J.Z.: Intellectual structure of cybersecurity research in enterprise information systems. Enterpr. Inf. Syst. 1–25 (2022)

10. Rezazade Mehrizi, M.H., van den Hooff, B., Yang, C.: Breaking or keeping the habits: exploring the role of legacy habits in the process of discontinuing organisational information systems. Inf. Syst. J. **32**(1), 192–221 (2022)

11. Villa, F., Bassani, E., Passaretti, F., de Ceglia, G., Viscuso, S., Zin, V., ... Villa, E.: Magnetron sputtering of Au-based alloys on NiTi elements: surface investigation for new products in SMA-based fashion and luxury accessories and watchmaking. Coatings, 12(2), 136 (2022)

12. Delavar, H., Gilani, H., Sahebi, H.: A system dynamics approach to measure the effect of information sharing on manufacturing/remanufacturing systems' performance. Int. J. Comput. Integr. Manuf. 1–14 (2022)

13. Sheketa, V., Poteriailo, L., Romanyshyn, Y., Pikh, V., Pasyeka, M., Chesanovskyy, M.: Case-based notations for technological problems solving in the knowledge-based environment. In: 2019 IEEE 14th International Conference on Computer Sciences and Information Technologies (CSIT), September, vol. 1, pp. 10–14. IEEE (2019)

14. Yu, R., Rashkevich, Y.M., Peleshko, D.D., Pasyeka, M.S., Stetsyuk, A.B.: Design of web-oriented distributed learning systems. Contr. Syst. Mach. **3**, 72–79) (2002)

15. Teslyuk, V., Kazarian, A., Kryvinska, N., Tsmots, I.: Optimal artificial neural network type selection method for usage in smart house systems. Sensors **21**(1), 47 (2021)

16. Boreiko, O.Y.: Developing a controller for registering passenger flow of public transport for the 'smart' city system/O. Y. Boreiko, V. M. Teslyuk. East-Eur. J. Enterpr. Technol. **6** 3(84), 40–46 (2016)

17. Boreiko, O.Y.: Development of models and means of the server part of the system for passenger traffic registration of public transport in the 'smart' city/O. Y. Boreiko, V. M. Teslyuk, A. Zelinskyy, O. Berezsky. East-Eur. J. Enterpr. Technol. **1** 2 (85), 40–47 (2017)

18. Medykovskyi, M.O., Tsmots, I.G., Tsymbal, Y.V.: Information analytical system for energy efficiency management at enterprises in the city of Lviv (Ukraine). Aktual'ni Problemy Ekonomiky= Actual Problems in Economics (175), 379 (2016)

19. Tsmots, I., Skorokhoda, O.: Methods and VLSI-structures for neural element implementation. In: 2010 Proceedings of VIth International Conference on Perspective Technologies and Methods in MEMS Design, April, pp. 135–135. IEEE (2010)

20. Nazarkevych, M., Hrytsyk, V., Voznyi, Y., Marchuk, A., Vozna, O. Method of detecting special points on biometric images based on new filtering methods. In: CEUR Workshop Proceedings 2923, pp. 243–251 (2021)

21. Liu, Z., Yang, Z., Yang, W., Sheng, J., Zeng, Y., Ye, J.: Optical management for back-contact perovskite solar cells with diverse structure designs. Sol. Energy **236**, 100–106 (2022)

22. Nazarkevych, M., Voznyi, Y., Hrytsyk, V., ...Havrysh, B., Lotoshynska, N.: Identification of biometric images by machine learning. In: 2021 IEEE 12th International Conference on Electronics and Information Technologies, ELIT 2021—Proceedings, pp. 95–98 (2021)

23. Kunanets, N., Vasiuta, O., Boiko, N.: Advanced technologies of big data research in distributed information systems. In: 2019 IEEE 14th International Conference on Computer Sciences and Information Technologies (CSIT), September, vol. 3, pp. 71–76. IEEE (2019,)
24. Boyko, N., Shakhovska, N.: Prospects for using cloud data warehouses in information systems. In: 2018 IEEE 13th International Scientific and Technical Conference on Computer Sciences and Information Technologies (CSIT), September, vol. 2, pp. 136–139. IEEE (2018)
25. Fitaa, R.O., Tawfiq, W.A.: Saudi consumers' perceptions of fashion brands' Applications for augmented reality. Int. Des. J. 12(1), 339–351 (2022)
26. Parvaze, S., Khan, J.N., Kumar, R., Allaie, S.P.: Public perception of flood risks and warnings in the flood-prone Kashmir Valley India. Curr. Sci. 122(5), 591 (2022)
27. Hawksbee, L., McKee, M., King, L.: Don't worry about the drug industry's profits when considering a waiver on covid-19 intellectual property rights. bmj, 376 (2022)
28. Šarauskis, E., Naujokienė, V., Kazlauskas, M., Bručienė, I., Romaneckas, K., Steponavičius, D., Jasinskas, A.: Precision variable rate seeding effectiveness for risks management in crop production (2022)
29. Krechowicz, M.: Towards sustainable project management: evaluation of relationship-specific risks and risk determinants threatening to achieve the intended benefit of interorganizational cooperation in engineering projects. Sustainability 14(5), 2961 (2022)
30. Halman, J.I., Hofer, A.P., Van Vuuren, W.: Platform-driven development of product families: linking theory with practice. J. Prod. Innov. Manag. 20(2), 149–162 (2003)
31. Granato, D., Barba, F.J., Bursać Kovačević, D., Lorenzo, J.M., Cruz, A.G., Putnik, P.: Functional foods: product development, technological trends, efficacy testing, and safety. Annu. Rev. Food Sci. Technol. 11, 93–118 (2020)
32. Vysotska, V., Lytvyn, V., Burov, Y., Berezin, P., Emmerich, M., Fernandes, V.B.: Development of information system for textual content categorizing based on ontology. In: COLINS, pp. 53–70 (2019)
33. Hrytsyk, V., Nazarkevych, M.: Real-time sensing, reasoning and adaptation for computer vision systems. In: International Scientific Conference 'Intellectual Systems of Decision Making and Problem of Computational Intelligence', May, pp. 573–585. Springer, Cham (2021)
34. Scotti, V.: Altmetrics, Beamplots, Plum X metrics and friends: discovering the new waypoints in the science metrics roadmap. AboutOpen 9, 1–2 (2022)
35. Antonyuk, N., Vysotska, V., Lytvyn, V., Burov, Y., Demchuk, A., Lyudkevych, I., ... Bobyk, I.: Consolidated information web resource for online tourism based on data integration and geolocation. In: 2019 IEEE 14th International Conference on Computer Sciences and Information Technologies (CSIT), September, vol 1, pp. 15–20. IEEE (2019)
36. Dronyuk, I., Nazarkevych, M., Fedevych, O.: Synthesis of noise-like signal based on Ateb-functions. Commun. Comput. Inf. Sci. 601, 132–140 (2016)
37. Coyago-Cruz, E., Corell, M., Moriana, A., Hernanz, D., Stinco, C.M., Mapelli-Brahm, P., & Meléndez-Martínez, A.J.: Effect of regulated deficit irrigation on commercial quality parameters, carotenoids, phenolics and sugars of the black cherry tomato (Solanum lycopersicum L.) 'Sunchocola'. J. Food Compos. Anal. 105, 104220 (2022)

A Real-Time Room Booking Management Application

Olena Shlyakhetko and Vitaliy Shlyakhetko

Abstract The paper presents a model of room booking management application. Booking and managing meeting rooms shouldn't be a hassle—it should just work. Our application allows logged in users to search available rooms according to parameters, make reservations and manage them. The practical part of work was made in C# as a Windows Forms Applications. As an IDE we used Microsoft Visual Studio Ultimate 2012. In order to create appropriate diagrams we used dedicated tools. The application "Facility management system" are designed and implemented. The application is the solution to make a workplace as smart and efficient as possible. It's allows real-time availability and booking. Using the system helps users find the right solution, optimize how their office is used, save money and make the workplace more flexible and accessible.

Keywords Health monitoring · Data consolidation · Analysis · Intelligent systems · Medical information · Big data · Data-driven medicine

1 Introduction

Facility Management System—definition

Facility management or facilities management [1–3] is a systematic approach based on advanced technologies, a level of service from the market leader, an experienced team of professionals, and comfortable conditions of cooperation. Also, this

O. Shlyakhetko (✉)
Information Systems Department, Faculty of Management, Comenius University Bratislava, Odbojárov 10, 25, 82005 Bratislava, Slovakia
e-mail: ivanochko2@uniba.sk

V. Shlyakhetko
Department of Financial and Economic Security, Accounting, and Taxation, Ukrainian Academy of Printing, Lviv, Ukraine
e-mail: vit_shl@ukr.net

N. Kryvinska et al. (eds.), *Developments in Information and Knowledge Management Systems for Business Applications*, Studies in Systems, Decision and Control 462, https://doi.org/10.1007/978-3-031-25695-0_9

management is a professional management discipline [4–6]. This discipline focuses on efficiently and effectively delivering organizations' support services [7–11].

According to the principle of organization, facility management is divided into:

– Internal facility management—the creation of own department in the company [12–14], which is responsible for all processes related to the operation of the building
– Outsourced facility management is the transfer of non-core activities to a professional company. With such cooperation, the contractor manages individual services [15–17] or several, depending on the terms of the contract
– Integrated facility management is a complete set of building management and maintenance measures.

Today, facility management is an essential function of any enterprise [18–20]. Competent operation of real estate improves the comfort and safety of people, thereby increasing productivity and optimizing all possible resources and systems [21–25].

Cooperation with a company specializing in home maintenance has many undeniable advantages that can simplify and improve the client's work [26–28]. Facility Management from the company is:

– Audit of the real estate object
– Optimization of resources due to the complexity of service provision
– Regular proposals for the implementation of innovations based on international best practices
– Transparent process management based on professional software [22, 29–31]
– Insurance of risks related to the operation of the building
– Time to solve strategic tasks
– Necessary, neat, qualified staff
– Quality service from the leader of facility management.

1.1 Main Problem

During the implementation of the planned application [32–35], we were thinking about the main problems related to the management of rooms or offices. We have assumed that our clients will be companies that use buildings equipped with a large number of conference rooms and offices. When undertaking the project, we conducted research to determine the most important needs of our application users [36–39]. The results of our analysis are presented below.

Main room/office management problems:

– The user who wants to reserve a room doesn't have access to information about its availability of this access is limited.
– Making and managing reservations requires the participation of third parties, which involves the need for additional employment and higher costs.

– Searching for available rooms in many buildings is complicated—booking rooms using parameters such as their number, capacity and chosen period would facilitate this process.
– Many companies decide to employee a facility manager, which is an effective but expensive solution.
– Limited access to booking history.
– People responsible for managing room reservations often don't know the details of the booking person, such as e-mail or phone, which would facilitate any contact.
– In many cases, you must be present to make a reservation. It's not possible to book a room outside the workplace.

Model building. Functional requirements:

1. Logging in using the login and password.
2. The fourth incorrect login sends mail to the administrator.
3. Searching available rooms by its number, capacity and chosen period of time.
4. Booking room for declared period of time.
5. Booking management, with the option of canceling reservation.
6. Viewing booking history.

Non-functional requirements:

1. The system will be easy to use and require a short time to learn how to use the system.
2. The system will run 24 h, 7 days a week.
3. The system should be able to expand with new functionalities.
4. The application is available in English language version.
5. The system should respond quickly to user actions.

1.2 Class Diagram

Our system is wrapped within six class (Fig. 1):

1. Logger
2. Connection
3. BookingPanel
4. Booking
5. User.

Room.

Class Logger has two private string-type fields: login and password and four public methods: send(), showError(), open() and close(), all of them have no entry parameters and their return type is void.

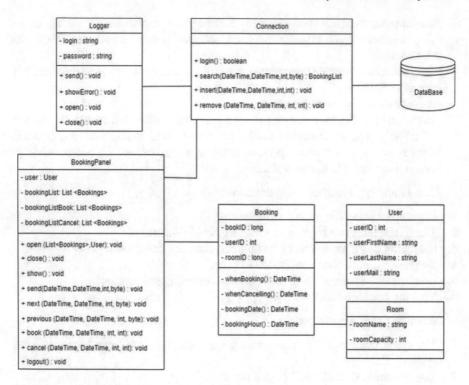

Fig. 1 Class diagram

Class Connection has no fields and four public methods:

```
login() - no entry parameters; return type: boolean,
search() - four entry parameters with type: DateTime, DateTime, int,
byte; return type: BookingList,
insert() - four entry parameters with type: DateTime, DateTime, int,
int; return type: void,
remove - same parameters and returning type as insert() method.
```

Class BookingPanel has four private fields: user, bookingList, bookingListBook, bookingListCancel. First has User type, rest of them have List < Bookings > type. This class has also nine public methods:

```
open() - two entry parameters: List<Bookings>, User,
close() - no entry parameters,
show() - no entry parameters,
send() - four parameters with type: DateTime, DateTime, int, byte,
next() - same as send() method,
previous() - same as send() method,
book() - same as send() method, but last parameter has int type,
cancel() - same as book(),
logout() - no entry parameters.
```

All methods in this class return void.

Class Booking has three private fields and four public methods. The fields are: bookID—type long, userID—type int, roomID—type long, and the methods are:

`whenBooking(), whenCancelling(), bookingDate(), bookingHour().`

All methods have no entry parameters and return type DateTime.

Classes User and Room have only four and two private fields, respectively:

`userID, userFirstName, userLastName, userMail, roomName, roomCapacity.`

First and last of them are int type, while rest are string.

Class Logger is responsible for process of logging user into system, exceptions handling and creating user session.

Class Connection is responsible for communication with database.

BookingPanel is responsible for user interface, it uses classes Booking, User and Room, which are needed for managing what user see in his GUI.

Data-flow diagram

This diagram (see Fig. 2) shows data flow during process of booking/cancelling, just after finished procedure of login without errors. Main view visible for user shows current availability of room number 1. This is possible because of default searching parameters (room no. 1, current week). System sends request to DB, get answer and refresh view. Described actions are made in behind, without user knowledge. Now user can select booking time for given parameters or change them. User can also cancel his booking. Each case is described below.

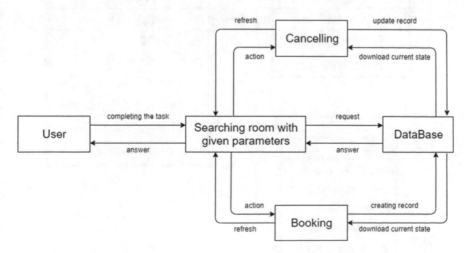

Fig. 2 Data-flow diagram

CASE: Change parameters

User is making his own parameters (for example change room or insert minimum room capacity) and press SEARCH button. Then new request is sent to DB and view is refreshed after getting an answer.

CASE: Booking

When user mark proper time for booking, BOOK button unlock. After pressing this button new request is sent to DB, new record is created, then current state is downloaded from DB and finally user view is refreshed.

CASE: Cancelling

When user select his booking on main view, CANCEL button became unlocked. After pressing this button new request is sent to DB, selected records are updated, current state is downloaded from DB and user view is refreshed.

Entity Relationship Diagram

Our database (Fig. 3) is as simple as it can be. It has only 3 tables: Rooms, Bookings and Users. Table Rooms is responsible for collecting information about room managed by our system. It has 3 columns:

- room ID (private key, type: small int),
- roomName (type: varchar)
- roomCapacity (type: int).

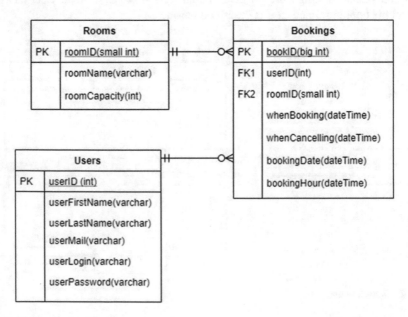

Fig. 3 Entity relationship diagram

Table User is responsible for collecting information about our system users. It has 6 columns:

- userID.
- userFirstName,
- userLastName,
- userMail,
- userLogin,
- userPassword.

First one is private key and has int type, rest of them are varchar type. Table Bookings is responsible for creating bookings. It has 7 columns. First one is bookID (private key, type: big int), second and third combined are foreign key based on userID from table Users and roomID from table Rooms. Rest of colums in this table (whenBooking, whenCancelling, bookingDate, bookingHour) are dateTime type.

Use case specification diagram

Our system (Fig. 4) give few possibilities to users, such as:

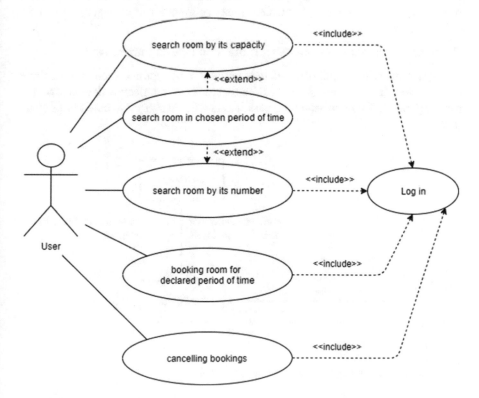

Fig. 4 Use case specification diagram

– searching room in chosen period of time,
– searching room by its capacity or number (this features extends previous one),
– booking room for declared period of time,
– cancelling existing user bookings.

All mentioned features included log in the system.

Communication diagram—booking by room number

At the beginning user chooses room. Then system (Fig. 5) send request to database, get answer and on that base refresh view. Now user can choose booking date on refreshed view. Next step made by system is creating proper record in database and refresh view again. If chosen parameters are unavailable, record won't be made. User will get information if booking is done or not.

Communication diagram—booking by room capacity

At the beginning user chooses minimum capacity of the room. Then system sends request to database, gets answer and on that base refreshes view. Difference between booking by room number and capacity is that in second version user gets list of all rooms fulfilled given capacity parameter. Rest of communication (Fig. 6) is the same in both cases.

Communication diagram—booking room with default parameters

Booking room with default parameters is nothing but simplified version of booking methods described at points 3.7 and 3.8. In that case communication does not include part responsible for input searching parameters and checking database at this stage (Fig. 7).

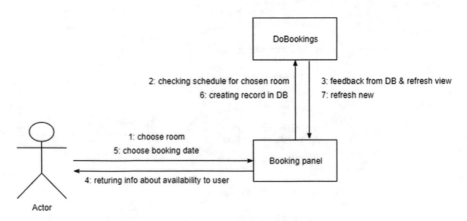

Fig. 5 Communication diagram—booking by room number

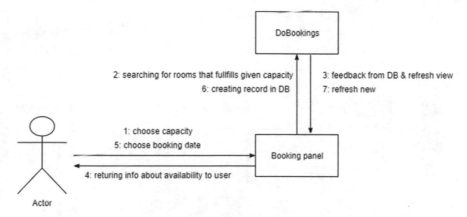

Fig. 6 Communication diagram—booking by room capacity

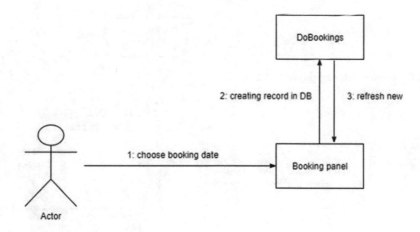

Fig. 7 Communication diagram—booking room with default parameters

Activity diagram—login to system

User opens system and sees logging windows with two field (Fig. 8): for login and password. After filling both of them and press LOG IN button inserted data are comparing with stored in database ones. If verification finished successful, user became logged in and activity is over. In other case system will check how many attempts of log in went wrong. If less than 4, user can insert data again. Consequences of fourth failing attempt are as follows: system sends mail to administrator and block user—activity is over.

Activity diagram—booking

User sets parameters (Fig. 9) on main window and press SEARCH button. When view is refreshed user is marking booking time and press BOOK button. If selected

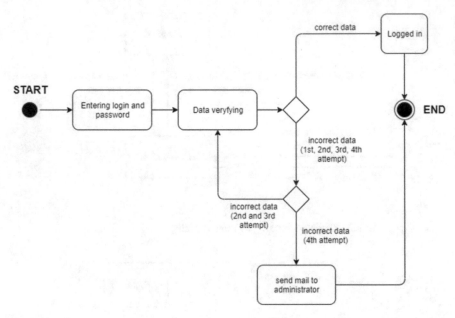

Fig. 8 Activity diagram—login to system

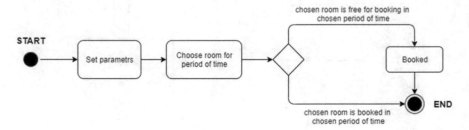

Fig. 9 Activity diagram—booking

room is available for booking in selected time, the room is booked and activity is over. In other case user marking is cancelled and activity is over. User can change parameters and try again or mark another period of time.

Activity diagram—cancelling

User highlights on main view bookings he wants to cancel and press CANCEL button (Fig. 10). If selected bookings belongs to him, cancelling will be processed and room will be available for every user—activity is over. In other case cancelling will be interrupted.

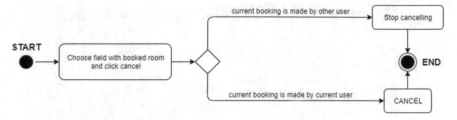

Fig. 10 Activity diagram—cancelling

2 Model Testing

2.1 Log In

After starting the application, a welcome page is displayed that allows you to log into the system (Figs. 11 and 12). Logging in requires a login and password (Fig. 13).

In case of providing incorrect data, the user will be informed by the system. If the user enters the data incorrectly more than three times, logging in will be impossible and the system will send mail to the administrator.

2.2 Booking Window

Logging in will enable the start screen showing the availability of room number 1 this week by default (Fig. 14).

The room's availability is clearly visible, logged-in user's reservations are green, others are orange (Fig. 15).

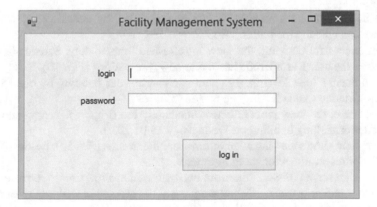

Fig. 11 Welcome page of application

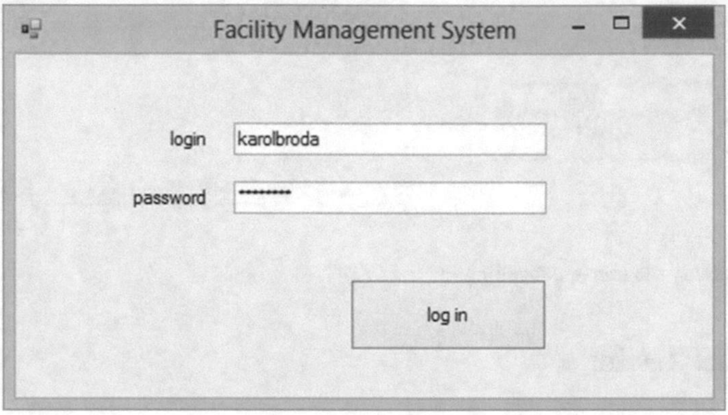

Fig. 12 Log into the system

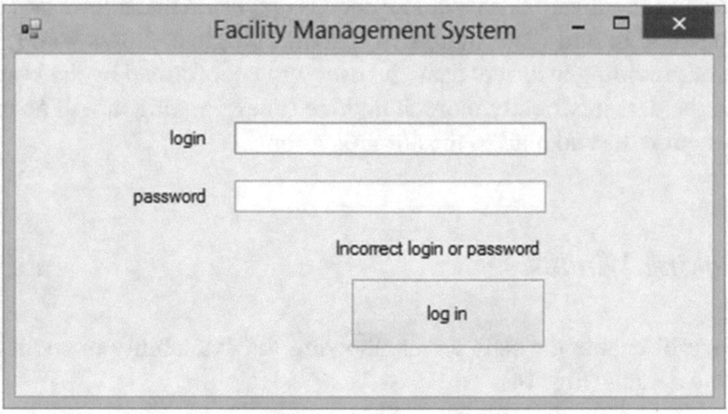

Fig. 13 Log into the system—informed by the system about incorrect data

The user can make a reservation at selected times, by checking the matching windows and clicking the button "Book".

After successful booking, the view is refreshed immediately. Screen shows the effect after the user Karol Broda has made new reservations (Fig. 16).

If you make a mistake (Fig. 17), you can cancel your booking by checking the matching windows and clicking the button "Cancel".

In this case, the view is refreshed immediately too (Fig. 18). Screen shows the effect after cancelling booking on Wednesday 15.01.2020.

The system allows viewing reservations for other rooms (Fig. 19). Below (Fig. 20) is the list of reservations for room number 9.

On the start screen (Fig. 21) there are two buttons that allow user to see reservation for previous and next weeks. Below we present the process of making a reservation by the user Bartłomiej Cota for the next Tuesday (Fig. 22).

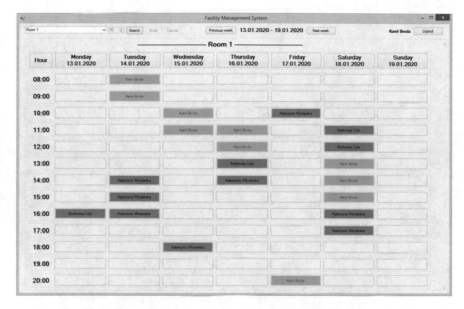

Fig. 14 Start screen (availability of room number 1 this week by default)

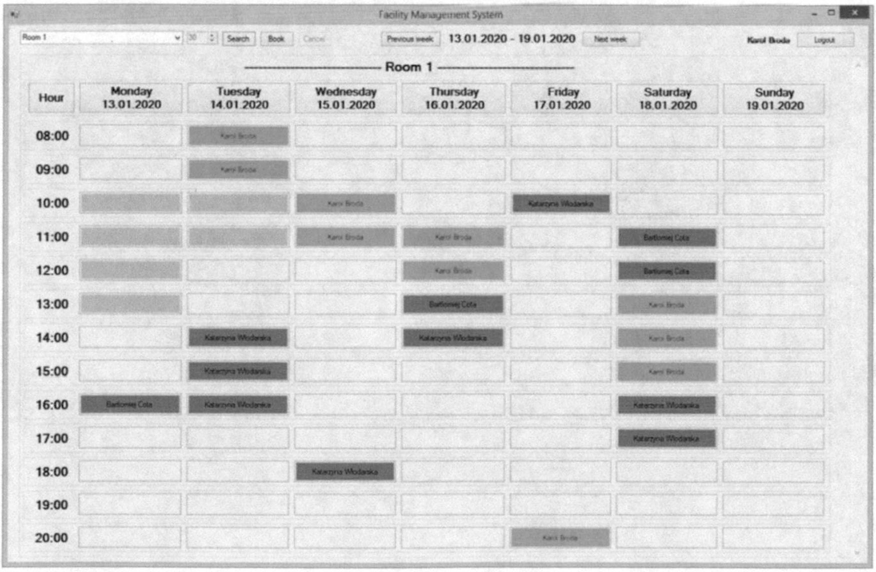

Fig. 15 Page with room's availability

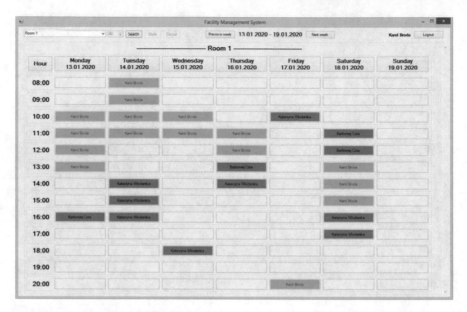

Fig. 16 Page with successful booking

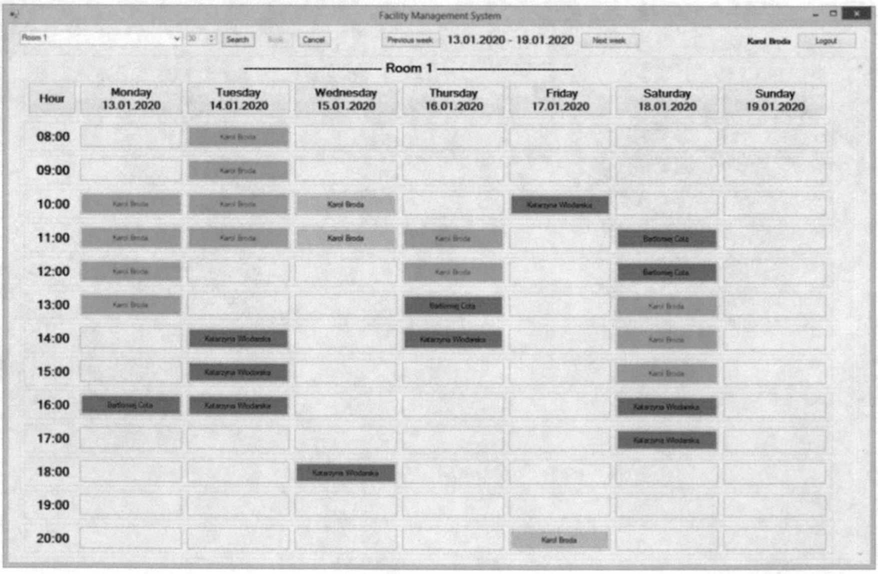

Fig. 17 Page with cancel your booking

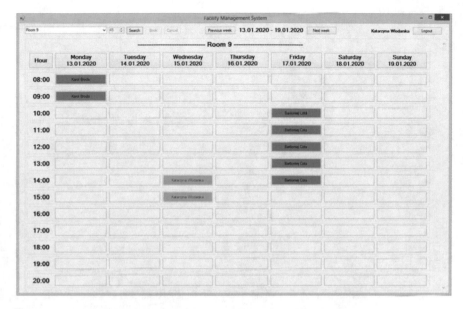

Fig. 18 Refreshed page

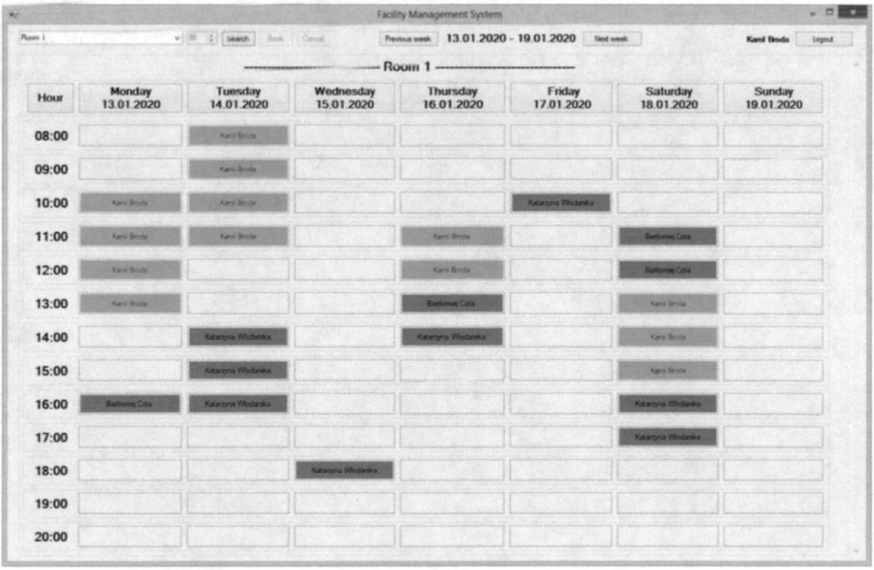

Fig. 19 The viewing reservations for other rooms

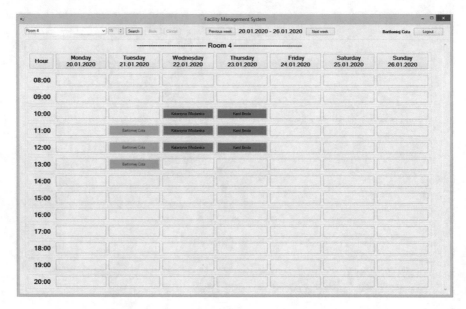

Fig. 20 The list of reservations for room number 9

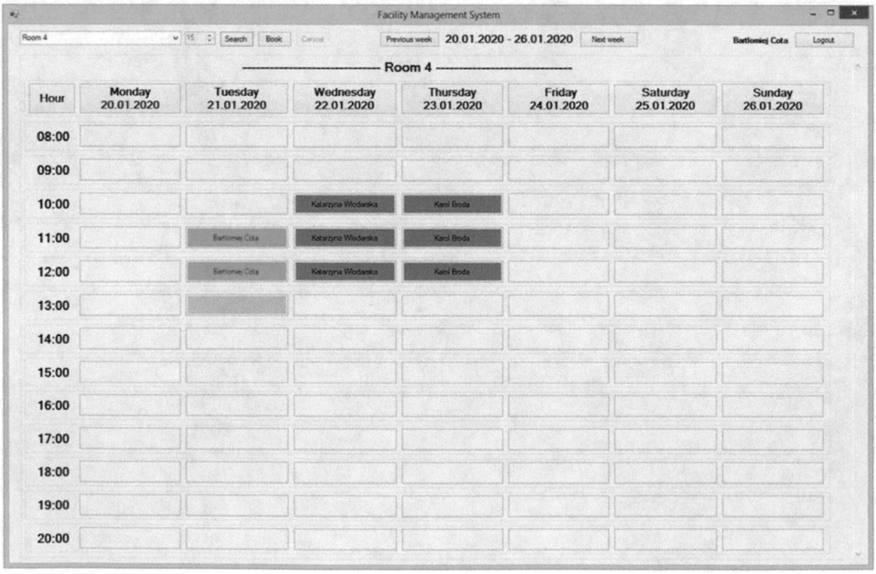

Fig. 21 The reservation for previous and next weeks

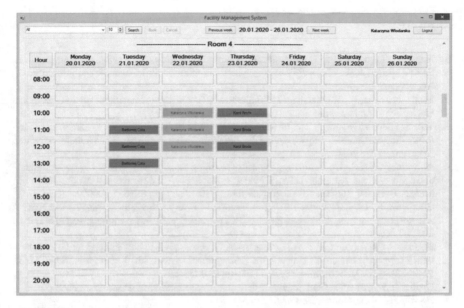

Fig. 22 The process of making a reservation by the user Bartłomiej Cota for the next Tuesday

The application allows searching room by its capacity. Below, the effect of searching a room with a at least ten people capacity, for the next week.

3 Conclusions

The "Facility management system" application we designed works following our expectations. Booking and managing rooms shouldn't be a hassle—it should just work. Our core platform is an extensible scheduling toolset that enables a variety of room booking use cases.

The different components of the same meeting room booking system work hand in hand to deliver solutions for every industry. We present the solution that helps to make an innovative and efficient workplace. It's easy and allows real-time availability and booking.

Using the system helps users find the right solution, optimize how their office is used, save money and make the workplace more flexible and accessible.

References

1. Facilities Manager Job Description|Telegraph Jobs Careers Advice. Telegraph Jobs. jobs.telegraph.co.uk. https://jobs.telegraph.co.uk/article/facilities-manager-job-description/. Accessed 27 Jan 2020
2. Broda, K., Cota, B., Włodarska, K., Problemowa, P.: Facility management system. Technical University of Łódź. Extramular studies. Academic year 2019/2020
3. What is Facilities Management Software?|ServiceChannel: The #1 Facilities Management Platform'. ServiceChannel. https://servicechannel.com/what-is-facilities-management-software/. Accessed 29 Jan 2020
4. Al-Mamary, Y.H.S.: Understanding the use of learning management systems by undergraduate university students using the UTAUT model: credible evidence from Saudi Arabia. Int. J. Inf. Manag. Data Insights **2**(2) (2022). https://doi.org/10.1016/j.jjimei.2022.100092
5. Goerlandt, F., Li, J., Reniers, G.: The landscape of safety management systems research: A scientometric analysis. J. Saf. Sci. Resil. **3**(3), 189–208 (2022). https://doi.org/10.1016/j.jnlssr.2022.02.003
6. Rosman, M.R.M., Aziz, M.A.A., Osman, M.A.F., Razlan, N.M.: Self-efficacy and user behavioral intention to use online consultation management system. Int. J. Eval. Res. Educ. **11**(3), 1240–1249 (2022). https://doi.org/10.11591/ijere.v11i3.22875
7. Facility management system. http://www.risatech.net/portal/portalimages/risa_FMS.pdf
8. Hollist, A.: Manage an Office Facility, p. 5
9. Ahmed, M.T., Kabir, M.H., Roy, S.: Web based student registration and exam form fill-up management system for educational institute. Int. J. Inf. Eng. Electron. Bus. (IJIEEB) **14**(2), 47–62 (2022). https://doi.org/10.5815/ijieeb.2022.02.04
10. Standardization. https://www.eurofm.org/index.php/standardization. Accessed 29 Jan 2020
11. Omri, M.N., Aonne, H.B.: Towards an intelligent approach to workflow integration in a quality management system. Int. J. Intell. Syst. Appl. (IJISA) **14**(3), 54–73 (2022). https://doi.org/10.5815/ijisa.2022.03.05
12. Lanza, V., Dubos-Paillard, E., Charrier, R., Provitolo, D., Berred, A.: An analysis of the effects of territory properties on population behaviors and evacuation management during disasters using coupled dynamical systems. Appl. Netw. Sci. **7**(1) (2022). https://doi.org/10.1007/s41109-022-00450-6
13. Chen, C.-C., Wang, K.-S., Hsiao, Y.-T., Chou, J.: ALBERT: an automatic learning based execution and resource management system for optimizing Hadoop workload in clouds. J. Parallel Distrib. Comput. **168**, 45–56 (2022). https://doi.org/10.1016/j.jpdc.2022.05.013
14. Dayan, V., Chileshe, N., Hassanli, R.: A scoping review of information-modeling development in bridge management systems. J. Constr. Eng. Manag. **148**(9) (2022). https://doi.org/10.1061/(ASCE)CO.1943-7862.0002340
15. Bhatnagar, A., Khatri, P., Krzywonos, M., Tolvanen, H., Konttinen, J.: Techno-economic and environmental assessment of decentralized pyrolysis for crop residue management: Rice and wheat cultivation system in India. J. Clean. Prod. **367** (2022). https://doi.org/10.1016/j.jclepro.2022.132998
16. Shakhovska, N., Fedushko, S., Gregušml. M., Melnykova, N., Shvorob, I., Syerov, Y.: Big Data analysis in development of personalized medical system. Procedia Comput. Sci. **160**, 229–234 (2019). https://doi.org/10.1016/j.procs.2019.09.461
17. Martínez-Zarzuelo, A., Rodríguez-Mantilla, J.M., Fernández-Díaz, M.J.: Improvements in climate and satisfaction after implementing a quality management system in education. Eval. Program Plan. **94** (2022). https://doi.org/10.1016/j.evalprogplan.2022.102119
18. Aljarallah, N.A., Dutta, A.K., Alsanea, M., Sait, A.R.W.: Intelligent student mental health assessment model on learning management system. Comput. Syst. Sci. Eng. **44**(2), 1853–1868 (2023). https://doi.org/10.32604/csse.2023.028755

19. Huang, H., Metawa, N., Rajendran G.: Integrated as a service in the construction of small and micro enterprise financial management platform system. In: Lecture Notes on Data Engineering and Communications Technologies, vol. 122, pp. 614–622 (2023). https://doi.org/10.1007/978-981-19-3632-6_72
20. Liu, X., Liu, G., Wan, H.-D.: Innovation of smart city management system based on computer application technology. In: Lecture Notes on Data Engineering and Communications Technologies, vol. 122, pp. 333–341 (2023). https://doi.org/10.1007/978-981-19-3632-6_41
21. Hu, Z., Khokhlachova, Y., Sydorenko, V., Opirskyy, I.: Method for optimization of information security systems behaviour under conditions of influences. Int. J. Intell. Syst. Appl. (IJISA) 9(12), 46–58 (2017). https://doi.org/10.5815/ijisa.2017.12.05
22. Gue, I.H.V., Lopez, N.S.A., Chiu, A.S.F., Ubando, A.T., Tan, R.R.: Predicting waste management system performance from city and country attributes. J. Clean. Prod. 366 (2022). https://doi.org/10.1016/j.jclepro.2022.132951.
23. Hu, Z., Odarchenko, R., Gnatyuk, S., Zaliskyi, M., Chaplits, A., Bondar, S., Borovik, V.: Statistical techniques for detecting cyberattacks on computer networks based on an analysis of abnormal traffic behaviour. Int. J. Comput. Netw. Inf. Secur. (IJCNIS) 12(6), 1–13 (2020). https://doi.org/10.5815/ijcnis.2020.06.01
24. Roy, S., Kabir, M.H., Ahmed, M.T.: Design and implementation of web-based smart class routine management system for educational institutes. Int. J. Educ. Manage. Eng. (IJEME) 12(2), 38–48 (2022). https://doi.org/10.5815/ijeme.2022.02.05
25. Khan, A.I., Alsolami, F., Alqurashi, F., Abushark, Y.B., Sarker, I.H.: Novel energy management scheme in IoT enabled smart irrigation system using optimized intelligence methods. Eng. Appl. Artif. Intell. 114 (2022). https://doi.org/10.1016/j.engappai.2022.104996
26. Lu, S.-H., et al.: Multimodal sensing and therapeutic systems for wound healing and management: a review. Sens. Actuators Rep. 4 (2022). https://doi.org/10.1016/j.snr.2022.100075
27. Chang, T.-Y., Ku, C.C.-Y., Cheng, T.-Y., Chung, C.-K., Chang Sanchez, E.: Modular counting management system for mall parking services. Comput. Ind. Eng. 171 (2022). https://doi.org/10.1016/j.cie.2022.108362
28. Yang, X., Altrjman C.: Application of big data in management information system. In: Lecture Notes on Data Engineering and Communications Technologies, vol. 122, pp. 594–603 (2023). https://doi.org/10.1007/978-981-19-3632-6_70
29. Zhang, S., Ou, W., Ren, G., Wang, H., Zhu, P., Zhang, W.: Risk model and decision support system of state grid operation management based on big data. In: Lecture Notes on Data Engineering and Communications Technologies, vol. 122, pp. 419–427 (2023). https://doi.org/10.1007/978-981-19-3632-6_51
30. Chang, Y., Zou, J., Fan, S., Peng, C., Fang, H.: Remaining useful life prediction of degraded system with the capability of uncertainty management. Mech. Syst. Signal Process. 177 (2022). https://doi.org/10.1016/j.ymssp.2022.109166
31. Zhang, S., Zhao, X., Li, X., Yu, H.: Heterogeneous fleet management for one-way electric carsharing system with optional orders, vehicle relocation and on-demand recharging. Comput. Oper. Res. 145 (2022). https://doi.org/10.1016/j.cor.2022.105868
32. Bedford, D., Bisbe, J., Sweeney, B.: Enhancing external knowledge search: the influence of performance measurement system design on the absorptive capacity of top management teams. Technovation 118 (2022). https://doi.org/10.1016/j.technovation.2022.102586
33. Jain, D.K., Neelakandan, S., Veeramani, T., Bhatia, S., Memon, F.H.: Design of fuzzy logic based energy management and traffic predictive model for cyber physical system. Comput. Electri. Eng. 102 (2022). https://doi.org/10.1016/j.compeleceng.2022.108135
34. Ahmed, F., Protik, R.C., Hasan, M.: Centralized library management system: an E-governance approach for improving accessibility of library resources of Bangladesh. In: Lecture Notes in Networks and Systems, vol. 401, pp. 741–750 (2023). https://doi.org/10.1007/978-981-19-0098-3_70

35. Hegde, S., Aishwarya, G., Hugar, A., Suneeta, V.B.: Bank management system using blockchain technology. In: Lecture Notes in Networks and Systems, vol. 401, pp. 209–218 (2023). https://doi.org/10.1007/978-981-19-0098-3_22
36. Iyanda, A.R., Ninan, O.D., Odejimi, D.J.: Students conversation management system. Int. J. Educ. Manage. Eng. (IJEME) **8**(4), 1–9 (2018). https://doi.org/10.5815/ijeme.2018.04.01
37. Tao, Q. et al.: Design and application of a public management system based on edge cloud computing. In: Lecture Notes on Data Engineering and Communications Technologies, vol. 122, pp. 736–745 (2023). https://doi.org/10.1007/978-981-19-3632-6_85
38. Hao, M., Nie, Y.: Hazard identification, risk assessment and management of industrial system: Process safety in mining industry. Saf. Sci. **154** (2022). https://doi.org/10.1016/j.ssci.2022.105863
39. Cheng, M., Tao, D., Xie, S., Cao, X., Yuen, A.H.: Exploring students' learning management system acceptance patterns: Antecedents and consequences of profile membership. Comput. Hum. Behav. **135** (2022). https://doi.org/10.1016/j.chb.2022.107374

Designing and Implementation of Online Judgment System

Iryna Ivanochko and Yurii Kostiv

Abstract In this paper we prepare extensive analysis of creation process for online judge solution. Design of architecture is inspired by methods used widely in professional applications and by experiences with existing systems of this kind. We present our vision of all system elements—web application, server application and executors. To create web application we use Material Design technology. Scalability is given by use of Google Kubernetes and Docker technologies. There is considered wide range of probable security threats, and after extensive testing eBPF solutions was chosen. In first version of system C++, Java and Python are supported. System offers also amazing extensibility by plugins, in these way even support of new programming language can be added.

1 Introduction

Research shows that one of the most efficient ways to learn or solidify programming skills is practice [1]. With the advancement of informational technologies, new learning methods are being developed. In aspects of the 4th Industrial Revolution and labour automation, student self-evaluation systems become an integrated part of modern education. Skill assessment platforms, like online judge, allow students to practice, check progress, and correct mistakes instantly without any external intervention or contribution. Most common services contain a huge base of various computer science-related tasks [2]. Obvious advantages of self-evaluation online environments [3–5] are time and money savings during the education process, however, appropriate design is highly demanding to achieve. The ability to send the source code or computer program, by user is the integral part to properly evaluate user input.

I. Ivanochko (✉) · Y. Kostiv
Lviv Polytechnic National University, Lviv, Ukraine
e-mail: irene.ivanochko@gmail.com

Y. Kostiv
e-mail: yurii.m.kostiv@lpnu.ua

© The Author(s), under exclusive license to Springer Nature Switzerland AG 2023   191
N. Kryvinska et al. (eds.), *Developments in Information and Knowledge Management Systems for Business Applications*, Studies in Systems, Decision and Control 462,
https://doi.org/10.1007/978-3-031-25695-0_10

However, this type of interaction is considered sensitive to exploitation and manipulation. Need for the execution of untrusted code challenges most of the commonly used security techniques. Despite the fact that a lot of sandboxes techniques can be found in this field [6–8], none of them satisfy the criteria of the modern cloud environment and microservices architecture, which are required to achieve scalability of a platform to handle web traffic [9–11]. This paper tries to solve scalability and security [12, 13] concerns in online judge platforms introducing design philosophy dedicated to cloud computing environment.

Main tasks raised to online judge system include: storage and categorize tasks, allow users to select, read, and solve exercises, automatically validate, and score received solutions. To meet requirements like automatic validation of solution, system receives computer program or its source code, executes it on prepared test cases and shows verdict. Additional validation can be made, for example: performance, memory usage or antiplagiat check. Execution of test cases has to be consistent, deterministic and easy to reproduce. This require to use containerized sandbox environment that can be easily reset to state before execution. Controlled environment allows to met some of security concerns related to execution of untrusted code [8, 11]. However more security constraints have to be made to make sure that no one is able to temper with system [7].

Proper testing of user submission requires running application over extensive amount of test cases. Some of sandboxing techniques also have negative impact on execution performance. To provide smooth user experience and do not force user to wait for verification too long, proper scalability has to be ensured. Modern web system with cloud-based architectures guarantee ability to handle even thousands of request per second and properly adjust infrastructure to handle that amount of traffic [14, 15]. Kubernetes orchestration is supported by most of cloud providers, reducing cost of implementation and maintenance at scale.

2 Architecture Design

2.1 Global Design

In high-level architecture several parts can be distinguished (Fig. 1). User communicates with the platform using web client in form of web page in his browser, following one of use cases described in Fig. 2. With web application as a main access point, user can make request to application server. Requests are processed and web server communicates with other parts of the system using asynchronous message queue. Several solution related to asynchronous messaging queues are discussed in [14], and Apache Kafka provides better scalability and throughput for heavy traffic systems [15–17]. Messages, categorized by topics, are read and processed by other part of systems like sandbox executor or database service, and after processing new messages

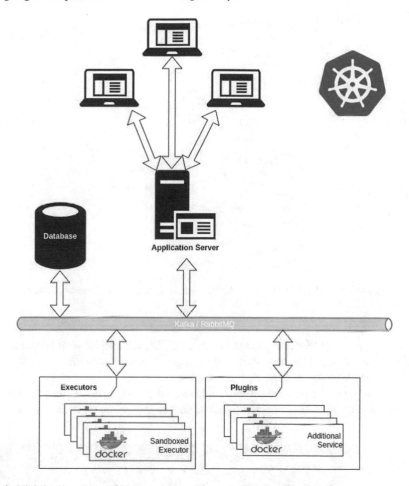

Fig. 1 High-level system architecture design. Chart was created using draw.io

are added to message bus. Whole system is orchestrated by Google Kubernetes software, allowing for scaling every part of the system independently, based on system needs. Asynchronous messaging queue also allows extensibility in form of plugins. Every part of the system has ability to read and modify message in which there are interested in. Specific part of the system was discussed in details in next sections.

2.2 Website Design

As for web platform, website displayed for an user specifies how they uses and experience system. For best user experience this web client has to be responsive and

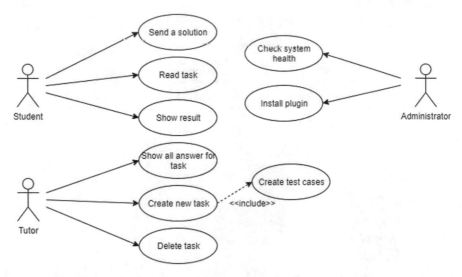

Fig. 2 Top level use cases for web service. Three roles are distinguished: Student, Tutor, and system administrator. Chart was created using draw.io

intuitive. Following Material Design Guidelines [17, 18] and Nielsen Heuristics for user interface design [18] helps with creating good looking website.

Provided views are on Figs. 3, 4, 5 and 6.

The most important of the provided views was the view presenting the content of the task (Fig. 5). It was decided to add a simple pre-checking code syntax editor

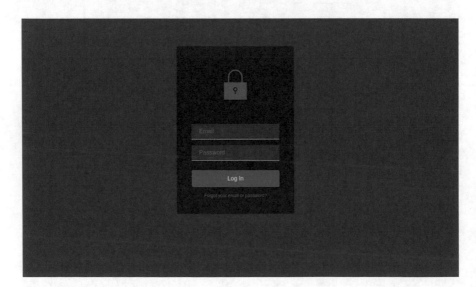

Fig. 3 Sample login page from provided website

Fig. 4 Sample register page from provided website

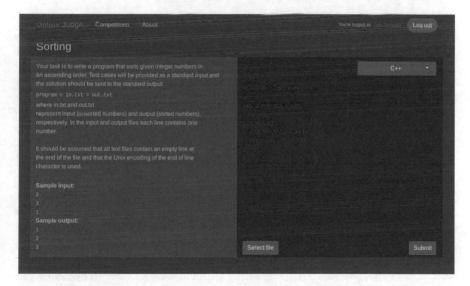

Fig. 5 Sample competition page with the task content and text editor for submission

right next to the task content. In the upper right corner of the editor there is a list with a selection of programming languages in which the user can prepare solutions to the problem. To give the user the freedom to provide a solution, it is also possible to select the appropriate file from the computer memory. In the lower right corner

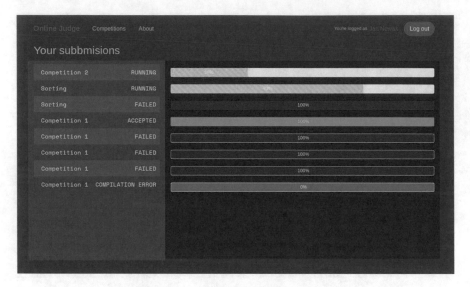

Fig. 6 Sample view of submitted answers for competitions with results

of the editor there is a button allowing the user to send a solution to be checked for correctness with the automatic tests prepared earlier.

In order to achieve the most intuitive user interface possible, the result of the tests carried out on the solution provided by the user was highlighted with the appropriate color. Red corresponds to a failure, green corresponds to a success, while yellow indicates that the tests are in progress (Fig. 6).

2.3 Main Web

Main web service (or application server) is considered as the simplest part of described system. Application has similar responsibility to typical CRUD systems widely used in the Internet. Processing user request, authentication, authorization and showing results can be done automatically by almost every modern web application framework. Using databus like kafka as an main communication channel with other part of the system allows scalability of main web service. Replicas of this component can independently gain requests, process it, write or read messages to or from databus, and communicating with the rest of the system with message queue. Architecture of this part is stateless, so it can be run in thousands of copies working independently, without extensive communication between replicas and having access to same data through message bus. Kubernetes allows to dynamically change number of replicas to handle traffic or limit resource usage when traffic is small. Conceptual division into subsystems is presented in Fig. 7. Entry gateway accepts request from clients and checks authorization rights using authentication module. This module

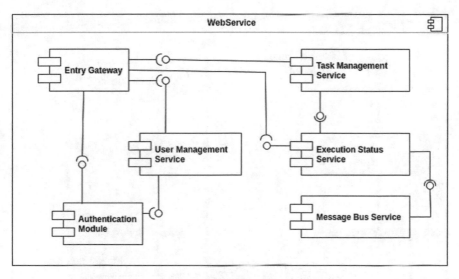

Fig. 7 Component diagram of web service. Interaction between all six components are done internally with application interfaces. Entry gateway is an entry point to the whole system from outside. Chart was created using draw.io

works together with user management service to check if user can access resource. Other request can be redirected from entry gate into user management service, if this request corresponds to change of user data, into task management service responsible for storing, accepting and displaying available tasks in the system, or into execution service responsible for executing task in one of the execution environment and informing task manager service about result.

2.4 Sandboxed Executors

Sandboxed executor comprises most challenging and difficult to properly design part of the system. This module is responsible for compiling and executing user provided programs. As mentioned in introduction, system cannot trust user code and has to ensure that executed code cannot temper with other part of the system. This means that application cannot read and write to files it is not supposed to, accessing memory of supervisor process it is running under, communicate with other part of the system or any entity in the web. Supervisor should also prevent from unauthorized resource exhaustion and therefore hanging up entire system. Executed application should not be allowed to allocate extensive amount of RAM, spawn threads, fork child processes, create new files, and should compile and execute tests in reasonable amount of time. Typical flow of execution request is shown in Fig. 8 and as followed:

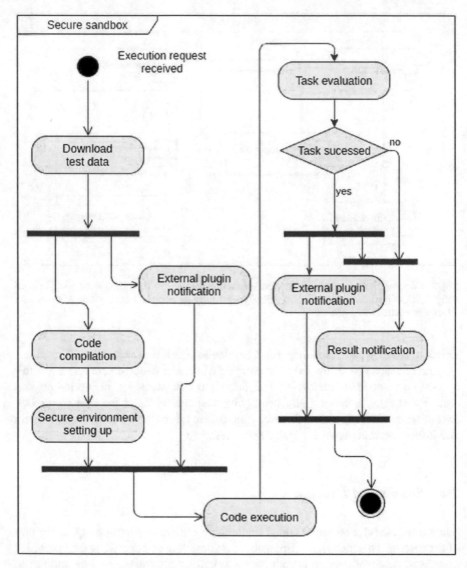

Fig. 8 Execution flow of a task sent to execution environment from web service. Chart was created using draw.io

1. Received source code is compiled.
2. Test data is downloaded into execution environment.
3. When sandbox is setup, external plugin are informed about execution.
4. Code is executed and tested against downloaded data.
5. Result is set in database.
6. When execution passes all tests, additional plugin notification is done.

Several considered security limitations are already forced by docker container-ization. Docker container provides resource isolation from rest of the system, mostly in terms of filesystem. Docker-based environment can also be easily and fastly restored to starting state. However it does not prevent from manipulating supervisor, exhausting resources and leaking other information to unauthorized user. Unfortu-nately cgroups limits provided by docker limits both supervisor and user solution, so it cannot be used for precise limitations and other mechanism has to be used. Linux Operating System provides several other solutions to limit application inter-action. Standard ulimits can limit resource usage in terms of opened files, memory and cpu interaction preventing resource exhaustion. File system related mechanisms like apparmor and disk quota can prevent unauthorized access to files and filling hard disk with extensive amount of data. To constrain system interaction in terms of creating websockets or spawning and forking childs Linux eBFP subsystem can by used to filter system calls made by application.

All those techniques helps to securely execute user solutions, however, following *the least privilege principle* forces to configure different security policies for different execution environment. Java application for example spawns additional threads for internal JVM usage and for correct functioning of Java-based application sandbox cannot prevent that. RAM requirements are also proportionally larger for this kind of execution environment. The most important part is to still provide secure environment even with relaxed policy required by Java programs. On the other hand, C/C++ application requires only one thread to function properly and sandbox can be more restrictive in term of opened file and used memory.

2.5   Plugins and Possible Extensions

Obviously, not all use cases can be covered upfront. To achieve flexibility, and allow to adapt the system into different environment and task types, extensive plugin system has to be included. Main architecture is prepared to allow as much extensibility as possible. Using centralized databus, plugins with simple API can subscribe for an event and process it. An event is defined like: sending a solution for a task, grading a task, running test for a solution, etc. With each event metadata are included, allowing to take substantial amount of information to realize additional tasks. Plugins can use this information to process additional checks, like anti plagiarism. Using subscrip-tion mechanizm, each plugin has to react for an subscribed event, before solu-tion is accepted. Plugins receive notification and process event following algorithm presented in Fig. 9.

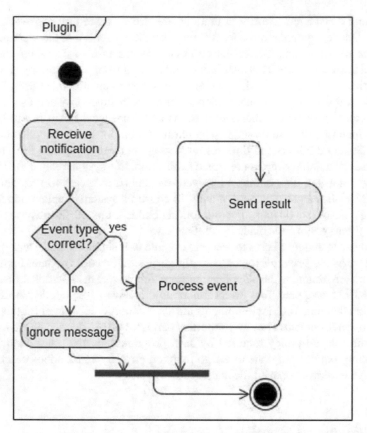

Fig. 9 Supported plugin architecture. Event is received by plugin. Extension can decide if the event needs to be processed or skipped. In case of event processing, result is passed to the next part of the system. Chart was created using draw.io

3 Testing and Evaluation

To perform security testing, several malicious application has been created. Applications covers most common attack vectors for that kind of system. This includes: memory exhaustion, disk space exhaustion, local file modification, forking run in background, creating threads, connecting into network. Each sandboxing techniques has been evaluated against this malicious applications. To evaluate sandboxing technique each malicious application is executed under the sandbox and it is checked if any of attack succeeded. A attack is called succeeded if any of computer resources described above are compromised. Flow of the test for each malicious code is shown in Fig. 10.

Operating memory is one of the most important resource for a computer system [19–22]. Malicious application may try to allocate as much as possible, to prevent

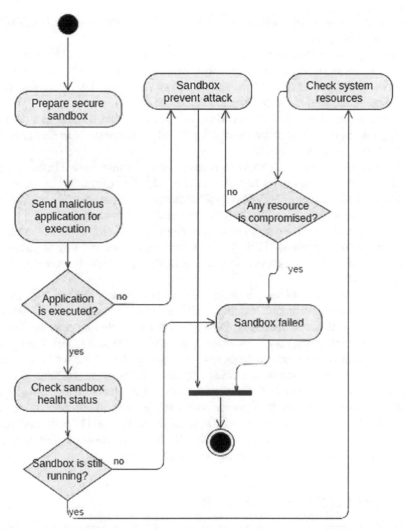

Fig. 10 Diagram describing how tests are executed. Malicious application can pass or fail test. Results are stored in table. Chart was created using draw.io

other, more critical software from running and even crushing the execution server. Therefor, proper limitation mechanism has to be used to prevent this kind of attacks.

Disk space is next important resource. Moreover, it is persistent between runs and simple restarting server does not solve the issue. Like in case of operating memory, filling whole space can lead to destabilization or shutdown of the executor.

Local file modification can help malicious application to gain control over the execution environment. Modifying files like input tests, which has to be available for judging software helps to cheat results. Via reading expected results that has to be stored in similar locations also allows to cheat. Other modification of critical files can be more dangerous.

In principle, user application is run against test cases and then it is killed to prevent resource leakage. However, creating a fork of user application can allow this fork to survive cleanup after run and then allow to run endlessly. That fork may not be considered sandboxed and then exceed other limitations.

Similar to memory and disk resources, cpu resource can be easy exhausted be creating a lot of threads or subprocesses. So called "fork bombs" can stall executors similarly to other security threats.

Last most important attack vector is connecting with other service in the Internet. Established connection allows to leak confidential information, test cases, system configuration allowing easier further exploitation, etc. Connection via a socket can be easily made with standard Linux API and this ability has to be limited.

Results of testing sandboxing techniques against those attack vectors are stored in the feature matrix below (Table 1). Note that application running in virtualized environment has different property than native applications that has to be considered during sandboxing.

Only one solution prevents all attack vectors considered. eBPF allows to filter all executed system interaction and then well written filter can prevent all malicious usage. However, this technique is expensive in both execution time and development time for a proper filtering code. The worst solution considered is a docker or cgroups mechanizm. Most of security options available are related to the whole container, where not only tested solution, but other important software are running. Ulimit is a basic UNIX mechanism that allows to do most of the work like limiting RAM, running processes and threads. However access to files are still not covered. Reading and writing to external files can be easily blocked by proper selinux policy, but in this case, files that are allowed to access by policy can be easily modified. This concludes that only mixing all of solutions can be considered secure.

Table 1 Feature matrix for sandboxing techniques

Sandbox	RAM	Disk	Files	Fork	Threads	Network
docker/cgroups	✗[a]	✗[a]	✗	✗	✗	✗[b]
ulimit	✔	✗	✗	✔	✔	✗
selinux	✗	✗	✔	✗	✗	✔
eBPF[c]	✔	✔	✔	✔	✔	✔

[a]Docker and cgroups even though allows to limit RAM and disk usage, this includes resources used be judging software. Extensive limitation of supervisor is not expected
[b]Similarly to RAM and disk, resources are limited on container basis, not process basis. Container need connection to the rest of the system and that disallow to use cgroups as a network sandbox environment
[c]Most powerful solution allows to capture and filter every system call application makes, but makes it expensive and hard to do right

Table 2 Native applications can be easily limited by most basic control mechanisms

Securing technique	C/C++	Java	Python
limiting memory via ulimit	✔	✗[a]	✔
limiting threads vis ulimit	✔	✗[a]	✔
limiting processes via ulimit	✔	✔	✔
file access control via selinux	✔	✔[b]	✔[b]

[a]Java virtual machine tends to allocate notable amount of memory for its internal usage and spawn several helper threads. Similarly to problem of distinguishing sandboxing software from sandboxed application, this ulimits cannot limit java application resources. However, in case of memory, JVM allows to manually limit memory accessible for executed application

[b]Docker and cgroups even though allows to limit RAM and disk usage, this includes resources used be judging software. Extensive limitation of supervisor is not expected

Following solution does not consider how to properly secure an asset, but only if an asset can be secured at all. Next table shows if specific techniques can be applied to specific execution environment and if it succeeded (Table 2).

Problems with techniques that requires runtime environment are more complex than with native code. Sandboxing must exclude requirements of runtime engine. eBFP may be required in several cases and specialized filter has to be developed for each of that case increasing complexity of sandboxed executor.

Other security related tasks can also be implemented via source code analysis and be provided into system using plugin mechanism described above. This may include for example anti-virus software check.

4 Conclusion

Securing online judge platform for different attack vectors are challenging. By far, easiest to execute sandboxed are native applications that not require any runtime engine like python or java virtual machine. Later technologies are more popular in industry and cannot be skipped in design of learning platform. Several mechanisms for sandbox environment can be found in Linux/Unix family of operating system, and all of them have different advantages and disadvantages. Only mixing most of them provides secure execution environment, but exception occurs depending on runtime used. Following tests are not extensive and may not include most sophisticated attack vectors. Further evaluation of sandbox techniques may be required and may shown additional flaws in those techniques when used in different runtime environment like C#/.Net for example or with more sophisticated attacks. Nethertheless used sandboxes allow to use provided online judge solution secured against most common attacks that can be performed against it. They also fit into cloud environment and runs inside docker container for fast, easy and scalable shiplement of sandboxed executors. Usage of message queues allows for easy extensibility of the system with plugin mechanizm and for scalability, as long as message queue can manage traffic. Solutions like kafka was tested in industry against millions messages per second [23], with far exceeds requirements of our system.

References

1. Fang, X., Zhang, H., Sun, Y.: A programming related courses E-learning platform based on online judge. In: Frontier and Future Development of Information Technology in Medicine and Education, pp. 3419–3423. Springer, Dordrecht (2014)
2. Wasik, S., Antczak, M., Badura, J., Laskowski, A., Sternal, T.: A survey on online judge systems and their applications. ACM Comput. Surv. (CSUR) 51(1), 1–34 (2018)
3. Yee, B., Sehr, D., Dardyk, G., Chen, J.B., Muth, R., Ormandy, T., Fullagar, N.: Native client: a sandbox for portable, untrusted x86 native code. In: 2009 30th IEEE Symposium on Security and Privacy, pp. 79–93 (2009)
4. Fedushko, S., Syerov, Y., Tesak, O., Onyshchuk, O., Melnykova, N.: Advisory and accounting tool for safe and economically optimal choice of online self-education services. In: Proceedings of the International Workshop on Conflict Management in Global Information Networks (CMiGIN 2019), vol. 2588, pp. 290–300. Ukraine, CEUR-WS.org (2020). http://ceur-ws.org/Vol-2588/paper24.pdf
5. Dolgikh, S.: Categorization in unsupervised generative self-learning systems. Int. J. Mod. Educ. Comput. Sci. (IJMECS) 13(3), 68–78 (2021). https://doi.org/10.5815/ijmecs.2021.03.06
6. Liu, D., Valdiviezo-Díaz, P., Riofrio, G., Sun, Y. M., Barba, R.: Integration of virtual labs into science e-learning. Procedia Comput. Sci. 75, 95–102 (2015)
7. Yi, C., Feng, S., Gong, Z.: A comparison of sandbox technologies used in online judge systems. Appl. Mech. Mater. 490, 1201–1204. Trans Tech Publications Ltd. (2014)
8. Pierścieniewski, L., Krzeszewska, U., Masłowski, P., Pietrusiak, F.: E-Learning system for computer science students-online judge. Student Seminar Work (2019)
9. Ahmad, A., Andras, P.: Scalability analysis comparisons of cloud-based software services. J. Cloud Comput. 8(1), 1–17 (2019)
10. Alla, M., Faryadi, Q., Fabil, N.: The impact of system quality in e-learning system. J. Comput. Sci. Inf. Technol. 1(2), 14–23 (2013)
11. Aeiad, E., Meziane, F.: An adaptable and personalised E-learning system applied to computer science Programmes design. Educ. Inf. Technol. 24(2), 1485–1509 (2019)
12. Hu, Z., Buriachok, V., Sokolov, V.: Implementation of social engineering attack at institution of higher education. In:1th International Workshop on Cyber Hygiene and Conflict Management in Global Information Networks. Kyiv, Ukraine. Available at: http://dx.doi.org/https://doi.org/10.2139/ssrn.3679106
13. Hu, Z., Odarchenko, R., Gnatyuk, S., Zaliskyi, M., Chaplits, A., Bondar, S., Borovik, V.: Statistical techniques for detecting cyberattacks on computer networks based on an analysis of abnormal traffic behavior. Int. J. Comput. Netw. Inf. Secur. (IJCNIS). 12(6), 1–13 (2020). https://doi.org/10.5815/ijcnis.2020.06.01
14. Bernstein, D.: Containers and cloud: from lxc to docker to kubernetes. IEEE Cloud Comput. 1(3), 81–84 (2014)
15. Rensin, D.K.: Kubernetes-Scheduling the Future at Cloud Scale. O'Reilly Media, CA, USA (2015)
16. Dobbelaere, P., Esmaili, K.S.: Kafka versus RabbitMQ: A comparative study of two industry reference publish/subscribe implementations: industry paper. In: Proceedings of the 11th ACM International Conference on Distributed and Event-Based Systems, pp. 227–238 (2017)
17. Kreps, J., Narkhede, N., Rao, J.: Kafka: a distributed messaging system for log processing. Proc. NetDB. 11, 1–7 (2011)

18. Design—Material Design. Accessed from: https://material.io/design/. Accessed 28 Nov 2019
19. Agbonifo, O., Boyinbode, O., Oluwayemi, F.: Design of a digital game-based learning system for fraction algebra. Int. J. Mod. Educ. Comput. Sci. (IJMECS) **13**(5), 32–41 (2021). https://doi.org/10.5815/ijmecs.2021.05.04
20. Hu, Z., Khokhlachova, Y., Sydorenko, V., Opirskyy, I.: Method for optimization of information security systems behavior under conditions of influences. Int. J. Intell. Syst. Appl. (IJISA) **9**(12), 46–58 (2017). https://doi.org/10.5815/ijisa.2017.12.05
21. Riahi, G.: E-learning systems based on cloud computing: a review. Procedia Comput. Sci. **62**, 352–359 (2015)
22. Siddiqui, A.T., Masud, M.: An e-learning system for quality education. Int. J. Comput. Sci. Issues (IJCSI) **9**(4), 375 (2012)
23. Nielsen, J.: 10 Usability Heuristics for User Interface Design (2020). Accessed from: https://www.nngroup.com/articles/ten-usability-heuristics/

E-commerce Drivers During the Pandemic and Global Digitalization: A Review Study

Solomiia Fedushko⬤, Olena Trevoho⬤, Oksana Hoshovska⬤,
Yuriy Syerov⬤, Natalia Mykhalchyshyn⬤, Denis Skvortsov⬤,
and Liudmyla Fedevych⬤

Abstract This paper investigates the current problems of e-commerce drivers during the pandemic and global digitalization. Digital transformation has become a tool for business diversification and competitiveness, but it is the reason for the whole business paradigm to shift toward digital reality. The several fundamental groups for contributing factor types were systematically structured based on countries and business sectors. The structured research employs the pandemic impact issue, technological innovations' impact on the post-pandemic e-commerce economy, and the relation between infrastructural changes, social networks, and companies' competitiveness. In this research, the authors provided the literature review based on analysis and developed indicative categorization according to the initial research objective stated by the authors. In all databases the articles were taken from, we worked on publications from Business, Computer Science, and Economics. In selecting relevant articles, we used files with criteria to provide appropriate parameters for further analysis. We have paid significant attention to identifying new trends in the literature that are discussed during pandemics. The publications for review were identified from

S. Fedushko (✉) · O. Trevoho · O. Hoshovska · Y. Syerov · N. Mykhalchyshyn · D. Skvortsov ·
L. Fedevych
Lviv Polytechnic National University, Lviv, Ukraine
e-mail: solomiia.s.fedushko@lpnu.ua

O. Trevoho
e-mail: olena.i.trevoho@lpnu.ua

O. Hoshovska
e-mail: oksana.v.hoshovska@lpnu.ua

Y. Syerov
e-mail: yurii.o.sierov@lpnu.ua

N. Mykhalchyshyn
e-mail: nataliia.l.mykhalchyshyn@lpnu.ua

D. Skvortsov
e-mail: denys.i.skvortsov@lpnu.ua

L. Fedevych
e-mail: liudmyla.s.fedevych@lpnu.ua

© The Author(s), under exclusive license to Springer Nature Switzerland AG 2023
N. Kryvinska et al. (eds.), *Developments in Information and Knowledge Management Systems for Business Applications*, Studies in Systems, Decision and Control 462,
https://doi.org/10.1007/978-3-031-25695-0_11

Business, Computer Science, and Economic databases. As the preferred reporting method, we chose meta-analyses (PRISMA) for the systematic literature review. We start by explaining how the relevant journal papers were defined and then explore the limits of our review strategy, present combined infographics and focus our attention on the 73 most relevant papers. An indicative categorization of the articles was developed due to the initial research objective stated by the authors. The analysis of the reports is then presented, highlighting current significant trends in the literature and research identified categories.

Keywords E-commerce · Review · Analysis · Pandemic · Global digitalization · E-commerce drivers · Scientific data base

1 Introduction

Is any company worldwide that overcame the COVID-19 pandemic without rethinking their existing strategies and adapting to the new situation? Digital platforms and technologies are being pulled to the frontline of every business toolset, and their role in business competitiveness continues to grow. Not only has digital transformation become a tool for business diversification and competitiveness, but it is the reason for the whole business paradigm to shift toward digital reality.

A boost in e-commerce resulted from the change in thinking and new business models brought about by the COVID-19 crisis. Despite the loosening of limitations in many states, the considerable increase in consumer e-commerce activity spurred by the COVID-19 epidemic was maintained recently, with online trade growing significantly in value. Online retail sales overgrew over time, with some differences across countries. The countries' and sectors' specifics are particularly interesting for the researchers in this respect. Global consumers continued to buy from websites and online marketplaces; the shift in consumer behaviour patterns was thus a centre of research conducted in this area.

The review is organised as follows. The publications for review were identified from Business, Computer Science and Economic databases. As the preferred reporting method, we chose meta-analyses (PRISMA) for the systematic literature review. We start by explaining how the relevant journal papers were defined and then explore the limits of our review strategy, present combined infographics and focus our attention on the 73 most appropriate papers. Then, an indicative categorisation of the articles was developed due to the initial research objective stated by the authors. The analysis of the reports is then presented, highlighting current significant trends in the literature and research identified categories. Finally, we make some conclusions and references guide applicable for your further study.

2 A Systematic Literature Review of Covid-19 E-commerce Drivers' Impact on Global Digitalization

To study the relevance of e-commerce among Internet users, the authors used statistics from Google Trends for the search queries "e-commerce", "ecommerce" and "electronic commerce" for the last 5 years (period 2018–2022) in all search categories. The result of this search demonstrates that the user interest over time in e-commerce is permanently increasing in the worldwide search on Google.

Statistical data of the Google Trends service for the last 5 years (see Fig. 1) show that the number of "e-commerce" searches on Google increased from 2017 to June 2022.

For 2018–2022, the peak of search queries was May 10–16, 2020. After that, the number of these user requests decreased, but from 2022, their frequency increased.

The Web of Science database was searched for scientific publications on the topic of electronic commerce using the following keyword: ALL (e-commerce OR ecommerce OR electronic AND commerce).

The search results for the following keyword are shown in Figs. 2 and 3. Figure 2 presents statistical data on publication activity on the subject of e-commerce for the period from 1999 to 2022.

34,709 publications were selected from the Web of Science Core Collection by the keyword ALL (e-commerce OR ecommerce OR electronic AND commerce). The most publications on this topic—3,040 works—were published in 2021. As we can see, the number of publications in this field is constantly growing with the development of global digitalization.

The statistics by year of publications about e-commerce in the Web of Science database are shown in Fig. 2.

The Statistics of selected 34,709 publications from the Web of Science Core Collection by Web of Science Categories for the keyword ALL (e-commerce OR ecommerce OR electronic AND commerce) is shown in Fig. 3.

We will use the following authoritative database of scientific publications for Scopus analysis. A search was made for scientific publications using the same keyword as in Web of Science—ALL (e-commerce OR ecommerce OR "electronic commerce"). 266,889 document results were found for this keyword (see Fig. 4); the year range to analyze is from 1930 to 2022.

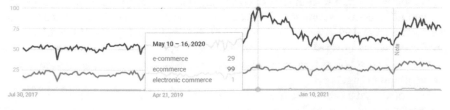

Fig. 1 Interest over time, search query "e-commerce" in Google Trends

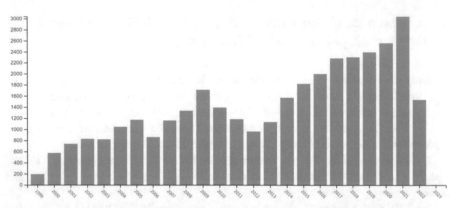

Fig. 2 The Statistics by year of publications about e-commerce in the Web of Science database

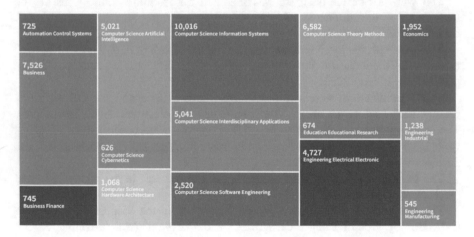

Fig. 3 The Statistics of publications about e-commerce by Web of Science Categories

The first scientific works in the Scopus database, in which the concept of "e-commerce" is mentioned, were published by Baer J. S., Rettig A. S., Cohen I. (1957), Sulayman M., Ur-quhart C., Mendes E., Seidel S. (1970), and Bodi A., Zeleznikow J. (1988).

A total of 231,913 documents for searching query ALL (e-commerce OR ecommerce OR "electronic commerce") have been published in the Scopus database for the period from 1992 to now. It is worth noting that in 2021, 24,721 works were published, which covered research on fundamental problems of e-commerce.

Since 1994, scholars have actively researched this subject area, as demonstrated by the graph in Fig. 5.

We conducted a systematic literature review to investigate the impact of covid-19 e-commerce drivers on global digitization.

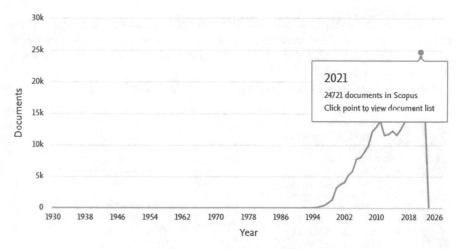

Fig. 4 Statistics by year of results of the searching query in Scopus database ALL (**e-commerce** OR **ecommerce** OR "**electronic commerce**")

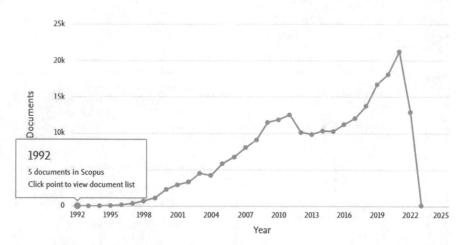

Fig. 5 Statistics by year of results of searching query in Scopus database ALL (e-commerce OR ecommerce OR "electronic commerce") from 1992 till today

The method of preferred reporting items for systematic reviews and meta-analyses (PRISMA) was chosen for the systematic literature review in this subject area.

Scientists use the preferred reporting items for systematic reviews and meta-analyses for a transparent and complete analysis of literary sources according to established criteria (type of publication, time period, language, location, etc.). This analysis is mainly used for a systematic review of the literature, which demonstrates the relevance of the studied subject area, highlighting the statistics of research conducted on this topic.

For a thorough analysis of all qualitative studies that highlight the impact of the covid-19 pandemic and e-commerce and global digitalization published in authoritative sources, the preferred reporting items for systematic reviews and meta-analyses (Fig. 6) were conducted.

PRISMA analysis was carried out according to the following stages:

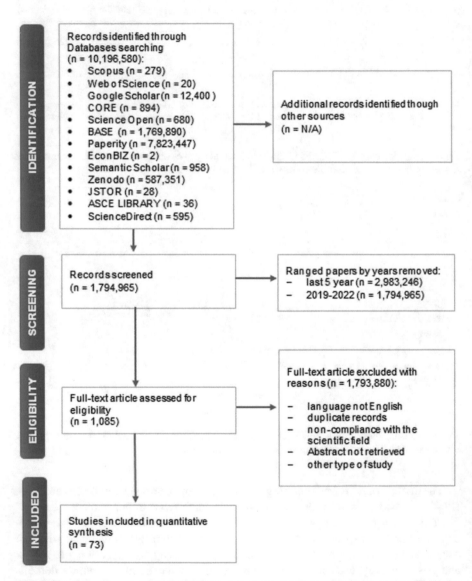

Fig. 6 The flow diagram of Preferred Reporting Items for Systematic Reviews (PRISMA) for systematic reviews, which included searches query "*ALL (e-commerce AND drivers AND covid AND digitalization)*" in 13 scientific database

Table 1 Search results for relevant articles on the query "ALL (e-commerce AND drivers AND covid AND digitalization)" in 13 scientific databases

Scientific database	Number of works
Scopus	n = 279
Web of Science	n = 20
Google Scholar	n = 12,400
CORE	n = 894
ScienceOpen	n = 680
BASE	n = 1,769,890
Paperity	n = 7,823,447
EconBIZ	n = 2
Semantic Scholar	n = 958
Zenodo	n = 587,351
JSTOR	n = 28
ASCE Library	n = 36
ScienceDirect	n = 595

- Identification
- Screening
- Eligibility
- Included.

At the first stage of the Identification of the PRISMA analysis, a search was made using the keyword "ALL (e-commerce AND drivers AND covid AND digitalization)" in 13 authoritative scientific databases:

- Scopus
- Web of Science
- Google Scholar
- CORE
- ScienceOpen
- BASE
- Paperity
- EconBIZ
- Semantic Scholar
- Zenodo
- JSTOR
- ASCE Library
- ScienceDirect.

A total of 10,196,580 publications were found for analysis in these 13 databases, which should be filtered according to specific criteria.

The results of the search for relevant articles with the query "ALL (e-commerce AND drivers AND covid AND digitalization)" in 13 scientific databases are presented in Table 1.

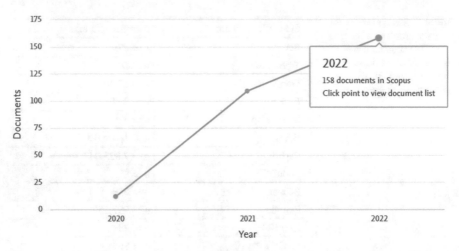

Fig. 7 Statistics of documents by years in the Scopus database for the search query "ALL (e-commerce AND drivers AND covid AND digitalization)"

According to the data of the table of results of the search for relevant articles for the query "ALL (e-commerce AND drivers AND covid AND digitalization)", the following number of articles were published in 13 scientific databases: Web of Science (n = 20), Google Scholar (n = 12,400), CORE (n = 894), Science Open (n = 680), BASE (n = 1,769,890), Paperity (n = 7,823,447), EconBIZ (n = 2), Semantic Scholar (n = 958), Zenodo (n = 587,351), JSTOR (n = 28), ASCE Library (n = 36), and ScienceDirect (n = 595).

In particular, 279 publications for the period 2020–2022 are placed in the Scopus database for this keyword "e-commerce drivers covid digitalization." Only 158 publications for several 7 months in 2022 have already been indexed in the Scopus database (see Fig. 7).

Statistics of documents in the Scopus database by subject area for the searching query "ALL (e-commerce AND drivers AND covid AND digitalization)" are shown in Table 2.

According to Table 2, the most significant number of documents have been published in the following subject areas: Business, Management and Accounting, Social Sciences, Computer Science, Economics, Econometrics and Finance, Energy, Environmental Science, Engineering, Decision Sciences, Psychology, and Mathematics.

In order to select relevant articles, they were filtered according to a set of criteria that provide grounds for excluding and including articles in the analysis parameters. Articles are analyzed according to some predefined categories, among them full-text articles excluded for the following reasons:

– language not English
– duplicate records
– non-compliance with the scientific field

Table 2 Statistics of documents by subject area in the Scopus database for the searching query "ALL (e-commerce AND drivers AND covid AND digitalization)"

Subject area	Documents
Business, Management and Accounting	148
Social Sciences	83
Computer Science	61
Economics, Econometrics and Finance	60
Energy	45
Environmental Science	45
Engineering	44
Decision Sciences	31
Psychology	18
Mathematics	17
Agricultural and Biological Sciences	8
Medicine	8
Arts and Humanities	5
Physics and Astronomy	3
Chemical Engineering	2
Materials Science	2
Chemistry	1
Earth and Planetary Sciences	1
Health Professions	1
Nursing	1

- abstract not retrieved
- other types of study.

The filtration reasons for excluding the articles for a systematic review are shown in Table 3.

According to data of the Table 3, ranged papers by years that we excluded from analysis are defined: last 5 year (excluded n = 2,983,246), 2019–2022 (excluded n = 1,794,965). In all databases, the analysis includes publications only from such areas of research (Fields) as Business, Computer Science and Economics.

In Scopus and Web of Science, only articles in the period 2020–2022 were published for this request; the number of articles did not change when the publication period was selected. In these two databases, the following publication types were selected for article analysis: Research article, Review article, Book, Book chapter, Conference.

In Semantic Scholar we excluded New, Editorial, Case Report, Letter and Comments for analysis, but included Review, Journal Article, Conference and Book In ScienceOpen, also we excluded Poster, Peer Review, Dataset. In services Zenodo, we excluded Image, Datasets, Software, Presentation, Posters, Video, Lessons, Physicalobject, and Other; in section, Publication includes only Article, Conference paper,

Table 3 The filtration reasons for excluding the articles for a systematic review

Scientific Database	Period	Number of relevant works	Number of papers (2019–2022)	Published in English	Open Access	Type—Article	Field
Scopus	2020–2022	279	279	244	108	96	84
Web of Science	2020–2022	20	20	18	12	12	9
Google Scholar	2018–2022	11,600	11,500	11,500	11,300	703	101
CORE	2019–2021	725	725	383	383	383	1
Science Open	2018–2022	672	666	666	271	270	4
BASE	2018–2022	1,389,972	343,566	78,888	64,390	3949	219
Paperity	2018–2022	991,103	849,349	549,349	234,646	483,569	254
EconBIZ	2021	2	2	2	1	1	1
Semantic Scholar	2018–2022	854	839	839	362	151	108
Zenodo	2021–2022	587,360	587,360	587,360	559,573	29,913	203
JSTOR	2020–2022	28	28	28	12	12	1
ASCE Library	2020–2022	36	36	36	36	30	19
Science Direct	2020–2022	595	595	595	168	159	81

and Book. In ASCE Library, we included Chapters/Proceedings Papers and excluded Case Study, Front Matter, Forum, Technical Paper, Technical Note, State of the Art Review. After filtering, we receive for analysis 1,085 articles. Based on the initial review, we included 73 articles in the quantitative synthesis of a systematic review.

In the next step, we conducted a detailed analysis of each of the 73 articles.

3 Discussion

We developed an indicative categorisation of the articles according to the initial research objective stated by the authors for identifying some significant trends in the literature on the changes during pandemics. In cases where the research objectives of the studies were not completely obvious, we used our assessment of the nature of the empirical findings to define them. The categorisation was developed from the initial author's content examination.

We consider that the suggested categorisation can serve as the basis for further studies of the investigated issue, as it provides a framework for a more detailed analysis of the more comprehensive range of literature, considering more broad definitions of the research objectives and a broader range of publication dates.

We provided several fundamental groupings for contributing factor types to give a systematic structure for the research findings. Furthermore, we classified the outcomes based on countries and industries/business sectors.

The COVID-19 pandemic crisis created unprecedented conditions for businesses worldwide, forcing many companies to adjust and replace traditional business models with new resilient ones. As a result, scientists devoted their studies to the impact of digital technologies and the effect of industrial digitalisation in pandemic crises based on countries' and industries' cases. The biggest part of the studies [1, 5, 7–11, 13–15, 20, 23, 24, 28, 32, 33, 35, 37, 39–42, 47, 50, 53, 55–61, 63, 70, 73] is rallied around the issue of the pandemic impact. The performed [5] pilot study's added value entails filling in a gap about the distinctions and overlaps across generational groups' use of mobile shopping during the Covid-19 pandemic. The article [8] investigates factors influencing digital transformation in companies due to the lockdown; the author [32] discusses COVID-19 as the reason for rapid technological development.

The paper [63] focused on theoretical aspects with exploratory cum distinct nature of the web-based business that represents Electronic Commerce-trade, recognising its difficulties and challenges during and after COVID-19.

One of the trends accompanying and gaining from the pandemic is e-commerce, and several researchers—[6, 7]—have investigated this issue. The research [6] looks into the causes of performance differences among international e-commerce SMEs. It demonstrates that marketing organisational agility, as indicated through market-driven and market-driving methods, is critical for maximising the effectiveness of online brand strength. The paper [7] provides insights into how consumers use place of sales/purchase and e-commerce payments methods, choose between local and imported goods, and other spending habits and travel preferences.

Digital transformation (DT) is a new reality that incorporates advancements in companies' strategy, infrastructure, and business processes related to contemporary information technologies [8, 12–14, 17, 18, 20, 23, 24, 29, 32, 33, 35, 36, 38, 44–46, 48, 50, 53, 56, 57, 59, 61, 65, 68]. Thus DT is inevitably influencing the activity of many modern businesses and, what is even more critical from the focus of our review, is closely related to the recent pandemic. The research [45] indicates the level of advancement of digitisation in small and medium service enterprises. The article [59] highlights key factors essential for retail banking success in post-pandemic conditions to manage with urgency and aptness for reassessing priorities and adjusted distribution model where physical and digital are combined. The article [13] investigates the influence of digital orientation and digital competence on digital transformation and the indirect impact of digital transformation on SMEs' revenues and business models. The research [29] enhances knowledge of the digital platforms' role in supplementing, replacing, or reconfiguring physical retail venues. Finally, the study [35] states that frequently used digital platforms should be estimated further for their impact on economic operations and business redemption.

A few researchers [19, 20, 36, 48, 56, 62] discussed technological innovations' impact in the post-pandemic e-commerce economy, mostly figuring out specific cases. In particular, the [56] study proposes a typology of rapid-response COVID-19 innovations from technological to frugal and social innovations. In addition, the research contributes a structural model of innovation activities that policymakers and innovators may implement as a systemic reaction to a crisis. The paper [36] stands at the crossroads of entrepreneurship and information systems studies, and its main contribution is to bring new perspectives on digital entrepreneurship eras from the perspective of history.

A number of studies [17, 20, 26, 37, 42, 52, 58, 66] is dedicated to the relation between internal/infrastructural changes, as well as relation to the social networks, and companies' competitiveness. The paper [58] provides an overview of the changed fashion companies' marketing and communication strategies on Instagram during a crisis according to the needs and feelings of their stakeholders, pointing out human–computer interaction limits and potential. The findings in [37] point to a new communications paradigm that establishes a better framework for thought leaders whose content is more engaging than traditional communication. The article [66] interrogates the speed of politicising the infrastructural space by advanced automation in pandemic conditions and the threat to labour outside of capital's circulations caused by automated infrastructure. The goods distribution methods are analysed in [52] with integrated criterion efficiency—generalized distribution utility, based on which vehicle capacity is substantiated and corresponds to the optimal consumer-driven logistics.

Examining methods, technologies and tools that form a basis for modern competitiveness is in the center of several publications: [1–4, 12, 41, 54, 60, 71, 72]. Technologies are considered the main drivers of many modern organisational changes. Thus, [1] is investigating the role of wireless communication technology in preventing the pandemic through monitoring automation, healthcare, virtual conferencing, and education. It discusses the limitations that wireless technologies present, as well as their possibilities in the post-pandemic era. Web analytics is the core of research in [2–4]. A web-based tool for collecting and comparing e-commerce site performance parameters is presented in [2]. The study [3] describes a method for creating web contact forms that incorporate textual analytics, commonly asked query processing, and a rule-based system. The article [4] presents tools for e-commerce site performance analysis by considering some key parameters (such tools as WebPageTest, PageSpeed Insight and GtMetrix are described as those that can collect the reports on the given parameters). E-commerce sites deal with input data from numerous users, and site responsiveness and stability play a crucial role in terms of user satisfaction. The article [12] shows that blockchain technologies seem to have the ability to alter the underpinnings of e-commerce by enabling trustless trade connections that work without specialised intermediaries or, in the case of permissionless blockchains, even central authority. [25] provides price prediction web tools to let users analyse and forecast product prices, allowing them to plan before making purchases. The Cohort analysis as a new practical method for e-commerce customers' research could answer various business questions and successfully solve real-world problems in e-commerce customer research; the obtained insights and the key e-business aspects from a customer point of view are analysed in [41]. The research [54] advances markerless augmented reality for Indian consumers who buy online furniture provided by mobile applications. Users can try virtual objects in a real environment with an easy-to-use user interface. The study [60] develops a bank clustering model in terms of the digitalisation level based on the corporate social responsibility principles in the face of new crisis threats; after testing 22 banks, results show the developed methodology increases customer loyalty, improves sustainability and makes banks attractive for investment. The paper [71] researches the cloud advantages and disadvantages of supporting the real-time service systems procedure using the Salesforce platform. The article [72] studies machine learning in e-commerce in pandemic conditions employing a daily updated model based on the Gaussian Process Regression approach to predict the daily delivery capacity. The study [26] aims to create a system that can assist organisations select and evaluating different third-party reverse logistics providers (3PRLPs) using a hybrid fuzzy multicriteria decision-making (MCDM) method; relevant economic, environmental, social, and risk-related criteria are incorporated and incorporated into the models.

Consumer behaviour is the focus of several studies conducted. A number of studies investigate [5, 12, 15, 16, 22, 27, 31, 34, 43, 60, 69] changes in the consumer behaviour patterns due to the influence of the pandemic [9, 11, 41]: the study [5] discusses how the pandemic situation may be handled using advanced techniques with the use of mobile phone technology, and [9] asserts that there is a vivid influence of the pandemic on shifts in consumer behaviour from offline to online shopping. Hamade [11] implies that consumers increased their use of e-commerce. At the same time, businesses/stores became increasingly reliant on their internet marketing to operate, mainly when long quarantine periods were relevant. The research [16] investigates consumer shopping/behaviour patterns via interviewing consumers about their online shopping experience. The authors of [references] concentrate on customer preferences within e-commerce in general; [12] demonstrates how blockchain technologies may affect several aspects of e-commerce in the domains of technological, legal and organisational and quality issues as well as consumer issues, while [27] investigates customers' online shopping motives, including the utilitarian, normative and hedonic factors. The authors of [31] examine motivation methods that can influence consumers to engage in more ecologically responsible online shopping behaviour and, as a result, to offer efficient solutions that can be applied in online stores. The aim of [34] is to prove that consumers' decisions on e-commerce platforms are influenced by demographic criteria such as age, gender, place of residence, and monthly income. The article [69] concludes that customer protection is suggested to reduce cyber threats during online shopping.

Organizations' strategies and processes are investigated in [24, 28–30, 36, 38, 40, 51, 68, 70]; part of the studies reviews business activity of SMEs in particular: [6, 9, 13, 45, 57]. Studies of the organisational developments resulting from the lockdown and digital transformation cover such issues as working conditions and perception of remote work. The article [24] underlines the fact that most workers face a negative economic-financial impact of remote work and studies the psychological–behavioural parameters influenced by working online. The paper [30] focuses on defining general conditions for natural and legal persons to meet to be an entrepreneur, as well as describing some particular conditions, such as qualification requirements; suggests the measures to be adopted to avoid the legal hurdles to business start-up. The trends of digital platforms in terms of working conditions are studied in the article [68], highlighting the relationship between platform capitalism and the pandemic complex. The article [70] is devoted to self–employed women and the challenges they have faced in pandemic times and possible initiatives to address those issues.

Sustainable development goals demand profound adjustments in each country, requiring coordinated initiatives from governments, society, and economic entities. [18] illustrates how digital transformation can contribute to ecological, economic, and societal sustainability, and [20] offers a robust, sustainable strategy solution for improving SMEs operations during a crisis in a developing country, fighting Cash Flow shortages and Supply chain disruptions. [21] analyses the correlation between environmental and functional aspects of SMEs and their growth and innovation potential; [28] studies modern economic concepts and their influence on business models. It concludes that companies should embrace the emerging Low-Touch Economy coupled with the digital economy and circular economy to redesign their business models, improve their positions, avoid possible recession, and sustain economic growth. The study [50] improves companies' digital transformation strategies to combat the pandemic consequences. The paper [57] employs a general literature review and insights to provide pandemic digitisation influence for sustainable development of Micro/Small Enterprises. It proposes ways to improve the successful digital transformation of the post-pandemic economy.

Quite a few articles are based on research within particular industries influenced by the pandemic or recent global digitalisation trends. The business sectors in focus include retail [15, 22, 23, 42, 44, 48, 54, 58, 64, 65], in particular fashion retail sector [15, 22, 42, 58], and furniture [54]. Thus, the study [15] exposes the COVID-19 pandemic's problems and prospects for the various distribution approaches in the fashion retail sector, such as shops where retailers showcase their products using a physical storefront, e-commerce, and multichannel sales. The article [22] detects and analyses customer behaviour patterns to manage the risk related to fashion e-commerce. The authors of [23] prove that the new retail industry chain will improve efficiency by incorporating digital technologies and social marketing. The research [42] covers changes in business models of fast fashion retailers in pandemic times and the e-commerce solutions exampled by the development of multi-channel and omnichannel based on Inditex. An extended marketing concept of place attachment to the study of the online retail environment provided by empirical evidence of consumer attachments both with traditional offline and online retail stores is illustrated in [44]. The article [48] investigates the unique characteristics of digitalisation-enabled retail business model innovation. It figured out digital influences on the retail business model innovation dimensions with the future research related to that model.

The researches [12, 33, 47, 53, 59] cover the financial sector including the area of fintech solutions and banking. In particular, [33] investigates the influence of the pandemic outbreak on the demand for digital transformation in fintech and fintech adoption worldwide. The research [47] analyses the key impact on the banking mortgage market in crisis, drawing up directions and digital algorithms for further development. [53] concludes that developing online technologies will be solutions for working with big data, creating personalised offers, managing production, and arising the economy.

The sector of agriculture is investigated in [14, 49, 51] with the focus on the digital grocery sector in the article [14]; the latter highlights critical aspects and technology used in the digital grocery sector and contributes by proposing a model of the digital grocery ecosystem to comprehend the grocery business's digital transformation better. The research [49] is dedicated to mechanisms underpinning open innovation success and sustainability in Italian agrifood businesses built on a single inductive case study method of data from an Italian agritech distributor.

There are some more narrow investigations—case studies within the built environment sector [32]—studying and transformative opportunities of DT and system thinking in the built environment, hotel business [67]—examining social commerce drivers in the hotel industry and their differential effects on social commerce acceptance with the help of a structural equation modelling procedure, and seaports [17]—investigating technological and marketing factors of seaports' strategies and business models advancement.

The service sector is in focus in the articles reviewing the communication sector [37], as well as public services [46] and real-time service systems [71]. The paper [10] analyses how SMEs in the service sector has dealt with the challenges created by the pandemic and intends to learn about which development factors they have prioritised and which technologies they have chosen to respond to changes.

Several articles [19, 20, 38, 46, 57, 62, 63] were constructed as academic literature reviews analysing e-commerce implementation or with elements of theoretical aspects of digital transformation, tools and methods or certain countries' cases and included a resulted evaluation of the materials; some mainly were literature reports. The article [19] discusses the factors influencing the revenues of online shops along the Digital Value Chain. The paper [38] reviews the academic literature analysing potential price differences for the same product depending on whether it is sold in a traditional establishment or through a digital platform. The research [62] classifies applied techniques in their advantages and disadvantages based on the articles review and categorisation types. The article [20] represents a Systematic Literature Review and A Case from an Emerging Economy. A systematic literature review [46] discusses future research directions on digitalisation, accountability and accounting in public services. The analyses consist of the critical digital accountability issues, dialogic and horizontal, multicentric responsibility, translators' roles, and digitalisation's social equity and inclusivity implications.

The topic of e-commerce development services explores by a number of researchers [2–6, 11, 12, 16, 19, 22, 23, 25–27, 31, 34, 38, 40–43, 51, 55, 61–64, 69, 72, 73] covering full advancement in technology, enhancing commercial and boosting economic growth. e-commerce The applied analysis, synthesis, and comparison of legal regulations in consumer protection processing in Romanian e-commerce are discussed in [43]. The research [61] studies systemic interactions with high-quality digital transaction platforms by e-commerce companies in crisis periods.

The research [69] focuses on e-commerce websites and their security vulnerabilities that may affect patronage of e-commerce services and client's trust and satisfaction. Within e-commerce industries, investigation in scientific articles the regional discussion and research appears specific topics like cross-border collaboration. The article [55] discusses the effect of the pandemic on B2B and B2C e-commerce (cross-border) all over the world and possible resilient e-commerce supply chain systems that can withstand any shocks in the future. The study [62] summarises applicable methods of the e-commerce domain due to the General Data Protection Regulation requirements.

Several studies provided research based on countries' cases as the whole or specific industry [38–40, 50, 51, 63, 65, 68, 73] (Table 4).

Table 4 The research based on countries' cases as the whole or specific industry

Country	Analysis
China	The article [39] attracts attention to the future economic problems within the context of the COVID-19 economy, discussing the upward trend in unemployment that changes the post-digital ecosystem in China The Chinese market is under the research [50] of deploying digital technology assets across their supply chains to mitigate the negative impact of COVID-19. The focus is on the breadth and depth of asset deployment dimensions, which reflect the scope and scale of the asset, respectively The proposed research model [63] examined the differences between e-commerce platforms used in Chinese retail, and the article proposes established results
India	The e-commerce business straightforward influenced buyers in India during and after the pandemic providing unprecedented changes. The challenges and opportunities faced by the Indian E-commerce industries in the post-Covid-19 era are explored in [40]
Indonesia	The study [51] is devoted to possible changes in the economic patterns where the digital economy helps to solve the problems of Indonesian agriculture faced in the Covid-19 pandemic through e-commerce
Kazakhstan	The paper [53] shows how Kazakhstan's fintech sector has gained results following technological progress, the growth of e-commerce, socio-demographic changes and banking restrictions
EU countries	The article [73] investigates e-commerce enterprises' activities and behaviour in EU countries using the hierarchical clustering method to identify specific economic and behavioural patterns of e-commerce activity due to analysing several variables
Spain	The paper [38] is estimating the development of e-commerce has nurtured business competition in Spain, reducing mark-ups The case of Uber is discussed in [68] and resulted in new trends and solutions emerging in the political and socio-economic identification
UK and Finland	The paper [65] highlights adapted digitalization in the perceptions and experiences of retail managers by developing existing theory, analyzing the qualitative data and studying opportunities in brick-and-mortar retail business models

4 Conclusion

The literature review provided in this research is based on analysis and developed indicative categorisation according to the initial research objective stated by the authors. In all databases the articles were taken from, we worked on publications from Business, Computer Science and Economics. In selecting relevant articles, we used files with criteria to provide appropriate parameters for further analysis. We have paid significant attention to identifying new trends in the literature that are discussed during pandemics. We have figured out several fundamental groups for contributing factor types, and they were systematically structured based on countries and industries/business sectors. Moreover, the structured research employs the pandemic impact issue, technological innovations' impact on the post-pandemic e-commerce economy, and the relation between internal/infrastructural changes, social networks, and companies' competitiveness. The reference matches make the review informative and easy to use by the researchers.

References

1. Saeed, N., Bader, A., Al-Naffouri, T., Alouini, M.-S.: When Wireless Communication Faces COVID-19: Combating the Pandemic and Saving the Economy (2020)
2. Hossain, M.T., Hassan, R., Amjad, M., Rahman, M.A.: Web performance analysis: an empirical analysis of E-commerce sites in Bangladesh. Int. J. Inf. Eng. Electron. Bus. (IJIEEB) 13(4), 47
3. Molnár, E., Molnár, R., Kryvinska, N., Greguš, M.: Web intelligence in practice. J. Serv. Sci. Res. 6(1), 149–172 (2014). https://doi.org/10.1007/s12927-014-0006-4
4. Amjad, M., Hossain, M.T., Hassan, R., Rahman, M.A.: Web application performance analysis of E-commerce sites in Bangladesh: an empirical study. Int. J. Inf. Eng. Electron. Bus. (IJIEEB) 13(2), 47
5. Wiścicka-Fernando, M.: The use of mobile technologies in online shopping during the Covid-19 pandemic—an empirical study. Procedia Comput. Sci. 192, 3413–3422 (2021). https://doi.org/10.1016/j.procs.2021.09.114
6. Tolstoy, D., Nordman, E.R., Vu, U.: The indirect effect of online marketing capabilities on the international performance of e-commerce SMEs. Int. Bus. Rev. 31(3), 101946 (2022). https://doi.org/10.1016/j.ibusrev.2021.101946
7. Borowski-Beszta, M.: The impact of the COVID-19 pandemic on the behavior of European consumers. In: Proceedings of The 3rd International Conference on Management, Economics and Finance (2021). https://doi.org/10.33422/3rd.icmef.2021.02.137
8. Smajlovic, M.O., Feng, S.: The Impact of the COVID-19 Lockdown on Digital Transformation in German Organizations (2021)
9. Kulsum, N.M., Salim, A., Sjuchro, D., Nugraha, A.R., Prastowo, A.A.: The impact of social media on personal shopper phenomenon in the covid-19 era (2020). https://doi.org/10.17762/PAE.V57I8.733
10. Gregurec, I., Furjan, M.T., Tomičić-Pupek, K.: The Impact of COVID-19 on Sustainable Business Models in SMEs (2021). https://doi.org/10.3390/SU13031098
11. Hamade, L.: The impact of COVID-19 on E-commerce use in lebanon: a quantitative study. In: Resilience and Economic Intelligence Through Digitalization and Big Data Analytics, Warsaw, pp. 88–97 (2021). https://doi.org/10.2478/9788366675704-010

12. Treiblmaier, H., Sillaber, C.: The impact of blockchain on e-commerce: a framework for salient research topics. Electron. Commer. Res. Appl. **48**, 101054 (2021). https://doi.org/10.1016/j.ele rap.2021.101054
13. Rupeika-Apoga, R., Petrovska, K., Bule, L.: The effect of digital orientation and digital capability on digital transformation of SMEs during the COVID-19 pandemic. J. Theor. Appl. Electron. Commer. Res. (2022). https://doi.org/10.3390/jtaer17020035
14. Abbu, H.R., Fleischmann, D., Gopalakrishna, P.: The Digital Transformation of the Grocery Business—Driven by Consumers, Powered by Technology, and Accelerated by the COVID-19 Pandemic (2021). https://doi.org/10.1007/978-3-030-72660-7_32
15. O\u27Connor, K.: The Challenges and Opportunities Created by a Global Pandemic\u27s Effects on Consumer Shopping Behavior Within the Fashion Retail Industry. Accessed 27 June 2022. [Online]. Available: https://core.ac.uk/reader/370425255
16. Bozzi, C., Neves, M., Mont'Alvão, C.: The 'Pandemic Effect' on e-Commerce (2021). https://doi.org/10.1007/978-3-030-85540-6_67
17. Henríquez, R., Martínez de Osés, F.X., Martínez Marín, J.E.: Technological drivers of seaports' business model innovation: an exploratory case study on the port of Barcelona. Res. Transp. Bus. Manag. **43** (2022). https://doi.org/10.1016/j.rtbm.2022.100803
18. Rosário, A.T., Dias, J.C.: Sustainability and the digital transition: a literature review. Sustainability (Switzerland) **14**(7) (2022). https://doi.org/10.3390/su14074072
19. Zumstein, D., Kotowski, W.: Success factors of e-commerce—drivers of the conversion rate and basket value. In: Proceedings of the 18th International Conference on e-Society (ES 2020) (2020). https://doi.org/10.33965/ES2020_202005L006
20. Hossain, M.R., Akhter, F., Sultana, M.M.: SMEs in Covid-19 crisis and combating strategies: a Systematic Literature Review (SLR) and a case from emerging economy. Oper. Res. Perspect. **9**, 100222 (2022)
21. SME Development in the Visegrad Area, springerprofessional.de. https://www.springerprofess ional.de/en/sme-development-in-the-visegrad-area/18809142. Accessed 27 June 2022
22. Sutinen, U.-M., Saarijärvi, H., Yrjölä, M.: Shop at your own risk? consumer activities in fashion e-commerce. Int. J. Consum. Stud. **46**(4), 1299–1318 (2022). https://doi.org/10.1111/ijcs.12759
23. Lv, C.: Research on the impact of COVID-19 epidemic on China's retail E-commerce industry. In: Proceedings of the 2021 6th International Conference on Social Sciences and Economic Development (ICSSED 2021) (2021). https://doi.org/10.2991/ASSEHR.K.210407.060
24. Battisti, E., Alfiero, S., Leonidou, E.: Remote working and digital transformation during the COVID-19 pandemic: economic–financial impacts and psychological drivers for employees. J. Bus. Res. (2022). https://doi.org/10.1016/j.jbusres.2022.06.010
25. Shahrel, M.Z., Mutalib, S., Abdul-Rahman, S.: PriceCop–price monitor and prediction using linear regression and LSVM-ABC methods for E-commerce platform. Int. J. Inf. Eng. Electron. Bus. (IJIEEB) **13**(1), 1
26. Wang, C.-N., Dang, T.-T., Nguyen, N.-A.-T.: Outsourcing reverse logistics for e-commerce retailers: a two-stage fuzzy optimization approach. In: Axioms (2021). https://doi.org/10.3390/axioms10010034
27. Koch, J., Frommeyer, B., Schewe, G.: Online Shopping Motives during the COVID-19 Pandemic—Lessons from the Crisis (2020). https://doi.org/10.3390/su122410247
28. Gigauri, I.: NEW ECONOMIC CONCEPTS SHAPING BUSINESS MODELS IN POST-PANDEMIC ERA. Accessed 27 June 2022. [Online]. Available: https://core.ac.uk/reader/389311735
29. Hardaker, S.: More than infrastructure providers—digital platforms' role and power in retail digitalisation in Germany. Tijdschr. Econ. Soc. Geogr. **113**(3), 310–328 (2022). https://doi.org/10.1111/tesg.12511
30. Srebalová, M., Horvat, M., Vačok, J., Vojtech, F., Filip, S.: Legal obstacles to freedom to conduct a business: experience of the Slovak Republic. Entrep. Sustain. Issues **7**(4), 3385–3394 (2020). https://doi.org/10.9770/jesi.2020.7.4(53)

31. Hollaus, M., Schantl, J.: Incentivizing consumers towards a more sustainable online shopping behavior: a study on nudging strategies in B2C E-Commerce. In: IC4E (2022). https://doi.org/10.1145/3514262.3514268

32. Cheshmehzangi, A.: From transitions to transformation: a brief review of the potential impacts of COVID-19 on boosting digitization, digitalization, and systems thinking in the built environment. J. Build. Constr. Plan. Res. **09**(01), Art. no. 01 (2021). https://doi.org/10.4236/jbcpr.2021.91003

33. Fu, J., Mishra, M.: Fintech in the time of COVID−19: technological adoption during crises. J. Financ. Intermediation **50**, 100945 (2022). https://doi.org/10.1016/j.jfi.2021.100945

34. Sneideriene, A., Beniusis, A.: Factors influencing the decision-making of users of lithuanian e-commerce platforms. Manag. Theory Stud. Rural Bus. Infrastruct. Dev. **44**(1), 72–83 (2022). https://doi.org/10.15544/mts.2022.08

35. Kushadiani, S.K., Nugroho, B., Mardian, S., Muhammad-Bello, B., Hermawan, A.: Evaluating digital economy in the Covid-19 pandemic era: a review. In: SIET (2021). https://doi.org/10.1145/3479645.3479653

36. Kollmann, T., Kleine-Stegemann, L., de Cruppe, K., Then-Bergh, C.: Eras of digital entrepreneurship: connecting the past, present, and future. Bus. Inf. Syst. Eng. **64**(1), 15–31 (2022). https://doi.org/10.1007/s12599-021-00728-6

37. Puntha, P., Jitanugoon, S., Lee, P.-C.: Engagement on social networks during the COVID-19 pandemic: a comparison among healthcare professionals, fitness influencers, and healthy food influencers. In: MISNC (2021). https://doi.org/10.1145/3504006.3504008

38. Lacuesta, A., Roldan, P., Serrano-Puente, D.: Effects of e-commerce on prices and business competition (2020). https://ideas.repec.org/a/bde/journl/y2020i12daan38.html. Accessed 27 June 2022

39. Civelek, M.E.: Effects of COVID-19 on China and the World Economy: Birth Pains of the Post-Digital Ecosystem. Accessed 27 June 27 2022. [Online]. Available: https://core.ac.uk/reader/327696405

40. Srivastava, A., Kumari, G.: E-COMMERCE IN THE POST COVID-19 ERA: CHALLENGES, OPPORTUNITIES AND, pp. 19–26 (2021)

41. Fedushko, S., Ustyianovych, T.: E-Commerce customers behavior research using cohort analysis: a case study of COVID-19. J. Open Innov.: Technol. Mark. Complex. **8**(1) (2022). https://doi.org/10.3390/joitmc8010012

42. Bilińska-Reformat, K., Dewalska-Opitek, A.: E-commerce as the predominant business model of fast fashion retailers in the era of global COVID 19 pandemics. Procedia Comput. Sci. **192**, 2479–2490 (2021). https://doi.org/10.1016/j.procs.2021.09.017

43. Peráček, T.: E-commerce and its limits in the context of the consumer protection:the case of the Slovak Republic. TBJ **12**(1) (2022). https://doi.org/10.24818/TBJ/2022/12/1.03

44. Horáková, J., Uusitalo, O., Munnukka, J., Jokinen, O,: Does the digitalization of retailing disrupt consumers attachment to retail places? J. Retail. Consum. Serv. **67** (2022). https://doi.org/10.1016/j.jretconser.2022.102958

45. Ingaldi, M., Klimecka-Tatar, D.: Digitization of the service provision process-requirements and readiness of the small and medium-sized enterprise sector. Procedia Comput. Sci. **200**, 237–246 (2022). https://doi.org/10.1016/j.procs.2022.01.222

46. Agostino, D., Saliterer, I., Steccolini, I.: Digitalization, accounting and accountability: a literature review and reflections on future research in public services. Financ. Account. Manag. **38**(2), 152–176 (2022). https://doi.org/10.1111/faam.12301

47. Yuzvovich, L., Sharafieva, M., Mokeeva, N., Nasyrova, G.: Digitalization of the Residential mortgage market in crisis conditions: main factors and drivers of development. In: SHS Web of Conferences (2021). https://doi.org/10.1051/SHSCONF/20219302004

48. Mostaghel, R., Oghazi, P., Parida, V., Sohrabpour, V.: Digitalization driven retail business model innovation: evaluation of past and avenues for future research trends. J. Bus. Res. **146**, 134–145 (2022)

49. Rialti, R., Marrucci, A., Zollo, L., Ciappei, C.: Digital technologies, sustainable open innovation and shared value creation: evidence from an Italian agritech business. British Food J. **124**(6), 1838–1856 (2022). https://doi.org/10.1108/BFJ-03-2021-0327

50. Ye, F., Liu, K., Li, L., Lai, K.-H., Zhan, Y., Kumar, A.: Digital supply chain management in the COVID-19 crisis: An asset orchestration perspective. Int. J. Prod. Econ. **245** (2022). https://doi.org/10.1016/j.ijpe.2021.108396

51. Rakhmadi, R., Junaidi.: Digital economy through E-commerce in agriculture in Indonesia. In: Proceedings of the Universitas Lampung International Conference on Social Sciences (ULICoSS 2021) (2022). https://doi.org/10.2991/assehr.k.220102.012

52. Galkin, A., Schlosser, T., Khvesyk, Y., Kuzkin, O., Klapkiv, Y., Balint, G.: Development of generalized distribution utility index in consumer-driven logistics. Energies **15**(3) (2022). https://doi.org/10.3390/en15030872

53. Sambetbayeva, A., et al.: Development and prospects of the fintech industry in the context of COVID-19 (2020). https://doi.org/10.1145/3410352.3410738

54. Raval, R., Sankhla, T., Shah, R., Nadkarni, S.: Developing markerless augmented reality for furniture mobile application. Int. J. Educ. Manag. Eng. (IJEME) **11**(1), 11

55. Kofi Mensah, I., Simon Mwakapesa, D.: Cross-border E-commerce diffusion and usage during the period of the COVID-19 pandemic: a literature review. In: 3rd Africa-Asia Dialogue Network (AADN) International Conference on Advances in Business Management and Electronic Commerce Research, pp. 59–65. New York (2021). https://doi.org/10.1145/3503491.3503500

56. Dahlke, J., Bogner, K., Becker, M., Schlaile, M.P., Pyka, A., Ebersberger, B.: Crisis-driven innovation and fundamental human needs: a typological framework of rapid-response COVID-19 innovations. Technol. Forecast. Soc. Chang. **169**, 120799 (2021). https://doi.org/10.1016/j.techfore.2021.120799

57. Bai, C., Quayson, M., Sarkis, J.: COVID-19 pandemic digitization lessons for sustainable development of micro-and small-enterprises. Sustain. Prod. Consum. **27**, 1989–2001 (2021). https://doi.org/10.1016/j.spc.2021.04.035

58. Noris, A., Cantoni, L.: COVID-19 Outbreak and Fashion Communication Strategies on Instagram: A Content Analysis (2021). https://doi.org/10.1007/978-3-030-78227-6_25

59. Kelecic, P.: Covid-19 crisis: opportunity for banks to reshape service models and foster digital transformation. Accessed 27 June 2022. https://core.ac.uk/reader/339165195

60. Kolodiziev, O., Shcherbak, V., Vzhytynska, K., Chernovol, O., Lozynska, O.: Clustering of banks by the level of digitalization in the context of the COVID-19 pandemic. Banks Bank Syst. **17**(1), 80–93 (2022). https://doi.org/10.21511/bbs.17(1).2022.07

61. Sanda, M.: Client's Quality Assessment of Digital Transaction Platforms Interactivenesses in a Covid-19 E-Commerce Business Environment (2021). https://doi.org/10.1007/978-3-030-79816-1_23

62. Koniew, M.: Classification of the user's intent detection in E-commerce systems–survey and recommendations. Int. J. Inf. Eng. Electron. Bus. (IJIEEB) **12**(6), 1

63. Patil, A.: Challenges and opportunities of e-commerce in India during covid-19. **6**(2), 8 (2021)

64. Li, S., Liu, Y., Su, J., Luo, X., Yang, X.: Can e-commerce platforms build the resilience of brick-and-mortar businesses to the COVID-19 shock? An empirical analysis in the Chinese retail industry. Electron. Commer. Res. (2022). https://doi.org/10.1007/s10660-022-09563-7

65. Donnelly, A.: Business Model Opportunities in Brick and Mortar Retailing Through Digitalization. Accessed 27 June 2022. [Online]. Available: https://core.ac.uk/reader/386287943

66. Lin, W.: Automated infrastructure: COVID-19 and the shifting geographies of supply chain capitalism. Prog. Hum. Geogr. **46**(2), 463–483 (2022). https://doi.org/10.1177/03091325211038718

67. Makudza, F., Sandada, M., Madzikanda, D.D: Augmenting social commerce acceptance through an all-inclusive approach to social commerce drivers. Evidence from the hotel industry. Malays. E Commer. J. **2** (2021). https://doi.org/10.26480/mecj.02.2021.55.63

68. Allegretti, G., Holz, S., Rodrigues, N.V.: At a crossroads: uber and the ambiguities of the COVID-19 emergency in Lisbon. Work. Organ., Labour Glob. (2021). https://doi.org/10.13169/workorgalaboglob.15.1.0085

69. Baako, I., Umar, S.: An integrated vulnerability assessment of electronic commerce websites. Int. J. Inf. Eng. Electron. Bus. (IJIEEB) 12(5), 24

70. Cheong, J.Q., Fadzlee, N. F. C. L. M., Mansur, K. H. M.: A systematic literature review of covid-19 impact to sme's adoption of e-commerce. J. BIMP-EAGA Reg. Dev. **6**(1), Art. no. 1 (2020)
71. Poniszewska-Maranda, A., Matusiak, R., Kryvinska, N., Yasar, A.-U.-H.: A real-time service system in the cloud. J. Ambient Intell. Human Comput. **11**(3), 961–977 (2020). https://doi.org/10.1007/s12652-019-01203-7
72. Bayram, B., Ülkü, B., Aydın, G., Akhavan-Tabatabaei, R., Bozkaya, B.: A Machine Learning Approach to Daily Capacity Planning in E-Commerce Logistics (2021). https://doi.org/10.1007/978-3-030-95470-3_4
73. Scutariu, A.-L., Susu, S., Huidumac-Petrescu, C.-E., Gogonea, R.-M.: A cluster analysis concerning the behavior of enterprises with e-commerce activity in the context of the COVID-19 pandemic. J. Theor. Appl. Electron. Commer. Res. 17(1), 47–68 (2022). https://doi.org/10.3390/jtaer17010003

Remanufacturing and Refurbishment of Electronic Devices—Their Future from a Business Perspective

Ann-Sophie Schweiger and Christine Strauss

Abstract Remanufacturing and refurbishment of electronic devices can be profitable business strategies for companies and a way to counteract the rapidly growing waste stream of electronic scrap. The industrial remanufacturing process provides a used device with a second life and a quality similar to a new device. However, companies and organizations that offer such devices still encounter difficulties. While earlier research has already partly covered the field of remanufacturing, it still lacks real-word examples and practical solutions to existing challenges. The identification and discussion of solutions for possible obstacles are thus a focus of this work. Furthermore, emerging trends in this industry as well as digitalisation and online trade with remanufactured devices are discussed. Attention is also paid to future developments in the field of remanufacturing, specifically to topics that could dominate this industry. The method applied to collect indicators for these various focal points was a Delphi study, which was conducted with experts in the field of remanufacturing. This study included a total of three survey rounds. It is noteworthy that the experts emphasized the importance of networking and various forms of cooperation in the reuse sector. Furthermore, the significance of remanufacturers' transparency, especially concerning the remanufacturing process as well as the condition and assessment of the offered devices, was emphasized.

1 Introduction

The remanufacture of electronic appliances is an industrial process that brings a used device through several steps to a condition similar to a new device [1, 2]. It is not only a possible way to mitigate the increasing amount of e-waste (electronic waste). For companies, remanufacturing can be described as a profitable [3] and promising endeavour [2]. From a business point of view, lots of valuable resources get lost when electronic devices are disposed of [4]. Gurita, Fröhling, and Bongaerts

A.-S. Schweiger · C. Strauss (✉)
University of Vienna, Oskar Morgenstern Platz 1, 1090 Vienna, Austria
e-mail: christine.strauss@univie.ac.at

© The Author(s), under exclusive license to Springer Nature Switzerland AG 2023   229
N. Kryvinska et al. (eds.), *Developments in Information and Knowledge Management Systems for Business Applications*, Studies in Systems, Decision and Control 462,
https://doi.org/10.1007/978-3-031-25695-0_12

[5] estimated the current market value of precious as well as critical metals in sold cell phones and smartphones in the German market of over € 600 million. Their model only included phones that were sold between 2004 and 2015, the potential of other devices and other time periods was not yet considered [5]. With an increasing shortage and a subsequent rise in the prices of some required critical materials or resources, remanufacturing can offer companies a crucial cost advantage [6].

Remanufacturing attracts the interest of researchers from various fields such as engineering, industrial design, and marketing. In addition, policy makers are increasingly acknowledging the potential of this industrial process especially in the context of circular economy efforts. Governments including the European Commission established laws and directives regulating the end of life, disposal, or recycling of products [6]. The European Commission communicated its ambitious new plan called the "European Green Deal" which should help amongst other goals to reach the sustainable development goals of the United Nations for the year 2030 by achieving a more sustainable economy [7, 8]. Within this context, a new Circular Economy Action Plan [9] targets topics such as transparency for consumers, sustainability and "enabling remanufacturing and high-quality recycling" [9]. Electronics as well as ICT (Information and Communication Technology) were identified as especially critical product groups [9]. This is an important aspect as government regulations are considered a main driver for remanufacturing activities in scientific literature [10].

A variety of papers was thus dedicated to remanufacturing as a business strategy. Many aspects have already been investigated, but other topics have not been sufficiently covered yet. Prior studies deal with aspects reaching from the product acquisition for remanufacturing [3] over the pricing of remanufactured products [11, 12]; to the consumer acceptance of remanufactured smartphones [2]. But even though the existing literature already covers a wide variety of different refurbishment topics, there is still a need for more in-depth research in this area. The business perspective is often missing i.e., theoretical statements are developed but they often lack empirical evidence. This study will therefore focus on a business point of view as well as practical recommendations for remanufacturers.

This work is composed as follows: First, relevant background information is given, and the context is presented. Subsequently the chosen method, i.e., the Delphi method, is explained in Sect. 3. After this the results of the empirical study are discussed and analysed in detail. Then the limitations and contribution of this work are pointed out, and finally a conclusion is drawn.

2 Background

Our society is facing a rapidly growing amount of e-waste. According to current data, almost 54 million metric tons of e-waste worldwide were counted in 2019 [13]. Globally, this was more than 7 kg of e-waste per person, less than 20% of which was collected and recycled. Although the number of e-waste collected and recycled is increasing, the problem is far from being halted on a global scale. During 2019, Asia recorded the largest amount of e-waste with nearly 25 million metric tons in total. Per capita, however, Europe is at the top of the statistics with more than 16 kg per person. Projections suggest a total of over 74 million metric tons of e-waste globally by 2030. The vast majority of e-waste generated is not appropriately collected and handled. Thus, it has a far-reaching negative impact on people and the environment caused by its harmful materials. Part of this problem concerning the handling of e-waste is the export of e-waste to developing countries. Electronic appliances are only classified as e-waste when they are disposed by their owners, but not when they are stored for example at home. This definition, therefore, includes almost any discarded everyday appliances such as smartphones, laptops, monitors, toners, lamps, refrigerators, etc. [13].

One reason for this growing waste stream is that consumers purchase more and more of these appliances today [13]. A further issue can be identified with regard to the average useful life. Many of these devices have a very limited lifetime. The average lifetime of a smartphone is estimated at less than 2, 5 years [4]. In Austria, a smartphone is currently used between 18 and 24 months [14]. Depending on the study, first-time use periods between four and six years are assumed for notebooks [15]. This indicates a clear trend. In addition, there is often a lack of options to repair the equipment [13]. The sharp increase of this waste stream has already been observed for decades and scientists have long been calling for a move towards more circular systems that would decrease the amount of scrap generated [16]. Especially, since a higher collection rate of e-waste alone does not mean that the total amount of e-waste is decreased [17].

In this context, the concept of a circular economy model has received more attention lately. Geissdoerfer, Savaget, Bocken, and Hultink [18] define circular economy "[...] as a regenerative system in which resource input and waste, emission, and energy leakage are minimized by slowing, closing, and narrowing material and energy loops. This can be achieved through long-lasting design, maintenance, repair, reuse, remanufacturing, refurbishing, and recycling". As part of the circular economy concept, research on refurbishment and remanufacturing is gaining momentum [19, 20].

The overall process starts with the acquisition of used devices that serve as cores [21]. Then, the used device is restored. The single steps of a simplified remanufacturing process can be found in Fig. 1 below. At this point it should be noted that remanufacturing means the product is restored into a "like-new condition" [2], but it cannot be sold as a new device [22], whereas refurbishment means the products are in a "satisfactory" [2] state when offered. However, in practice and in scientific literature few authors distinguish between these terms [2]. Following this approach, the expressions will be used synonymously in this text. After the remanufacturing or refurbishing, the renewed devices are marketed and sold to new customers. Each of these phases brings its own challenges, some of which are addressed in the subsequent sections [21].

As indicated above remanufacturing can be seen as a kind of product reuse [23] or more precisely as a kind of product recovery [6]. Gallo et al. [6] distinguish between

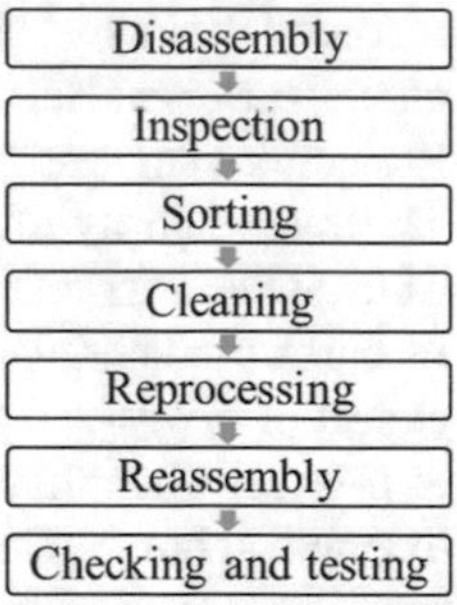

Fig. 1 Simplified
remanufacturing process
(adapted from [6])

Disassembly
↓
Inspection
↓
Sorting
↓
Cleaning
↓
Reprocessing
↓
Reassembly
↓
Checking and testing

different forms of product recovery such as restoration, refurbishing, remanufac-turing, and cannibalization. The latter can be defined as only certain still usable components of a used product are taken. Additionally, it is possible that used products are directly reused without modifications. Recycling products "allows to recover only raw materials but not the added value of production cycle" [6]. Therefore, Guide and Wassenhove [3] classify recycling as "material recovery", whereas the other forms of product recovery are seen as "value-added recovery".

The focus of this work will be on the remanufacturing of electronics, especially consumer electronics. This sector seems particularly suitable, as remanufacturing electronics allows to retain a high value compared to the initial price or the market value [6]. Additionally, researchers predict high growth in this sector [3]. Yet, reman-ufacturing can be applied in many industries. It is already established in the auto-motive industry, the arms industry, the space sector, medical appliances, and many other areas worldwide [3, 24, 25].

Remanufacturing in general offers multiple benefits for society, the environment, consumers, and producers. Firstly, it can be argued that remanufacturing creates job opportunities for skilled workers because it is currently more work-intensive than manufacturing. This might be especially appealing for developing and emerging countries. Society can further benefit from the reduction of e-waste, the energy savings, and the conservation of raw materials [25, 26]. In a similar way, the envi-ronments profits when less devices that possibly contain hazardous materials end up in landfills [25]. When less new devices are needed, the resulting energy savings go hand in hand with lower carbon dioxide emissions. However, a major environmental advantage of remanufacturing is the preservation of raw materials [4], especially non-renewable resources [6]. All these points allow remanufacturing to be under-stood as a contribution to environmental protection. Moreover, another party that possibly benefits from remanufactured devices is the party of consumers. They can profit from less expensive products compared to new products [4] in better conditions than normal second-hand products [25]. In the European Union these products also come with a one-year warranty [27].

For companies remanufacturing devices and their sale offers further advantages, yet it has its pitfalls, too. At this point, it can be distinguished between manufac-turers who usually produce new devices and actors such as independent or contract remanufacturers whose core business is remanufacturing. Researchers assume that Original Equipment Manufacturers (OEMs) are in a superior position compared to independent remanufacturers, but they need to overcome other challenges [21]. When OEMs do not engage in remanufacturing activities, this leaves room for independent remanufacturers to enter the market [6]. Manufacturers often hesitate because they are afraid that offering remanufactured products could decrease their new product sales [23]. Selling new products usually allows higher profit margins compared to remanufactured devices [21]. These concerns about possible cannibalization effects when both new and remanufactured products are offered were already addressed in prior research [2, 28, 29]. The effects are not always clear. Gallo et al. [6] state that "the company can profit from new products on the primary market and from remanufactured ones on the secondary one, and if these markets are separated, there is no risk of demand substitution". Cannibalizing one's product sales of new products may even help to prevent the loss of market share [6]. Atasu et al. [30] conclude that in the segment of green consumers cannibalization can be found. Accordingly, these green consumers attribute an equal value to new and remanufactured devices. Thus, the remanufactured option is chosen when the product price is lower. Guide and Li [29] point out that cannibalization is rather an issue in the Business-to-Business (B2B) area and barely concerned consumer products in their auction experiments.

In particular for OEMs remanufacturing offers some not immediately recognizable benefits. When manufacturers take back their used products, they can collect data on different types of defects of the devices [6]. This could be used to improve the products and to enhance customer satisfaction if this is desired by the manufacturer. Furthermore, the manufacturers have a higher level of control along the value chain when remanufacturing their own products. When an independent remanufacturer, for example, refurbishes the devices of an OEM, consumers still associate these devices with the OEM due to the branding of the products [6]. Furthermore, some researchers argue that manufacturers may demonstrate corporate citizenship by remanufacturing their used devices.

In addition, both OEMs and remanufacturers can profit from saving energy and raw materials [25, 31]. The energy needed to remanufacture a product is estimated at about 60% of the energy required to manufacture it. The effects of saving materials in this context are even greater than the energy savings [26]. This may turn out to be a competitive advantage for businesses when prices for resources such as raw materials or energy rise due to increasing shortages. However, it only holds true under the assumption that raw materials are a major cost driver of the product [6]. From a business perspective the main argument for remanufacturing might still be that it can be profitable for companies [3, 6].

In this work, several different topics are covered. Firstly, suitable solutions to current barriers that remanufacturers encounter when offering remanufactured devices on the market are discussed. Beside potential barriers, upcoming trends and hot topics are assessed. Moreover, an outlook into remanufacturing in the year 2026 is taken. The growth perspective and market forecasts are of particular interest for this outlook. Although it is impossible to predict the future, specific emerging trends in the remanufacturing industry can be identified. It is no less important to address suitable adaptation strategies for remanufacturers for possible industry developments. In addition, one focus is on the supporting and driving role of digitalisation, especially the e-commerce of remanufactured goods.

3   Delphi Study on Remanufacturing

Even though many different refurbishment topics were already investigated, there is still a need for more research efforts in this field. The business point of view is often underrepresented. This work will therefore focus on a business perspective and practical recommendations for remanufacturers.

Due to the explorative character of the topic the Delphi method is particularly suitable for the empirical study. This group decision tool encompasses multiple survey rounds where industry experts answer relevant questions. One characteristic is the feedback mechanism by which the experts get insights into the other experts' assessments as of the second survey round. Moreover, this method was chosen due to its flexibility and its suitability as a forecasting tool [32].

Twelve to 15 experts participated in each round of the Delphi survey. An expert on this panel had at least three years of professional experience in the field of remanufacturing and brought relevant technical knowledge as well as marketing and sales know-how. In addition, many of the experts worked in a managerial position. The experts were associated with organizations based exclusively in Austria and Germany. Overall, the panel consisted of, among others, remanufacturing firms, associations representing reuse companies, wholesalers, and an online marketplace.

The Delphi study included three rounds in total and was conducted online from March to September 2021. The study delivered quantitative and qualitative results. Quantitative results were achieved through rating questions. As part of the feedback mechanism mentioned above, previously reported statements were ranked or rated by all experts from the second round onwards. These statements were mainly rated on a Likert scale [33]. Furthermore, it was always possible to choose between an English and a German version of the questionnaires.

In round one, answers were received from 13 experts from twelve different companies. This questionnaire included mostly open questions to capture the diversity of opinions. In round two, the questionnaire was filled out by 15 experts from 13 different companies. In the last round, twelve questionnaires were received and thus twelve experts from ten different organizations participated. Table 1 summarizes the overall results categorized in three main areas of interest. All results will be discussed in greater detail in the following sections.

Table 1 Summary of key results of the Delphi study

Topic		Results
Current barriers to remanufacturing	Consumer awareness and knowledge about remanufactured alternatives	There is significant room for improvement. The importance of cooperation in the reuse sector and with other players e.g., mobile operators was underlined. Moreover, single lighthouse projects serve as positive examples and encourage imitation in the industry
		In general, vendors should approach consumers in different ways and via different channels online and offline
		In addition, consumer knowledge can be enhanced by communicating the peculiarities of remanufacturing processes
	Consumer perception of remanufactured devices	Remanufacturers can set the right framework to evoke more positive associations with their devices: Offering remanufactured devices of a high quality, the service quality, the company reputation, extend warranties, transparency with regard to the remanufacturing process, and the grading of the devices and their condition. They can further invest in advertising and constantly emphasize the benefits of remanufactured devices. The overall goal is to gain the trust of consumers
Current barriers to remanufacturing	Consumers' willingness to pay	This aspect was assessed less challenging than expected. Good communication from the remanufacturer or vendor to the consumers might suffice to counteract many of the consumers' uncertainties and reservations

(continued)

Table 1 (continued)

Topic	Results
Quality and design of electronic products	To further support the remanufacture of these devices the experts considered the measures listed below among others as the most suitable ones. Further, more governmental support for these demands is expected to be achieved by gaining more influence, e.g., with the help of lobbying and advocacy groups (1) An enhance level of repairability of electronic devices (2) Access to transparent product information and repair instructions for remanufacturers (3) An increased level of durability of electronic devices
Technical advancement and obsolescence	At this point no consensus was reached. However, there are many casual users that do not need the latest models to satisfy their needs. These consumers may form the major target group for remanufactured devices

(continued)

Table 1 (continued)

Topic		Results
Trends and developments in the remanufacturing industry	Green marketing	Environmental benefits are already communicated by remanufacturers. However, the experts supported more marketing and educational efforts to highlight the environmental benefits of remanufactured devices Transparent and simplified information on the resources saved with the purchase of a remanufactured device instead of a new one should be provided. This tackles the issue when the environmental problems related to the manufacturing of electronics and the disposal of the devices are not recognized by consumers. In addition, remanufacturers may communicate the positive environmental or social impact of their organization to consumers in order to highlight the sustainability mindset
	Quality marks	A consensus was not reached. Many experts criticized the possible costs for remanufacturers to obtain a quality certificate and the difficulties when implementing such certifications. A quality certification issued by an independent (third) party received the highest approval of the experts. The legal warranty would suffice according to many experts

(continued)

Table 1 (continued)

Topic	Results
Leasing models	The experts had usually a positive attitude towards leasing models for remanufactured devices. However, they raised concerns if these models would be attractive from a business perspective due to the lower prices of remanufactured devices. The consumers' attitude towards these models might also be challenging. Again, many experts supported the idea that the legal warranty gives customers enough security. In that case, leasing models were proposed to give consumers more security
Industry developments	It is possible that remanufacturing of electronic devices will gain importance and market share. However, it could take longer than five years to reach this state. According to the experts, the focus of remanufacturing should be on sustainability. Market growth may be accompanied by more standardization and increasing regulation. Further, the market might become more attractive for manufacturers. The latter can have negative effects on the business activities of independent remanufacturers, such as more competition for suitable cores. Moreover, customers might prefer to buy the devices directly from the manufacturers and hereby squeeze out independent companies. In such a situation, remanufacturers can look for market niches or seek cooperation with manufacturers

(continued)

Table 1 (continued)

Topic		Results
Digitalisation	Digitalisation	Many experts would like to focus more on digital support in the future. There is potential in areas such as online marketing and e-commerce
	e-commerce	The online trade of electronic devices is booming. Remanufacturers can take advantage of this and offer their devices online. It does not have to be their own store other platforms are also suitable. The devices can thus be offered to a larger audience

4 Barriers to Remanufacturing

Obstacles and barriers to remanufacturing companies and companies that offer remanufactured electronics were a central part of the Delphi study. This section first focuses on the issues related to consumer behaviour, namely the awareness and the perception of remanufactured products and the consumers' willingness to pay for these goods. Subsequently, the barriers for remanufacturing companies resulting from the quality, functionality and design of new products are discussed. As these points can rather not be controlled by remanufacturers, a look will be taken at government support and regulation. Lastly, the experts' assessment of the impact of the technological progress on remanufacturing is introduced. These are all considered external factors, but especially the consumer-related barriers can be influenced to a certain extent through targeted marketing.

4.1 Consumer Awareness

Studies showed, that consumers were often not aware of refurbished alternatives to new devices and did not know what exactly refurbishing means [2]. These essential questions were therefore to be discussed in the Delphi study, and possible strategies to address them were identified over the course of three rounds. Initially, nine out of 13 experts saw "room for improvement" regarding the consumer awareness and their understanding of a refurbishment process. A minority of three experts stated that potential customers are aware of remanufactured products and that they would know what remanufacturing stands for. Most of the experts surveyed thus partly confirm the scientists' findings by stating that there is a need for improvement. However, the majority did not agree with the thesis that customers would know too little about the availability of refurbished products or remanufacturing processes. Nevertheless, a need for action can be derived from the evaluations received. Van Weelden [2] therefore stated that "attracting consumers starts with building awareness". In general, companies can advertise to raise awareness for their products.

In the subsequent rounds, more creative ideas besides advertising were gathered and assessed. An often-mentioned idea was connecting remanufacturing companies. Ten out of twelve experts supported the idea of networking in the reuse sector on a supra-regional or local level. According to eleven out of twelve experts these networks should organize fairs or events together. The experts unanimously agreed that networks of remanufacturers should organize awareness campaigns together and seek common communication channels in (specialized) media. Working together in certain areas allows organizations to bundle resources such as expertise, personnel, online or offline reach or financial resources to achieve greater common goals. A positive example of cooperation named by one of the experts is the cooperation between reuse companies and a Berlin shopping centre called Re-Use-Superstore, where a variety of second-hand products are easily accessible for consumers [34].

This even counteracts the problem that consumers stated in studies that they did not know where to find refurbished equipment or that they would not want to visit separate stores [2, 35]. The high level of approval for various collaborations is one aspect that particularly stands out. To the best of the authors knowledge, cooperation between refurbishers was not specifically targeted in earlier studies. However, individual comments can be found. It is stated that there are many different forms of cooperation, independent refurbishers could work together in some areas and compete in others [6]. In practice there are some initiatives in Austria and Germany that focus on connecting reuse and repair organizations [36, 37].

Further, all experts considered transparency as a key to improve the consumers' understanding and knowledge of a refurbishment process. This means e.g., communicating what steps their remanufacturing process encompasses. In the next step, this transparency could possibly help to minimize the ambiguity that consumers associate with refurbished products [1] and to reduce prejudices. Overall, investing in public relations was endorsed by all participants. Public relation efforts are according to the panel a reasonable way to convey a positive image of refurbished devices and raise awareness for negative consumption habits and related environmental issues. Ten out of eleven experts agreed that in general, events such as talks or showing documentary films can help to reach and inform consumers. Also, projects such as repair cafes should be supported by refurbishers. These events enable direct contact and exchange between organizations and interested consumers. Using online channels to inform consumers was endorsed by eleven out of twelve experts. The online activities must not be limited to promote one's products. Objective content that informs users on relevant topics, possibly in an entertaining way, may be well received by potential consumers. Remanufacturers can provide assistance for searching or undecisive users by offering for example test results, testimonials or product comparisons. In addition, films and reports in television were considered as appropriate means by ten out of twelve experts.

Again, ten out of twelve experts agreed that remanufacturers should organize campaigns together with mobile operators and refurbished devices. Some companies already offer refurbished smartphones with contracts and promote this as resource- and emission-saving [38]. In view of the experts' assessments, this can probably be regarded as a sensible development, but it is unclear what proportion of the total volume of equipment placed on the market by this company is accounted for by refurbished equipment. Additionally, other contracts by mobile providers that promise new equipment at short intervals invite customers to engage in unsustainable consumer behavior.

Moreover, using print media to inform consumers was confirmed by nine out of twelve experts. In general advertising positively influences the sales of a company. The results of a study showed, however, that online advertisement of an electronic company led to over 20% more sales, whereas advertising in printed media only increased sales by 3.7%. The study only investigated the role of Facebook as an online medium. The authors assumed that great potential also lies in other popular online media such as Instagram, YouTube and Twitter [39]. Other tools to promote one's products are endorsements by influential people. Ten out of twelve experts supported the idea that companies should aim at high-profile customers who could serve as role models. This does not necessarily mean that celebrities or influencers should be chosen, but maybe companies or public institutions.

Additionally, sponsoring events was only endorsed by eight out of twelve experts and thus received the lowest support. On the one hand, when a company decides to sponsor events, positive effects for its brands and corporate image are expected. Sponsorships are popular in the areas of sports, cultural life, or charity. Nowadays a company's sponsorship strategy usually encompasses multiple sponsorships, also referred to as a portfolio strategy [40]. Researchers underlined the importance of "the fit with the company" [41] with regard to the sponsored artist and the event organizer. Moreover, when pursing a portfolio strategy companies should engage in different sponsorship categories to prevent the negative effects of being overrepresented in one category [41]. On the other hand, sponsoring multiple events might not be desirable for smaller or medium sized companies. Besides, a connection between refurbished electronics and traditional sports is difficult to establish. There are however many opportunities in the field of social or environmental sponsorship. Moreover, e-sports (electronic sports) are characterized by increasing popularity and attract big players such as Vodafone or Adidas [42, 43].

Many popular promotional tools were named in the course of the study, especially different facets of classic advertising. The vast majority of experts endorsed networking and cooperation in the reuse sector or with other players in the market such as mobile operators. This might not be surprising as amongst the experts some already were part of such networks, nevertheless the idea of networking and cooperation was also considered positive by other experts. The introduced Re-Use Superstore is a good way to draw attention to remanufactured products and at the same time it can serve as an example project for other cities. Further, the experts highlighted the transparency with regard to their remanufacturing process and the grading of their remanufactured devices. Companies can, for example, allow better insights by providing a look behind the scenes or other additional information material.

4.2   Consumer Perception

Consumer awareness is only one aspect of the consumer-related issues. A decade ago, Gallo et al. [6] described the consumer's attitude towards refurbished products as "the most difficult obstacle in this business". The majority of experts stated that in their experience customers perceive refurbished devices as "rather positive" or "positive". However, this seems not to hold true for consumers in general. In the second round, 14 out of 15 experts agreed that consumers' associations with refurbished devices have to change. One expert referred to used cars as an example where it is very common to purchase previously used products. In Austria, for example, 1,096,177 s-hand vehicles have been registered over 371,252 new motor vehicles in the year 2021 [44, 45]. The same year in Germany, 3.2 million new motor vehicles were registered, and 7.79 million motor vehicles changed ownership [46, 47]. This shows that in some cases consumers prefer the second-hand option over the new alternative.

The next step was to discuss how organizations could evoke positive associations with their refurbished devices. The most frequently mentioned proposals can be categorized into marketing and legal measures. Many experts stated that the quality and the economic and ecological benefits of refurbished devices should be not only communicated but emphasized. It may help to attach more importance to the wording and the terms used to describe these second-hand products. One expert pointed out that, on the one hand, the term "refurbished" does not have a negative connotation in German, and companies could put more effort in giving it a positive meaning. On the other hand, the term might be inconvenient as it is an English word, and it has not yet become customary in the German language. In this context, eleven out of 15 experts supported the call for a third separate category besides "used" and "new". Terms such as "pre-loved" device or "second-life product" were suggested. Over the past years, organizations and researchers promoted multiple different names such as "certified pre-owned" or "reconditioned" that all refer to remanufacture or refurbishment of products [23].

Additionally, a strong company image and good reputation were considered crucial factors to evoke positive associations with the products offered. Many experts saw transparency as an important factor. More precisely, transparency regarding the refurbishing process. This seems particularly reasonable as consumers often misinterpret remanufacturing or refurbishing processes [2]. Besides this, some of the experts named the warranty of refurbished devices as a signal for quality and reliability. According to these experts the warranty could either be longer or more comprehensive than legally required or similar to the warranty of new devices. Earlier research underlined the importance of warranty on the perceived risk of consumers [2]. In the European Union these second-hand devices come already with a one-year warranty [27].

As 14 out of 15 experts confirmed that it is necessary to earn the consumers' trust to make remanufacturing mainstream, trust building measures were evaluated. Again, transparency was seen as a critical point by some of the experts. Consumers should have the opportunity to gain insights in the refurbishment process and organizations should clearly communicate the condition of the refurbished device. The above-mentioned prolonged warranties were also in this context considered a trust-building measure. According to the experts this holds true for a good company reputation and high-quality products and services. These criteria such as reputation and service quality were found to be important factors in the consumer decision-making process. They allow companies to mitigate the "product performance-related risks" [2]. Another simple action at aims at transparency would be showing reviews of customers on the online product pages. Further, the researchers underlined the importance of a bond between the customer and the company. On the basis of this relationship customers could serve as "ambassadors" and the company would benefit when customers called attention to refurbished products and recommended them to others [2]. In the study one expert acknowledged this so-called word of mouth in this context.

4.3   Consumers' Willingness to Pay

The lower perceived quality and functionality of refurbished products compared to new products can lead to a lower willingness to pay (WTP) of consumers [6, 48]. Various studies focused on this topic and thus additional reasons for a lower willingness to pay have been identified such as poor seller reputation [49] or disgust due to previous use [23]. [1] investigated the relationship between ambiguity tolerance and the WTP. Their findings suggested that people who have a low level of ambiguity tolerance are only willing to pay lower prices for remanufactured products [1]. In addition, other researchers indicated that the market for remanufactured products is limited by people who only considered purchasing refurbished products when heavily discounted [48]. Thus, vendors encounter difficulties if the remanufacturing process is costly, and consumers are not willing to pay reasonable prices for remanufactured products [50]. Other ways to enhance the consumers' WTP that were introduced are eco-certifications [48] or quality certifications [2], they are discussed below. Overall, the low WTP is in scientific literature still seen as an issue that needs to be tackled. It can be assumed that this question is highly relevant for companies that sell remanufactured devices, therefore this topic was included in the study.

Later, nine out of eleven experts agreed that good communication is key to increase the willingness to pay of consumers. This is consistent with the assumptions of Gallo et al. [6]. Some experts noted that offering products of good quality must be considered a prerequisite. 14 out of 15 experts confirmed that vendors should emphasize the environmental benefits and the high quality of refurbished products to enhance the consumers' WTP. The idea to provide transparent information on the classification and the condition of the refurbished equipment was supported by all but one expert. This also holds true for the proposal to underline the price difference of the refurbished product to a comparable new device, not the cheapest new device. Eight out of 15 experts doubted that the legal warranty alone would be a convincing argument for potential customers. Lund [26] summarized that the warranty is usually not in the focus of consumers when evaluating purchase decisions, but the lack of warranty would rise doubts and signal a high risk to consumers.

The lowest agreement in this section yielded the statement that there was enough demand, it would be therefore not necessary to target these consumers with a low WTP. The majority of ten out of 15 experts disagreed with this assertion. An attempt was made to dig deeper and find out at which part of the statement the experts would express their disagreement. If there was enough demand in their opinion or if they would still try to target this group of consumers that is characterized by a low WTP. Despite this statement being unpopular in the Delphi study, researchers showed that, depending on the product type, a certain percentage of consumers might be unreachable. Through experiments and follow-up interviews Guide and Li [29] concluded that some people would never consider buying a remanufactured product and would always decide for the new product regardless of the price. Abbey et al. [23] came to similar results in their experiments with products such as laptops. They estimated that 20–40% of consumers would never buy a remanufactured device. This, however, leaves 80 60% of consumers who would possibly purchase remanufactured devices [23].

One expert pointed out another difficulty related to the pricing: According to this expert, cheap products often have a bad image among consumers. When the products are too heavily discounted many people get suspicious. Researchers assumed that the critical threshold of price discounts for remanufactured products is 40%. The level of discount should be carefully chosen [23]. Another expert stated in this context that the prices of remanufactured products should actually be higher. However, it must be taken into account that studies have identified upper limits for the willingness to pay of consumers. In an experiment, subjects were not willing to pay higher prices for remanufactured products than for new devices [51]. The results of another survey suggested that consumers are willing to pay higher prices only for eco-certified remanufactured products compared to non-eco-certified remanufactured products [48].

Yet, the communication measures mentioned above could be far more beneficial for companies than certifications which are associated with higher costs. Overall, the willingness to pay was rated as less challenging than suggested earlier by other researchers. On a five-point Likert scale from 1 = "not challenging at all" to 5 = "very challenging", the mean value was 2.2. The answers ranged from 1 to 4. It could therefore be described as one problem among many, but not the most serious one. However, ranking issues according to their importance was not part of this Delphi study. In another study, ranking questions in the context of a Delphi study led to unsatisfying results [52], they were thus eluded in this work.

4.4 Quality, Functionality, and Product Design of Electronic Devices

The abovementioned three barriers are all related to consumer behavior. From another point of view the question arose if electronic (consumer) products are suitable for remanufacturing based on their quality and the product design. Hatcher et al. [53] fundamentally doubted that these devices are suitable for remanufacturing at all based on the results of a case studies with Chinese recyclers. In Europe, the devices are often only used for a short time which does not indicate that the devices would be particularly durable. Today's average lifetime of a smartphone is less than 2, 5 years [4], more precisely in Austria between 18 and 24 months [14]. Used products still need a certain level of quality and functionality when they should serve as cores in remanufacturing processes [3, 5]. Additionally, consumers' taste might change and after a while, some exterior product designs could no longer be desirable [2, 24]. Some authors therefore focused on design related aspects of remanufacturing [54]. It was stated that a timeless design is regarded as an asset in a closed-loop supply chain of a more circular economy, therefore Apple products seemed to be especially suitable for remanufacturing [2]. Improving the product quality and their durability

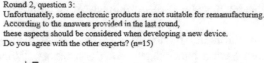

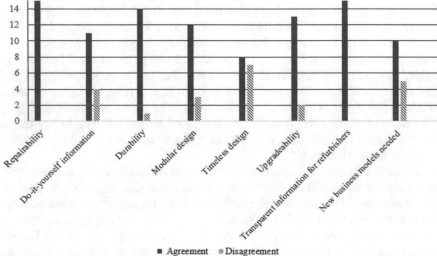

Round 2, question 3:
Unfortunately, some electronic products are not suitable for remanufacturing.
According to the answers provided in the last round,
these aspects should be considered when developing a new device.
Do you agree with the other experts? (n=15)

■ Agreement ⧹Disagreement

Fig. 2 Supporting remanufacturing in the product development process

would be an important step along with a more timeless design [54]. The overall product design has further the highest influence on the environmental performance of a product later [55].

Figure 2 above displays the experts' opinions on what key aspects should be considered to better support the remanufacturing of electronic products. According to the experts, multiple points should be considered in the product development process to make products more suitable for remanufacturing. All agreed that repairability is a critical aspect. It should be possible to easily open the device, replace components, and have access to spare parts. This was also criticized in the case study with Chinese recycling companies. Disassembling the device without destroying it was often not possible and thus identified as a major issue [53]. The claim for transparent information for refurbishes was supported by all experts. Ideally, access to information such as repair instructions should be available. In fact, according to the experts, there is a deficit with longevity. 14 out of 15 experts endorsed the demand for durable devices. The products must be designed to be durable which requires, for example, high-quality components. According to an expert, consumer products are, in contrast to high-end business IT (Information Technology) equipment, of poor quality which impedes remanufacturing and remarketing. That the durability of devices and the use of quality components is an essential aspect of remanufacturing is also underlined by researchers [53, 56]. 13 out of 15 experts agreed that upgrades of the devices should be facilitated e.g., adding memory which can improve the speed. Modular design is seen by twelve out of 15 experts as a way to simplify remanufacturing. Eleven out of 15 experts supported the idea of making do-it-yourself information such as repair instructions accessible to everyone.

In this context, there are some advances from manufacturers: Apple has announced to provide their customers with original spare parts and instructions for repair in the future [57].

In general, it is often due to the product design when the equipment cannot be reused. Manufacturers should therefore already ask at the beginning of the development phase what can be done with the product after its useful life. A holistic perspective that considers reuse is necessary. If products are to be developed specifically for remanufacturing, all individual steps of the remanufacturing process must be included [56]. In [58] a more detailed definition of this approach is provided. Accordingly, "design for remanufacture is a combination of design processes whereby an item is designed to facilitate remanufacture. Design for remanufacture is guided by an assessment of product or component value over time. This value may vary depending on the market and material demand and supply, legislation, and technological improvements" [58]. Sundin and Bras [59] proposed the "remanufacturing property matrix" for product designers that introduced pertinent product characteristics that should be considered during the different stages of a remanufacturing process. During the cleaning, for example, access to the components should be easy and the parts should be wear-resistant [59]. Another guideline proposed by Charter and Gray [58] included various "design for X" approaches such as "design for disassembly", "design for upgrade" and "eco-design". In their comprehensive work, they underlined the environmental benefits of this design approach, but also stated that there exist serious obstacles such as a lack of awareness and related to this a lack of demand for the remanufactured products. At the time of the study, over ten years ago the "design for remanufacture" approach had not yet been taken up by OEMs to any significant extent [58]. This seems to be valid until today and more research in this area would be desirable [56].

In the Delphi study, the results for implementing a timeless design of electronic products as a means to support remanufacturing are ambivalent. Only eight out of 15 experts saw timeless design as an important factor. In contrast to the majority of the experts, researchers underlined "the power of aesthetic durability to extend the product life cycle" [2]. In this context they rather focused on the consumer perception than on the general suitability of products for remanufacturing. Wallner et al. [54] interviewed consumers to investigate potential design issues of consumer products. While they did not consider electronic devices such as laptops or smartphones, they concluded that a timeless design of refurbished products makes them more appealing to consumers. Additionally, consumers assessed repairable products such as modularly designed devices as "more durable and trustworthy" [54] compared to non-repairable products. This is based on the assumption that repairable products can be better remanufactured [54].

It is not surprising that the experts unanimously confirmed that repairability and repair information for refurbishers should be considered in the product development of new products to support remanufacturing. The critical role of repairability for remanufacture has been emphasized in earlier scientific work [59]. Relevant information on the repair of devices would undoubtedly simplify remanufacturing. However, legislators set the parameters and some demands would probably meet with great resistance from manufacturers. Even in other cases of individual projects of the European authorities that are not related to circular economy efforts such as a uniform charger for electronic devices, the resistance of manufacturers delayed more serious interventions for a long time [60]. Currently, batteries are in the focus of the European Parliament. In the course of the circular economy idea, rechargeable batteries should be, for example, easier to replace [61].

In the specific case of remanufacture, manufacturers fear declining sales of their new equipment. There are, however, many reasons for engaging in remanufacturing activities and many well-known manufacturers in various industries such as BMW or Hewlett-Packard offer remanufactured products. With these products, they can serve the segment at the lower end of the market, which might otherwise resort to cheaper new products [23]. OEMs that do not produce themselves are facing difficulties due to remanufacturing [62]. They might not even be able to remanufacture their own devices. Contract companies, which take over the production for these OEMs, would be offered a new source of income through the remanufacturing of these products. Zhou et al. [62] summarized three possible strategies for OEMs faced with remanufacturing: Conducting remanufacturing activities, let a contract remanufacturer perform it or do not enter the business at all.

Governments, on the other hand, might have two possible ways to support remanufacturing: A regulation of the market that forces manufacturers to implement a more favourable product design or subsidies for remanufacturers. While in the European Union existing guidelines such as the "eco-design directive" are targeted at energy efficiency and the environmental impact of devices. This directive theoretically also aims at the "ease for reuse" [63] that encompasses remanufacturing [63]. During this study twelve out of 15 participating experts agreed that legislators must set the right framework. It was stated that the prices of new devices should more comprehensively include negative externalities such as such as detrimental emissions, pollution, or consumption of critical raw materials. The "European Green Deal" is meant to support remanufacturing of electric and electronic devices [7]. In the first round of the Delphi study, most experts were aware of these plans, but the feelings about these plans were mixed. Concerns regarding the implementation were mentioned. One expert was afraid that smaller, local refurbishers could be discriminated and larger players will control the market. With regard to government support, 13 out of 15 experts saw government subsidies for refurbishers as essential to make remanufacturing mainstream. In game theoretic models, subsidies were in most scenarios beneficial and helped to develop the remanufacturing industry [64]. Government support is not limited to subsidies, it can imply setting tax incentives for remanufacturers or remanufactured goods [20, 65].

The final round of the Delphi study was dedicated to the question how these demands for repairability, durability, transparent information for remanufacturers and the subsidies could be best pushed through. Many experts mentioned lobbying or organizing oneself in advocacy groups. One expert mentioned that the reuse sector is large, and many players have different or divergent concerns; every group should thus have its own representation. At this point, it should be noted that smaller advocacy groups while being more homogeneous might have weaker bargaining power. Additional suggestions were that remanufacturers should participate in standardization committees to gain influence on product standards through legislative processes or seek allies such as manufacturers. On the downside, these manufacturers might have higher bargaining power compared to independent remanufacturers. Concerning a practical implementation of these demands the experts were rather optimistic. It was stated that it might be a lengthy process, but some demands are already being implemented.

Indeed, the European Parliament is willing to face some of the issues. A label for products displaying the level of durability is aspired on a European level [66]. France is pioneer and introduced the "repairability index" for electronic and electric products in 2021: A number between zero and ten indicates the repairability of the device based on five categories: documentation, dismantlability, availability of spare parts, their price, and product-specific aspects such as remote maintenance. In some cases, the lack of traceability is criticized, and that the same measures are taken for a wide range of very different devices. Nevertheless, it can be considered a good first aid to orientation for consumers, but some reuse experts call for more comprehensive, regulatory measures. Moreover, uniform standards for the European market would be desirable [67].

4.5   Technical Advancement and Obsolescence

The last barrier that was tackled in this context was the technical advancement. When innovative products are introduced to the market, they often replace existing devices over time. Mobile phones, for example, where quickly replaced by smartphones [5]. Additionally, smartphones have much more functionalities than mobile phones [68] and new models are introduced to the market at always shorter intervals [5]. This plays an important role as a potential buyer might fear that the remanufactured device becomes obsolete more quickly after the purchase than a new product [2], although the devices such as smartphones are nowadays used less than 2,5 years on average [4]. It is a peculiarity of remanufactured products that they were sold on the primary market before being refurbished and offered to new customers [2, 24]. Further, it is possible that smartphones become redundant sooner or later [69]. Lund [26] referred to this as the "technological life". This section is dedicated to the technological progress, the issues with short life cycles and their implications for the remanufacturing business.

When asked for the impact of technological change on remanufacturing, 14 out of 15 experts considered it critical for remanufacturing when software and security updates are discontinued. Even if the devices are technically ok, they can often no longer be used by many users as a lack of security updates threatens the security of the computer. After ten years, for example, Microsoft announced to discontinue the security updates for Windows 10 in 2025 [70]. Further, all experts supported the statement that in many cases new devices do not contain any innovative features compared to the previous models. This means that few innovation boosts were observed when it comes to devices like smartphones and laptops. Here, the critical attitude of the participating experts with regard to announcements by manufacturers became clear. The divergent interests of remanufacturers in contrast to manufacturers who mainly count on the sales of their new devices also stand out. The experts rather focused on sustainability. 14 out of 15 experts would prefer durable devices for environmental reasons. The vast majority of 13 out of 15 experts consented to the statement that some innovative features make the products more prone to malfunction. 13 out of 15 experts observed that there are a significant number of consumers who do not need the latest models due to various reasons. Eight out of ten experts confirmed later that these people can be considered the major target group of remanufacturers.

Regarding the general impact of the technological advancement, the experts' opinions varied more widely. One expert stated that the technological advancement today is almost insignificant for remanufacturing. However, nine out of 15 experts disagreed with this point of view. One expert stated that the technological advancement is favourable for remanufacturing as a wide range of different models can be offered to the customers. This was also noted by another expert who immediately criticized this development as favourable for remanufacturing but opposed to sustainable development. This is considered an interesting aspect, as remanufacturing itself contributes to a more sustainable way of consumption, while at the same time it can profit from unsustainable short product life cycles in some ways. It was tried to delve at this point, but no consensus amongst the experts was reached. Some experts thought that these short life cycles might be beneficial for refurbishers, but they rejected this assumption for sustainability reasons. It was further argued that refurbishing electronic devices can mitigate environmental challenges, but not resolve them. When the devices would be durable and easy to repair in the first place, one would be less reliant on refurbishment.

Another expert claimed that the second life cycle is mostly independent from the first life. It would thus have little influence on the "second life" of a device if it had only been used for a short period of time during its "first life". This might be questionable with regard to the consumers' perception, as they fear that the remanufactured devices would become obsolete more quickly than new products [2]. The older a device would be, the stronger this fear could be. Moreover, no agreement was reached regarding the statement that some technological developments such as cloud services would be favourable for remanufacturing. Cloud computing enables the use of online storage space or software. It was argued by one expert that cloud services could therefore facilitate the replacement of devices when the data is stored in the cloud. However, this could be neither confirmed nor rejected in the third round.

One of the key aspects of this section is that the interests of manufacturers and remanufacturers can diverge widely, especially with regard to sustainability. As mentioned earlier, many devices are not designed for a second or even third "life". The manufacturers benefit when consumers demand new models at short and regular intervals. In addition, remanufacturing used equipment can mitigate some of the environmental issues, but in the long run, more durable and repairable equipment may be the more sustainable solution.

5 Future Trends and Industry Developments in Remanufacturing

The following section addresses possible future trends and hot topics in the remanufacturing industry. The first one that was examined in more detail was green marketing, which is supposed to help remanufacturers to better promote their products. Further, quality certifications and leasing models were seen as ways to accommodate the consumers' needs in scientific literature.

5.1 Green Marketing

The environmental benefits of remanufacturing activities and the contribution to a more circular economy have been widely discussed in scientific literature [2]. As mentioned in the beginning, remanufacturing of electronic devices can help to reduce the increasing amount of e-waste [5, 6]. In addition, when a product is remanufactured, the resources previously used in the manufacturing process, such as raw materials or energy, can be saved to a certain extent [3]. Further, the need for new devices is reduced and thus the consumption of resources is decreased when a new device is substituted by a refurbished one [71]. However, in studies even "green" consumers were surprisingly unaware of the environmental benefits of remanufactured products [2, 23]. Reaching more environmentally conscious consumers by highlighting the environmental benefits could be in the interest of companies that market remanufactured equipment. Great potential lays in people's increasing interest in environmental protection. In a representative study, 65% of the Germans attributed very great importance to environmental and climate protection. In this study citizens were considered to be aware of the need to change their own consumption behaviour [72]. In Austria, a majority of 76.2% respondents said they paid attention to environmental and social issues when buying certain products [73]. However, the actual purchasing behaviour can of course deviate from such declarations of intent. Moreover, in this context electronics were not specifically targeted [72, 73]. Consumers can nevertheless be made aware of the environmental impact of electronic products. Refurbished equipment can be promoted as a more environmentally friendly alternative to new equipment.

Therefore, in all three survey rounds, approaches to reach more environmentally conscious consumers were discussed. On the one hand, an expert initially stated that there are already lots of green marketing activities, and it would only be a matter of time until remanufactured products will receive more attention of consumers. Another expert added that, on the other hand, it is difficult for consumers to fully understand the environmental issues related to our consumer habits, for example, resource consumption or e-waste. This was named as one of the reasons why people do not consume in a more responsible manner. The three most popular approaches supported by all 15 experts were: *(i)* generally more marketing and educational work (this is in opposition to the above introduced statement that there are already suffi-cient marketing measures), *(ii)* providing transparent and simplified information on the resources saved when purchasing a refurbished (in this context, establishing a label would be a viable option), and *(iii)* informing customers what positive social and/or environmental impact one's organization has was considered as a good way to reach more environmentally conscious consumers. Additionally, providing trans-parent and simplified information for consumers on the environmental impact of a new device received support by 13 out of 15 experts. Equally popular was the idea to generally emphasize the negative aspects of purchasing electronic devices to reach environmentally conscious consumers. Marketing campaigns with a social aspect such as donating to charity for every sold device were endorsed by eleven out of 15 experts. Marketing campaigns in conjunction with regional environmental activities such as planting a tree in Germany or Austria for every sold device were rejected by six out of 15 experts.

The experts' call for more marketing and educational efforts suggests that current efforts are either insufficient and/or not targeted enough to reach the right consumers and/or that environmental issues are not emphasized enough. Uninformed consumers might not even be fully aware of the issues related to e-waste or the resource consump-tion of new devices. Consequently, these consumers might not see any need for action to change their consumption behaviour. In this study, the approaches where consumers are informed about ecological issues related to the manufacturing of electronic devices were thus considered helpful by the vast majority of the experts. This underlines the assumption that there is a lack of awareness amongst consumers in the first place or that the environmental benefits of remanufactured devices are rather abstract for consumers. To tackle this issue organizations can consider the following aspects that are based on the experts' verbal answers: First of all, the environmental benefits must be communicated more comprehensively and based on scientific findings and valid data. Second, the way how this information is commu-nicated is equally important. The experts advised using simple language, practical examples, reference variables and context information. Environmental organizations, for example, estimated that in total almost 13,000 L of water or 160 bathtubs are used for one smartphone, from mining to packaging. This information should be put in context, like for example water shortage or contamination in mining countries [74]. Reliable data is hard to find and often offered like in this example by environmental organizations. It is difficult to verify these proposed numbers for consumers.

In previously conducted studies, interesting dynamics were observed. Informing consumers about the environmental benefits of remanufactured products led in an experiment to a lower willingness to pay for the new product. The authors under-lined the importance of informing and educating consumers. Consumers that were informed during the experiment did not underestimate the value of the remanufac-tured product [51]. In addition, Wang et al. [35] advised remanufacturers to promote the environmental benefits of their devices and to select a marketing strategy that specifically targets environmentally conscious consumers. The experts' ideas related to promoting the environmental benefits or addressing the environmental issues as well as the ideas that are generally aimed at informing consumers, can be therefore considered promising. At this point, however, it should not be forgotten that even with sustainably manufactured or remanufactured products, advertising can tempt consumers into excessive consumption [22].

Moreover, the question arose if an eco-label for refurbished electronics could be useful. Some researchers stated that an eco-label could be used to inform consumers about the environmental benefits of remanufactured devices [35]. Harms and Linton [48] investigated the impact of an eco-certification to mitigate the lower willingness to pay of consumers for products with remanufactured components. They concluded that remanufacturing has a negative impact on the WTP for the product, but eco-labelling has a positive influence and it "appears to be a signal of lower risk" [48]. Van Weelden et al. [2] cautiously suggested that the environmental benefits of these products have a lower influence on the consumer's decision-making process than often assumed. In surveys many consumers reported that they would use well-known national eco-labels such as *Blauer Engel* in Germany or *Grüner Strom* label for green electricity in Austria for better orientation [72, 73]. From another perspective, consumers do not always profit from labels. In some areas there is a variety of different labels, but not all of them are meaningful. Some non-obligatory labels can be misleading, and the evaluation criteria are not transparently comprehensible which is criticized by consumer advisors [75]. Labels that are intended purely as a marketing tool and do not contain any information relevant to consumers should be avoided if the idea of environmental protection is to be credibly communicated.

5.2 Quality Marks

Not only eco-labels could possibly be informative for consumers and beneficial for remanufacturers, but also quality marks for remanufactured goods would signal trustworthiness to the consumers when they associate a rather high risk with the purchase of a remanufactured electronic device. Especially when uniform standards for refurbishing electronics are currently lacking [2, 4, 25]. This quality certification could be issued by the brand-owner [76], or by an independent authority, or the remanufacturer could establish its own quality certification.

Initially, the experts were asked if in their opinion quality certifications were a reasonable way to meet the customer's needs. Hereby, the experts assessed the issue rather from a business perspective. The answers suggested that six out of twelve experts endorsed any kind of quality certification. With regard to quality certifications the costs and implementation difficulties for remanufacturers were criticized. In the second round, the highest rated alternative to minimize the perceived risk of consumers was an independent quality certification. This was endorsed by twelve out of 15 experts. Nine out of 15 experts approved of quality certifications issued by remanufacturers or sellers. Quality labels issued by the manufacturer or OEM were considered to be of little significance and not independent enough. They were therefore not discussed in the subsequent rounds. In general, eight out of 15 experts found that quality certifications would be too expensive for refurbishers. Moreover, seven out of 15 experts thought that quality certifications are too difficult to implement for refurbishers. After all, nine out of 14 experts also agreed with the statement that a warranty or extended warranty would give consumers sufficient security.

Since independent quality certification was the most popular alternative, the experts were conclusively asked how these certificates should look like in order that refurbishers would benefit from it. These are not so much specific plans as factors that should be considered in the development process. In this process refurbishers and consumers should be integrated. Among other things, the demands were related to the costs incurred for refurbishers. Ideally, the prices for the certificates are reasonable. One expert stated that it should be free from charges for refurbishers. The experts mentioned that one could start from scratch and certify single steps of a refurbishment process such as data erasure. Hence, it would be favourable if processes, or in the end, the refurbished devices are checked by an independent party. Certifying single steps of the process is already possible. There is, for example, a certified data erasure software [77] and if the data medium must be destroyed, this can also be certified [78]. One expert stated that a quality certification should not be issued by the big players in industry. Additionally, it would be a prerequisite that the issuing party has a good reputation and is known by consumers. Only this would guarantee the necessary transparency and would thus be beneficial for refurbishers and consumers. Some organizations already established their own labels for certified remanufactured equipment [79, 80]. Apple is also selling its certified refurbished devices [81].

In a game theoretic model, Niu and Xie [76] investigated if a remanufacturer who suffers from a disadvantageous quality image would benefit from quality certification of the brand-owner. Their results suggested that the brand-owner benefits from additional revenues, and when competition is intense the remanufacturer also benefits from this certification [76]. Therefore, for an OEM issuing a quality certification is an additional source of income and a way to earn money on the secondary market with the sale of refurbished products from other refurbishers. Most experts in the study thus considered a quality label issued by a manufacturer as detrimental for remanufacturers. This could strengthen the already powerful positions of OEMs in the market. A truly independent label was rated more positively and could be beneficial for remanufacturers when it is not too costly. However, many experts considered the existing warranty of one year in the European Union [27] or an extension of the statutory warranty to be sufficient to mitigate the perceived risk of consumers.

5.3 Leasing Models

In earlier research, leasing models were proposed as another possible way to mitigate the perceived high risk of consumers [4]. As part of the servitization concept they are further a trending topic in manufacturing industries. They allow companies to remain competitive apart from product innovations [82]. Moreover, rental models with a sustainable approach can be seen as part of a circular economy [17]. However, in a study on refurbished smartphones leasing options did not convince customers [4]. The study focused only on smartphones and individual consumers, therefore not considering opportunities in the B2B segment or other equipment. Theoretically, the introduced tools offer potential benefits for remanufacturers and sellers of remanufactured products. Practically, they are not common or not widely used in the remanufacturing industry.

Initially, the results showed that six out of twelve experts had initially a positive attitude towards leasing models. In addition, in the context of what would be necessary to further support the remanufacturing of electronic products, ten out of 15 experts saw the need for new business models. Leasing models or more generally speaking product-service-systems (PSS) were named as promising approaches. The topic was therefore subsequently discussed. On the one hand, eleven out of 15 experts supported the idea of leasing models in the second round. On the other hand, nine out of 15 experts raised concerns that the lower prices of refurbished devices make leasing models unattractive from a business perspective. When evaluating the impact of the lower prices of refurbished equipment the opinions differed. While one expert stated that the leasing models make sense when the prices of the electronic products are sufficiently high, another expert saw the lower prices of refurbished devices as a benefit. In the case that the refurbished devices are significantly less expensive than new devices, the leasing models for refurbished products could be offered at lower prices. In this context, it was claimed that lower prices might attract consumers with insufficient capital and low credit-rating, and this would make the system more prone to fraud according to an expert. For another expert the lower prices of leasing models are the key aspect of this concept. Accordingly, the benefits of leasing models are the lower prices and not more security for consumers. An expert doubted further that the B2B sector will offer many business opportunities in the near future. The expert justified this with the assumption that companies would see their IT equipment rather as a store of value and would thus be hesitant to lease this infrastructure. However, IT equipment ages and therefore becomes more susceptible to safety defects over time, or it is just no longer up to the new challenges. Depending on the area of application and the workload, devices must be replaced from time to time.

Additionally, nine out of 14 experts considered the warranty or the opportunity to extend the legal warranty as sufficient to provide security for consumers. Nevertheless, in the last round, the majority of the experts concluded that leasing models could still be applied reasonably and sustainably. One expert considered leasing models as especially interesting for larger refurbishers. Whether leasing models are attractive from a business perspective is according to the experts depending on the product category and the product prices. Suitable product categories could be smartphones, laptops or even electronic toys, e-scooters, or e-bikes.

The above discussed leasing models do fall under the category of product-service-systems. Theoretically, servitization allows companies in the end to offer PSS where product and service are inseparably combined, or the product could even be made redundant by service offerings. Besides, there are multiple different forms that can be differentiated whether the focus is on services or the product [82]. PSS are considered favourable for companies since they can enhance the relationship with the customer and allow to bind customers stronger to the company. Further, the company is able to collect more data about customer behaviour and especially their needs which could lead to more innovative offerings [83]. Following Tukker [83] one possibility is to distinguish three types of PSS:

(1) Product-oriented services where the focus is on selling products, but services can be purchased additionally. The major source of revenues remains the sale of products. Take-back agreements for products that are no longer used would be an example as well as maintenance arrangements.
(2) Use-oriented services mean that the product itself is still an important part of the concept, but the revenues are not generated by product sales. The product is owned by the company and can be used by customers. It would be possible that the same product is used by more than one customer. Leasing and rental models are part of this category.
(3) Result-oriented services, hereby a third party performs a task, and the outcome is the central part. Examples for this category are booking a caterer or outsourcing the cleaning of facilities.

However, similar to remanufactured devices, consumer acceptance is the major issue for establishing PSS [84]. In experiments with Belgian consumers between 15 and 30 years, less than one third of the consumers considered leasing a smartphone. Consumers feared the financial consequences, possible risks of signing a leasing contract and a lack of control in the situation [85]. It is therefore possible that leasing contracts that are suggested as a way to minimize the perceived risks of consumers when buying refurbished devices [4], do not meet these expectations. Mashhadi et al. [84] identified in their study previous consumption behaviour as the main driver for future consumption behaviour. Consumers who purchased their phone, are likely to purchase a phone again in the future. While a consumer who used a leased phone, would be open to a new leasing contract for a phone. In the Delphi Study, the experts were rather optimistic about offering leasing contracts for remanufactured equipment. Building on the findings of the studies cited above, however, a lack of acceptance of these models on the part of consumers could pose a greater challenge.

5.4 Industry Developments in the Near Future of Remanufacturing

This section is dedicated to the future of remanufacturing, in particular the year 2026. Possible upcoming industry developments are discussed. The experts shared their assessment of the future of the industry. In the focus were increasing competition, regulation, and standardization of the industry.

The assessments about the near future of the remanufacturing market were developed over all three survey rounds. It is, of course, impossible to make concrete assumptions about the future and the results should rather be used as a basis to make predictions [32]. Moreover, the experts' assessments are not entirely unbiased [86]. It is assumed that they will believe in the growth of the industry and the success of their businesses. Nevertheless, their assessments are usually based on market knowledge and thus considered valuable. Further, as mentioned at the beginning, Delphi studies try to make use of self-fulfilling prophecies to manipulate the future [32].

In the focus were first the market growth and the future market share. The market size of remanufactured electronics is rather difficult to capture due to the market structure and e.g., players that also manufacture new devices [6]. It was estimated that in Europe approximately 15% of all smartphones are collected for recycling or remanufacturing activities. The secondary market for smartphones including refurbished smartphones was estimated at 6% of the primary market volume [4]. However, this data could be outdated by now and was thus not presented as a reference point in the survey. In the first round, twelve out of 13 experts thought that remanufacturing will be more important in five years than it is today. Amongst these experts are three experts that stated that remanufacturing will be more important than today, but still a niche market. One expert expressed concerns that this development of market growth might have some negative side effects. Accordingly, when the number of refurbishers increases, the number of companies who offer poor quality also increases and the image of all refurbishers might suffer. It is thus argued that uniform standards for this market will be needed to gain the consumers' trust. Additionally, another expert addressed the issue of "supply constraints" [30]. Before a product can be remanufactured it has to be offered and sold on the primary market first. This way the market is limited [30].

The different assumptions and opinions varied widely, therefore the experts were subsequently asked to assess the different statements. Overall, the assumptions were rather optimistic. 13 out of 15 experts agreed that remanufacturing of electronic devices will no longer be a niche market in five years. Twelve out of 15 thought that remanufacturing will be more important in five years, but still only an additional offer to new electronic devices. Eight out of 15 experts estimate that remanufacturing of electronic devices will no longer be a niche market, but in ten years. Ten out of 15 experts disagreed that remanufactured electronics will be a larger niche in five years. This is the highest disagreement in this section. The slightly less optimistic assessment, that the market for remanufactured electronic devices will grow moderately in the next five years, is supported by nine out of 15 experts. The vast majority of experts thought that remanufacturing will be more important in five years than it is now. Regarding the growth of the market the majority of the experts expected a moderate level of growth. Based on the collected responses, it must be concluded that some experts supported contradictory statements. For example, it is assumed that remanufactured products will no longer be a niche product in five years' time, but at the same time it is stated that the market will grow only moderately. This can be seen below in Fig. 3.

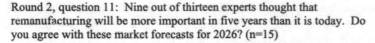

Round 2, question 11: Nine out of thirteen experts thought that remanufacturing will be more important in five years than it is today. Do you agree with these market forecasts for 2026? (n=15)

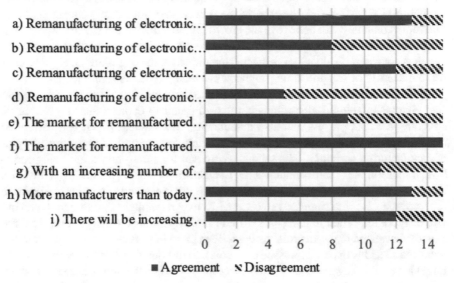

Fig. 3 The future importance of remanufacture and industry growth

When asked for the market share of remanufactured devices in percent of the total market of electronics in the European Union, the average value was 18.44%. While 5% being the lowest estimate and 50% the highest. There was no comparison value given in order to collect unbiased results and due to the lack of current, reliable, and accessible market data. The future market share is difficult to estimate, and often separate markets are considered e.g., the market for smartphones or the market for laptops in a specific region. The intention was to capture a rough estimate of the future market share and derive the optimism or pessimism of the panel. Compared to the assumed market share of 6% [4], a market share of 18.44% represents more than a threefold increase and not "moderate growth". For the global market of used smartphones analysts expected a compound annual growth rate of 11.2% from the year 2019 to 2024 [87]. This is in line with the expectations of some researchers. Gallo et al. [6] saw high growth potential especially in the European Union. Following this assessment, Harms and Linton [48] as well as Guide and van Wassenhove [3] predicted an increasing importance of refurbishment. This leaves notable room for growth.

Moreover, according to the majority of the experts, sustainability will play an important role in the industry in the future. Increasing competition in terms of which refurbisher is more sustainable was considered realistic by twelve out of 15 experts. Some experts placed their hopes above all on increasing environmental awareness among consumers and on a greater importance of a circular economy. Eleven out of 15 experts thought that with an increasing number of refurbishers in the market, uniform quality standards will be needed. All experts supported the idea that the market for remanufactured electronic devices will be more professional in five years than it is today. This could according to the experts manifest itself in various aspects. More providers could mean a broader variety of products that are offered on the market. Again, regulations and certifications are named as a facet of a more professional market. The importance of uniform standards for the grading of the devices were highlighted by an expert. Some experts thought that the companies will increase their marketing activities and become more present. In this context, an expert noted that the quality of the online appearances of the companies will be enhanced. Some experts predicted that there will be larger companies, and more manufacturers will offer remanufactured devices. The companies might rely more on modern technology and logistics in the future. One expert stated that the processes will be increasingly standardized. Further, it would be possible that the refurbishment processes will be increasingly automated. This is an interesting point, which is supported by Gallo et al. [6]. This process might take longer than five years, but the beginnings could already be visible.

The idea that more manufacturers than today will engage in remanufacturing activities in five years was endorsed by 13 out of 15 experts. This development could have far-reaching effects on the existing (independent) refurbishers. Some experts feared a negative impact for refurbishers. One possible consequence could be more competition and maybe even increased rivalry with regard to the number of available cores for remanufacturing. However, today only a small part of the total available amount of used devices in Europe is remanufactured due to various reasons [4]. Another expert assumed that consumers will then prefer buying refurbished devices from the manufacturers. The expert concluded that customer contact and service will thus gain importance for independent refurbishers. One strategy for the independent refurbishers would be filling in niches left by the manufacturers. More precisely, specializing on certain product categories or services that manufacturers do not cover. Hence, some experts thought that market cleansing mechanisms are likely to occur soon. This would reduce the number of players in the market after a state of excess supply. As OEMs who operate as remanufacturers often gain a dominating position in these industries [88], the scepticism of independent remanufacturers does not appear unfounded. From another perspective, cooperation between independent refurbishers and manufacturers might be possible in different forms. The OEM as a brand-owner could certify the remanufactured products of refurbishers [76]. Some manufacturers prefer outsourcing the refurbishment of their used devices which could be carried out by contract remanufacturers. This aspect was briefly addressed by Gallo et. al. [6]. The authors suggested that for manufacturers cooperation might be problematic when resources or components are scarce and manufacturers risk becoming too pending on other (remanufacturing) companies [6]. Another issue for manufacturers could be that third parties collect too much confidential information on the products [6]. In addition, some OEMs that outsource their manufacturing may not be able to remanufacture their own equipment themselves or may show no interest in starting to remanufacture [62]. Overall, it would be possible that the market for remanufactured electronics grows, and the players operate in a more professional manner. This development could attract more manufacturers that enter the market. This might have both positive and negative consequences for the existing (independent) remanufacturers. Adapting to these challenges is part of the business strategy.

6 The Role of eBusiness and Digitalisation in the Remanufacturing Industry

Digitalisation and electronic processing of business operations theoretically offers various opportunities and possibilities for businesses [89]. They have the potential to significantly change the way organizations conduct business, it is therefore expected that sooner or later the impact on the remanufacturing industry will further increase.

Based on the answers in the first round displayed in Fig. 4 below, potential areas which might currently not profit from digitalisation were identified and discussed in the subsequent rounds. The results show that all participating companies used digital solutions for marketing and customer communication as well as internal communication. Ten out of eleven experts stated that their organizations used digital solutions in their procurement departments. Eight out of eleven experts answered that their logistics departments were supported digitally. In distribution only seven out of eleven experts indicated they their company relied on digital solutions. An expert named additionally "accounting analytics & business intelligence" as other areas in the organization which used digital support. Overall, nine out of twelve participating experts

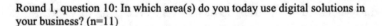

Round 1, question 10: In which area(s) do you today use digital solutions in your business? (n=11)

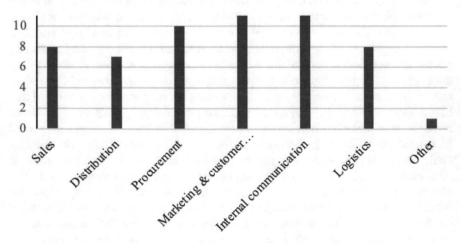

Fig. 4 Status quo of digital support in the participating organizations

said that their company had planned to make more use of digital support in the future, or they already implemented their plans for more digitalisation. One expert stated that almost all processes in their company were digital today.

They are further planning to implement a dynamic pricing system which could consider the prices of competitors, the demand, and the supply. When interpreting these answers, the structure of the panel has to be taken into consideration. Most organizations, but not all, can be described as smaller or medium-sized companies. Therefore, further assessments of the degree of digitisation are difficult in this framework.

As mentioned above, the majority of the experts planned to make more use of digital support in the future or already implemented their plans. When asked for specific areas of interest, it was stated that their companies would further make use of online marketing and strengthen public relations via the internet. Additionally, distribution with the focus on e-commerce of refurbished electronics was named. An expert pointed out that their company would like to optimize and automate business processes. Another expert thought that the process documentation could increasingly benefit from digitalisation. It was mentioned that the online condition monitoring of electronic devices could become more relevant. Until now this is mainly used for machinery and equipment, it is questionable whether there are no legal restrictions on remote monitoring of privately owned electronics. However, too few responses have been collected on this point to make more meaningful statements. Based on the answers given, however, it can be assumed that there are differences or major differences in the degree of digitalisation among the organizations surveyed.

Although "smart remanufacturing" is a new field of research, it has not been discussed in depth in this work due to a lack of publications. The reasons for that are manyfold, such as missing implementations of theoretical approaches in practice [90], lack the financial resources to implement expensive projects such as investments in big data forecasting methods examined by Niu and Zou [71] (which applies particularly to small and medium-sized remanufacturers), or the simple current absence of need for the use of such methods. However, this might change in the future.

The e-commerce of electronic products is currently booming in European countries such as Austria not only due to the coronavirus pandemic [91]. The Austrian Retail Association refers to a recent study conducted by European market research institutes. Accordingly, one quarter of consumers will continue purchasing more products online after the coronavirus pandemic [92]. In addition, more and more people use their smartphones to shop online [93]. Websites and online shops should thus be mobile-friendly. Not all participating organizations use online opportunities to sell their products, e.g., offering and selling products via a virtual marketplace (marketspace) or their own online shop. Due to the structure of the panel this is difficult to interpret, as for example, some of the participating companies only operated in the B2B sector where online trade might be very specific and highly depending on product and sector.

However, to reveal the underlying reasons why organizations in industry would possibly decide not to have an online store or sell their devices online via a marketspace every expert of the panel was consulted. It was stated that first of all, maintaining an additional online store is a huge effort, especially for companies which usually manage a stationary store. Second, refurbished devices are sometimes individual items, therefore every device would require its own photos and descriptions according to some experts, which implies another huge workload. Moreover, the assortment changes constantly. It was stated that sometimes the products sell very well offline, so there would not be enough devices to be offered online. Additionally, for some remanufacturers who focus on store-based retailing, it could be difficult to organize their internal processes in order to maintain an online shop. With regard to online marketspaces an expert pointed out that larger platforms or marketspaces might require specific quality standards than some refurbishers would not be able to meet. Beside the online shop, long-term investments of companies in online marketing would be needed to make it a success. If companies decided to offer their products online, they would need ideally a comprehensive strategy.

Choosing an online shop system or offering the products via a marketspace plays a key role within this online strategy. However, there are more aspects to consider, such as target group, online product range and online assortment, design of the online shop and in general the design of the company's website. A company could decide to design the product pages in order to dispel or ease possible doubts or insecurities of consumers in a way that underlines the above discussed benefits of refurbished devices. The payment options and terms, customer support as well as the distribution of purchased devices further need to be defined. Organizations can complement their online appearance with content marketing such as blog articles, podcasts, or videos. Companies may be present on social or professional networks and use these channels to interact and network with various stakeholders such as potential customers, other companies, or advocacy groups. Therefore, companies need to decide on a large number of issues in order to formulate a comprehensive e-commerce strategy.

7 Contribution and Limitations

The Delphi study at hand focused on essential topics in the remanufacturing industry that concern remanufacturing companies and vendors of remanufactured electronics located in Austria and Germany. It was implemented from a business perspective, and practical recommendations for remanufacturers were developed. Therefore, possible solutions to eliminate those obstacles currently faced by remanufacturers were of great importance in this context. Furthermore, it was examined how trends brought up in the literature are perceived and assessed in practice. Finally, a look into the future was ventured, which developments could occupy remanufacturers in the mid-term future. The matter of digitalisation and, above all, online trade with remanufactured devices was addressed.

One of the overall goals of this study was that industry experts evaluate issues and industry developments from their point of view. Emphasis was usually not placed on the company's internal processes, such as on details of their remanufacturing processes. The focus was rather outwardly oriented on the interaction with other stakeholders such as consumers, customers, and partly also governments. The technical perspective was addressed to a lesser extent as this work was predominantly written from a business perspective.

Moreover, the choice of research questions necessitates a broad discussion on a large variety of topics. This allowed to collect lots of valuable insights in the first round. Since not all of them could be investigated in the course of this study, this leaves room for further research for example in the form of in-depth interviews. However, the questions of the subsequent two rounds of this study had to be limited to essential or ambiguous points. Otherwise, the survey rounds would have been too long and the willingness of panellists to participate would have declined. Additionally, the evaluation of statements on a Likert scale was chosen to shorten and simplify the questionnaires, even though this has some drawbacks such as no textual feedback on every single statement. Giving written feedback on each statement would have taken too much time for the respondents and could have ultimately led to a higher dropout rate or even a termination of the study if only few responses had been collected. Instead, the experts were able to leave optional comments under each question in case one wanted to explain her/his choice or to express other thoughts.

With regard to the individual topics, it should be mentioned that in the area of e-business and digitisation, fewer responses were collected and therefore evaluated. In addition, these responses are partly very company-specific like the use of digital support, to name an example. They are therefore not suitable for generalisations. Elsewhere, the focus was narrower. One expert critically noted that other devices, not only information and communication technology infrastructure, should have been addressed in the study. This focus was chosen because consumers replace devices such as smartphones more frequently than larger devices. When purchasing a smartphone, consumers sometimes do not replace a broken device; they simply desire to own a newer device [15]. In addition, there could be too much variation in terms of consumer perception and the underlying mechanisms between a smartphone and a washing machine, to name an example. It is certainly very useful to examine remanufacturing and the sale of other appliances in future research.

At this point, the bilingual nature of the study should be addressed. Potential experts should have the choice between German and English questionnaires to additionally address non-German speaking experts and thereby broaden the circle of available experts. It was determined in advance that this work would be written in English. Later, however, the experts responded predominantly in German. The questionnaires in English were nevertheless retained because the basic design of the study was not to be changed. Hence, it was necessary to translate statements from German to English and vice versa with great care. As multilingual surveys are quite common, no disadvantages should result from this circumstance [52].

The total duration of the three survey rounds exceeded the recommended maximum of five months [94] by one month. However, as the recommendation rather aims at keeping the drop-out rate low and the study at hand showed a very good respondents' loyalty, this time lag did not have any negative effect. As all experts in the panel came either from Austrian or German organisations the results cannot be generalised. Nevertheless, they can also serve as impulses, inspiration or impetus for practitioners and scientists in other markets or future studies.

8 Conclusion and Future Research Areas

Remanufacturing has recently gained importance in scientific literature and in practice. It is considered as an alternative approach reducing the rapidly growing amount of e-waste. By purchasing a remanufactured device without buying a new one, energy and raw materials may be saved. While some obstacles for remanufacturers already have been addressed, there is still a need for practical solutions to overcome these obstacles.

The Delphi study presented in this work included three survey rounds, collected in spring through summer of 2021. This was only made possible by the experts' willingness to participate in this study. Various aspects of the remanufacturing of electronic devices were collected and discussed. Building on the experience and knowledge of the participating experts, it was possible to develop insightful assessments and approaches to solve challenging problems. Hereby, the experts emphasized the importance of the sustainability concept in remanufacturing. The potential of cooperation and networking between remanufacturers and other stakeholders was further underlined. The organizations can jointly create more awareness amongst consumers about resource consumption in the manufacture of electronic devices and the difficulties associated with their disposal. At several points, the importance of the remanufacturer's transparency became evident. On the one hand, transparency with regard to the remanufacturing process should be provided. On the other hand, the condition, in which the equipment is offered for sale and the criteria based on which this assessment was made should be indicated. Thus, remanufacturers themselves can take some courses of action to drive growth in their industry. However, not all these difficulties can be overcome by the efforts of remanufacturers alone. Legislative institutions should become more active to facilitate a sustainable consumption of electronic devices.

Many relevant issues of the broad topic of remanufacturing were discussed from a business perspective. Future research may focus on consumer behaviour, consumers' expectations and willingness to pay, and observations of real sales situations. Furthermore, companies might provide insights into best practice examples from their own operations. Solution approaches developed in this study could be subjected to a practice text or tested in experiments. In addition to obtaining practice-relevant results, another aim of this study was to draw attention to the topic of remanufacturing and the underlying environmental problems.

Acknowledgements The authors would like to thank Andreas Mladenow for his valuable comments and suggestions on an earlier version of this chapter.

References

1. Hazen, B. T., Overstreet, R. E., Jones-Farmer, L. A., Field, H. S.: The role of ambiguity tolerance in consumer perception of remanufactured products. Int. J. Prod. Econ. **135**(2), 781–790 (2012). https://doi.org/10.1016/j.ijpe.2011.10.011
2. Van Weelden, E., Mugge, R., Bakker, C.: Paving the way towards circular consumption: exploring consumer acceptance of refurbished mobile phones in the Dutch market. J. Clean. Prod. **113**, 743–754 (2016). https://doi.org/10.1016/j.jclepro.2015.11.065
3. Guide, V.J., Wassenhove, L.N.: Managing product returns for remanufacturing. Prod. Oper. Manag. **10**(2), 142–155 (2001). https://doi.org/10.1111/j.1937-5956.2001.tb00075.x
4. Mugge, R., Jockin, B., Bocken, N.M.P.: How to sell refurbished smartphones? an investigation of different customer groups and appropriate incentives. J. Clean. Prod. **147**, 284–296 (2017). https://doi.org/10.1016/j.jclepro.2017.01.111
5. Gurita, N., Fröhling, M., Bongaerts, J.: Assessing potentials for mobile/smartphone reuse/remanufacture and recycling in Germany for a closed loop of secondary precious and critical metals. J. Remanufacturing **8**(1–2), 1–22 (2018). https://doi.org/10.1007/s13243-018-0042-1
6. Gallo, M., Romano, E., Santillo, C.L.: A perspective on remanufacturing business: issues and opportunities. In: Bobck, V (ed.) International Trade from Economic and Policy Perspective. InTech (2012). https://doi.org/10.5772/48103
7. European Union: European Commission (2019). Communication from the Commission to the European Parliament, the European Council, the Council, the European Economic and Social Committee and the Committee of the Regions on the European Green Deal. 11/12/2019 [Press release]. COM(2019) 640 final. Retrieved 15/05/22, from: https://eur-lex.europa.eu/legal-content/EN/TXT/?qid=1596443911913&uri=CELEX:52019DC0640#document2
8. United Nations: THE 17 GOALS|Sustainable Development (2022). Accessed 15 May 2022, from United Nations: https://sdgs.un.org/goals
9. European Union: European Commission (2020). Communication from the Commission to the European Parliament, the European Council, the Council, the European Economic and Social Committee and the Committee of the Regions on a new Circular Economy Action Plan for a cleaner and more competitive Europe. 11/03/2020 [Press release]. COM(2020) 98 final. Retrieved 15/05/22, from: https://eur-lex.europa.eu/legal-content/EN/TXT/?qid=1583933814386&uri=COM:2020:98:FIN
10. Zlamparet, G. I., Ijomah, W. L., Miao, Y., Awasthi, A. K., Zeng, X., Li, J. [Jinhui]: Remanufacturing strategies: a solution for WEEE problem. J. Clean. Prod. **149**, 126–136 (2017). https://doi.org/10.1016/j.jclepro.2017.02.004
11. Neto, F., Bloemhof, J., Corbett, C.: Market prices of remanufactured, used and new items: Evidence from eBay. Int. J. Prod. Econ. **171**, 371–380 (2016). https://doi.org/10.1016/j.ijpe.2015.02.006
12. Vorasayan, J., Ryan, S.M.: Optimal price and quantity of refurbished products. Prod. Oper. Manag. **15**(3), 369–383 (2006). https://doi.org/10.1111/j.1937-5956.2006.tb00251.x
13. Forti, V., Baldé, C. P., Kuehr, R., Bel, G.: The Global E-waste Monitor 2020: Quantities, flows and the circular economy potential. Bonn/Geneva/Rotterdam: United Nations University/United Nations Institute for Training and Research, International Telecommunication Union, and International Solid Waste Association (2020). Accessed 15 May 2022, from: https://collections.unu.edu/view/UNU:7737
14. Arbeiterkammer: Handy-Lebensdauer (2021). Accessed 15 May 2022, from https://www.arbeiterkammer.at/interessenvertretung/umweltundverkehr/umwelt/abfall/Handy-Lebensdauer.html
15. Prakash, S., Dehoust, G., Gsell, M., Schleicher, T., Stamminger, R.: Einfluss der Nutzungsdauer von Produkten auf ihre Umweltwirkung: Schaffung einer Informationsgrundlage und Entwicklung von Strategien gegen "Obsoleszenz". Dessau-Roßlau (2016). Accessed 15 May 2022, from: https://www.umweltbundesamt.de/sites/default/files/medien/378/publikationen/texte_11_2016_einfluss_der_nutzungsdauer_von_produkten_obsoleszenz.pdf

16. Steinhilper, R.: Remanufacturing: The Ultimate Form of Recycling. Fraunhofer-IRB-Verl, Stuttgart (1998)
17. Hobson, K., Lynch, N., Lilley, D., Smalley, G.: Systems of practice and the circular economy: transforming mobile phone product service systems. Environ. Innov. Soc. Trans. **26**, 147–157 (2018). https://doi.org/10.1016/j.eist.2017.04.002
18. Geissdoerfer, M., Savaget, P., Bocken, N.M.P., Hultink, E.J.: The circular economy—a new sustainability paradigm? J. Clean. Prod. **143**, 757–768 (2017). https://doi.org/10.1016/j.jclepro.2016.12.048
19. Abbey, J.D., Meloy, M.G., Atalay, S., Guide, V.J.: Remanufactured products in closed-loop supply chains for consumer goods. Prod. Oper. Manag. **24**(3), 488–503 (2015). https://doi.org/10.1111/poms.12238
20. Hazen, B.T., Mollenkopf, D.A., Wang, Y. [Yacan]: Remanufacturing for the circular economy: an examination of consumer switching behavior. Bus. Strat. Environ. **26**(4), 451–464 (2017). https://doi.org/10.1002/bse.1929
21. Matsumoto, M., Umeda, Y.: An analysis of remanufacturing practices in Japan. J. Remanufacturing **1**(1), 1–11 (2011). https://doi.org/10.1186/2210-4690-1-2
22. Esmaeilian, B., Onnipalayam Saminathan, P., Cade, W., Behdad, S.: Marketing strategies for refurbished products: survey-based insights for probabilistic selling and technology level. Resour., Conserv. Recycl. **167**(1), 1–10 (2021). https://doi.org/10.1016/j.resconrec.2021.105401
23. Abbey, J.D., Meloy, M.G., Blackburn, J., Guide, V.J.: Consumer markets for remanufactured and refurbished products. Calif. Manage. Rev. **57**(4), 26–42 (2015)
24. Chen, J., Venkatadri, U., Diallo, C.: Optimal (re)manufacturing strategies in the presence of spontaneous consumer returns. J. Clean. Prod. **237**, 117642 (2019). https://doi.org/10.1016/j.jclepro.2019.117642
25. Sharma, V., Garg, S.K., Sharma, P.B.: Identification of major drivers and roadblocks for remanufacturing in India. J. Clean. Prod. **112**, 1882–1892 (2016). https://doi.org/10.1016/j.jclepro.2014.11.082
26. Lund, R.T.: Remanufacturing: the experience of the United States and implications for developing countries (1. print). Integrated resource recovery: Vol. 2. Washington. The World Bank, DC (1984)
27. European Union: European Parliament and Council of the European Union. Directive 1999/44/EC. On certain aspects of the sale of consumer goods and associated guarantees
28. Atasu, A., Guide, V.J., van Wassenhove, L.N.: So What if remanufacturing cannibalizes my new product sales? Calif. Manage. Rev. **52**(2), 56–76 (2010). https://doi.org/10.1525/cmr.2010.52.2.56
29. Guide, V.J., Li, J. [Jiayi]: The potential for cannibalization of new products sales. Remanufactured Prod. Decis. Sci. **41**(3), 547–572 (2010). https://doi.org/10.1111/j.1540-5915.2010.00280.x
30. Atasu, A., Sarvary, M., van Wassenhove, L.N.: Remanufacturing as a marketing strategy. Manage. Sci. **54**(10), 1731–1746 (2008). https://doi.org/10.1287/mnsc.1080.0893
31. Özer, H.S.: A review of the literature on process innovation in remanufacturing. Int. Rev. Manag. Mark. **2**(3), 139–155 (2012)
32. Cuhls, K.: Die Delphi-Methode—eine Einführung. In: Niederberger, M., Renn, O. (eds.) Lehrbuch. Delphi-Verfahren in den Sozial-und Gesundheitswissenschaften: Konzept, Varianten und Anwendungsbeispiele, pp. 3–31. Springer Fachmedien Wiesbaden, Wiesbaden (2019). https://doi.org/10.1007/978-3-658-21657-3_1
33. Skulmoski, J.G., Hartman, F., Krahn, J.: The delphi method for graduate research. J. Inf. Technol. Educ.: Res. **6**, 1–21 (2007). https://doi.org/10.28945/199
34. Senatsverwaltung für Umwelt, Mobilität, Verbraucher- und Klimaschutz. Re-Use-Superstore im Karstadt am Hermannplatz öffnet wieder - Berlin.de. 13/08/2021 [Press release] (2021). Accessed 15 May 2022, from: https://www.berlin.de/sen/uvk/presse/pressemitteilungen/2021/pressemitteilung.1095537.php

35. Wang, Y. [Yacan], Wiegerinck, V., Krikke, H., Zhang, H.: Understanding the purchase intention towards remanufactured product in closed-loop supply chains. Int. J. Phys. Distrib. Logist. Manag. **43**(10), 866–888 (2013). https://doi.org/10.1108/IJPDLM-01-2013-0011
36. Repanet: Wir über uns-Repanet (2021). Accessed 15 May 2022, from RepaNet–Re-Use- und Reparaturnetzwerk Österreich: https://www.repanet.at/ueberuns/
37. ReUse e.V.: https://reuse-verein.org/der-verein (2022). Accessed 15 May 2022, from ReUse e.V.: https://reuse-verein.org/der-verein
38. T-Mobile Austria: Refurbished Handy–Nachhaltig erneuerte Smartphones (2022). Accessed 15 May 2022, from T-Mobile Austria: https://www.magenta.at/handytarife/refurbished
39. Lin, Y., Ahmad, Z., Shafik, W., Khosa, S.K., Almaspoor, Z., Alsuhabi, H., Abbas, F.: Impact of facebook and newspaper advertising on sales: a comparative study of online and print media. Comput. Intell. Neurosci. **2021**, 5995008 (2021). https://doi.org/10.1155/2021/5995008
40. Chien, P.M., Cornwell, T.B., Pappu, R.: Sponsorship portfolio as a brand-image creation strategy. J. Bus. Res. **64**(2), 142–149 (2011). https://doi.org/10.1016/j.jbusres.2010.02.010
41. Bruhn, M., Holzer, M.: The role of the fit construct and sponsorship portfolio size for event sponsorship success. Eur. J. Mark. **49**(5/6), 874–893 (2015). https://doi.org/10.1108/EJM-09-2012-0517
42. Adidas, A.G.: Annual Report 2019 (2020). Accessed 15 May 2022, from: https://report.adidas-group.com/2019/en/
43. Nguyen, T., Galinski, L.: Digital-marketing, sportsponsoring und E-Sports–Warum E-Sports ein Blue Ocean ist. In: Terstiege, M. (ed.) Digitales Marketing—Erfolgsmodelle aus der Praxis: Konzepte, Instrumente und Strategien im Kontext der Digitalisierung, pp. 259–277. Wiesbaden, Heidelberg: Springer Gabler (2020). https://doi.org/10.1007/978-3-658-26195-5_16
44. Statistik Austria: Kraftfahrzeuge—Gebrauchtzulassungen (2022). Accessed 15 May 2022, from Statistik Austria: https://www.statistik.at/statistiken/tourismus-und-verkehr/fahrzeuge/kfz-gebrauchtzulassungen
45. Statistik Austria: Kraftfahrzeuge – Neuzulassungen (2022). Accessed 15 May 2022, from Statistik Austria: https://www.statistik.at/statistiken/tourismus-und-verkehr/fahrzeuge/kfz-neuzulassungen
46. Kraftfahrt-Bundesamt: Besitzumschreibungen (2022). Accessed 15 May 2022, from Kraftfahrt-Bundesamt: https://www.kba.de/DE/Statistik/Fahrzeuge/Besitzumschreibungen/FahrzeugklassenAufbauarten/2021/2021_u_fzkl_tabellen.html?nn=3527070&fromStatistic=3527070&yearFilter=2021&fromStatistic=3527070&yearFilter=2021
47. Kraftfahrt-Bundesamt: Neuzulassungen (2022). Accessed 15 May 2022, from Kraftfahrt-Bundesamt: https://www.kba.de/DE/Statistik/Fahrzeuge/Neuzulassungen/FahrzeugklassenAufbauarten/2021/2021_n_fzkl_tabellen.html?nn=3524574&fromStatistic=3524574&yearFilter=2021&fromStatistic=3524574&yearFilter=2021
48. Harms, R., Linton, J.D.: Willingness to pay for eco-certified refurbished products: the effects of environmental attitudes and knowledge. J. Ind. Ecol. **20**(4), 893–904 (2016). https://doi.org/10.1111/jiec.12301
49. Subramanian, R., Subramanyam, R.: Key factors in the market for remanufactured products. Manuf. Serv. Oper. Manag. **14**(2), 315–326 (2012). https://doi.org/10.1287/msom.1110.0368
50. Debo, L.G., Toktay, L.B., van Wassenhove, L.N.: Market segmentation and product technology selection for remanufacturable products. Manage. Sci. **51**(8), 1193–1205 (2005). https://doi.org/10.1287/mnsc.1050.0369
51. Michaud, C., Llerena, D.: Green consumer behaviour: an experimental analysis of willingness to pay for remanufactured products. Bus. Strat. Environ. (2010). https://doi.org/10.1002/bse.703
52. Lego, T.: Assessment of eCall's effects on economy and the automotive industry as well as its perception and future (2018)
53. Hatcher, G.D., Ijomah, W.L., Windmill, J.F.C.: Design for remanufacturing in China: a case study of electrical and electronic equipment. J. Remanufacturing **3**(1) (2013). https://doi.org/10.1186/2210-4690-3-3

54. Wallner, T.S., Magnier, L., Mugge, R.: An exploration of the value of timeless design styles for the consumer acceptance of refurbished products. Sustainability **12**(3), 1213 (2020). https://doi.org/10.3390/su12031213

55. Ijomah, W.L., McMahon, C.A., Hammond, G.P., Newman, S.T.: Development of design for remanufacturing guidelines to support sustainable manufacturing. Robot. Comput. Integr. Manuf. **23**(6), 712–719 (2007). https://doi.org/10.1016/j.rcim.2007.02.017

56. Lindkvist Haziri, L., Sundin, E.: Supporting design for remanufacturing—a framework for implementing information feedback from remanufacturing to product design. J. Remanufacturing **10**(1), 57–76 (2020). https://doi.org/10.1007/s13243-019-00074-7

57. Apple: Apple kündigt Self Service-Reparatur an (2021). Accessed 15 May 2022, from: https://www.apple.com/at/newsroom/2021/11/apple-announces-self-service-repair/

58. Charter, M., Gray, C.: Remanufacturing and product design. Int. J. Prod. Dev. **6**(3/4), 375–392 (2008)

59. Sundin, E., Bras, B.: Making functional sales environmentally and economically beneficial through product remanufacturing. J. Clean. Prod. **13**(9), 1–25 (2005). https://doi.org/10.1016/j.jclepro.2004.04.006

60. Becker, L.: USB-C im iPhone per Dekret: Apple hält EU-Vorhaben für innovationsfeindlich. Heise Online (2021). Accessed 15 May 2022, from: https://www.heise.de/news/USB-C-im-iPhone-per-Dekret-Apple-haelt-EU-Vorhaben-fuer-innovationsfeindlich-6200029.html

61. European Union: European Parliament: New rules on batteries: Meps want more environmental and social ambition. 10/03/2022 [Press release] (2022). Accessed 15 May 2022, from: https://www.europarl.europa.eu/news/en/press-room/20220304IPR24805/new-rules-on-batteries-meps-want-more-environmental-and-social-ambition

62. Zhou, Q., Meng, C., Yuen, K.F., Sheu, J.: Remanufacturing authorization strategy for an original equipment manufacturer-contract manufacturer supply chain: cooperation or competition? Int. J. Prod. Econ. **240**, 108238 (2021). https://doi.org/10.1016/j.ijpe.2021.108238

63. European Union: European Parliament and Council of the European Union. Directive 2009/125/EC. Establishing a framework for the setting of ecodesign requirements for energy-related products

64. Qiao, H., Su, Q.: Impact of government subsidy on the remanufacturing industry. Waste Manag. **120**, 433–447 (2021). https://doi.org/10.1016/j.wasman.2020.10.005

65. Zhou, Q., Yuen, K.F.: Analyzing the effect of government subsidy on the development of the remanufacturing industry. Int. J. Environ. Res. Public Health **17**(10) (2020). https://doi.org/10.3390/ijerph17103550

66. European Union: European Parliament: Parliament wants to grant EU consumers a "right to repair". 25/11/2020 [Press release] (2020). Accessed 15 May 2022, from: https://www.europarl.europa.eu/news/en/press-room/20201120IPR92118/parliament-wants-to-grant-eu-consumers-a-right-to-repair

67. Brand, R.: Ein Reparaturindex soll in Frankreich leicht zu reparierende Elektronik fördern. Heise Online (2021). Accessed 15 May 2022, from: https://www.heise.de/hintergrund/Ein-Reparaturindex-soll-in-Frankreich-leicht-zu-reparierende-Elektronik-foerdern-5994578.html

68. Dabove, P.: The use of smartphones in the 21st century. In: Dabove, P. (ed.) Smartphones: Recent innovations and applications. Nova Science Publishers, Inc., New York (2019)

69. Sappin, E.: Three Tech Innovations Leading The Smartphone To Its Grave. Forbes (2018). Accessed 15 May 2022, from: https://www.forbes.com/sites/forbesnycouncil/2018/03/22/three-tech-innovations-leading-the-smartphone-to-its-grave/?sh=226db4b16a07

70. Microsoft: Windows 10 Home und Pro—Microsoft Lifecycle: Supportzeiträume (2022). Accessed 15 May 2022, from https://docs.microsoft.com/de-de/lifecycle/products/windows-10-home-and-pro.

71. Niu, B., Zou, Z.: Better demand signal, better decisions? evaluation of big data in a licensed remanufacturing supply chain with environmental risk considerations. Risk Anal.: Off. Publ. Soc. Risk Anal. **37**(8), 1550–1565 (2017). https://doi.org/10.1111/risa.12796

72. Gellrich, A.: 25 Jahre Umweltbewusstseinsforschung im Umweltressort: Langfristige Entwicklungen und aktuelle Ergebnisse (2021). Accessed 15 May 2022, from: https://www.umweltbundesamt.de/sites/default/files/medien/5750/publikationen/2021_hgp_umweltbew usstseinsstudie_bf.pdf

73. Statistik Austria: Umweltbedingungen, Umweltverhalten: Ergebnisse des Mikrozensus 2019. Wien (2020). Accessed 15 May 2022, from: https://www.statistik.at/services/tools/services/publikationen/detail/1059?cHash=a2afdf8219722018e157232f69972710

74. Burley, H.: The land and water footprints of everyday products: Mind your step (2015). Accessed 15 May 2022, from: Friends of the Earth Trust website: https://catalogue.unccd.int/587_mind-your-step-report-76803.pdf

75. Verbraucherportal Baden-Württemberg: Labels: Labels auf Produkten hinterfragen (2021). Accessed 15 May 2022, from https://www.verbraucherportal-bw.de/,Lde/Startseite/Verbrauch erschutz/Labels

76. Niu, B., Xie, F.: Incentive alignment of brand-owner and remanufacturer towards quality certification to refurbished products. J. Clean. Prod. **242**, 118314 (2019). https://doi.org/10.1016/j.jclepro.2019.118314

77. Blancco Germany: Datenlöschsoftware|Sichere Datenvernichtung—Blancco Germany (2021). Accessed 15 May 2022, from https://www.blancco.com/de/

78. TÜV Süd: TÜV SÜD Zertifizierte Datenträgervernichtung–Information zu Zertifikaten und Prüfzeichen (2022). Accessed 15 May 2022, from https://www.tuvsud.com/de-de/dienstleistu ngen/auditierung-und-zertifizierung/pruefzeichenuebersicht/datentraegervernichtung

79. Bb-net media: Refurbished IT von tecXL—Technik wie neu | BB-NET (2022). Accessed 15 May 2022, from https://bb-net.de/tecxl/

80. Green Panda: RETEQ CERTIFIED – Green Panda—Gebrauchte Notebooks (2022). Accessed 15 May 2022, from https://www.greenpanda.de/refurbished/de/infos/ueber-reteq/

81. Apple: Warum refurbished Produkte? Zertifiziert refurbished (2022). Accessed 15 May 2022, from https://www.apple.com/at/shop/refurbished/about

82. Kryvinska, N., Kaczor, S., Strauss, C., Gregus, M.: Servitization strategies and product-service-systems. In: Zhang, L.-J. (ed.) 2014 IEEE World Congress on Services (SERVICES 2014): Anchorage, Alaska, USA, 27 June–2 July 2014, pp. 254–260. IEEE, Piscataway, NJ (2014). https://doi.org/10.1109/SERVICES.2014.52

83. Tukker, A.: Eight types of product–service system: eight ways to sustainability? experiences from SusProNet. Bus. Strateg. Environ. **13**(4), 246–260 (2004). https://doi.org/10.1002/bse.414

84. Mashhadi, A.R., Vedantam, A., Behdad, S.: Investigation of consumer's acceptance of product-service-systems: a case study of cell phone leasing. Resour. Conserv. Recycl. **143**, 36–44 (2019). https://doi.org/10.1016/j.resconrec.2018.12.006

85. Rousseau, S.: Millennials' acceptance of product-service systems: leasing smartphones in Flanders (Belgium). J. Clean. Prod. **246**, 118992 (2020). https://doi.org/10.1016/j.jclepro.2019.118992

86. Okoli, C., Pawlowski, S.D.: The Delphi method as a research tool: an example, design considerations and applications. Inf. Manag. **42**(1), 15–29 (2004). https://doi.org/10.1016/j.im.2003.11.002

87. IDC: IDC Forecasts Worldwide Market for Used Smartphones to Reach 351.6 Million Units with a Market Value of $65 Billion in 2024 (2021). Accessed 15 May 2022, from IDC: https://www.idc.com/getdoc.jsp?containerId=prUS47258521

88. Toffel, M.W.: Strategic management of product recovery. Calif. Manage. Rev. **46**(2), 120–141 (2004). https://doi.org/10.2307/41166214

89. Wirtz, B.W.: Electronic Business (3., vollst. überarb. und erw. Aufl.). Gabler, Lehrbuch, Wiesbaden (2010).

90. Kerin, M., Pham, D.T.: Smart remanufacturing: a review and research framework. J. Manuf. Technol. Manag. **31**(6), 1205–1235 (2020). https://doi.org/10.1108/JMTM-06-2019-0205

91. Statista: Österreich—Nutzung von Online-Shopping 2020|Statista (2021). Accessed 15 May 2022, from https://de.statista.com/statistik/daten/studie/298302/umfrage/nutzung-von-online-shopping-in-oesterreich/.

92. Austrian Retail Association: E-Com Report DACH 2020 (2021). Accessed 15 May 2022, from https://www.handelsverband.at/presse/presseaussendungen/e-com-report-dach-2020/
93. Austrian Institute for SME Research: E-Commerce-Studie Österreich 2021: Die 12. Studie zum Konsumentenverhalten im Distanzhandel (2021). Accessed 15 May 2022, from: https://www.handelsverband.at/fileadmin/content/images_events/eCommerce_Day_2021/Praesentationen_ECD2021/07_KMUForschung_eCommerceStudie2021.pdf
94. Delbecq, A.L., van de Ven, A.H., Gustafson, D.H.: Group Techniques for Program Planning—A Guide to Nominal Group and Delphi Processes. Green Briar Press, Middleton, Wisconsin (1986)

Nuclear Waste Potential and Circular Economy: Case of Selected European Country

Mária Srebalová, Tomáš Peráček, and Boris Mucha

Abstract The scientific study focuses on analysis of baselines for the use of spent fuel and radioactive waste in circular economy within the broader framework of state energy security. On the basis of both European and national strategies and existing legislation, it defines the essence of the circular model of the economy and within it the concept and meaning of waste management. It analyses the approach and regulation for the management of a specific type of waste, which is nuclear waste compared to other wastes. On this basis, it examines "de lege lata" possibilities for further use of spent nuclear fuel and radioactive waste in the increasing demands of the company for energy sources. It assesses the direction and impact of strategic documents at both international and national level in relation to the analyzed topic. When processing the study, the method of primary analysis of legislation and strategic documents was applied. As a result of our research, we provide answers to questions about the possibility of using nuclear waste and proposals de lege ferenda. The added value of our scientific study is especially in a broader context to critically highlight selected legislative problems and to propose ways to address them.

M. Srebalová
Faculty of Law, Comenius University in Bratislava, Šafárikovo Námestie 6, Bratislava, Slovak Republic
e-mail: maria.srebalova@flaw.uniba.sk

T. Peráček (✉)
Faculty of Management, Comenius University in Bratislava, Odbojárov 10, Bratislava, Slovak Republic
e-mail: peracek2@uniba.sk

B. Mucha
Ministry of the Interior of the Slovak Republic, Pribinova 2, Bratislava, Slovak Republic

1   Introduction

The Slovak Republic is one of the countries with high waste production of various types [3]. The main reason is the current economic model, which can be called linear. Within its framework, the raw materials used for the production of products become waste in landfills or incineration plants. As the materials of the Ministry of the Environment of the Slovak Republic show, this causes extensive environmental risks and raw materials are rapidly depleted with worldwide increasing consumption [48]. On the contrary, the economic model, which is as far as possible waste-free, ideally closed, is called circular economy or circular economy [44]. This model tends to be seen as part of the so-called green economy. It already consumes less natural resources and less energy in the production of products. Materials are reused. The green economy, as the broadest of these concepts, represents a shift in economic development towards sustainable business and infrastructure practices [6]. It focuses on a number of areas, the most important of which are:

– adaptation to climate change;
– circular economy and sustainable use of resources;
– sustainable transport,
– energy efficiency,
– green buildings and housing,
– sustainable management in the country;
– a sustainable bioeconomy [50].

The circular economy is based on resource efficiency. It covers all links in the value chain. From production through consumption, repair and remanufacturing, waste management to the return of raw materials back to the economy and their use in the production cycle as secondary raw materials [46]. It is a trend in modern European legislation. It focuses on ambitious targets for recycling municipal waste, packaging waste, landfilling, sorting of bio-waste, textiles and hazardous municipal waste, as well as extended producer responsibility [1]. Waste and management is therefore a key factor in the circular economy. In the Slovak Republic, for a long time, the centre of interest of the public, state institutions and legislation has been primarily municipal waste, the issue of collection of kitchen biowaste and the return of disposable beverage packaging.

However, waste is generated on a much wider scale and in specific areas of the economy [2]. The question of whether waste management is aimed at all types of waste and whether development documents and subsequent legislation will remember them will be examined on the basis of current policies, strategies and legislation at the level of the European Union and the Slovak Republic. We will focus in particular on the recovery of waste arising from the peaceful use of nuclear energy [20].

According to Filip et al. [29] the operation of nuclear installations and the use of nuclear energy in the Slovak Republic is an important factor for energy security, which is now crucial for the overall level of security of states [29]. On nuclear energy, we find conflicting views among scientists, in the public and between the different states of the European Union. However, a number of actors consider it to be a very environmentally friendly way of achieving the appropriate operating conditions of nuclear installations, including safe transport of nuclear fuel or the introduction of accident prevention plans. For example, this way of generating electricity does not generate greenhouse gases. Čajková and Gogová [5] argues that, however, the production of spent nuclear fuel and radioactive waste is an integral part of this way of generating energy [5]. These are, on the contrary, extremely dangerous for the environment and human health. Horvat and Lysá consider that energy security and environmental safety have a number of common points. Environmental safety prefers the use of energy resources that are environmentally friendly and at the same time accentuate its preservation for future generations [38].

2 Theoretical Background

According to Nováčková and Vnuková [54], the European Union's environment policy in relation to the circular economy has a number of objectives in general:

1. to maintain, protect and improve the quality of the environment;
2. to protect human health,
3. prudent and rational use of natural resources;
4. promote measures at international level to address regional or global environmental problems and, in particular, action against climate change [54].

In the most general sense, the legal basis for the creation and functioning of the circular economy is the Treaty on the Functioning of the European Union (TFEU) [4]. In Articles 11 and 191 to 193, it favours the reuse and recycling of things and promotes sustainable and renewable resources and materials. In December 2019, the European Commission adopted the 'European Green Deal', which sets out the Green Economy Action Plan, which is an important part of the European Commission's strategy to deliver on the 2030 Agenda and the Sustainable Development Goals formulated by the United Nations [23].

The Green Deal is a new growth strategy. It aims to transform the European Union's economy for a sustainable future. The circular economy is one of the key areas of this document and the European Commission has set out to prepare a new Circular Economy Action Plan focusing on the sustainable use of resources as well as a new industrial strategy based on circular economy principles [40, 43].

The Green Deal expresses Europe's ambition to be carbon neutral by 2050. The adoption of the above-mentioned document is expected to accelerate and support the transformation of all manufacturing sectors, with an emphasis on protecting and enhancing natural resources and protecting the health of the population from environmental risks and impacts. One of the ways to prevent, eliminate or reduce them is consistent risk management [7].

Ensuring the transition to a circular economy and the use of smart solutions are crucial. The circular economy offers an opportunity to increase sustainability and competitiveness and creates space for green investments [41]. This Circular Economy Action Plan also includes measures to promote green goods and services [28]. The Green Convention's climate headline target is enshrined in the 2020 European Climate Law (the so-called European Climate Law). As part of the Ecological Convention, this law enshrines in the EU legal order the legally binding objective of achieving EU climate neutrality by 2050. It also commits the European Commission to review the current 2030 climate target. As a follow-up, in September 2020, the European Commission published a set of measures, the 'Green Package', aimed at increasing climate and energy targets [8].

The Green Deal is also a new growth strategy focused on resource efficiency. It aims to achieve a climate-neutral and clean circular economy and the decarbonisation of energy [67]. This will require a full modernisation of industry, in particular energy-intensive industries such as steel, chemical and cement industries, resource-intensive industries such as textiles, construction, electronics and plastics. Key documents for achieving these objectives are the new European Industrial Strategy 2020 and the new Circular Economy Action Plan. For a cleaner and more competitive Europe, adopted in March 2020 [74]. The new EU Industrial Strategy sets out a set of measures to ensure that the European Union's industry can cope with the green and digital transitions, reap their benefits and maintain global competitiveness, and strengthen the EU's strategic autonomy. The new Circular Economy Action Plan includes measures to contribute to reducing pressure on the environment and natural resources, changing consumer behaviour, promoting recycling and the use of secondary raw materials, reducing waste generation and moving towards sustainable production and consumption patterns. This includes the establishment of a sustainable product policy framework, consumer empowerment, the introduction of "circularity" in production processes, the creation of a well-functioning EU secondary raw material market, the introduction of minimum binding criteria for green public procurement, a proposal for harmonisation of separate waste collection systems and a comprehensive monitoring framework for the circular economy, and other measures. This Action Plan sets out a new long-term strategic framework that underpins the industrial transition and one of the key tools to achieve the climate neutrality objective by 2050 [77].

In point 2.1.3. The Green Deal particularly highlights access to resources, which is a strategic issue of security. In particular, it aims to ensure the supply of sustainable raw materials from both primary and secondary sources. Key sectors of EU industry need new breakthrough technologies in the field of climate, which could be commercially applied. It prioritises clean hydrogen, fuel cells and other alternative fuels, energy storage and carbon capture, storage and utilisation.

For the sake of completeness, according to Clark et al. [9], it should be noted that, as part of the transition towards a sustainable economic model, Europe is building on resource efficiency activities [9]. In 2015, the European Commission adopted an EU action plan for the circular economy. It contains concrete measures covering all elements of the value chain from production to consumption, repair and remanufacturing, waste management to the return of raw materials to the return to the economy and their use in the production cycle already as secondary raw materials [23].

The transition towards a circular economy also includes a revision of a number of European waste directives, which include ambitious recycling targets for both municipal and packaging waste, landfilling, sorting of biowaste, textiles and hazardous municipal waste, and extended producer responsibility. As a follow-up to the Circular Economy Action Plan, the Commission has also developed a European Strategy for Plastics in a Circular Economy. Following this document, for example, a directive has been adopted which prohibits or restricts certain product groups of single-use plastics, which make up the majority of waste on European beaches.

3 Strategic Elements of the Slovak Republic

Already in 2016, during the Slovak Presidency of the Council of the EU, the Ministry of Environment of the Slovak Republic launched theso-called "Transition to green economy"process as a platform for discussing the benefits of a green and circular economy. According to the Programme Declaration of the Government of the Slovak Republic for 2020–2024, Slovakia is to actively promote national interests and, together with other states, shape the EU's environmental and climate policy. It shall act actively and responsibly at the international level and support the achievement of the Sustainable Development Goals through the use of the experience of the transformative and accession processes [35].

The transition to a circular economy for the Slovak Republic is also defined as one of the strategic priorities of the Environmental Policy Strategy 2030 (hereinafter referred to as "the 2030 Environment Strategy"). Its core vision is to achieve better environmental quality and a sustainable circular economy. It should use as few non-renewable natural resources as possible and toxic or otherwise hazardous substances. On the basis of internationally comparable indicators, which measure the level of performance achieved in individual areas of the environment, Slovakia lags behind developed countries especially in waste management and air quality. For example, the Slovak Republic achieves one of the lowest levels of municipal waste recycling in the EU. Landfilling is predominant and, on the contrary, its rate is one of the highest in the EU [36]. Let us outline the circular economy objectives under Chap. 10 of the 2030 Enviro Strategy, 'Towards a circular economy'.

By 2030, the recycling rate of municipal waste, including preparing for re-use, will increase to 60% in Slovakia and the landfilling rate will be reduced to less than 10% by 2035. Emphasis should be placed on waste prevention, the creation of conditions for priority use of residues and recovered industrial waste to reduce the use of natural materials, the eco-design of products and the lack of technologies to treat certain types of waste so far. The measures for the above objectives are: increasing landfill fees, introducing incentives for quantity collection, enforcing the polluter-pays principle in particular for the prevention of black landfills and preventing the generation of biodegradable and food waste. In particular, according to the Enviro Strategy 2030, the recycling rate of municipal waste, including its preparation for re-use, is to be increased.

The Waste Management Programme and the Waste Prevention Programme are important waste management planning documents which Member States are required to draw up under Directive 2008/98/EC of the European Parliament and of the Council of 19 November 2008 on waste and repealing certain Directives (the so-called Framework Directive). The Slovak Waste Prevention Programme 2019–2025, as a strategic document, identified as the main objective a shift from material recovery to waste prevention [35]. It also sets specific targets for nine types of waste: mixed municipal waste, biodegradable waste, food waste, paper waste, bulk waste, plastics and packaging, construction waste, hazardous waste and extractive waste. Indicators are set for each type of waste to evaluate the achievement of the target, together with the possibilities for their financing.

The circular economy is also covered by the Economic Policy Strategy of the Slovak Republic up to 2030 [42]. In addition to the Ministry of Economy, it also includes tasks for the Ministry of the Environment. According to paragraph 16, this is about creating the conditions for encouraging the closure of material waste streams in selected industries through the re-use of recyclates.

As part of this task, it is necessary to:

– to promote resource efficiency in the economy of the Slovak Republic by identifying material waste streams;
– by means of instruments to support the reuse of recyclates, then close them;
– when placed on the market, establish a mechanism to promote incentives for the marketing of products from recycled materials in the form of tax and environmental benefits, the mandatory procurement of such products by national and local authorities and the prior use of such materials in investment activities.

The Slovak Republic's 2030 low carbon development strategy adopted on the basis of Article 15 of the EP and Council (EU) 2018/1999 provides a long-term strategic outlook for the transition to a low-carbon economy. It ensures consistency with other strategic documents and action plans within the national economy (energy, industry, transport, agriculture and forestry, waste). It introduces binding and indicative targets for individual areas. It also offers a list of measures and funding options for achieving climate neutrality by 2050 [25, 26].

The Waste Management Programme of the Slovak Republic 2021–2025 is an important strategic document of the Government of the Slovak Republic, which determines the direction of waste management for a specified period. It reflects the national waste management policy and covers the entire territory of the Slovak Republic [34]. It is developed in line with the requirements of sustainable growth under the Circular Economy Action Plan, the European Green Deal and the new Circular Economy Action Plan. It complies with the requirements of European Union and Slovak law. In this context, according to Žofčinová et al. [73] it is necessary to add that in February 2021, the European Parliament approved non-legislative material called the EU Circular Economy Action Plan, recommending measures to Member States to reduce waste production and achieve the transition to a circular economy. For example, it supports short local recovery chains, supports local governments in developing circular economy centers and sharing expertise in waste management [75].

In relation to the topic analyzed, we will deal more closely with the links between circular management and waste management from the point of view of legislation [33].

4 Legislative Baselines for Waste Management in the Slovak Republic

Based on the analysed link between the economic model of circular waste management as an important part of it, we will come closer to the legislative anchoring of waste management in the following text [32]. Waste management is generally defined as a set of activities aimed at preventing, reducing and managing waste generation and environmental hazards. Waste management has fixed targets, with binding order of its priorities. The main objective is to prevent or reduce the adverse impacts of waste generation and waste management and to reduce the overall impacts of resource use and to increase the efficiency of such use of resources. Waste management priorities are applied within the binding waste hierarchy:

- prevention of waste generation,
- preparing for re-use,
- recycling,
- other recovery, such as energy,
- disposal.

A departure from a specified waste hierarchy is possible only for certain types of waste, and this is also a reason considering the life cycle in relation to the overall impacts of the generation and management of such waste. The basis of the legislation on waste management is primarily from Act No. 79/2015 Col. on waste and amending certain acts, as amended ('the Waste Act'). The Waste Act has the status of general legislation when regulating waste management. Special legislation must be followed for specific types of waste. This also follows from the legal definition of waste and the negative definition of waste under the Waste Act, which is not covered by the Waste Act [53].

According to Paragraph 2(1) of the Waste Act, waste is a movable thing or a substance which the holder discards or intends to discard or is obliged to dispose of it in accordance with the Waste Act or other special regulations [31]. However, waste is not a substance or movable thing which is a by-product and thus fulfils the following conditions:

(a) is the result of a production process the primary objective of which is not the production of that substance or object;
(b) its continued use is ensured,

(c) may be used directly without further processing other than normal industrial processes,
(d) it is formed as an integral part of the production process,
(e) its further use complies with this Act and with specific regulations laying down requirements for the product, the protection of the environment and the protection of human health with regard to its particular use and does not lead to overall adverse impacts on the environment or human health;
(f) it fulfils specific criteria if they have been laid down for the substance or object, and
(g) consent has been given that a substance or movable object is considered as a by-product and not as waste.

The Waste Act does not apply to a listed category of waste, among which radioactive waste is listed. According to Sect. 1 (2), the Waste Act also does not apply to, for example, natural agricultural or forestry material, the management of air pollutants, the management, capture, transport and permanent storage of carbon dioxide in the geological environment, the management of precious metal waste and the like. Keiveillant et al. (2021) adds that the management of radioactive waste is carried out in accordance with Sect. 2 (k) of Act No 541/2004 Col. on the peaceful use of nuclear energy (the Atomic Act) and amending certain acts. Waste management means the collection, transport, recovery and disposal of waste, including the supervision of these operations and the subsequent care of disposal sites [45]. This includes the conduct of the trader and the intermediary. Preparing waste for re-use means a recovery operation related to inspection, cleaning or repair in which a product or part thereof that has become waste is prepared for re-use without any other pre-treatment [39]. Waste recovery is an activity the main result of which is the beneficial use of waste in order to replace other materials in a manufacturing activity or in the wider economy. Disposal of waste is the opposite of recovery, even if the secondary result is the recovery of substances or energy.

5 Legislative Baselines of the European Union in the Field of Spent Nuclear Fuel and Radioactive Waste Management

The legislation on the management of spent nuclear fuel is formed at several levels. At the level of international law, European Union law and national law [37]. As an EU Member State, the Slovak Republic is bound by its legislation. The field of nuclear energy, including the management of spent nuclear fuel and radioactive waste, regulates a significant number of sources, out of which the most essential ones, consisting of primary sources but also secondary EU law.

- Treaty establishing the European Atomic Energy Community (1957),

- Council Regulation (Euratom) No 1493/93 of 8 June 1993 on shipments of radioactive substances between Member States, as amended, OJ L 148 [17],
- Commission Regulation (EC) No 1209/2000 of 8 June 2000 on the notification under Article 41 of the Treaty establishing the European Atomic Energy Community, as amended by Commission Regulation (Euratom) No 1352/2003 of 23 July 2003, OJ L 138, 9. 6. 2000 [18],
- Commission Regulation (Euratom) No 302/2005 of 8 February 2005 on the application of the Euratom safeguards scheme, OJ L 54, 28. 2. 2005,
- Council Directive 2006/117/Euratom of 20 November 2006 on the supervision and control of shipments of radioactive waste and spent fuel, OJ L 337, 5 December 2006 [12],
- Council Regulation (EC) No 428/2009 of 5 May 2009 setting up a Community regime for the control of exports, transfer, brokering and transit of dual-use items—Last amendment—Commission Delegated Regulation (EU) 2020/1749 of 7 October 2020 amending Council Regulation (EC) No 428/2009 setting up a Community regime for the control of exports, transfer, brokering and transit of dual-use items, OJ L 134, 29 May 2009 [16],
- Council Directive 2009/71/Euratom of 25 June 2009 establishing a Community framework for the nuclear safety of nuclear installations—as amended by Council Directive 2014/87/Euratom of 8 July 2014 amending Directive 2009/71/Euratom establishing a Community framework for the nuclear safety of nuclear installations, OJ L 172, 2 July 2009 [13],
- Council Directive 2011/70/Euratom of 19 July 2011 establishing a Community framework for the responsible and safe management of spent fuel and radioactive waste, OJ L 199, 2 August 2011 [14],
- Consolidated version of the Treaty establishing the European Atomic Energy Community (2016/C 203/01) OJ C 203, 26. 10. 2012,
- Directive 2012/18/EU of the European Parliament and of the Council of 4 July 2012 on the control of major-accident hazards involving dangerous substances, amending and subsequently repealing Council Directive 96/82/EC, OJ L 197, 24 July 2012 [27].
- Council Directive 2013/59/Euratom of 5 December 2013 laying down basic safety standards for protection against the dangers arising from exposure to ionising radiation and repealing Directives 89/618/Euratom, 90/641/Euratom, 96/29/Euratom, 97/43/Euratom and 2003/122/Euratom [15],
- Commission Recommendation 2006/851/Euratom of 24 October 2006 on the management of funds for the decommissioning of nuclear installations and the management of spent fuel and radioactive waste, OJ L 330, 28 November 2006, [11]
- Commission Recommendation 2008/956/Euratom of 4 December 2008 on criteria for the export of radioactive waste and spent fuel to third countries, OJ L 338 [10].

European Union legislation thus forms a safety framework for the operation of nuclear installations, lays down basic safety standards for protection against the dangers arising from ionising radiation and in the management of radioactive waste and spent nuclear fuel. It also accentuates responsible and safe handling of them [23, 24].

6 Sources of Legislation of the Slovak Republic in the Field of Peaceful Uses of Nuclear Energy

Slovak legislation in the field of peaceful uses of nuclear energy is being built in accordance with European legislation and in accordance with international recommendations and requirements [30]. The strategic documents expressing the policy, principles and development strategy in this area are those relating to energy security and environmental security:

1. The Enviro Strategy 2030;
2. The Economic Policy Strategy of the Slovak Republic up to 2030,
3. The Government Programme Statement,
4. Integrated National Energy and Climate Plan 2021–2030,
5. Energy Security Strategy of the Slovak Republic,
6. Strategy for the Final Part of Nuclear Energy [35].

The policy, principles and strategy for the further development of the use of nuclear energy shall be based on and develop fundamental development documents in the field of the environment in accordance with the principles of nuclear safety [47]. Similarly, they are based on the fundamental objectives of the transition towards a green, low-carbon and inclusive economy. They emphasise the need to use both economic and green energies that meet sustainability criteria. They gradually foresee a decline in coal-based electricity generation and a reduction in dependence on fossil fuel imports, with a shift towards more environmentally acceptable alternatives [19].

According to the Nuclear Regulatory Authority of the Slovak Republic, the main objective of the 2014 Strategy of the Final Part of Peaceful Uses of Nuclear Energy is to further develop the state policy on the final part of the peaceful use of nuclear energy into specific objectives and activities in order to protect the environment more effectively from the long-term consequences of the use of nuclear energy [49].

However, it should be noted that developing strategies, particularly in the area of nuclear energy, is not a closed topic for EU Member States. National positions on how to achieve the objectives of the Ecological Convention vary [73]. The current situation points to the fact that the energy security of the Slovak Republic is not stable. The energy raw materials that form the energy mix of the Slovak Republic are in the vast majority imported from abroad, gas, oil and nuclear fuel from the Russian Federation. Domestic production is minimal. There are no major deposits of coal, oil, natural gas or uranium ores in Slovakia. In the Slovak Republic, energy sources are of different origins, but more than 50% are produced from the core. According to Canyas et al. (2021) fossil sources (20%), water sources (16%) and electricity also come from renewable sources other than water (6%) [65].

With regard to the use of nuclear energy, it should be noted that this item of pre-accession negotiations with the European Union was a particularly important point for several of its Member States. In the field of "Energy", the Slovak Republic therefore committed to shut down two units of the Jaslovské Bohunice Nuclear Power Plant. The closure of these blocks took place in 2006 and 2008, when more than 25 years of commercial operations were closed. This moment was important from the point of view of the energy security of the Slovak Republic, because until that point the Slovak Republic was an exporter of electricity, after which, on the contrary, it became its importer [70]. Within the Jaslovské Bohunice nuclear power plant, two more units are still in operation.

There is also the Mochovce Nuclear Power Plant on the territory of the Slovak Republic. Currently, its two units are in operation, for which it is planned to shut down in about 2060. The other two units at the Mochovce nuclear power plant are under construction. (UJD) Nuclear installations on the territory of the Slovak Republic are also:

(1) technologies for the treatment and treatment of radioactive waste,
(2) intermediate storage of spent nuclear fuel,
(3) the final treatment of liquid radioactive wastes,
(4) the Republican Radioactive Waste Repository,
(5) integral radioactive waste warehouse (UJD).

After completion of the third unit of the Mochovce nuclear power plant, nuclear power sources in Slovakia should exceed 70%. Such a high current and planned share puts us second in the European Union countries after France. We also include the Czech Republic among the states with a high share of nuclear energy. These countries associate their energy security with the use of nuclear energy for the future. In line with the requirements of carbon neutrality, this source needs to be low-emission and stable in the long term. It is not realistic in the future to reduce energy consumption, especially electricity [76]. Whether the process of obtaining nuclear energy will be classified as a sustainable technology and whether it is actually possible to count on the transition to the green economy depends on strategic decisions. One of these should be the outcome of the procedure called upon by the European Commission. One of the preconditions for the necessary decisions is the addition of the so-called taxonomy of sustainable investment. This followed up on the Action Plan on Financing Sustainable Growth, which aims to mobilise the private sector to move towards a low-carbon, more resource-efficient and sustainable economy. The taxonomy is a means of identifying investment opportunities for projects and economic activities that contribute significantly to climate and environmental objectives. This is intended to help investors and companies change their activities to make them more environmentally sustainable or in line with the objectives of the [72]. This document identified as environmentally sustainable activities the use of water and marine resources, the circular economy, pollution prevention and control, and the protection and restoration of biodiversity and ecosystems. However, the taxonomy already encoded its addition and extension to other objectives. Once the full text is finalised, the taxonomy is intended to form an integral part of the sustainability reporting by both financial and non-financial corporations. According to the European Green Deal, the taxonomy will also he taken into account in the guidance documents on tracking climate and environmental progress and on sustainability verification under Invest EU. Despite the opposition of most Member States to the production and use of nuclear energy, the European Commission presented on 2 February 2022, at the insistence of several EU Member States, a complementary climate delegated act supplementing the Taxonomy Regulation. This was preceded by intense political negotiations in Brussels supported by the French presidents and prime ministers of six EU Member States (Czech Republic, Hungary, Poland, Romania Slovakia and Slovenia). These political efforts have been underpinned by the concerns of the nuclear sector and citizens' initiative about the possible far-reaching negative impacts of non-integration of nuclear energy into the so-called taxonomic framework. The European Parliament and the Council of the EU have set up four months to examine the delegated act and raise objections. The delegated act is the result of efforts to mitigate and adapt to climate change, covering in particular gas and nuclear activities. It responds to the fundamentally different positions of the EU Member States [52].

This process leading to the presentation of a taxonomy is intended to direct the flow of funds towards sustainable activities in the European Union. With its rules, the European Commission should help meet the objectives of the Green Deal also in relation to the EU's commitment to carbon neutrality by 2050 [71].

Changing the approach to the use of nuclear energy can address the issue of the safety of nuclear installations and the safe management of nuclear waste.

EU legislation already aims at improving safety standards for the operation of nuclear power plants, ensuring the safe management and disposal of nuclear waste. The need for a change in the European energy policy outlined above should provide EU Member States and its citizens with greater guarantees of safety in the use of nuclear energy, including its final stages. Coordination is essential as nuclear safety transcends national borders [59].

7   Nuclear Waste Management in the Slovak Republic

The legislation on the management of radioactive waste in the Slovak Republic is based on three basic pillars:

- international treaties and conventions by which the Slovak Republic is bound as a member of the International Atomic Energy Agency (IAEA);
- directives and regulations of the Council of the European Atomic Energy Community, by which the Slovak Republic is bound by its membership of the EU;
- strategy and development documents.

The legislative bases form within the framework of national legislation laws and other generally binding legislation in the field of nuclear safety. The basis of the legislation for the peaceful use of nuclear energy at national level is Act No 541/2004 Coll. on the peaceful use of nuclear energy (the Atomic Act) and, on the basis thereof, the decrees issued. It regulates basic concepts and requirements for the safe management of radioactive waste as a fundamental piece of legislation for the management of radioactive waste. The Atomic Act is followed by Act No. 308/2018 Coll. on the National Nuclear Fund and on amendments to the Atomic Act. Both fundamental laws stem from a broader strategy for the management of radioactive waste, which includes, for example, procedures for its treatment into a form suitable for disposal or long-term storage, as well as disposal and long-term storage in surface or deep storage sites.

Nuclear safety means the state and ability of a nuclear installation or transport facility and its operation to prevent the uncontrolled development of fissile chain reaction or illicit release of radioactive substances or ionising radiation into the working environment or into the environment, and to limit the consequences of accidents and accidents of nuclear installations or the consequences of events in the transport of radioactive materials. Nuclear safety, physical security and emergency preparedness are primarily the responsibility of the permit holder, who is required, in particular, to comply with the documentation assessed or approved, to comply with and evaluate the fulfilment of the principles of peaceful use of nuclear energy, in particular the principle that the use of nuclear energy must be fully justified by the benefits which manifestly outweigh any risks arising from such activities, in particular in comparison with the different ways in which the same objective can be achieved. The competent administrative body for the assessment of documentation is the Nuclear Regulatory Authority of the Slovak Republic.

The production and use of nuclear energy is linked to the concept of nuclear fuel cycle. This represents a series of activities and processes related to the use of uranium as a basic fuel for electricity production. According to Galamboš and Krajňák [30] by fission of 235U by thermal neutrons, heat is obtained from the fresh nuclear fuel in the power reactor, which is converted into electricity in the turbogenerator [30]. The nuclear fuel cycle consists of several stages:

– extraction and treatment of uranium ore,
– conversion and enrichment,
– production of fuel,
– use of fuel in the reactor,
– spent fuel storage
– reprocessing of fuel
– definitive storage of spent fuel.

The conversion of nuclear energy into electricity includes processes associated with the production of spent nuclear fuel and radioactive waste.

Nuclear law distinguishes between spent nuclear fuel and radioactive waste. On the other hand, it does not regulate the concept of nuclear waste as a legislative concept, nor does it provide a legal definition of nuclear waste. For the purposes of our research into the use of these radioactive materials in the circular economy, we will, on the contrary, work with the notion of nuclear waste in the sense that it must be understood as both spent nuclear fuel and radioactive waste. Spent fuel is a nuclear fuel that has been irradiated in the core of a nuclear reactor and has been permanently removed from it.

According to the Atomic Act, spent nuclear fuel (hereinafter referred to as "NPP") is of a dual legal nature. Where possible, it can be considered as a usable resource that can be reprocessed or as radioactive waste that can be identified for disposal. Radioactive waste means any unused material in gaseous, liquid or solid form which cannot be introduced into the environment because of its radionuclide content or its level of contamination by radionuclides. Thus, the VPS cannot automatically be generalized and characterized as waste under the Waste Act [22].

Radioactive waste is generally a material produced as a by-product of a nuclear reaction without further use. According to the Atomic Act, it is an unusable material in liquid, solid or gaseous form, which cannot be introduced into the environment because of the radionuclide content. Decree of the Nuclear Regulatory Authority of the Slovak Republic No. 30/2012 Coll. laying down details on requirements for the management of nuclear materials with radioactive waste and spent nuclear fuel divides radioactive waste into classes.

Transient radioactive waste—wastes which, due to their very short half-conversion period, fall below the limit for their safe introduction into the environment. Very low-activity radioactive waste, the activity of which is slightly higher than the limit value.

Low-activity radioactive waste whose mass activity of radionuclides is less than 400 Bq/g, does not produce residual heat and, after treatment, meets the limits and conditions of safe operation for the surface type of radioactive waste disposal site.

Medium-active radioactive wastes with a radionuclide mass activity of more than 400 Bq/g may produce residual heat and the measures to extract it are lower than for highly active radioactive waste after treatment do not meet the limits and conditions of safe operation for the surface type of radioactive waste disposal site.

High-activity radioactive wastes which can only be deposited in a deep type of radioactive waste disposal site, the residual heat recovery measures being an important factor in the design of these disposal sites [56].

The current strategy for radioactive waste management in the Slovak Republic is based on the following steps:

- treatment of radioactive waste into a form suitable for disposal or long-term storage;
- disposal of low and medium radioactive radioactive waste in a surface repository and long-term storage of radioactive wastes unacceptable in a surface repository;
- research and development of a deep storage site for the disposal of radioactive waste unacceptable in a surface storage site [57].

Requirements for the management of radioactive waste are laid down in the Atomic Act and Decree of the Nuclear Regulatory Authority of the Slovak Republic No. 30/2012 Coll. laying down details of requirements for the management of nuclear materials, radioactive waste and spent nuclear fuel. Radioactive waste management is a key term for an integrated system for the collection, sorting, storage, treatment,

treatment, handling and disposal of radioactive waste from nuclear installations, as well as for the treatment and disposal of so-called institutional radioactive waste. The management of radioactive waste means its disposal, including the final stages of the management of institutional radioactive waste [59]. Currently, disposal is provided in the framework of the surface disposal of radioactive waste on the basis of the consent issued by the Nuclear Regulatory Authority of the Slovak Republic. Shipments of radioactive waste are subject to the issue of a permit for shipments of radioactive material upon request. We include the term radioactive waste in connection with its shipment under the broader term radioactive material [68]. The shipment of radioactive materials is the loading of nuclear material, radioactive waste from a nuclear installation, spent fuel, IRAO, orphan and unused sources, radioactive waste of unknown origin at the place of loading, their transport and unloading at the place of destination. These activities are carried out both within a nuclear installation and between nuclear installations. In addition to the general conditions, the applicant must also meet specific conditions. It is about:

– documentation of the quality management system for the authorised activity,
– provisional physical protection plan or physical protection plan;
– categorisation of selected devices into safety classes,
– the internal emergency plan of the nuclear installation, the plan for the protection of the population and the emergency traffic order.

The storage of radioactive waste or spent fuel is to be understood as meaning only the temporary placement of radioactive waste or spent fuel in certain premises or facilities enabling their isolation, control and protection of the environment with a view to removing them in the future [59]. On the other hand, when we are talking about the disposal of radioactive waste or spent fuel, we are talking about the permanent placement of radioactive waste or spent fuel in a radioactive waste disposal site or a spent fuel disposal site, in which case there is no intention to remove it [63].

The disposal of radioactive waste or spent fuel is defined in the Atomic Act as permanent storage. Its safety is achieved by isolation of treated RAO from the environment using engineering and natural barriers. For surface storage RAO is in operation in the Slovak Republic Repository RAO in Mochovce, which obtained approval for permanent operation in November 2001. In order to ensure the storage of RAOs, which cannot be stored in the surface type of storage site, alternatives are currently being considered, one of which is the construction of a deep storage site in the Slovak Republic and another is the export of spent nuclear fuel for storage abroad. The Slovak Republic is also involved in projects to build an interregional deep-water disposal site for radioactive waste [58]. (UJD) Long-term storage of spent nuclear fuel, decommissioning and management of radioactive waste from decommissioning may be carried out only by a legal person established, established or authorized by the Ministry of Economy of the Slovak Republic, subject to the authorisation of the Authority [64]. According to the previous sentence, the legal entity must be the holder of a permit for the operation of the storage site, the holder of a permit for the operation of a storage facility or the holder of a permit for the decommissioning and management of radioactive waste from decommissioning, the Slovak Republic must hold a participation in that person and at the same time that person cannot hold a permit for the operation of a nuclear facility for the production of electricity or nuclear energy research involving nuclear reactors which are used for controlled fissile energy. An intermediate storage facility is a nuclear facility that serves the safe interim storage of spent fuel prior to further processing in a processing plant or definitive composition. Spent nuclear fuel is stored in storage pools [61, 62].

When defining the concept of spent fuel storage, it is necessary to highlight the difference between the above concepts of spent fuel storage and spent fuel storage. The storage of spent nuclear fuel, including long-term storage, differs from its disposal by the intention of subsequently removing spent fuel or radioactive waste from the site of disposal. On the other hand, the disposal of spent nuclear fuel and radioactive waste is their permanent placement in a repository [66].

According to the Atomic Act, the disposal site of spent fuel and radioactive waste is legally defined as a nuclear installation and is therefore covered by all provisions relating to the construction of nuclear installations under the conditions of the Slovak Republic, the main purpose of which is the disposal of radioactive waste or spent nuclear fuel, enabling their isolation, control and protection of the environment [21]. The construction and placement of nuclear installations, the commissioning, operation and decommissioning or closure of a storage site constitute one of the groups of activities defined as the nuclear energy use process, to which the general prohibition applies, with exceptions if the consent or permit of the relevant public authority is granted [60]. The Atomic Act regulates several successive permit procedures in connection with the construction and operation of nuclear installations, namely:

– location
– the building
– commissioning
– operation [69]
– decommissioned
– closure of storage
– institutional control [51].

The permanent storage of spent fuel is the final part of nuclear energy as an activity of peaceful uses of nuclear energy. In the conditions of the Slovak Republic, the legal aspects of the construction of the storage site are regulated mainly by the Atomic Act, but also include aspects of environmental protection, in particular the mandatory nature of the EIA process, as well as several processes falling within the field of construction law [55]. Although the law of the Slovak Republic allows the construction of a (deep) spent nuclear fuel storage site in its territory, we do not currently have such a nuclear facility to permanently resolve the disposal of spent nuclear fuel without an alternative to its further relocation. According to Clark et al. [9] as the constant production of spent nuclear fuel and radioactive waste increases the need for a permanent solution to this problem, the deep storage site is the subject of qualified assessment and search for the most suitable location, in cooperation with the Czech Republic [9].

8 Conclusion

As we stated in the introduction, this scientific study focused on the analysis of the baselines for the use of nuclear waste in the circular economy within the broader framework of state energy security. The analysis showed that strategic or legislative assumptions of a transnational or national nature do not include the intention to treat spent nuclear fuel and radioactive waste as a secondary raw material that should be part of the circular economy. On the other hand, they pay attention to the protection of the environment, life and health of the population and emphasise the safety of their management.

The acquisition of energy is one of the basic conditions for the existence and development of civilisation, and the demand for its volume increases worldwide. Unlike other types of energy, there are different views on nuclear energy. Those that point out that it poses too much risk and also those that show that its production is among the most environmentally friendly, unlike, for example, the burning of fossil fuels, which generates a large amount of polluting substances. Alternative energy sources, which we also include among the ecologically advantageous ones, are not yet obtained in sufficient volume. In the production of nuclear energy, the only waste is spent nuclear fuel. It is therefore a product of the process of generating electricity in nuclear power plants. Spent nuclear fuel can be considered a usable raw material. In terms of environmental protection, spent nuclear fuel management is one of the most complex areas of the nuclear sector.

We are of the opinion that the current global fragility is, however, forcing individual states and their international groups to think about their energy security from a new perspective. They review the future development of energy production, the methods of energy distribution and storage, as well as the elimination of undesirable outputs of these processes. As a Member State of the European Union, the Slovak Republic coordinates its development objectives with its partners. One of the most up-to-date joint discussions was the Brussels European Council on 17 December 2021. However, it is clear that, in particular in the context of the conflict in Ukraine, the sanctions against Russia and its response to them and the soaring energy prices, a number of negotiations will be needed to increase Europe's energy self-sufficiency. Whatever the conflict, the European Union will have to work much more to secure alternative energy supplies and find ways to achieve this, even in the longer term.

This also paves the way for a new approach to the use of nuclear waste. However, this will require further scientific research and new legal solutions. From the available sources of information, we can conclude that current trends in the disposal of nuclear waste are directed towards their disposal and final disposal in deep-water storage sites, without presupposing further management of them.

The management of spent nuclear fuel and radioactive waste is regulated separately from the management of other wastes in the Slovak Republic's legal order. Unlike waste management, which envisages the reuse of waste as a secondary raw material, nuclear waste is not yet treated with this stage. It will therefore be necessary to promote research in the field of research and support the idea of including nuclear waste as waste that can be re-used in the circular economy. Also in the Slovak Republic, several scientific activities in the field of natural sciences and law have long examined the possibility of disposal of spent nuclear fuel and radioactive waste either on the territory of our state or in cooperation with other states in a common repository.

It will also be necessary to prioritise, in particular, the interest in promoting ever higher safety standards in the management of nuclear waste and thoroughly preparing the conditions for its re-use. Common positions, a common policy of the Member States and informing the public about a safe final phase of the core have a high potential to have a positive impact on the public opinion of the citizens of the European Union. On the contrary, the negative public relationship also results from the lack of up-to-date and high-quality information on scientific progress, technical and economic readiness in this area.

Acknowledgements This scientific study was created thanks to the support of the Slovak Research and Development Agency grant under contract no. APVV-18-0534 "Legal regulation of spent nuclear fuel management".

References

1. Aksamovic, D., Simunovic, L.: The EU scheme for the state aid rules in the air transport sector during the Covid-19 crisis. Juridical Tribune **12**(1), 51–67 (2022). https://doi.org/10.24818/TBJ/2022/12/1.04
2. Androniceanu, A., Popescu, C.R.: An inclusive model for an effective development of the renewable energies public sector. Administratie si Manag. Public **2017**(28), 81–96 (2017)
3. Čajka, P., Abrhám, J.: Regional aspects of V4 countries' economic development over a membership period of 15 years in the european union. Slovak J. Polit. Sci. **19**(1), 89–105 (2019). https://doi.org/10.34135/sjps.190105
4. Čajková, A., Čajka, P., Elfimova, O.: Personality and Charisma as Prerequisites for a leading position in public administration. Springer Proc. Bus. Econ. 199–211 (2018). https://doi.org/10.1007/978-3-319-74216-8_21
5. Čajková, A., Gogová, A.: Case study of the knowledge management process in selected department of state administration in Slovakia. Stud. Syst. Decis. Control **420**, 533–545 (2022). https://doi.org/10.1007/978-3-030-95813-8_21
6. Cajková, A., Jankelová, N., Masár, D.: Knowledge management as a tool for increasing the efficiency of municipality management in Slovakia. Knowl. Manag. Res. Pract. In press (2021). https://doi.org/10.1080/14778238.2021.1895686
7. Čajková, A., Šindleryová, I.B., Garaj, M.: The covid-19 pandemic and budget shortfalls in the local governments in Slovakia. Sci. Pap. Univ. Pardubic. Ser. D: Fac. Econ. Adm. **29**(1), 1243 (2021). https://doi.org/10.46585/sp29011243
8. Chochia, A., Nässi, T.: Ethics and emerging technologies–facial recognition. Revista de Internet, Derecho y Politica **34** (2021). https://doi.org/10.7238/idp.v0i34.387466
9. Clark, T.A., Guarrieri, T.R.: Modeling terrorist attack cycles as a stochastic process: analyzing chemical, biological, radiological, and nuclear (CBRN) incidents. J. Appl. Secur. Res. **16**(3), 281–306 (2021). https://doi.org/10.1080/19361610.2020.1761743
10. Commission Regulation. Commission Regulation (EC) No 956/2008 of 29 September 2008 amending Annex IV to Regulation (EC) No 999/2001 of the European Parliament and of the

Council laying down rules for the prevention, control and eradication of certain transmissible spongiform encephalopathies (Text with EEA relevance). [online] [cit. 1.6.2022] Available from: https://eur-lex.europa.eu/legal-content/EN/TXT/?uri=CELEX%3A32008R0956& qid=1654849393744

11. Council of the European Union. Commission Regulation (EC) No 851/2006 of 9 June 2006 specifying the items to be included under the various headings in the forms of accounts shown in Annex I to Council Regulation (EEC) No 1108/70 (Codified version) (Text with EEA relevance). [online] [cit. 1.6.2022] Available from: https://eur-lex.europa.eu/legal-content/EN/ TXT/?uri=CELEX%3A32006R0851&qid=1654849277954

12. Council of the European Union. Council Directive 2006/117/Euratom on the supervision and control of shipments of radioactive waste and spent fuel. [online] [cit. 1.9.2021] Available from: https://eur-lex.europa.eu/legal-content/SK/TXT/?uri=CELEX%3A32006L0117& qid=1632817755207

13. Council of the European Union. Council Directive 2009/71/Euratom establishing a Community framework for the nuclear safety of nuclear installations. [online] [cit. 1.9.2021] Available from: https://eur-lex.europa.eu/legal-content/SK/TXT/?uri=CELEX%3A32009L0071& qid=1632817810413

14. Council of the European Union. Council Directive 2011/70/Euratom establishing a Community framework for the responsible and safe management of spent fuel and radioactive waste. [online] [cit. 1.9.2021] Available from: https://eur-lex.europa.eu/legal-content/SK/TXT/?uri= CELEX%3A32011L0070&qid=1632817851569

15. Council of the European Union. Council Directive 2013/59/Euratom laying down basic safety standards for protection against the dangers arising from ionizing radiation and repealing Directives 89/618/Euratom, 90/641/Euratom, 96/29 / Euratom, 97/3/Euratom and 2003/122/Euratom. [online] [cit. 1.9.2021] Available from: https://eur-lex.europa.eu/legal-content/SK/TXT/?uri=CELEX%3A32013L0059&qid=1632817877880

16. Council of the European Union. Council Regulation (EC) No 428/2009 of 5 May 2009 setting up a Community regime for the control of exports, transfer, brokering and transit of dual-use items. [online] [cit. 1.6.2022] Available from: https://eur-lex.europa.eu/legal-content/EN/ TXT/?uri=CELEX%3A32009R0428&qid=1654849124889

17. Council of the European Union. Council Regulation (Euratom) No. 1493/93 of 8 June 1993 on shipments of radioactive substances between Member States, as amended, OJ L 148. [online] [cit. 1.6.2022] Available from: https://eur-lex.europa.eu/legal-content/EN/TXT/?uri=celex% 3A31993R1493

18. Council of the European Union.: Commission Regulation (EC) No 1209/2000 of 8 June 2000 determining procedures for effecting the communications prescribed under Article 41 of the Treaty establishing the European Atomic Energy Community. [online] [cit. 1.6.2022] Available from: https://eur-lex.europa.eu/legal-content/EN/ALL/?uri=CELEX%3A32000R1209

19. Aliyev, A.G., Shahverdiyeva, R.O.: Scientific and methodological bases of complex assessment of threats and damage to information systems of the digital economy. Int. J. Inf. Eng. Electron. Bus. (IJIEEB) 14(2), 23–38 (2022). https://doi.org/10.5815/ijieeb.2022.02.02

20. Dănișor, D.C., Dănișor, M.-C.: Modern solidarity and administrative repression. Juridical Tribune 11(3), 472–496 (2021). https://doi.org/10.24818/TBJ/2021/11/3.04

21. De Geer, L.E., Wright, C.M.: The 22 september 1979 vela incident: radionuclide and hydroacoustic evidence for a nuclear explosion. Sci. Glob. Secur. 26(1), 20–54 (2018). https://doi.org/10.1080/08929882.2018.1451050

22. Dufala, M., Šmelková, J.: Respository of Spent Nuclear Fuel in the Legal Order of the Slovak Republic. Wolters Kluwer, Bratislava (2020)

23. European Commision.: Communication from the Commission to the European Parliament, the Council, the European Economic and Social Committee and the Committee of the Regions, [online] [cit. 1.6.2022] Available from: https://eur-lex.europa.eu/legal-content/SK/ TXT/HTML/?uri=CELEX:52015DC0614&from=NL

24. European Parliament and the Council.: Directive (EU) 2018/851 of the European Parliament and of the Council of 30 May 2018 amending Directive 2008/98/EC on waste. [online]

[cit. 1.6.2022] Available from: https://eur-lex.europa.eu/search.html?scope=EURLEX&text=2018%2Г851&lang=sk&type=quick&qid=1654356944054

25. European Parliament and the Council: Directive (EU) 2018/852 of the European Parliament and of the Council of 30 May 2018 amending Directive 94/62/EC on packaging and packaging waste. [online] [cit. 1.6.2022] Available from: https://eur-lex.europa.eu/search.html?scope=EURLEX&text=2018%2F852&lang=sk&type=quick&qid=1654356990451

26. European Parliament and the Council: Directive 2008/98/EC of the European Parliament and of the Council of 19 November 2008 on waste and repealing certain Directives. [online] [cit. 1.6.2022] Available from: https://eur-lex.europa.eu/legal-content/SK/TXT/?uri=CELEX%3A32008L0098&qid=1654356718145

27. European Parliament and the Council: Directive 2012/18/EU of the European Parliament and of the Council of 4 July 2012 on the control of major-accident hazards involving dangerous substances, amending and subsequently repealing Council Directive 96/82/EC Text with EEA relevance. [online] [cit. 1.6.2022] Available from: https://eur-lex.europa.eu/legal-content/EN/TXT/?uri=CELEX%3A32012L0018&qid=1654849690671

28. Filip, S., Šimák, L., Kováč, M.: Manažment rizika. Bratislava, Sprint (2011)

29. Filip, S., Kabát, L., Filipová, L.: The safety measurement in EU Countries and regional units and Slovakia. Conterporary Res. Organ. Manag. Adm. **5**(2), 19–27 (2017)

30. Galamboš, M., Krajňák, A., Viglašová, E.: Zadná časť jadrovej energetiky v Slovenskej republiky. Bezpečnosť jadrovej energie **22**(60) (2014)

31. Gale, R.P., Armitage, J.O., Hashmi, S.K.: Emergency response to radiological and nuclear accidents and incidents. Br. J. Haematol. **192**(6), 968–972 (2021). https://doi.org/10.1111/bjh.16138

32. Gauntlett, L., Amlôt, R., Rubin, G.J.: How to inform the public about protective actions in a nuclear or radiological incident: a systematic review. Lancet Psychiatry **6**(1), 72–80 (2019). https://doi.org/10.1016/S2215-0366(18)30173-1

33. Government of the Slovak Republic.: Environmental Policy Strategy of the Slovak Republic up to 2030. [online] [cit. 1.6.2022] Available from: https://www.minzp.sk/iep/strategicke-materialy/envirostrategia-2030.html

34. Government of the Slovak Republic.: Programme Declaration of the Government of the Slovak Republic for 2020–2024. [online] [cit. 1.6.2022] Available from: https://www.mpsr.sk/programove-vyhlasenie-vlady-slovenskej-republiky-na-obdobie-rokov-2020-2024/800-17-800-15434/

35. Government of the Slovak Republic.: Programme Declaration of the Government of the Slovak Republic for 2020–2024, [online] [cit. 1.6.2022] Available from: https://www.nrsr.sk/web/Dynamic/DocumentPreview.aspx?DocID=494677

36. Government of the Slovak Republic.: Resolution of the Government of the Slovak Republic No. 87/2019 Coll. Environmental Policy Strategy of the Slovak Republic up to 2030. [online] [cit. 1.6.2022] Available from:

37. Groza, A.: The principle of mutual recognition: from the internal market to the European area of freedom, security and justice. Juridical Tribune **12**(1), 89–104 (2022). https://doi.org/10.24818/TBJ/2022/12/1.07

38. Horvat, M., Lysá, A.: Energy Law in Slovakia. Warszawa, C. H. Beck (2017)

39. Ismaila, A., Md Kasmani, R., Ramli, A.T.: Numerical evaluation of the severity of consequences of external fire and explosion incident at a nuclear power plant. Nucl. Eng. Des. **355**, 110314 (2019). https://doi.org/10.1016/j.nucengdes.2019.110314

40. Ivanov, A.: Computational model of incidents at Russian WWER nuclear power units. In: Proceedings of the 2021 IEEE Conference of Russian Young Researchers in Electrical and Electronic Engineering, ElConRus, pp. 421–424 (2021). https://doi.org/10.1109/ElConRus51938.2021.9396637

41. Jančíková, E., Pásztorová, J.: Promoting eu values in international agreements. Juridical Tribune **11**(2), 203–218 (2021). https://doi.org/10.24818/TBJ/2021/11/2.04

42. Jankurová, A., Ljudvigová, I., Gubová, K.: Research of the nature of leadership activities. Econ. Sociol. **10**(1), 135–151 (2017). https://doi.org/10.14254/2071-789X.2017/10-1/10

43. Joamets, K., Chochia, A.: Access to artificial intelligence for persons with disabilities: legal and ethical questions concerning the application of trustworthy AI. Acta Baltica Historiae et Philosophiae Scientiarum 9(1), 51–66 (2021). https://doi.org/10.11590/ABHPS.2021.1.04

44. Joamets, K., Chochia, A.: Artificial intelligence and its impact on labour relations in Estonia. Slovak J. Polit. Sci. 20(2), 255–277 (2020). https://doi.org/10.34135/sjps.200204

45. Kerveillant, M., Lorino, P.: Dialogical and situated accountability to the public. The reporting of nuclear incidents. Account., Audit. Account. J. 34(1):111–136 (2021). https://doi.org/10.1108/AAAJ-04-2019-3983

46. Poniszewska-Maranda, D., Matusiak, R., Kryvinska, N., Yasar, A.-U.-H.: A real-time service system in the cloud. J. Ambient Intell. Human Comput. 11(3), 961–977 (2020). https://doi.org/10.1007/s12652-019-01203-7

47. Poniszewska-Maranda, D., Kaczmarek, N.K., Xhafa, F.: Studying usability of AI in the IoT systems/paradigm through embedding NN techniques into mobile smart service system. Computing 101(11), 1661–1685 (2019). https://doi.org/10.1007/s00607-018-0680-z

48. Ministri of Enviroment of Slovak Republic. Waste Management Programme of the Slovak Republic 2016–2020. [online] [cit. 1.6.2022] Available from: https://www.minzp.sk/files/sek cia-enviromentalneho-hodnotenia-riadenia/odpady-a-obaly/registre-a-zoznamy/poh-sr-2016-2020_vestnik.pdf

49. Ministri of Enviroment of Slovak Republic.: Waste Prevention Programme of the Slovak Republic 2019–2025. [online] [cit. 1.6.2022] Available from: https://www.minzp.sk/files/sek cia-enviromentalneho-hodnotenia-riadenia/odpady-a-obaly/registre-a-zoznamy/ppvo-sr-19-25.pdf

50. Ministri of Enviroment of Slovak Republic.: Waste prevention program. [online] [cit. 1.6.2022] Available from: https://www.minzp.sk/odpady/program-predchadzania-vzniku-odpadu/

51. National Council of the Slovak Republic. Act no. 541/2004 Coll. on the Peaceful Use of Nuclear Energy (Atomic Act) and on Amendments to Certain Acts, as amended. [online] [cit. 1.6.2022] Available from: https://www.slov lex.sk/pravne-predpisy/SK/ZZ/2004/541/20191001

52. Matúšová, S.: Nováček, P: New generation of investment agreements in the regime of the European Union. Juridical Tribune 12(1), 21–34 (2022). https://doi.org/10.24818/TBJ/2022/12/1.02

53. National Council of the Slovak Republic.: Act No 79/2015 on waste and amending certain acts, as amended, and amending certain acts. [online] [cit. 1.6.2022] Available from: https://www.slov-lex.sk/pravne-predpisy/SK/ZZ/2015/79/20220101.html

54. Nováčková, D., Vnuková, J.: Competition issues including in the international agreements of the eropean union. Juridical Tribune 11(2), 234–250 (2021). https://doi.org/10.24818/TBJ/2021/11/2.06

55. Nuclear Regulatory Authority of the Slovak Republic Official website. [online] [cit. 1.6.2022] Available from: https://www.ujd.gov.sk/ujd/www1.nsf

56. Nuclear Regulatory Authority of the Slovak Republic: Decree No. 30/2012 Coll. laying down details of requirements for the management of nuclear materials, radioactive waste and spent fuel. [online] [cit. 1.6.2022] Available from: https://www.slov-lex.sk/pravne-predpisy/SK/ZZ/2012/30/20160301

57. Nuclear Regulatory Authority of the Slovak Republic: supervision of nuclear waste management. [online] [cit. 1.6.2022] Available from: https://www.ujd.gov.sk/cinnosti/dozor-nad-rao-jm/radioaktivne-odpady-rao/

58. Oláh, J., Aburumman, N., Popp, J., Khan, M.A., Haddad, H., Kitukutha, N.: Impact of industry 4.0 on environmental sustainability. Sustainability (Switzerland), 12(11), 4674 (2020). https://doi.org/10.3390/su12114674

59. Pavlovič, M.: Zborník z XXI. ročníka medzinárodnej vedeckej česko-poľsko-slovenskej konferencie práva životného prostredia. Bratislava (2020)

60. Poiană, O.: An overview of the European energy policy evolution: from the European energy community to the European energy union. Online J. Model. New Eur. 22(1), 175–189 (2017). https://doi.org/10.24193/ojmne.2017.22.09

61. Prokopenko, O., Mishenin, Y., Mura, L., Yarova, I.: Environmental and economic regulation of sustainable spatial agroforestry. Int. J. Glob. Environ. Issues **19**(1–3), 109–128 (2020)
62. Săraru, C.S.: Considera ţii cu privire la limitele libertătii contractuale în dreptul public impusc de integrarea în uniunea europeană. Transylvanian Rev. Adm. Sci. **1**, 131–140 (2008)
63. Săraru, C.S.: The European groupings of territorial cooperation developed by administrative structures in Romania and Hungary. Acta Juridica Hungarica **55**(2), 150–162 (2014)
64. Shimada, Y., Nomura, S., Ozaki, A., Higuchi, A., Hori, A., Sonoda, Y., Yamamoto, K., Yoshida, I., Tsubokura, M.: Balancing the risk of the evacuation and sheltering-in-place options: a survival study following Japan's 2011 Fukushima nuclear incident. BMJ Open **8**(7), e021482 (2018). https://doi.org/10.1136/bmjopen-2018-021482
65. Canyaş, O., Bayata Canyaş, A.: Approach towards the right to be forgotten under turkish law in comparison with eu and us laws: a need for a reform? Juridical Tribune **11**(2), 174–202 (2021). https://doi.org/10.24818/TBJ/2021/11/2.03
66. Števček, M., Ivančo, M.: The conception and institutional novelties of recodification of private law in the Slovak Republic. In: The Law of Obligations in Central and Southeast Europe: Recodification and Recent Developments, pp. 33–49 (2021)
67. Stratone, M.E., Vataamanescu, E.M., Treapat, L.M., Rusu, M., Vidu, C.-M.: Contrasting traditional and virtual teams within the context of COVID-19 pandemic: from team culture towards objectives achievement. Sustainability **14**, 4558 (2022). https://doi.org/10.3390/su14084558
68. Fedushko, S., Ustyianovych, T., Gregus, M.: Real-time high-load infrastructure transaction status output prediction using operational intelligence and big data technologies. Electronics **9**(4), 668 (2020). https://doi.org/10.3390/electronics9040668
69. Tokareva, V., Davydova, I., Adamova, E.: Legal problems of the use of orphan works in digital age. Juridical Tribune **11**(3), 452–471 (2021). https://doi.org/10.24818/TBJ/2021/11/3.03
70. Troitino, D.R., Chochia, A., Kerikmäe, T.: The incapacity of the union to act as a reliable actor in the international arena. Eur. Union: Polit., Econ. Soc. Issues 11–32 (2017)
71. Troitiño, D.R., Kerikmäe, T., Chochia, A.: Foreign affairs of the European Union: How to become an independent and dominant power in the international arena. In: The EU in the 21st Century: Challenges and Opportunities for the European Integration Process, pp. 209–230 (2020). https://doi.org/10.1007/978-3-030-38399-2_12
72. Tsubokura, M., Murakami, M., Takebayashi, Y., Nomura, S., Ono, K., Ozaki, A., Sawano, T., Kobashi, Y., Oikawa, T.: Impact of decontamination on individual radiation doses from external exposure among residents of Minamisoma City after the 2011 Fukushima Daiichi nuclear power plant incident in Japan: a retrospective observational study. J. Radiol. Prot. **39**(3), 854–871 (2019). https://doi.org/10.1088/1361-6498/ab280e
73. Žofčinová, V., Čajková, A., Král, R.: Local leader and the labour law position in the context of the smart city concept through the optics of the EU. TalTech J. Eur. Stud. **11**(2), 3–26 (2022). https://doi.org/10.2478/bjes-2022-0001
74. Žofčinová, V., Horváthová, Z., Čajková, A.: Selected social policy instruments in relation to tax policy. Social Sci. **7**(11), 241 (2018). https://doi.org/10.3390/socsci7110241
75. Žofčinová, V., Košíková, A.: Selected legislative instruments of family policy supporting work–life balance: a comparison of Italy and the Slovak Republic. Online J. Model. New Eur. **39** (2022)

Knowledge Management and Local Government

Martin Molčan and Andrea Čajková

Abstract We feel the need for knowledge management in the context of ongoing and necessary changes in public administration institutions, which are constantly undergoing processes associated with modernization, the use of new technologies and comprehensive streamlining of the provision of relevant public services for citizens. In this chapter, we focus on the level of quality in the use of knowledge management at the local government level. The aim of the article is to identify the level of preparedness of local governments for changes resulting from digital transformation, both in the context of the relevant strategy, its implementation in connection with the action plan or vision, but especially the real practice of local government in the use and management of knowledge for strategic management of the municipality. We analyze the situation in selected municipalities in Slovakia, and the research problem is the readiness of local government for challenges arising from the need for continuous improvement of public services, especially in terms of ability of responsible municipal staff to use knowledge management to meet this goal. At the same time, the fact whether employees increase their qualifications and knowledge through vocational training or education.

1 Introduction

The art of working with knowledge is not essential only for the private sector. It is crucial for revenue-oriented companies. We feel the same need in institutions of public government, where there are running processes connected with modernization, using the new technologies, and comprehensive streamlining of service delivery to citizens.

It would be possible to state that local government application is currently considered one of the basic features of a democratic state. One of the essential tasks of the

---

M. Molčan · A. Čajková (✉)
Faculty of Social Sciences, Department of Political Sciences, University of St. Cyril and Methodius in Trnava, Bučianska 4/A, Trnava 917 01, Slovak Republic
e-mail: andrea.cajkova@ucm.sk

© The Author(s), under exclusive license to Springer Nature Switzerland AG 2023
N. Kryvinska et al. (eds.), *Developments in Information and Knowledge Management Systems for Business Applications*, Studies in Systems, Decision and Control 462, https://doi.org/10.1007/978-3-031-25695-0_14

local government should be viewed as caring for the overall development of its territory and its inhabitants' needs. The European Charter of Local Self-Government (1985) defines local self-government as a natural-legal institute, which is an expression of the rights and ability of the population to manage that part of public affairs which belongs to its own. The Local Self-Government then represents the authority of towns and municipalities not only to make independent decisions but also to carry out a legally defined range of matters that are directly related to the interests of the municipality's population. Self-governing responsibility is practically the duty of a city, municipality, or higher territorial unit to ensure all tasks related to meeting the basic needs of the population living in the relevant territory, which are defined by law. *"The ability of knowledge of self-governing and the corresponding competence possibilities, the true essence of territorial self-government is manifested, the result of which should be a public benefit, resp. For the benefit of its citizens,"* points out Kráľ. As independent economic and legal entities, cities, municipalities, and regions can perform tasks beyond their competencies as long as they have the meaning and the necessary level of professional and active human potential. According to Hajšová [8], the quality of human capital is an assumption for the development of society and an assumption for local government management. However, according to Molitoris [28], the demands on the public administration in terms of the quality of its performance are escalating, and the complexity of its tasks is increasing. We were satisfied with lawfully administered public administration in the past, but today we require "good public administration" or even cooperative self-government [35]. There is an uprising question, is it possible to await from public administration constant improvement way of performing it? Modern companies are aware of the need for clear classification and categorization of business knowledge, so they compile financial statements and "knowledge" statements.

This chapter aims to identify the level of preparedness of local authorities for changes resulting from digital transformation, both in the context of the relevant strategy and its implementation and about the action plan or vision, but especially the actual practice of local government in the use and management of knowledge in favor of strategic village management. We analyze the situation in selected municipalities in Slovakia. The research problem is the readiness of the local government for the challenges arising from the need to improve public services constantly. Mostly in terms of the ability of municipal officials to use knowledge management to meet this goal. At the same time, whether employees increase their qualifications and knowledge through vocational training or education.

We want to achieve the stated objectives, in particular through a thorough study of relevant legislation, in combination with both academic and scientific literature.

Due to the nature of the scientific article, we use several scientific methods of knowledge suitable for knowledge management and public administration. This concerns, in particular, the use of a analysis method to examine legal and regulatory environment and abstraction. Using the comparative method, we make available different opinions mainly from foreign experts who have experience with the knowledge management in public administration. In this way we strive for not only a administrative but also a managerial view of the examined issue. Due to our own years of experience in knowledge management and public administration, we also use expert approach in combination with qualitative research. This scientific study represents a further continuation of our research in the field of knowledge management.

The study suggests that a holistic knowledge management strategy promotes the use of information at the municipal level by providing a systematic management framework for the collection and use of data. However, the primary factor in the success of this strategy will always be the people—employees and elected representatives of local governments. In the context of the researched issues, the research questions are set as follows:

RQ1: Data quality at the required level is guaranteed—What are the small towns' and villages' educational structures and knowledge assets?

RQ2: Is the knowledge management strategy carefully integrated into the general management system, especially in upgrading the qualifications of competent employees?

RQ3: There are transparent processes and responsibilities for the constant addition of data for the knowledge management strategy—what is the flow of information flow and verification?

RQ4: Document management—is the knowledge management strategy followed by the city/municipality strategy?

The survey was done among the mayors of Slovak municipalities in the form of structured interviews. Due to the limitations resulting from the pandemic situation, the interviews were conducted by telephone from January to March 2022.

2 Theoretical Backgrounds

Legislatively, the competencies of local self-government are determined primarily by Act no. 369/1990 Coll. on general establishment, as amended, as well as Act 416/2001 Coll. on the transfer of certain powers from state administration bodies to municipalities and higher territorial units, as amended. At the same time, from the point of view of the exercise of competencies and their financing, it is necessary to distinguish between the original competencies and the exercise of state administration transferred to municipalities.

In the Slovak environment, M. Vernarský [44] deals with the legislative definition of self-government, the position of various bodies, and the results of their activities, for example, D. Klimovský and J. Nemec [19]. The authors D. Klimovský and J. Gašparík [18] point out that the interactions that arise in implementing policy between entities operating at the municipal level can be of different quality. Therefore, it is necessary to consider the importance of participatory public policy-making, its potential benefits, and at the same time limits. N. Kováčová [24], for example, also deals with the importance of the participation aspect and specific obstacles to political participation at the municipal level. This is related not only to the fact that different entities have different positions but also to the fact that each of these entities defends its interests. As for the relations between the bare political-administrative bodies of the municipal self-government, several approaches can be used in their classification. One of the most widely used approaches is comparing political-administrative relations at the local level with political-administrative ties at the national level.

Many authors, such as Balážová et al. [1], draw attention to the issue of the ability of specifically smaller municipalities to perform both original and transferred competencies. The state has moved more than 200 powers to the municipality without conducting serious research into whether the municipality is able to provide these competencies. Municipalities have so many transferred competencies that they do not have time to ensure the performance of their own original competencies. These are set for both a municipality with 100 inhabitants and a municipality with 100,000 inhabitants. It is also necessary to realize that the fundamental problem is the ability of the municipality to implement these competencies, both in terms of financial, often, especially personnel. Democracy is much higher in our country than expertise in elected positions. In this context, ZMOS [47] further draws attention to another important fact: that the personnel and technical provision of both types of competencies are, in principle, dependent on the municipalities. According to the constitution and the law, the state must pay the municipality for the delegated performance of state administration. However, we still have the rigid approach chosen many years ago. The state pays the municipalities for the delegated performance of state administration, usually based on the number of inhabitants. In the case of smaller municipalities, with the best will of finance for the continuous education of representatives or employees, they are simply absent.

The informatization, digitization, and informatization of processes and jobs have made learning and the use of advanced digital competencies necessary for us to be able to face constant change and use it in the context of knowledge management. Not only expertise and experience in management and planning but nowadays, in addition to communication and system management skills, technical and digital knowledge and skills are also essential for municipalities at the level of their employees. If we take into account the opinion of T. Peráček et al. [33] that the manager's mandatory equipment includes strategic thinking, the ability to anticipate and entrepreneurial thinking, even in the context of the processes of creating a knowledge environment and the ability to use ICT as a carrier of information and knowledge in the form of e-government. At the same time, it is necessary, as Vojtech et al. [45], changes were introduced to ensure a functioning information society and the building of wise government in the context of strategic knowledge management. This should generally be done by making information technology an integral part of our daily lives, including at the level of local government management and strategic decision-making. According to Mazur (2020), many cities in the world have long used the potential of their inhabitants to gain incentives to address or improve public services. And it is the digitization and automation of the provision of these stimuli through mobile applications that have great potential in improving the motivations and reducing barriers to providing feedback to public authorities, especially at the local level.

The European Union's challenges are also moving towards a sustainable society, the digitization of services, smart solutions, and the strengthening of environmental responsibility, and it is, therefore, necessary for them to be reflected in the settings and policies of local and regional authorities [11]. At the same time, Kaliňák [12] points out that decisions on the digitization of services and the use of intelligent technologies should not be about spontaneous decisions or risky pilot projects but about starting points on which, we can reliably anticipate impacts. It is another step toward modernizing local government, disseminating examples of good practice, and promoting practical inter-municipal cooperation [6].

3 Knowledge Management

Knowledge management is about building organisational intelligence by enabling people to improve the way they work in capturing, sharing and using knowledge. It involves using the ideas and experience of employees, customers and suppliers to improve the organisation's performance. Building on what works well leads to better strategy, practice and decision-making (Improvement and Development Agency for Local Government, UK, 2009). The main goal of knowledge management is to find a way to facilitate access to information as well as institutional knowledge for people who are looking for it in relation to the people who have it. Based on this process, individual knowledge becomes knowledge collectively, the expertise of employees, their work performance and the efficiency of work procedures increase. It is also important to make it clear that knowledge management is not just about managing information and communication technologies (ICT), but rather about one of its many areas, as we are talking about an issue that can be seen from different angles due to its wealth of ideas.

According to UN ECLAC [13], there is no blueprint for implementing knowledge management, it cannot be achieved overnight and requires analysis and planning. Governments that aim to implement knowledge management practices normally learn from others that have done so already, but in the end, the final outcome and strategy will be tailor-made to the needs, wants, goals and budget of the implementing government. Also, at the level of local governments, according to Ngcam [29], knowledge management should be perceived and used as a strategic tool to increase the efficiency of the organization in accelerating the provision of services and achieving operational objectives. The rise and growth of the importance of knowledge management are one of the managerial answers to the empirical trends associated with globalization and post-industrialism. The author considers the impact of biographical profiles to be crucial given the diverse workforce and points out that knowledge management (KM) is also recognized as a core activity for attracting, growing, and retaining intellectual capital in organizations.

Brooks [3] states that the easiest way to explain the difference between information and knowledge management is to argue that knowledge management focuses on people, takes their views and intuition into account, while information focuses on processes and hard facts. Other major differences include:

(a) Information management: in addition to being technology-based, it explicitly characterizes facts, figures and other hard data and can be measured at short intervals.

(b) Knowledge management: leads to unique innovations, includes communication, management processes, and organizational culture, and can be measured by what changes in the behavior and work of individuals and teams have taken place over a period of time.

According to Sedláková and Válovská [39]: "Knowledge, and skills help to increase the growth rate of state performance and competitiveness, while it is necessary to support them at all levels of public administration, not excluding self-government. The current conditions in which Slovakia's self-government requires continuous improvement, adaptation, and development of the educational level of elected and executive representatives and all its employees."

First, each organization must identify the way the documentation is used, the necessary knowledge for its preparation, and where it can be obtained. How the individual processes take place and based on what information and documents. Pioneers in the field of knowledge management Nonaka and Takeuchi [30] describe knowledge assets as those that are the output of the knowledge creation process, recognizing four basic types of knowledge assets (Fig. 1).

Knowledge capital is also spoken of when knowledge is concentrated, organized, purposefully integrated into the productive value-creating process [9]. knowledge management has already been addressed in many research studies, both from the point of view suitability of tools to support knowledge management, application of knowledge management in business environment, and development of knowledge management in different types of organizations [20, 34], dealing with specifics of knowledge management in public administration and government [14], as the prerequisite or barrier of cooperative relations of municipalities [6, 13], the development of the knowledge society, from the point of view of leadership and creativity [17, 26, 40] the necessarily related digital transformation [37] or the importance of knowledge sharing [25].

Žárska et al. [46] brings its own definition of the factors of knowledge self-government determining the general self-government as a knowledge institution in the process of building a knowledge economy by generalizing the starting points and approaches to local self-government (Table 1).

According to Žárská (2012), the scope of competencies is vital because the self-government can carry out activities only within the area of its delegated powers—the strength of the self-government grows with the number of competencies. However, it also draws attention to the importance of the overall climate and the level of support for building a knowledge society in the country. It also illustrates maturity in equipment and use of ICT and the financial potential of self-government, which projects can strengthen from national and international grant schemes, factors of technical and economic potential, as well as the level of development and quality of planning documents.

Experiential Knowledge assets

Tacit knowledge shared through common experiences

- Skills and know-how of individuals
- Care, love, trust, and security
- Energy, passion and tension

Conceptual Knowledge Assets

Explicit knowledge articulated through images, symbols, and language

- Product concepts
- Design
- Brand equity

Routine Knowledge Assets

Tacit knowledge routinized and embedded in actions and practices

- Know-how in daily operation
- Organizational routines
- Organizational culture

Systematic Knowledge Assets

Systemised and packed explicit knowledge

- Documents, specifications, manuals
- Database
- Patent and licenses

Fig. 1 Four types of knowledge assets. *Source*: Nonaka and Takeuchi [30]

The interconnection of municipalities and the possibilities and strategies of the region is also more than necessary in this area of research. It is also reflected in Slovak realities in economic and social development programs—whether at the level of a self-governing region or municipalities. In these documents, this issue is a central part of the analytical-strategic part, which contains, in particular, the evaluation and analysis of the economic situation and social situation, including the state of public health, environmental condition, and the situation in the field of culture, facilities, and service. The authors Terem, Čajka, Rýsová [41] focus on the Europe 2020 Strategy in the context of Slovakia. Rodríguez-Posea and Garcilaza [38] analyze the impact of investments on cohesion in European regions without understanding the regional structure in the territory. Our knowledge of the conditions of regional development mechanisms and factors is minimal, and this is an area of fundamental research for understanding regional development.

Table 1 Factors of knowledge self-government

Local government as an object	Local government as a subject	
1	2	3
Qualifications and education employees and elected representatives	Financial capacity of the municipality	Range of competencies
Flexibility and motivation of executive management	Level of ICT use and computer literacy of executive management	Business structure
Quality management and management processes	Communication intensity with actors in the territory	Infrastructure equipment
An informed and proactive citizen	Obtaining projects successfully	Pressure from local entrepreneurs and interest of investors
Number and activities of civic associations	Government support for knowledge society	Social and economy climate
Intensity of inter-municipal and international cooperation	Knowledge level of countryside society	Existence and success of implementation of strategic and planning documents

4 Data, Information and Knowledge in "Knowledge Management"

To clarify the process of knowledge management, it is necessary to clarify the very concept of knowledge through the so-called typology of knowledge:

- Explicit knowledge—are available on the information carrier, e.g., databases, information systems [7], text files in computers, paper documents. It exists in the form of words and numbers such as data, documents, formulas, specifications, or manuals. It can be expressed in letters, words, characters, they can be transmitted, archived. Explicit knowledge is actually information. Examples are e.g., laws, internal regulations, directives, regulations, applications, forms.
- Implicit knowledge—is not yet expressed knowledge of employees acquired through education or training. Most are tied to people's awareness. They represent procedures and processes in handling data that are in public administration registers and are used to deal with citizens' requests, they help to create legislation.
- Tacit knowledge—personal knowledge carried by a person. They are a hidden kind of knowledge that we normally acquire by socializing with the environment and about the existence of which we often do not even know (intuition, experience, judgment, empathy). This represents knowledge, which is in the form of experience, opinions, values or emotions insight, and mental models that employees obtained through social interaction and organisational learning processes [5].

Table 2 The difference between data, information and knowledge

Data	Information	Knowledge
Text, that does not give the answer to a specific issue	Text, that answers to questions who, what, when and where	Text, that answers to questions how and why
Facts and news	Data that makes sense	Legitimate, true beliefs
Non interpreted but recorded facts	Flow of meaningful messages	The beliefs and opinions formed based on these reports
Simple observation	Relevant data for a specific purpose	Valuable information from the human mind
Group of discontinuous data	A message that changes the recipient's perception	Tested, evaluated information in context

Source Knowledge Management for Library and Information Science (Majerová 27; In Kianicová Jašová 2020)

Emphasis in knowledge management is placed more strongly on tacit knowledge, which is a means for the organization to gain a competitive advantage. They are therefore what creates values and decides on the performance of organizations [43]. Knowledge management is a system that manages collective knowledge across the whole organization (explicit and implicit knowledge) and is a spiral process that involves identifying, validating, storing, and refining knowledge for users to access it, and follows the following results:

- Reuse knowledge by others for similar needs.
- Delete knowledge due to lack of credit.
- Changing the form of knowledge and creating it in a new form.

However, it is always necessary to take into account the impacts of contextual moderators, such as national culture, economy, and industry, on the KM leadership–organisational performance relationships. Also, it is essential to realize and take into account the difference between data, information and knowledge in "knowledge management" (Čajková et al. [4]) (Table 2).

5   Digitalization a Knowledge Management

According to Sous a Rocha (2019) and other authors (e.g., [10, 16], strategic knowledge management offers a distinct approach to the management of people and systems and a response to the changes occurring in a turbulent environment, a means of improving organizations performance. Kouřilová et al. [23], for example, point out the importance of digitization as a prerequisite for regional competitiveness in the context of knowledge management.

Agenda 2030 for sustainable development

The 2030 Agenda for Sustainable Development was endorsed at the UN Special Summit in New York in 2015 and sets the general framework for the world's countries to eradicate poverty and achieve sustainable development by 2030. It builds on the Millennium Development Goals adopted in 2000. The agenda also includes the Addis Ababa Action Program, approved by the United Nations in July 2015, which sets out the various means needed to implement the 2030 agenda, including domestic resources, private finance and official development assistance. The agenda contains 17 sustainable development goals and 169 related sub-goals, which balance three aspects of sustainable development—economic, social and environmental.

However, the authors Barrantes Briceño and Almada Santos [2] point to Knowledge management as the missing piece in the 2030 agenda and SDGs puzzle. With the enormous potential and vision of the sustainable development goals (SDGs), there is still a barrier in its progress and development: the knowledge use, in both the local knowledge aspects and general knowledge management. It is therefore more than necessary to link Knowledge Management to SDGs, which will help and promote its use to educate and involve all those interested in meeting these goals.

Also, already within the Draft National Priorities for the Implementation of Agenda 2030 (2018), six priority areas have been declared, which take into account the specifics of Slovakia and which will be the basis for further strategic and conceptual work. They are: 1. Education for a dignified life. 2. Moving towards a knowledge and environmentally sustainable economy in a context of demographic change and a changing global environment. 3. Poverty reduction and social inclusion. 4. Sustainable settlements, regions and countries in the context of climate change. 5. Rule of law, democracy and security. 6. Good health.

Strategy of digital transformation of Slovakia 2030

The Government of the Slovak Republic approved Resolution No. 206/2019 of 7 May 2019 framework supra-ministerial Strategy of Digital Transformation of Slovakia 2030, which aims to transform Slovakia into a modern country by 2030 with an efficient public administration ensuring intelligent use of land and infrastructure. Also, a country with an information society whose citizens will reach their full potential and live a quality and safe life in the digital age in the context of respecting and building digital humanism.

This document defines the policy and specific priorities of Slovakia in the context of the ongoing digital transformation of the economy and society under the influence of innovative technologies and global megatrends of the digital age. The strategy emphasizes current innovative technologies, such as artificial intelligence, the Internet of Things, 5G technology, big data, analytical data processing, blockchain, and super-powerful computers, which will become a new engine of economic growth and strengthen Slovakia's competitiveness.

The target entity of the digital transformation is the citizen, who should have a simpler and better everyday life in the workplace and in private, as well as the citizen-entrepreneur, to whom the state should reduce the administrative burden as much as possible and support it with appropriate incentives.

The strategy follows the creation of the new EU multiannual financial framework for 2021–2027, including cohesion policy instruments and the directly managed Digital Europe and Connecting Europe Facility programs, where the need to develop the digital economy receives special attention. The strategy analyzes the current initial state of Slovakia and offers a vision for the future. At the same time, it sets out the framework tasks and areas in which Slovakia urgently needs to multiply the potential in the digital transformation. One of these areas is public administration.

Slovakia plans to approach the development of digital transformation on three levels:

(1) Concepts and policies—This level ensures innovation in selected sectors and industries. Policies and legislative frameworks will be adapted to support the digital transformation, either by simplifying, removing obsolete rules or adopting completely new concepts.
(2) Innovation laboratories—This level will be a tool for experimenting with new ways of performing public administration. Innovative laboratories will be set up for selected sectors to experiment with new policies, business models and technologies and to help manage the digital transformation process.
(3) The new approach to projects—At this level, it presupposes a shift in the perception of project creation and an orientation not only towards grants from cohesion policy instruments, but also towards directly managed EU programs.

As stated in the document, the purpose of the strategy is not to set specific measures, but to define the vision on which the measures will be based. At the same time, the success of the fulfillment of the vision depends on broad political support beyond the mandate of the current government.

The action plan of digital transformation of Slovakia for 2019–2022

This plan includes short-term measures to vision a digital transformation strategy. It also includes improving the capacity of public administrations to innovate and use data for the benefit of citizens. The strategic goal consists of two themes: Data in Public Administration and Innovation in Public Administration. Under the first theme, the Action Plan addresses methods to significantly improve public administrations' use and processing of data for analytical purposes so that public administrations can provide quality services and the state can make decisions based on the best available knowledge. The second topic deals with implementing innovations in public administration in practice. The first wave of informatization and creation of public administration through digital technologies occurred from 2010 to 2015 within the Operational Program Informatisation of Society.

The ambition of introducing innovations into public administration significantly reduced the time needed to put innovation into practice. Several measures were to ensure this. It should become commonplace in public administration to test pilot solutions. Experimentation, organization of competitions, and involvement of companies should become the standard for the functioning of public administration institutions.

Within the ambitions, the concept of an innovative public–private partnership was envisaged, based on which it will be possible to support the implementation of innovative solutions. The private partner was to be the driving force, as it is often able to provide better and more exciting services at lower costs and usually has the knowledge needed for successful innovation. The concept of the partnership was to be implemented in practice, especially in emerging laboratories, and was aimed at supporting innovations such as artificial intelligence. However, this measure has not yet been implemented.

Other measures of the Action Plan concern the local government directly. One of them is establishing a methodological and evaluation unit and an expert platform to support the development of smart cities and regions. Therefore, this measure's main objective is to create a methodological and evaluation team for smart cities and regions and to create an expert platform to support the development of smart cities and regions. The inspiration for this platform is the URBIS platform set up by the European Commission (DG REGIO) and the EIB. Once created, the methodological and evaluation unit would focus on the following activities:

• Elaboration of the National Strategy for the Development of Smart Regions (including a benchmark for the area of smart cities and regions in Slovakia).
• Development of recommended methodology and procedures for creating strategies and concepts of smart regions.
• Analysis and evaluation of the impact of the implementation of intelligent region projects. The expert platform will support the municipality in preparing and implementing projects and attract investment in intelligent development.

The aim is not to replace the municipality's activities but to complement the activities of the cities and regions themselves and to bring expert support in specific areas. The Expert Platform and the Methodological and Evaluation Unit will support local governments that do not meet the conditions for providing support from pan-European initiatives. Although this measure was supposed to be implemented last year, it has not yet started.

An essential measure in the action plan for local governments is the establishment of a digital innovation office in local governments. Many public administration agendas are carried out at the level of self-government, which has its own specifics. These are larger and smaller cities and municipalities with different needs and priorities, as well as self-governing regions facing other challenges. In particular, digital technologies can be described as a central tool for innovation in the twenty-first century, thanks to which cities, municipalities, and regions can improve the lives of their inhabitants, improve the provision of public services and streamline decision-making and local government. This approach is referred to in the world as smart cities, but it is also about smart regions, smart cities, and smart municipalities or communities. This approach is being implemented relatively slowly in our country, as Slovakia is still struggling with a certain digital divide between the level of informatization and the use of smaller self-government technologies.

6 The Results of Research and Discussion

Local self-government in Slovakia currently consists of 2,890 municipalities. Within the methodology, the Statistical Office of the Slovak Republic divides them into two primary size groups according to the number of inhabitants: municipalities with a population of up to 1,999 inhabitants and less (2,452 municipalities), municipalities with more than 2,000 inhabitants (438 municipalities). The largest share in the size group of municipalities with less than 1,999 inhabitants consists of municipalities with 500 to 999 inhabitants (749), the second-largest share of municipalities with 200 to 499 inhabitants (707), and the third-highest share are municipalities with 1,000 to 1,999 inhabitants (584). Municipalities with a population of 2,000 to 4,999 (303) have the largest share in the second size group (Statistical Office of the Slovak Republic, 2022). These municipalities, following §22 of Act no. 369/1990 Coll. We can still classify the municipal establishment as municipalities because they do not meet one of the basic requirements for declaring a municipality a city: the population of 5,000 and more. For this scientific work, we have worked with these size groups of municipalities, except for the municipality of Jarovnice with 7,000 inhabitants, which, however, does not yet have the status of a city.

The research objective of this work was to determine the current state, needs, and shortcomings of local government in knowledge management. Due to the time constraints, we have narrowed the range of examined objects to the lowest possible number to provide a sufficient sample of Slovakia about demography and territoriality. We determined the territoriality based on the methodology of the Ministry of the Interior of the Slovak Republic on the territorial and administrative organization of the Slovak Republic (MV SR, 2022), within which Slovakia is divided into eight regions and 79 districts. We geographically determined the selection of the surveyed subjects so that each of the eight regions or regions is represented. In each area, we selected municipalities, which we divided into three size groups for our research in terms of population:

1. size group—villages up to 999 inhabitants (small municipalities)
2. size group—villages from 1,000 to 1,999 inhabitants (middle size municipalities)
3. size group—villages with more than 2,000 inhabitants (big municipalities).

For our research, we selected 18 municipalities, with each region represented in each size group. For the sake of clarity, based on the division of Slovakia at the NUTS II level, we have created four basic geographical groups talking about the Bratislava region, western Slovakia (Trnava, Nitra and Trenčín region), central Slovakia (Žilina and Banská Bystrica region), and eastern Slovakia (Prešov, and Košice region). In the Bratislava region, we selected the municipalities of Zálesie (Pezinok district), Vištuk (Senec district) and Suchohrad (Malacky district). The Trnava region is represented by the municipalities of Špačince (district of Trnava), and the Tomášikovo and Trenčín and Nitra regions are represented by the municipality of Hajná Nová Ves (district of Topoľčany), which is located in the imaginary geographical intersection of both regions. Central Slovakia is represented in our research by municipalities from the Žilina region Svätý Kríž (district Liptovský Mikuláš), Kolárovice (district Bytča) and Terchová (district Žilina). The Banská Bystrica region is in the village of Tajov (Banská Bystrica district), Vyhne (Žiar nad Hronom district) and Slovenská Ľupča (Banská Bystrica district). Eastern Slovakia is represented by the municipalities of the Prešov region Úzovské Pekľany (district Bardejov), Kojatice (district Prešov) and Jarovnice (district Sabinov) and the Košice region municipalities Dargov (district Trebišov), Vinné (district Michalovce) and Poproč (district Rimavská Sobota).

We chose a semi-structured telephone interview based on eight open-ended questions to provide relevant data for our research. The respondents were the mayors of these municipalities, in one case, the head of the municipal office. The questions focused on the following areas of data collection:

1. What is the knowledge needs of self-government?
2. What knowledge assets or resources does it have?
3. What gaps exist in her knowledge?
4. How does knowledge flow within, to and from the organization?
5. What are the barriers to information flow?
6. How do people currently support or prevent the efficient flow of knowledge of technologies?
7. How does document management work?
8. What decision support systems do they use?

Research evaluation

Educational structure and knowledge assets

Based on the obtained data, we came to the not surprising conclusion that very few university-educated people participate in the running of self-government. The size of the municipality does not only condition this according to the number of inhabitants of the range of services that the municipality or municipal office provides to citizens or the public. For example, two municipalities with the same number of 600 inhabitants, Suchohrad and Dargov, have a diametrical difference in employees, namely six, respectively two. This is related to the scope of competencies that the municipality exercises. While Dargov, the municipality, only takes care of the running of the kindergarten, in Suchohrad, the municipality employees are also in charge of the registry office, the building office, and the building administration. More than a third of the addressed municipalities have problems employing university-educated people, and the most common reason is to provide such employees with adequate financial rewards. This is related to lower table salaries in public administration than in the private sector. Another reason why municipalities fail to fill specialized jobs is the lack of labor market experts. In addition to economists or IT specialists, they are experts in public procurement, creating projects to obtain subsidies from Euro funds or other sources, and experts for work at the cadastral or land office. This is one of the reasons why mayors rely on employing the most capable people they have at their disposal or focus on the paid services of external contractors from the ranks of private workers (Tables 3, 4 and 5).

Table 3 Number of university-educated employees in small municipalities

Name of the village	The population	Number of employees of the municipal office	
		Total	With university education
Hajná Nová Ves	350	1	0
Úzovské Pekľany	460	1	1
Suchohrad	600	6	0
Dargov	600	2	0
Tajov	650	2	1
Svätý Kríž	800	3	0

Source own research.

Table 4 Number of university-educated employees in medium-sized municipalities

Name of the village	The population	Number of employees of the municipal office	
		Total	With university education
Vyhne	1 100	7	3
Kojatice	1 200	3	0
Vištuk	1 400	5	0
Vinné	1 700	11	5
Tomášikovo	1 700	4	2
Kovárovce	1 800	4	2

Source own research

Table 5 Number of university-educated employees in large municipalities

Name of the village	The population	Number of employees of the municipal office	
		Total	With university education
Zálesie	2 200	5	1
Špačince	2 400	7	0
Poproč	2 700	5	0
Slovenská Ľupča	3 200	9	4
Terchová	4 000	11	4
Jarovnice	7 000	11	4

Source own research

Improving qualifications and their forms

Slovak municipalities have a wide and diverse range of employees to ensure the fulfillment of their own and state-delegated competencies. However, not every professional
or public service can acquire an employee with the appropriate specialization. And
so they try to replace this need for the field by training employees, most often in
the form of courses or training. We have not encountered a case where an employee

is studying at a university in a given field in addition to employment. Still, in one municipality, the employees are studying at a university.

The most common forms of increasing qualifications and acquiring new knowledge and skills are training, courses, or retraining of employees. Furthermore, it is a self-study focused on resources on the Internet in the literature. The exchange of information and knowledge between respective municipal authorities, through personal acquaintances or friends, and only then in direct communication with superior public administration bodies can be considered a particular specificity. Of course, it is a matter of passing on experience and knowledge to younger employees by older ones, not only when retiring.

Information flow and verification

The mayors of the municipalities we contacted often pointed to the lack of financial resources they have to deal with. This is associated with a lower number of employees and especially with the accumulation of several tasks entrusted to existing employees. That is why they point to the time-consuming nature of their work and are not always willing to engage in further education at the expense of their time off work. Money also plays a role in choosing the training courses or training itself. Small and medium-sized municipalities, in particular, cannot finance them themselves and the state rarely contributes to education.

Of the technical obstacles, this is mainly the poor coverage of the quality Internet network, outdated municipal offices 'equipment, or employees' low skill when working with a computer or new communication technology. In one case, even municipal employees resist digitizing their work or using modern forms of communication.

Another most common obstacle was the growth of new information on the Internet, not all of which were relevant, and employees had difficulty working with selecting and filtering authentic sources. The issue of verifying the data obtained is also related to this obstacle. The employees of the addressed municipalities verify these directly through other internet sources, relevant laws, or professional literature. An important part is a direct communication with superior public administration bodies, whether telephone or electronic, contacting other mayors who are already solving the problem or consulting with experts in law or construction. In two cases, the mayors told us they also follow simple common sense.

Document management

Each of the addressed municipalities archives documents by the relevant law, both in physical and electronic form. Documents are archived in files or on electronic media. The documents the law must publish are available on the municipalities' websites, which also serve as a freely available electronic archive. In the municipalities, this task is entrusted to a trained worker. In Jarovnice, they have a specialist employed to archive—a trained archivist.

7 Conclusion

Most of the mayors of the municipalities we contacted are also satisfied with the staffing of the municipal offices they currently have at their disposal. Although they would like to employ workers with adequate vocational training, they have to make do with what the current labor market offers. Hiring a specialist in project creation or public procurement is not only beyond their economic capabilities. In most municipalities, the specialized competencies transferred from the state are performed by employees with the most necessary training or further education. From a financial point of view, only the largest municipalities, which also have higher incomes, can afford university-educated employees.

Despite this educational handicap, it cannot be claimed that municipalities do not fulfill the tasks or functions entrusted to them. They can also do this thanks to systematic training in the form of courses, training or retraining, but the most common form of increasing education and expertise is self-study and finding relevant information on the Internet or through information flows between individual municipalities. According to mayors, obtaining information from the state administration is lengthy and often confusing. Therefore, through the mentioned information channels, they also verify a lot of information.

The most common obstacles to improving knowledge management are the lack of financial resources for employees and their further education, the lack of professionals in the labor market, and insufficient internet coverage in Slovakia. The mayors or employees of municipalities try to bridge these shortcomings by obtaining relevant information from sources other than the primary ones, such as communication, especially with state administration bodies. It is lengthy, and they would not be able to meet the set deadlines, for example, when submitting applications for calls. Thanks to this effort, it can be stated that the local self-government in Slovakia, although underestimated in terms of the educational census of employees, can provide entrusted tasks and services to its citizens.

Acknowledgements The paper was prepared within the project VEGA no. 1/0320/21 "The role of universities in building the knowledge economy".

References

1. Balážová, E., Papcunová, V., Pospišová, L., Knežová, J.: Výkon kompetencií miest v SR. Slovenská poľnohospodárska univerzita v Nitre. (2020) ISBN: 978-80-552-2281-3. 184 s. doi:https://doi.org/10.15414/2020.9788055222813.
2. Barrantes Briceño, C.E., Almada Santos, F.C.: Knowledge management, the missing piece in the 2030 agenda and SDGs puzzle. Int. J. Sustain. High. Educ. **20**(5), 901–916. https://doi.org/10.1108/IJSHE-01-2019-0019 (2019)
3. Brooks, R.: Difference between Information Management and Knowledge Management [online] [cit. 2021-11-10]. Dostupné z: https://blog.netwrix.com/2020/01/21/information-management-vs-knowledge-management/ (2020)

4. Cajková, A., Jankelová, N., and Masár, D.: Knowledge management as a tool for increasing the efficiency of municipality management in Slovakia. Knowl. Manage. Res. Pract (2021): 1–11. https://doi.org/10.1080/14778238.2021.1895686
5. Cheng, Q., Chang, Y.: Influencing factors of knowledge collaboration effects in knowledge alliances. Knowl. Manag. Res. & Pract. 18(4) 380–393 (2020). https://doi.org/10.1080/147 78238.2019.1678412
6. Dušek, J.: Evaluation of development of cooperation in South Bohemian Municipalities in the years 2007–2014. Eur. Ctryside 9(2), 342–358 (2017). https://doi-1org-158ybleog08fe.han proxy.cvtisr.sk/. https://doi.org/10.1515/euco-2017-0021
7. Fedushko, S., Ustyianovych, T., Gregus, M.: Real-time high-load infrastructure transaction status output prediction using operational intelligence and big data technologies. Electronics 9(4), 668 (2020). https://doi.org/10.3390/electronics9040668
8. Hajšová, M.: Význam ľudského kapitálu pre rozvoj samosprávy. In: Sociálno-ekonomická revue, 02/2014. Fakulta sociálno-ekonomických vzťahov, Trenčianska univerzita Alexandra Dubčeka v Trenčíne (2014). ISSN 1336-3727
9. Jirásek, J.A.: Znalostní kapitál. Moderní řízení 35(6), 4–7 (2000)
10. Joyce, W.F., Slocum, J.W.: Top management talent, strategic capabilities, and firm performance. Organ. Dyn. 41(3), 183–193 (2012)
11. Jánošková, B.: Future of slovak regional self-government or quo vadis self-governing region? Politické Vedy. [online] 23(4), 228–236 (2020). ISSN 1335-2741. https://doi.org/10.24040/pol itickevedy.2020.23.4.228-236
12. Kaliňák, M.K.: Analýza možností na zvýšenie potenciálu miestnej územnej samosprávy pri realizácii hospodárskych politík miest a obcí. [cit. 2021-10-11]. Dostupné na: (2017). https:// www.zmos.sk/download_file_f.php?id=1220820
13. Kavan, Š., Brehovská, L.: Spolupráce Jihočeského kraje a přeshraničních regionů se zaměřením na ochranu obyvatelstva. In 19th International Colloquium on Regional Sciences. Conference Proceedings, pp. 907–914. Brno: Masarykova univerzita, (2016). ISBN 978-80-210-8273-1. https://doi.org/10.5817/CZ.MUNI.P210-8273-2016-117
14. Khilji, N.K., Roberts, SA.: An exploratory study of knowledge management for enhanced efficiency and effectiveness: the transformation of the planning system in the UK local government. J. Inf. Knowl. Manag. 14(1) (2015)
15. Kianicová Jašová, K.: ZNALOSTNÝ MANAŽMENT PRE KNIŽNIČNO - INFORMAČNÚ VEDU (2020). ISBN: 978-80-554-1706-6
16. Kianto, M., Vanhala, P., Heilmann, P.: The impact of knowledge management on job satisfaction. J. Knowl. Manag. 20(4), 621–636 (2016)
17. Kim, S.S.: Exploitation of shared knowledge and creative behavior: the role of social context. J. Knowl. Manag. 24(2), 279–300 (2020). https://doi.org/10.1108/JKM-10-2018-0611
18. Klimovský, D., Gašparík, J.: Základný rámec participatívnej tvorby verejných politík a súvisiace pojmy. In: Klimovský, D. (ed.) Participatívne procesy v praxi čítanka participatívnej tvorby verejných politík. 2. Vyd, pp. 12–26. Bratislava: MV SR/ÚSR ROS (2020). ISBN 978-80-89051-63-2
19. Klimovský, D., Nemec, J.: Local self-government in Slovakia. In: Brezovnik, B., Hoffman, I., Kostrubiec, J. (eds.), Local self-government in Europe, pp. 355–382. Maribor: Institute for Local Self-Government Maribor (2021). ISBN 978-961-7124-00-2. https://doi.org/10.4335/ 978-961-7124-00-2
20. Kluge, J., Stein, W., Licht, T.: Knowledge unplugged: The Mckinsey & Company Global Survey on knowledge management. Palgrave Publishing (2002). https://doi.org/10.1080/109 7198X.2002.10856333
21. Kompetencie pre digitálnu dobu – AMAVET. [cit. 2021-10-21]. Dostupné na: https://www. amavet.sk›kompetencie-pre-digitalnu-dobu
22. Koncepcia ďalšieho vzdelávania zamestnancov verejnej správy. [cit. 2021-10-21]. Dostupné na: https://www.minv.sk›implementacia-systemu-dal
23. Kouřilová, J., Kubíková, M., Pělucha, M.: Digitalizace jako předpoklad regionální konkurenceschopnosti? Analýza disparit na příkladu ČR. In Klímová, V., Žítek, V. (eds.)

24th International Colloquium on Regional Sciences. Conference Proceedings, pp. 1–5. Brno: Masarykova univerzita (2021). ISBN 978-80-210-9896-1
24. Kováčová, N.: Analysis of Municipal Elections 2018 in context of K. Reif and H. Schmitt´s theory. Politické vedy. **22**(4), 149–171 (2019). ISSN 1335–2741. https://doi.org/10.24040/pol itickevedy.2019.22.4.149-171
25. Lei, H., Ha, A.T.L., Le, P.B.: How ethical leadership cultivates radical and incremental innovation: the mediating role of tacit and explicit knowledge sharing. J. Bus. & Ind. Mark. **35**(5), 849–862 (2019)
26. Liu, G., Tsui, E., Kianto, A.: Revealing deeper relationships between knowledge management leadership and organisational performance: a meta-analytic study. Knowl. Manag. Res. & Pract. (2021) DOI: https://doi.org/10.1080/14778238.2021.1970492
27. Majerová, I.: Measurement of Innovative Performance of Selected Economies of the European Union and Switzerland. Int. J. Inf. Educ. Technol. **5**(3), 228–232 (2015). https://doi.org/10.7763/IJIET.2015.V5.506.
28. Molitoris, P.: Verejná správa v súčasnom demokratickom a právnom štáte. In: Verejná správa v súčasnom demokratickom a právnom štáte (Časť 1.): recenzovaný zborník príspevkov z medzinárodnej vedeckej konferencie/Paluš, Igor; Mital, Ondrej; Žofčinová, Vladimíra, eds. – 1. vyd. – Košice (Slovensko) : Univerzita Pavla Jozefa Šafárika v Košiciach (2018). ISBN 978-80-8152-700-5
29. Ngcamu, B.S.: Perceptions of knowledge management: A local government perspective. J. Gov.Ance Regul. **1**(4–1), 133–138 (2012). doi: https://doi.org/10.22495/jgr_v1_i4_c1_p2
30. Nonaka, I., Takeuchi, H.: The Knowledge Creating Company: How Japanese Companies Create
31. Návrh národných priorít implementácie Agendy 2030. Uznesenie vlády SR **273**(13) (2013)
32. Operačný program Informatizácia spoločnosti. [cit. 2021-10-21]. Dostupné na: https://www.minv.sk
33. Peráček, T., Kocisova, L., Mucha, B.: Importance of the e-government act and its impact on the management and economy of the enterprise in the Slovakia. In Management and Economics in Manufacturing. Global Scientific Conference on Management and Economics in Manufacturing, pp. 90–96 (2017)
34. Peráček, T.: The Perspectives of European Society and the European Cooperative as a Form of Entrepreneurship in the Context of the Impact of European Economic Policy. Online J. Model. New Eur. **34**, 38–56 (2020). doi: https://doi.org/10.24193/OJMNE.2020.34.02
35. Picker, C.: Cooperative governance. In: Blome-Drees J., Göler von Ravensburg N., Jungmeister A., Schmale I., Schulz-Nieswandt F. (eds.), Handbuch Genossenschaftswesen. Springer VS, Wiesbaden (2022). doi: https://doi.org/10.1007/978-3-658-18639-5_27-1
36. Podpora inovatívnych riešení v slovenských mestách. [cit. 2021-10-21]. Dostupné na: https://www.mhsr.sk › uploads › files
37. Poniszewska, D., Matusiak, R., Kryvinska, N., Yasar, A.-U.-H.: A real-time service system in the cloud. J Ambient Intell Human Comput **11**(3), 961–977 (2020). https://doi.org/10.1007/s12652-019-01203-7
38. Rodríguez-Pose, A., Garcilazo, E.: Quality of government and the returns of investment: examining the impact of cohesion expenditure in european regions. Regional Studies **49**(8), 1274–1290 (2015). ISSN 1360-0591. doi: https://doi.org/10.1080/00343404.2015.1007933
39. Sedláková, S., Válovská, Z.: Riadenie organizácií verejnej správy v procese transformácie. Verejná správa a spoločnosť: vedecký časopis. **15**(1), 75–82 (2014). ISSN 1335-7182
40. Szostak, M.: Post-communist burden influence on the perception of creative identities: consequences for managers and leaders. Eur. Res. Stud. J. **XXIV**, 282–302 (2021). doi: https://doi.org/10.35808/ersj/2354
41. Terem, P., Čajka, P., Rýsová, L.: Europe 2020 Strategy: Evaluation, Implementation, and Prognoses for the Slovak Republic. Econ. Sociol. **8**(2),154–171 (2015). doi: https://doi.org/10.14254/2071-789X.2015/8-2/12. ISSN 2071-789X
42. UN ECLAC. Knowledge Management: informing decisions to realise good governance," Sede Subregional de la CEPAL para el Caribe (Estudios e Investigaciones) 38381, Naciones Unidas Comisión Económica para América Latina y el Caribe (CEPAL). Policy Brief (2013). August 2013 LC/CAR/L.413

43. Van Stigt, R., Driessen, P.P.J., Spit, T.J.M.: A user perspective on the gap between science and decision-making. Local administrators' views on expert knowledge in urban planning. Environ. Sci. & Policy. **47** (2015). doi: https://doi.org/10.1016/j.envsci.2014.12.002
44. Vernarský, M.: Current state's attitude towards municipal self-government in Slovakia. Slovak J. Polit. Sci. **19**(2) (2019). doi: https://doi.org/10.34135/sjps.190203
45. Vojtech, F., Srebalová, M., Majerčáková, D. Elektronizácia územnej samosprávy v slovenskej republike dnes. In: Medzinárodné a vnútroštátne právne aspekty priamej účasti obyvateľov na miestnej samospráve. Lívia Trellová – Ľudovít Máčaj (zost.) 1. Vyd, p. 98. Bratislava: Univerzita Komenského v Bratislave, Právnická fakulta (2020)
46. ŽÁRSKA, E., – HALÁSKOVÁ, M., – GALATA, M.: Indicators of knowledge autonomy : case study of Slovak and Czech municipalities. In ECON : journal of economics, management and business. Ostrava : Faculty of Economics VŠB - Technical university of Ostrava, **21**(1), (2012). ISSN 1803-3865.
47. ZMOS. Výkon a financovanie prenesených kompetencií. Bratislava (2019)
48. Žofčinová, V. Osobitné kategórie zamestnancov. Košice 214 (2018). ISBN 978-80-8152-645-9

Comparison of Methodologies Used in Cybersecurity Reports

Ondrej Čupka⊙, Ester Federlova⊙, and Peter Vesely⊙

Abstract New legislation in the field of cybersecurity makes public and private institutions legally responsible for keeping the nation's IT infrastructure and data safe under the threat of legal consequences. This sudden shift saw many companies and public institutions scrambling to meet the new legal demands but have often found a lack of professionals, institutions, systems, but mainly information. The laws do oblige most organizations to report incidents and breaches. How to turn this data into information is still up to discussion. It is this that has led us to our research, which consists of defining, comparing, and assessing the methodologies of four major cybersecurity reports, characterizing their interactions with cybersecurity legislation in the Slovak republic to create a base for developing a methodology for Slovakia. The reports we used are Verizon's Data Breach Investigation Report, IBM's X-Force Threat Intelligence Report, Ponemon Institute's Separating Truth from Myth in Cybersecurity, and e-Governance Academy's National Cybersecurity index. Despite the difference in their approaches, we were able to discover unifying best practices and issues that should not be repeated in the development of a new methodology. Three key best practices are sound statistical methods, timeliness of data, and transparency. Three key issues were lack of transparency in methodology, limitations, bias and reasoning, inconsistency, and loss of relevance due to late publication. Additionally, we have identified that reports of this nature are in line with cybersecurity legislature and national strategies. Such reports may be used as a part of preventative measures by creating broad awareness, increasing interest in the topic, warning organizations of any dangers, and finding ways of creating and improving cybersecurity strategies to build capacities and resilience.

O. Čupka · E. Federlova · P. Vesely (✉)
Faculty of Management, Comenius University Bratislava, Odbojarov 10, Bratislava 831 04, Slovakia
e-mail: peter.vesely@fm.uniba.sk

O. Čupka
e-mail: cupka9@uniba.sk

E. Federlova
e-mail: federlova2@uniba.sk

Keywords Cybersecurity · Security · Information · Attack · Threat · Reports

1 Introduction

Digital security has become an important issue since the internet became broadly available. But it has only recently become a matter of national security and an issue that touches most if not all areas of human activity. While public and private services moved steadily to the internet, especially with the proliferation of social media, cloud technology, or communications technologies such as messengers and video calling, digital security naturally gained in importance. Because as our dependence grew so did the threats. This was only accelerated by the introduction of legislation, most notable the General Data Protection Regulation and the Cybersecurity Act of the European Union, and the Covid-19 pandemic, which saw a large shift of public, private, and commercial life to the digital sphere. From malware attacks causing the loss of confidentiality for millions of users, ransomware attacks shutting down oil pipelines, crippling the energy supplies of a nation, to nation-state attackers conducting espionage, influencing elections, and seeding doubt and conflict in the public with disinformation campaigns. The stakes are high and ever-growing. To take appropriate measures that will prevent or at least mitigate these threats we require good sources of information on the always evolving cybersecurity landscape. After several discussions, we realized that the Slovak Republic had access to plenty of data, but it does not have a good enough way to turn it into information yet. This article is therefore meant as the first step in developing a cybersecurity report for Slovakia. This first step is to find out how the global and respected players are conducting their research, and what the interactions with the main cybersecurity legislation are.

In this paper, we are going to compare the methodologies used in different papers on the state of cybersecurity and identify the best practices used in them. This is to serve as a base for further research in the future and the development of a methodology for the Slovak republic. The reports we will discuss are Verizon's Data Breach Investigation Report, IBM's X-Force Threat Intelligence Report, Ponemon Institute's Separating Truth from Myth in Cybersecurity, and e-Governance Academy's National Cybersecurity Index. And the legislation we analyzed are the aforementioned General Data Protection Regulation, the Cybersecurity Act, and the soon-to-be ratified National Cybersecurity Strategy for the years 2021 to 2025. Due to the complexity of this issue, we have divided the main goal into the following sub-goals:

- Appropriately characterize the current state of literature in the security sciences and especially cybersecurity
- Identify and gain access to the most trusted and used cybersecurity reports
- Analyze the methodology used in the cybersecurity reports
- Compare the methodologies for their advantages and disadvantages
- Identify the best practices for cybersecurity report methodologies
- Analyze the interactions with current legislation and upcoming national strategy on cybersecurity

The research questions we are looking to answer in this paper are:

- What are the best practices in the methodology used in cybersecurity reports?
- What mistakes should be avoided?
- What are the advantages and disadvantages of the different types of methodologies?
- What are the interactions of such research with national legislation?

2 Conceptual Background

The emergence of modern society, based on technological progress and the discovery of new raw materials and energy sources, has brought with it unprecedented prosperity for mankind, but also unprecedented risks. Each system naturally carries instruments for its prosperity as well as for its demise. Improved means of travel have brought the whole world closer together with all the economic and social benefits, but it also allows for global pandemics and crime. The Internet has enabled the emergence of the world's richest businesses, and made enormous amounts of information available to the world, but has also enabled the emergence of cybercrime, hackers, and constant monitoring and analysis of digital users. Ulrich Beck, therefore, called the current society a society of risk. Security sciences, security management, and particularly cybersecurity are therefore becoming crucial for correctly identifying and preventing these risks.

3 Methodology

For this paper we define the following methodological approaches:

- Selection of relevant literature based on the consultant's recommendations and the basis of own research
- Study of selected literature
- Creating the literature overview with the help of analysis, deduction, comparison, synthesis, and induction
- Analyzing and deducing the methodologies of cybersecurity reports
- Comparing the reports
- Selection of legislation and national strategies on cybersecurity and analyze their interaction with this kind of research.

Data Sources and Collection Methods

Several primary and secondary sources of information were used in creating this thesis. These were mainly from foreign sources. The primary sources of information were the reports that were analyzed, namely the X-Force Threat Intelligence Index 2021 from IBM, Verizon's 2020 Data Breach Investigations Report, Ponemon

Institutes Separating the Truths from the Myths in Cybersecurity, and the National Cybersecurity Index from the e Governance Academy, as well as laws and regulation such as GDPR and the Cybersecurity Act.

Methods Used for Evaluating and Interpreting Results

The following methods were used for the evaluation and interpretation of the literature:

- Analysis—the analysis was necessary, especially when evaluating the theoretical literature, as well as during the identification and characterization of the various methodologies of the selected reports.
- Synthesis—was used in formulating the outputs of our work and making conclusions.
- Comparison—was the most used method and necessary in compiling a comprehensive view of the issue and in fulfilling the main goal of the thesis.
- Deduction—was used to draw specific conclusions related to the set goals.
- Induction—with the induction we combined different opinions of relevant authors to draw our conclusions.

Characteristic of the Investigated Object

The investigated objects of the thesis are first and foremost the approaches to conducting research and especially the methodologies used for compiling the chosen four cybersecurity reports. Afterward, they are the institutions conducting the research and the legislation on cybersecurity of the Slovak republic.

The reports are as varied as the institutions that published them. The first two are the reports from multinational companies IBM and Verizon, the X-Force Threat Intelligence Report and the Data Breach Investigation Report, then is a report from the research institute Ponemon, the Separating the Truths from the Myths in Cybersecurity, and the state-supported e-Governance Academy with its National Cybersecurity Index. The reports were chosen as they are some of the most trusted and widely used reports by IT professionals. They are very different in how they approach defining the cybersecurity landscape which was done intentionally to understand why and how such research may be conducted. Leaning on the goals set out in the thesis, it is done to support the creation of a similar framework, methodology, and report for the Slovak republic.

The X-Force Threat Intelligence Report and Data Breach Investigation Report are similar in nature. Both companies leverage their large customer base using their IT products and services to gather direct data on incidents and breaches they experienced as well as gathering data from external sources. The reports focus on a statistical analysis of the gathered incidents and breaches on aggregated data, broken down based on industry, geography, attack type, and vector. Based on the findings they shed light on when, why, and how incidents and breaches happen and offer measures that can be taken to prevent, mitigate or limit the damage caused by them.

The Ponemon Institute's Separating the Truths from the Myths in Cybersecurity on the other hand focuses this report on gathering data via a survey among IT professionals to find out their views on key statements surrounding cybersecurity. Such a study, especially when conducted multiple times may offer valuable information about the shifting views of IT professionals. The report then statistically analyses the responses and offers explanations as to why this may be so and what it means for the field.

The e-Governance Academy's National Cybersecurity Index offers a very different approach to understanding the state cybersecurity in a country. As the name suggests it is first and foremost an index, which measures the capacity of a nation to deal with cybersecurity threats based on a set of 46 indicators. The indicators are made up of legislative, institutional, and policy conditions. How a nation fulfills these conditions the e-Governance Academy assesses based on publicly available information, which is sent to the team responsible for the NCSI or gathered from the team themselves. The result is a transparent comparison between 160 different countries and a list of recommendations to the countries, which may help guide them in creating a more resilient state.

Legislation in the Slovak republic on cybersecurity is mainly made up of the General Data Protection Regulation, the Cybersecurity Act, and the resulting National Cybersecurity Strategy. GDPR for the most part does not interfere with research, as it specifically protects the personal information of natural persons. Although it offers some opportunities as the enforcement of this legislation may result in more available information on how data is protected. The Cybersecurity Act offers even more opportunities as it establishes the Computer Security Incident Response Team, a unified information system, and obliges organizations offering basic and digital services to report incidents as well as making it available to all organizations voluntarily. This way it has a growing database of the incidents happening in Slovakia. The act also creates a need for a National Cybersecurity Strategy. Here the goals are set for the country in the field of cybersecurity, which calls for improved data gathering, analysis, assessment, and distribution of results.

4 Results

In this chapter we are going to analyze the methodology of four renowned cybersecurity reports; IBM's X-Force Threat Intelligence Index, Verizon's Data Breach Investigation Report, Ponemon Institute's Separating the Truths from the Myths in Cybersecurity, and the National Cybersecurity Index of the e-Governance Academy. Afterward, we are going to compare these methodologies to identify their advantages and disadvantages. Then we are going to list any best practices we can identify and rank them. The chapter will afterward end with an examination of legislation surrounding cybersecurity in the Slovak republic and how they interact with cybersecurity reports.

X-Force Threat Intelligence Index 2021

IBM or International Business Machines Corporation is an American multinational company with its core business in information technology. IBM produces and sells computer hardware, middleware, and software, but also provides consulting and hosting services [1]. The company is also highly active in research, receiving more than 9 100 patents in 2018 alone, 110 000 in total [2]. The headquarters are located in Armonk, New York, USA. The company was founded in 1911 as the Computing Tabulating Recording Company by Charles Ranlett Flint and Thomas J. Watson Sr., and in 1924 renamed IBM. The current CEO is Virginia Marie Rometty [1]. On the fortune 500 list IBM is ranked 38th with almost 74 bn. USD in revenues made, assets of more than 123 bn. and employing 352 600 employees worldwide in 2020. IBM is therefore also one of the 30 companies included in the Dow Jones Industrial Average index. [3] IBM is one of the world's largest employers with about 70% of its employees based outside of the United States of America in more than 170 countries [4].

IBM is one of the largest and most recognized companies in the IT sector. Offering a multitude of services and IT solutions from data storage and management, cloud solutions, an AI personal assistant, internet of things solutions, and much more, to companies all over the world. To keep offering these services IBM needs above all to make sure that it and its products are safe. Which requires it to have the highest standards of cybersecurity. IBM also soon recognized, that many of their clients struggle with cybersecurity and that cybersecurity is an issue that cannot be solved by a single actor. Which is one of the reasons the IBM X-Force Threat Intelligence was created [5]. X-Force is a branch of IBM Security since 2006 but has been in existence since 1996 as a part of Internet Security Systems, which was acquired by IBM. The X-Force Exchange platform was created in 2015 to support collaborative defense in cybersecurity. This cloud platform offers threat intelligence collected by X-Force since its founding in 1996 and is steadily updated with new information gathered globally by IBM and its clients. It offers information on IP and URL reputation, vulnerabilities, malware, web applications, and spam. This way the users of this service can access potentially vital information on current risks and threats in the cybersecurity landscape and can take appropriate actions to mitigate them. Benefits of the platform are the size of the database, regular updates, high security with ISO27001 and IBM Privacy Shield certification, and limited free access. Access is free for up to 5 000 records per month, above that, users have to pay a fee of 2000 USD per 10 000 records [6].

From the information gathered, IBM X-Force also compiles a yearly report, the X-Force Threat Intelligence Index. This index was first published in 2017 by IBM as a combination of the X-Force Threat Intelligence Report and the Managed Security Services Cyber Security Intelligence Index [7].

Published methodology of research

Unfortunately, IBM has yet to release the methodology used for this report. While understandable as it may be considered a part of the know-how of the company, it makes it difficult to assess its accuracy, verify and confirm its findings or hint at its possible weak points. The only mention of any methodology can be found in the first X-Force Threat Intelligence Index from 2017:

"To better understand the security threat landscape, X-Force uses both data from monitored security clients and data derived from non-customer assets such as spam sensors and honeynets. X-Force runs spam traps around the world and monitors more than eight million spam and phishing attacks daily. It has analyzed more than 37 billion web pages and images. IBM Security Services monitors billions of events per year from more than 8 000 client devices in more than 100 countries. This report includes data IBM collected between the 1st of January 2016 and the 31st of December 2016. In this year's report, IBM X-Force Threat Research adopted the MITRE Corporation's Common Attack Pattern Enumeration and Classification (CAPEC) standard for attack categorization. The top five attacked industries were determined based on data from a representative set of sensors from each industry. The sensors chosen for the index had to have event data collected throughout the entire year of 2016. The insider/outsider identification utilized in this report includes all source and destination IP addresses identified in the attacks and security incidents targeting the representative set of sensors. A single attack may involve one or many attackers [8]."

While being itself is a quite short explanation of the methodology used, there is no mention of how specifically the data has been gathered or analyzed and assessed. We can therefore only guess that a similar approach has been made when compiling the report for 2021. Although, when comparing the two reports we can see a difference in the structure and contents of the reports and can assess that the methodology has gone through some changes since its inception. Nevertheless, the methodology may be partially identified based on the structure and content of the report.

Report structure and summary

After the executive summary, which highlights the most important points of the report, comes the section Top Attacks of 2020. Here we also learn what is considered an attack in this report and how information on it is gathered: "Attacks and incidents are used interchangeably in this report. An incident refers to an organization's hotline call to the X-Force Incident Response team that results in the investigation and/or remediation of an attack or suspected attack" [9]. Two distinctions are immediately drawn between attack types, for example, ransomware, data theft, etc., and initial attack vectors, such as scan-and-exploit, phishing, credential theft, etc. Or said differently, between the goal of the attack—to steal data, extort money, etc.—and how the attack was realized—by scanning for vulnerabilities, misusing the trust of employees, stealing credentials, etc. Importantly, these are compared with data from the report from last year to put it into perspective, track changes in the behavior of malicious actors, and help point to what the focus of cyber security should be [9].

The top three attack types and attack vectors are subject to deeper analysis within the report. For the X-Force Threat Intelligence Index 2021, the most common attack types were Ransomware, Data Theft, and Server Access. A unifying, common metric was not used for analysis rather the authors go into the specifics of each type.

Ransomware was analyzed based on its prevalence; what types of ransomware were most commonly used. Interestingly, an estimation of the criminal profits of the most popular type of ransomware was also made with a deeper analysis of which industries and countries were targeted as well as recommendations on how to respond to such an attack. Next was Data Theft and Server Access attacks, where only basic statistics were stated regarding the affected geographic regions and industries. Next comes a list of the top 10 most exploited vulnerabilities in 2020 continuing into Top attack vectors. Both parts offer basic statistics about the types, share of the total, and some comparisons to the previous year [9].

The next chapter is dedicated to Advanced Threat Actors and as such to organized and coordinated cybercriminal groups. The 2021 report was about a unique situation where X-Force was able to get insights into the working of a state-sponsored threat group from Iran. Apart from finding clear evidence of attacks conducted by this group, it was also able to access a misconfigured server, gaining access to a large database of educational videos and data files used by the threat group. Additionally, several advanced threat actors were found exploiting the ongoing pandemic for phishing campaigns. Unlike most phishing attacks, these also point to possible state actors based on their focus and sophistication [9].

Operational technology is hardware and software that detects or causes change to equipment, assets, processes, and events. Here we can picture automatic systems that monitor and control dams, chemical plants, or industrial machinery in a car factory. Any successful hacking attempt would have the potential to cause great damage to health and property. Therefore, a chapter is dedicated to this topic within this index. This chapter contains statistics on the most common attack types and the change in registered attacks. Similarly, to previous chapters, the top three threats to operational technology are analyzed deeper. In this case Ransomware, Remote Access Trojans, and Insider Threats [9].

Special mentions within the X-Force Threat Intelligence Index 2021 are Spoofed Brands and Financial Cybercrime. Spoofed brands are a tool of social engineering strategies and relate to malicious domains on the internet, which pretend to be original and branded. They are used to trick users and misuse their trust in the brands and services they know to steal information or download malicious software via links or emails. Top 10 spoofed brands are listed, although it is unclear what brands are monitored as the 10 brands make a total of 100% and there is no "other" item on the list. Similar to the chapter on Operational technology, Financial cybercrime is also specifically mentioned as it is the most common type of cybercrime. The top 10 banking trojan families are listed with the top three, Ramit, Trickbot and QakBot analyzed deeper [9].

Probably one of the most important chapters is about New Malware Threats. These are the new threats that were identified in 2020. For this year it was mostly new code innovations in Linux malware, which was probably caused by malicious actors adapting to the move of many companies into the cloud. These malware codes are specialized in ransomware and crypto-mining. The new most used programming language for malware was Go or Golang, which allows attackers to adapt code to other platforms automatically. As such, malicious code written for Windows may quickly be adapted for OS X, Linux or others. Emotet spam malware was detected to be in use once again but improved with new features and better protected against scans [9].

The largest chapter of the report contains a breakdown of the aggregated data based on geography and industry. Based on geography attacks are shown for Europe, North America, Latin America, Asia, and together with the Middle East, and Africa. First regions are ranked based on the share of attacks conducted in them. After that each region is separately listed with information regarding the attack volume, attack types, which countries were the most attacked within the region, and a breakdown of the attack types. The breakdown of industries is made of the top 10 industries by attack volume. First, they are ranked, where not only their current rank is listed, but also the change from last year and the specific percentage share of attacks. Each sector has its page with a selection of the most important information about the attack share, most used attack types, and vectors. It also contains detailed information on how the situation changed in previous years. The top ten sectors are Finance and Insurance, Manufacturing, Energy, Retail, Professional Services, Government, Healthcare, Media, Transportation, and Education [9].

The chapters until now were focused on the past, data on what was recorded the last year. The last two chapters Looking Ahead and Recommendations for Resilience are focused on the future. Looking Ahead lists four key takeaways for the next year. For 2021 the takeaways are to remain vigilant as it is expected that thousands more vulnerabilities will be reported on old and new systems. Double extortions are going to persist for ransomware attacks. Threat actors are likely to continually shift to other attack vectors and based on the big changes to the shares of attacked industries, there is not a single one safe. Recommendations for resilience has 8 key recommendations on how to prepare for the possible threats in the next year. For 2021 it is for example to implement multifactor authentication, have ready plans in case of a ransomware attack, or backup data offline [9].

Common Attack Pattern Enumeration and Classification (CAPEC).

CAPEC is a public catalog that contains common attack patterns of malicious attackers. Its purpose is to further understand how weaknesses in applications and other systems are exploited for both the purposes of education and prevention. This catalog was first established by the U.S. Department of Homeland security as an initiative to increase cybersecurity. It was first released in 2007 [9].

Attack patterns are defined by CAPEC as descriptions of the common attributes and approaches used by malicious actors to exploit known weaknesses. They also define the obstacles that an attacker may have to solve before gaining access and how they may solve them. These descriptions are the result of an in-depth analysis of real-world exercises of so-called blue and red teams. These exercises simulate how an attacker would act when trying to gain access to a system. The exercise features two teams, the blue team trying to defend a system and the red team trying to breach it. The catalog currently defines 546 attack patterns. For each attack pattern, there is a description of what it is, what weaknesses an attacker would exploit, or what weaknesses make this attack pattern possible. Next is the execution flow which includes step-by-step instructions on how this attack is conducted. The consequences of a successful attack are listed afterward as well as examples from the real world which are either real or at least theoretically possible [9].

Summary

The base of the X-Force Threat Intelligence Index is the incidents gathered by X-Force. Those incidents are investigated for their source and categorized based on the attack type and attack vector and with the use of CAPEC. It is uncertain what specific types are monitored or how CAPEC was integrated into the methodology. Only those types are listed which are with a bigger share than 5%, which for 2020 means the "other" category type contains 29%. The initial vectors do not contain any "other" category but makeup seven types, which together make up 101%, pointing to calculation errors. These are Scan and Exploit, Phishing, Credential Theft, Remote Desktop, Brute Force, Removable Media, and BYOD. The overall top three attack types and vectors are deeply analyzed, listing specific methods and types, further breakdowns, and overall information specific to that type. This information is then broken down based on geographic regions and industries. Regions are Europe, North America, Asia, Latin America, and the Middle East with Africa. The most attacked countries in a given region are also listed separately. For industries only the top 10 most attacked regions are listed, it is unknown if more are analyzed. These industries are Finance and Insurance, Manufacturing, Energy, Retail, Professional Services, Government, Healthcare, Media, Transportation, and Education. All of these regions and industries are then ranked and compared based on the share of attacks that were recorded in the current and previous year. Each is then separately investigated for specific information in them, the biggest threats, and changes in the behavior of malicious actors they experienced [9].

This core is then expanded with topics that are deemed key areas of concern for cybersecurity. For the 2021 report, it was Advanced Threat Actors. This topic is dedicated to highly organized and possibly state-sponsored cybercriminal groups. Threats to Operational Technology, which contains information on threats not only to digital assets but also physical ones. Brand Spoofing is a very common tactic of social engineering attacks. And New Malware Threats that have been identified during the year, may indicate a shift in the structure of future attacks. The report then concludes with recommendations and key takeaways [9].

Unfortunately, with most data showing it is unclear how it was analyzed if there are more data points than is shown if some incidents have several categories or just one, or even how the individual categories or topics of interest were chosen. In most graphs showing shares of a given phenomenon, the percentages counted together do not make 100% and there is no indication of the missing data. It is also unclear why some data was chosen to be included in the report, for example, an estimate of the revenue of a specific ransomware type, but not others. Trust is therefore created not by transparency and solidity of its methodology, but mostly by the brand of IBM.

4.1 Verizon: Data Breach Investigation Report

Verizon Communications is a multinational company with its core business in communication technology. Verizon is one of the world's leading providers of technology, communications, information, and entertainment products, and services [10]. The company was founded in the year 2000 by Bell Atlantic Corp. and GTE Corp., which was one of the greatest mergers in the history of the USA [11]. On the Fortune 500 list, Verizon is ranked the 20th largest company in the US with 128 bn. USD in revenues, around 290 bn. USD in assets and around 132 thousand employees worldwide in 2020. It is also included in the Dow Jones Industrial Average index [12].

Similarly to IBM, Verizon is a highly respected company in its field, which includes the field of cybersecurity. As one of the largest communications technology companies in the world cybersecurity has always been an important issue for the company. The company is in a unique position, as many attacks can be directly tracked within its network and technology provided to its customers. This grants insights into security threats that not many companies have. It has according to its website 25 years of experience in incident data analysis, processes 61 billion security events per year on average, and has more than 500 thousand networks and hosting devices which it has under management [13]. The data gathered is then compiled into an extensive, highly detailed, and freely available report, the Data Breach Investigation Report.

Published Methodology of Research

The Verizon Data Breach Investigation Report is in stark contrast with the X-Force Threat Intelligence Index from IBM in its transparency of the methodology used. The 2020 report features several attachments detailing the methodology of how the data was analyzed and categorized full acknowledgments of biases and a link to a page containing mentions of all mistakes that were discovered within the report after publishing [14].

The methodology of the Data Breach Investigation Report is mostly unchanged compared to previous years, which makes the report more valuable for research as the data can also be better compared with previous years. All incidents included in the report were individually reviewed and converted into the VERIS framework, which will be further explained later in this chapter. This was done to create a common, anonymous aggregate dataset. The collected data came in various forms and therefore underwent different steps of conversions for this report. Three basic methods were used. First, direct recording of paid external forensic investigations and related intelligence operations conducted by Verizon using the VERIS WebApp. Second via a direct recording by contributors using VERIS. And third, by converting contributors' existing schema into VERIS. To safeguard anonymity all contributors had received instructions to omit any information that might identify organizations or individuals involved. Reviewed spreadsheets and VERIS WebApp JavaScript Object Notation (JSON) were ingested by an automated workflow that converts the incidents and breaches into the VERIS JSON format as necessary, adds missing enumerations, and then validates the record against the VERIS schema. The automated workflow subsets the data and analyses the results. Based on the results of this analysis, the validation logs from the workflow, and discussions with the contributors providing the data, the data is cleaned and reanalyzed. This process runs for roughly three months as data is collected and analyzed [14].

The data used was not exclusively categorized and can have multiple values (i.e., "Social," "Malware" and "Hacking"). This means that percentages do not always add up to 100%. For example, if there are five botnet breaches, the sample size is five. But if each botnet used Phishing, installed Keyloggers, and Used stolen credentials, there would be five Social actions, five Hacking actions, and five Malware actions, adding up to 300%. But this is expected by the analysts and handled correctly in the analysis. Another point is that a part of the presented data has the tag unknown, this is equivalent to unmeasured. This is to say that if a record contains elements that have been marked as unknown, it means that statements about such an element could not be made because of a lack of information. Such data is also not included in the sample size. The data tagged as other is counted as it means the value was known but not part of VERIS or not included, for example with lists featuring a list of top 5 or top 10 in a category. Finally, not applicable, (normally NA), may be counted or not counted depending on the hypothesis. For this year's report, Verizon used confidence intervals more often to allow it to analyze smaller sample sizes which count as containing 30 or fewer samples. Appropriate measures were taken to avoid bias. Firstly, sample sizes smaller than five were not analyzed at all. The count or percentage for small samples is available. This goes for figures too and is why some figures lack the dot for the median frequency. And thirdly, small samples may include information about the value being in a range, or values being greater or less than each other. These all follow the hypothesis testing and confidence interval approaches [14].

For a potential entry to be eligible for the incident and breach database, a couple of requirements were defined. The entry must be a confirmed security incident, defined as a loss of confidentiality, integrity, or availability. After being confirmed as a security incident, the entry is assessed for quality. Verizon created therefore a subset of incidents that pass the quality filter. The requirements for an incident are first, the incident must have at least seven enumerations (e.g., threat actor variety, threat action category, variety of integrity loss, et al.) across 34 fields or be a DDoS attack. Exceptions are given to confirmed data breaches with less than seven enumerations. And secondly, the incident must have at least one known VERIS threat action category (hacking, malware, etc.). In addition to having the level of details necessary to pass the quality filter, the incident must be within the time frame of analysis, which was November 1st, 2018, to October 31st, 2019, for this report. Incidents and breaches affecting individuals that cannot be tied to an organizational attribute loss are also excluded. For example, if a person's private device was targeted by an attack, then it would not be included in the report. Lastly, for something to be eligible for inclusion in the Data Breach Investigation Report, Verizon needs to be aware of it happening, which it acknowledges may lead to several potential biases [14].

Verizon admits, that many breaches go unreported and many more are unknown by the victims themselves and thereby unknown for this report. Therefore, until an exhaustive census of every breach that happens in the entire world each year can be conducted or on the whole study population, sampling must be used. Unfortunately, sampling introduces bias. The first type of bias is the random bias introduced by sampling. For this year's report, the maximum confidence is $\pm1.5\%$ for breaches and $\pm0.5\%$ for incidents, which is related to the sample size. Any subset with smaller sample size is going to have a wider confidence margin. The second source of bias is sampling bias. This is mitigated by collecting breaches from a wide variety of contributors, but it does not eliminate bias. For instance, some breaches, such as those publicly disclosed, are more likely to enter the database used for this report, while others, such as classified breaches, are less likely. See the picture below for a visualization of potential sampling bias. Each radial axis is a VERIS enumeration and the stacked bar charts represent data contributors. The ideal is an equal distribution between all contributors along all axes. Axes only represented by a single source are more likely to be biased. However, contributions are inherently thick-tailed, with a few contributors providing a lot of data and many contributors providing a few records within a certain area. Still, most axes have multiple large contributors with small contributors adding appreciably to the total incidents along that axis. It can be noticed that there is a large single contribution on many of the axes. This may seem like a large source of bias, but it represents a contribution aggregating several other sources, so not an actual single contribution. It also occurs along most axes, limiting the bias introduced by that grouping of indirect contributors. The third source of bias is confirmation bias. Because the entire dataset is used for both exploratory analysis as well as hypothesis testing, the hypotheses are therefore tested on the same data which was used to make them. But still, this is limited by the use of a variety of contributors and by a consistent multiple-review process. Additionally, the findings are often consulted with subject matter experts in the specific areas ahead of release (Fig. 1).

There are other instances where subsets of data are defined for analysis beyond passing the quality requirements. These subsets consist of legitimate incidents that would eclipse smaller trends if left in. These are removed and analyzed separately. This year, there were two subsets of legitimate incidents that were not analyzed as part of the overall database. A subset of web servers that were identified as secondary targets, such as taking over a website to spread malware. And second, botnet-related incidents. Both subsets were separately analyzed for the last three years as well. Finally, some subsets were created to help further the analysis. In particular, a single subset is used for all analysis within the Data Breach Investigation Report unless otherwise stated. It includes only quality incidents as described earlier and excludes the aforementioned two subsets [14].

Since the 2015 issue, the Data Breach Investigation Report includes data that requires analysis that did not fit into the usual categories of incident or breach. Examples of non-incident data include malware, patching, phishing, DDoS, and other types of data. The sample sizes for non-incident data tend to be much larger than the incident data but from fewer sources. An effort is made to normalize the data,

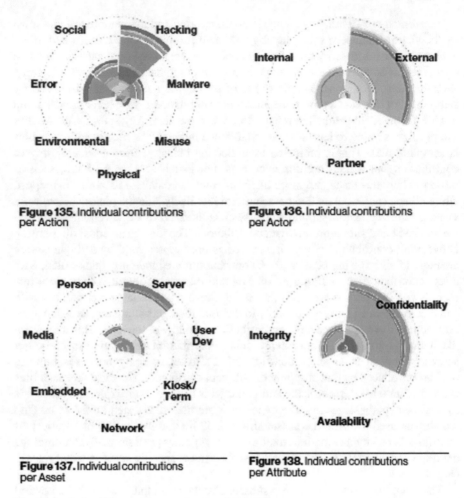

Figure 135. Individual contributions per Action

Figure 136. Individual contributions per Actor

Figure 137. Individual contributions per Asset

Figure 138. Individual contributions per Attribute

Fig. 1 Analysis of bias—sources and types of data [14]

for example, by weighting records by the number contributed from the organization so all organizations are represented equally. Partners with similar data are also put together to conduct the analysis wherever possible. Once the analysis is complete, the findings are discussed with the relevant partners to validate them against their knowledge of the data [14].

The Vocabulary for Event Recording and Incident Sharing (VERIS) is a set of metrics designed to provide a common language for describing security incidents in a structured and repeatable manner. VERIS is a tool to standardize how cybersecurity incidents are categorized and shared. VERIS thereby helps organizations to collect useful incident-related information and to share that information anonymously and responsibly with others. For this reason, the Veris Community website was set up to help guide users on how to use VERIS and share information [15]. The framework gathers information based on the 4A's. On the actor who causes the incident, the action or the methods used, the affected assets and how they were affected or the attribute. In Fig. 2 we can see the 4A's and how they intersect. The intersection represents association within an incident, not a direct connection. VERIS is rooted in the examination of evidence and post-incident analysis but is designed to provide metrics useful to risk management [16].

Apart from using VERIS for this report, it was used also in combination with the ATT&CK framework developed by the company MITRE. The Adversarial Tactics, Techniques, and Common Knowledge (ATT&CK) framework were developed in 2013 as a means of codifying adversarial behavior and released publicly in 2015. ATT&CK has become a well-established way of describing the tactical actions used by attackers. The 260+ Techniques in ATT&CK for Enterprise are logically grouped with their corresponding 11 Tactics, which describe the different objectives an adversary might take as part of their intrusion. While there is some overlap between VERIS and ATT&CK, they also complement each other. VERIS was created to codify incidents and ATT&CK to codify adversary techniques. Therefore, Verizon developed a framework, which combines the two, the VERIS Common Attack Framework (VCAF) [14].

Report structure and summary of findings

Since the Verizon report contained such a detailed methodology, we will conduct a shorter examination of its structure than the report from IBM. As mentioned before, the Verizon 2020 Data Breach Investigation Report is highly detailed in covering all aspects of the cybersecurity landscape of the year in-depth on 119 pages. Apart from the appendices, it is divided into 6 chapters. The first is the foreword and summary of

Fig. 2 VERIS overview [16]

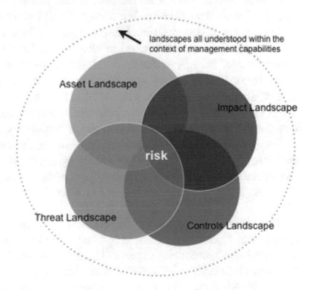

the report, after that are Results and Analysis, Industry Analysis, Small and Medium Businesses Overview, Regional Analysis, and Closing Remarks [14].

Results and Analysis contain subchapters on the topics of Actors, Actions, Threat action varieties, Error, Malware, Ransomware, Hacking, Social, Assets, Attributes, How many paths must a breach walk down? Timeline and Incident classification patterns and subsets. Actors and Actions are two of the 4A's and show data on the actors, their motives, and varieties, while Actions are about the methodology of attacks, and varieties in both incidents and breaches. Next, the attack varieties are analyzed within their subchapters. These subchapters contain breakdowns of the subvarieties of the given attack types, the vectors of attack used, and the motives of these types of attacks. And at the end of this chapter are the rest of the 4A's, assets, and attributes. Asset sales are dedicated to the affected devices and systems in the recorded attacks which were targeted and why. Attributes as mentioned before are about how the systems and companies were affected. How many paths must a breach walk down? Is an interesting chapter analyzing how many and what kind of different steps are taken by an attacker during an incident or breach? These have implications for creating an effective response as the behavior changes during the attack. The Timeline shows how the average time to discover breaches and contain them changed over the years. Similarly, the chapter Incident classification patterns and subsets offer a unique overview of how the patterns of the attacks changed over the years of data gathering and how last year compares to the aggregated data over several years [14].

Chapter 3 goes under the heading of Industry Analysis. Over the year Verizon was able to gather data on more than 157 thousand incidents and 108 thousand breaches. While the previous chapter analyzed them based on the 4A's this chapter breaks it down to the different industries that are defined. In total, 20 industries are listed as well as one category for unknown. For each of these industries, we can find information on the frequency, patterns, threat actors, actor motives, compromised data, and even recommendations on how to best limit the threats faced by that industry. This creates a base for what is analyzed for each industry. Apart from that, various topics specific to the threat landscape are also available. The fourth chapter is focused on Small and Medium-Sized Businesses and is also the smallest chapter in the report. It has a similar structure to the last chapter, except expanded by specific breakdowns of data for the threats faced by these kinds of businesses. The next chapter continues with Regional Analysis. The data is split based on the four defined regions as well as the category unknown. Similarly, to the previous chapters, the unknown category is part of the aggregated analysis, but not part of the specific analysis undertaken in the chapters. The data contained in these chapters is consistent with the previous ones [14].

The last chapter before the Appendices is Wrap-up. Here we can find the last recommendations and the Year in Review. Next are 5 appendices. The first one was the most important for this paper as it detailed the methodology used for this report. The next discusses the VERIS Common Attack Framework, which we have already discussed. After that a paper on financially motivated cybercrimes, which were the most common motives for cybercrime. Closing the report are the last two appendices, one is a short paper on the adoption of VERIS by the American state of Idaho and lastly the list of contributing organizations [14].

Summary

The Data Breach Investigation Report is a highly detailed report on cybersecurity incidents and breaches around the globe. Thanks to Verizon being one of the largest telecommunications companies in the world, it also has access to a large sample of data from its customers and various other contributors. The data comes specifically from companies, not private persons. The gathered data is then filtered based on completeness and classified by the VCAF framework, on the 4A's, and the attack type used by malicious actors. After that, the data is analyzed as a whole to assess the prevalence of certain attack types, motivations, or specific targets for that year and afterward put into contrast with the previous years for analysis if any changes in patterns of behavior of malicious actors occurred. Next, the data is broken down based on the defined areas of interest, mostly based on geography, industry type, with a special mention and of the situation in small to medium-sized businesses.

Over the years, the methodology used for this report has been consistent and has only sparsely changed. The great advantage of this is the ability to compare the data during the years and identify trends and patterns. The whole methodology is transparent, freely available, and open to criticism. Verizon has made efforts to address any mistakes found in the reports as well as any limitations or biases that could be observed. It is largely thanks to this that the report has become one of the most important sources of information on global cybercrime for professionals and the broad public.

The large sample and consistency of the methodology also pose a disadvantage. The field of cyber security is very dynamic and prone to fast changes which require a similarly quick source of accurate information to adequately protect against new threats or methods of attack. The 2020 report gathered information from November 2018 to October 2019 and was released in May 2020. An obvious point of criticism is, therefore that the information comes too late. This means that between the end of information gathering on cybercrime and the publishing of the findings is about 7 months. This becomes apparent especially for the 2020 report as the 7-month period saw a huge shift in the cyber security landscape because of the global Covid-19 pandemic. As important as those changes were, there is little to no mention of them within the report, making it in some ways irrelevant to publication.

4.2   Ponemon Institute: Separating the Truths from the Myths in Cybersecurity

The Ponemon Institute is an American research institute founded in 2002 by Dr. Larry Ponemon and Susan Jayson. Based on the website, the Institute is dedicated to independent research and education that advances the responsible use of information and privacy management practices within business and government. The institute is also the parent organization of the Responsible Information Management (RIM) Council. The RIM Council has members from leading multinational corporations and government organizations. It focuses on the development and execution of responsible information practices for the collection and use of personal, sensitive, or confidential information. Members are invited to monthly calls to be briefed on the Institute's research. Each year, the Institute hosts an event called RIM Renaissance to bring together smart individuals in privacy and security who are passionate about the work they do and the challenges they face [17].

Based on the above-mentioned and their published research, we can see that the Ponemon Institute has a wide range of topics that are covered under cybersecurity. In 2018 it conducted a one-time report called Separating the Truths from the Myths in Cybersecurity, which was sponsored by BMC. The purpose of this study was to better understand the security myths that can be barriers to a more effective IT security function and to determine the truths that should be considered important for the overall security posture [18].

Published Methodology of Research

Participants in the research were asked if a selection of statements is considered truthful or if they are based solely on conjecture or gut feeling, in other words, a myth. Specifically, respondents rated each statement on a five-point scale from -2 = absolute myth, -1 = mostly myth, 0 = can't be determined, $+1$ = mostly truth and $+2$ = absolute truth. All myths and truths are not equal and range from -1.04 to $+0.78$. Drawing upon nonparametric statistical methods, statements were separated that had a statistically significant positive value that was above 0 considered truth, from those statements that had a statistically significant negative value at or below 0 considered myth [19].

The sample was intentionally composed of 40 194 IT and IT security practitioners located in North America, the United Kingdom, and the EMEA region and who have various roles in IT operations and security. As shown in Table 1, 1 517 respondents completed the survey. Screening removed 191 surveys. The final sample was 1 326 surveys or a 3.3 percent response rate.

Next, the respondents are divided based on their answers to several demographic questions. Based on their position within the company. For this report, about 55% of

Table 1 Sample response [19]

	NA	UK	EMEA (excl. UK)	Global
Total sampling frame	17 500	10 093	12 601	60 194
Total returns	679	402	436	1 517
Rejected or screened surveys	74	57	60	191
Final sample	605	345	376	1 326
Response rate (%)	3.5	3.4	3.0	3.3

respondents had supervisory roles within their company. Based on who the person primarily reports to within the company. In this case most respondents, about 40%, report to the chief information officer or head of IT. Next, based on the number of employees. About 72% were from companies with more than 1 000 employees. And lastly, based on the industry sector the company operates. The largest sectors are financial with 17%, then public services with 11%, and the services sector with 10% [19].

The report also states all caveats that need to be taken into account when dealing with its findings. Four caveats are enumerated. First is the non-response bias. As can be gathered from Table 1 and the low response rate it may be possible, that the respondents are not a representative sample and are very different from those that did not respond. The second is the sampling frame bias. This may be caused by selectively choosing potential respondents that are not representative of the majority. Additionally, subjects were compensated if they responded within a specified period, which also may have caused some bias. The third is self-reported results. The respondents answered the questions themselves and for that reason, great trust has to be placed on the integrity, accuracy, and truthfulness provided in the responses [19].

The statements to be assessed by the respondents based on their truthfulness are as follows:

- There is a skills gap in the IT security field.
- Security patches can cause a greater risk of instability than the risk of a data breach
- The cloud is cost-effective because it is easier and faster to deploy new software and applications than on-premises
- Greater visibility into all applications, data, and devices and how they are connected lowers an organization's security risk.
- Malicious or criminal attacks are the root cause of most data breaches.
- A strong security posture enables companies to innovate and take risks that can lead to greater profitability.
- IT security and IT operations work closely to make sure the resolution and remediation of security problems are completed successfully.
- Many organizations are suffering from investments in disjointed, non-integrated security products that increase cost and complexity.
- Too much security diminishes productivity.
- A strong security posture does not affect consumer trust. (In other words, a strong security posture is considered beneficial to improving consumers' trust in the organization.)
- Automation is going to reduce the need for IT security expertise.
- Artificial intelligence and machine learning will reduce the need for IT security expertise
- It is difficult or impossible to allocate the time and resources to patching vulnerabilities because it leads to costly business disruptions and downtime.
- Insider threats are costlier to detect and contain than external attacks.
- Nation-state attacks are mainly a threat to government organizations.

- Security intelligence tools provide too much information to be effective in investigating threats[19].

Report structure

The report structure is standard for a report of this type. The first chapter is the Introduction. Here the reader can find out about the circumstances that led to the report being created, basic information on the methodology, the respondents, and also the results. The second chapter is dedicated to key takeaways where all assessed statements are listed and each discussed separately based on the responses and the meaning for cybersecurity in companies. The last chapter before the already discussed Methods chapter is the Findings. The structure of this chapter is similar to the second one but is more deeply discussed with the addition of graphs showing the percentages of answers given. Additionally, the report has one appendix where the detailed survey results are listed [19].

Summary

The Ponemon Institute report Separating the Truths from the Myths in Cybersecurity is very different from the past two reports we have analyzed. The report takes a different approach to find out the situation in cybersecurity. It is not about analyzing incidents and breaches that have happened over the course of a year, but about finding out the posture towards issues in this field by IT professionals in the form of a survey. While it may not provide the reader with information on specific changes in the cybersecurity landscape or give examples, recommendations, and facts about how a good cybersecurity policy should be carried out and what to focus on, it still fills a gap in the understanding of the field that the other reports mostly leave out. Specifically, it is the view of the IT professionals on the problem of cyber security, finding out how they view and therefore shape cybersecurity in their companies, what they view as a problem, what they consider a myth, and what they see as a priority from their perspective. This way the report grants valuable insights. With such information, it may be possible to better understand and therefore better improve the field by focusing efforts based on the views of actual IT-professional in the field. It makes it possible that recommendations formulated based on reports and research may be given from a place of understanding of the professionals in the field.

4.3 e-Governance Academy: National Cybersecurity Index

The e-Governance Academy was founded in 2002 as a non-profit foundation. It was created as a joint initiative of the government of Estonia, the Open Society Institute, and the United Nations Development Programme. The foundation assists public sector and civil society organizations in the field of digital transformation. This is offered through consultancy, training, networking, research, and assisting in the implementation of e-government solutions. The management system for its project management, study visits, and consulting services have been independently certified to ISO 9001:2015 standard. The academy also plays a key role in policy planning and implementation including the formation of the lead organizations responsible for developing a nation's interoperability framework and related legal and technical framework development and management. The partners of the e-Governance Academy include the Global Forum on Cyber Expertise, is a member of the Estonian Information Security Association, an Advisory Observer of the Secure Identity Alliance, the global identity and secure e-services advisory body, a member of the Estonian Roundtable for Development Cooperation and a member of the Estonian Network of Estonian Non-profit Organizations. It has assisted several countries in their digital transition mostly in Central and Eastern Europe, Asia, and Africa [20].

The e-Governance Academy is also the creator of the National Cybersecurity Index or NCSI. This index ranks 160 countries around the globe based on their cybersecurity policy and their preparedness for a cybersecurity attack. It acts as a tool for countries to more easily identify gaps in policy and strategy and to help improve and guide them. The index also prides itself on being one of the most consistent and transparent indices on the globe [21].

The purpose of this global index is to measure the preparedness of countries to prevent cyber threats and manage cyber incidents. The NCSI is also a database with publicly available evidence materials and a tool for national cyber security capacity building. The vision is to develop a comprehensive cyber security measurement tool that provides accurate and up-to-date public information about national cyber security. In the coming years, the NCSI team has the ambition to develop new applications for national cyber security analysis and development [22].

Published Methodology of Research

The base of the framework of NCSI is the fundamental cyber threats. First, it is a denial of e-services, when services are not accessible. Second, data integrity is breached when data is modified by unauthorized users. And thirdly, data confidentiality breach, when confidential data is exposed. These threats directly affect the normal functioning of national information and communication systems and electronic services, including critical e-services. According to NCSI, to manage these cyber threats, a country must have appropriate capacities for baseline cyber security, incident management, and general cyber security development [22] (Fig. 3).

The index has been developed using a five-step process. Identification of national-level cyber threats, identification of cyber security measures and capacities, selection of important and measurable aspects, development of cyber security indicators, and grouping of cyber security indicators. The NCSI focuses on measurable aspects of cyber security implemented by the central government, which are legislation in force, meaning legal acts, regulations, orders, etc. Next, established units, such as existing organizations, departments, etc. Then cooperation formats, committees, working groups, etc. And outcomes of policies, exercises, technologies, websites, programs, etc. [22].

The NCSI is organized into categories, capacities, and indicators, of which there are 3 categories, 12 capacities, and 46 indicators in total. Each indicator has a value,

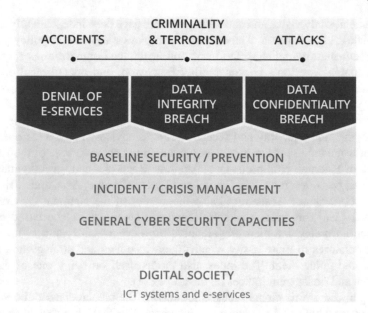

Fig. 3 The NCSI framework [22]

which shows the relative importance of the indicator in the index. The values are given by an expert group awarding 1 point for a legal act that regulates a specific area, 2–3 points for a specialized unit, 2 points for an official cooperation format, and 1–3 points for an outcome or product. Country ratings are based on public evidence, which may be provided as legal acts, official documents, and official websites [22]. Opinions, studies, news articles, reviews, etc. are generally not accepted as evidence materials. The material provided has to be public and publicly accessible, submitted in English, or at least understandably translated. The material is assessed as yes or no, also meaning that either all points or none are awarded for that indicator [23]. Based on the points awarded the NCSI Score is calculated and shows the percentage the country received from the maximum value of the indicators. The maximum NCSI Score is always 100 or 100% regardless of whether indicators are added or removed [22].

In addition to the NCSI Score, the index table also shows the Digital Development Level or DDL, which is calculated according to the ICT Development Index and Networked Readiness Index. The DDL is the average percentage the country received from the maximum value of both indexes. Also calculated is the difference showing the relationship between the NCSI score and DDL. A positive result shows that the country's cyber security development is by, or ahead of, its digital development. A negative result shows, that the country's digital society is more advanced than the national cyber security area [22].

Several ways are used to collect data for the index. Sources of data are the country's government officials, an organization or individual, or the NCSI team that conducts a public data collection. When data collection is complete, the provided information is reviewed by at least two NCSI experts. After inspection, the dataset is published on the NCSI website. Data collection, review, and publication is a continuous process, meaning it is not published annually, but as new data is provided and reviewed the change to the list will be made immediately [22].

Report structure

The report structure for any of the 160 countries is similarly simple. First, the overall ranking and awarded points in the NCSI are shown. After that, we can see a comparison with the other indexes, the Global Cybersecurity Index, the ICT Development Index, and the Networked Readiness Index. Each shows how large the cybersecurity capabilities fare from the perspective of government and institutions as well as the technological development of the country on a social level. This aims at helping to better assess the cybersecurity capabilities lack behind its digital development. Some basic information is also provided on the country, specifically its population, area, and GDP. Then the report returns to data concerning only the NCSI. Three graphs show the percentage of fulfillment of the indicators in the 12 capacities, which are Policy, Threats, Education, Global, Digital, Essential, eID and TS, Personal, CIRC, Crisis, Police, and Military. The other two graphs show the percentages of fulfillment in time and the timeline of a countries country's ranking [23] (Fig. 4).

The last and most important part of the country report is a list of the indicators the country fulfills. These are divided into three categories, General Cybersecurity Indicators, Baseline Cybersecurity Indicators, and Incident, and Crisis Management Indicators. Each is divided into four capacities and then further into a varying number of indicators totaling 46. Each indicator has its weight in points as described in the previous chapter on the methodology [23].

The categories, indicators, and their weight in points are as follows:

1. Cyber Security Policy Development—7:

   - Cyber security policy unit—3
   - Cyber security policy coordination format—2
   - Cyber security strategy—1
   - Cyber security strategy implementation plan—1

2. Cyber Threat Analysis and Information—5

   - Cyber threats analysis unit—3
   - Public cyber threat reports are published annually—1
   - Cyber safety and security website—1

3. Education and Professional Development—9

   - Cybersafety competencies in primary or secondary education—1
   - Bachelor's level cyber security program—2
   - Master's level cyber security program—2
   - Ph.D. level cyber security program—2
   - Cyber security professional association—2

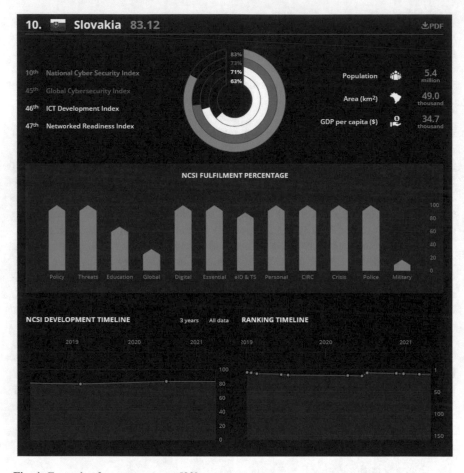

Fig. 4 Example of a country report [23]

4. Contribution to Global Cyber Security—6

   - Convention on Cybercrime—1
   - Representation in international cooperation formats—1
   - International cyber security organization hosted by the country—3
   - Cyber security capacity building for other countries—1

5. Protection of Digital Services—5

   - Cyber security responsibility for digital service providers—1
   - Cyber security standard for the public sector—1
   - Competent supervisory authority—3

6. Protection of Essential Services—6

   - Operators of essential services are identified—1
   - Cyber security requirements for operators of essential services—1

- Compctent supervisory authority—3
- Regular monitoring of security measures—1

7. E-identification and Trust Services—9

- Unique persistent identifier—1
- Requirements for cryptosystems—1
- Electronic identification—1
- Electronic signature—1
- Timestamping—1
- Electronic registered delivery service—1
- Competent supervisory authority—3

8. Protection of Personal Data—4

- Personal data protection legislation—1
- Personal data protection authority—3

9. Cyber Incidents Response—6

- Cyber incidents response unit—3
- Reporting responsibility—1
- Single point of contact for international coordination— 2

10. Cyber Crisis Management—5

- Cyber crisis management plan—1
- National-level cyber crisis management exercise—2
- Participation in international cyber crisis exercises—1
- Operational support of volunteers in cyber crises—1

11. Fight Against Cybercrime—9

- Cybercrimes are criminalized—1
- Cybercrime unit—3
- Digital forensics unit—3
- 24/7 contact point for international cybercrime—2

12. Military Cyber Operations—6

- Cyber operations unit—3
- Cyber operations exercise—2
- Participation in international cyber exercises—1 [23]

Summary

The National Cybersecurity Index is one of the most transparent and therefore also trusted indices. It is used to quickly assess the state of cybersecurity in any of the 160 countries that are listed in it. The index is very different from the other reports that have been listed in this paper in both methodology, purpose, and structure. Unlike the other reports, the NCSI is meant and focused on countries and not businesses or

IT professionals as individuals or employees. It is a tool to assess on the strategic level if a country has the legal, institutional, and crisis management capabilities to respond to cybersecurity threats within its borders and offer support outside of them. This is done based on publicly available and accessible governmental information in forty-six defined indicators divided into four capacities and twelve categories. Each indicator is assessed in a binary way, assigning all or no points based on assigned weights. From the points, a percentage is calculated by dividing granted points and the total, which constitutes the index.

Each country report contains the NCSI, ranking among 160 countries as well as the change in ranking and awarded points over time. Each indicator is listed with all indicators and the total points gained. If points are gained, evidence is also presented as well as all the contributors who provided the data. Realizing that the NCSI on its own does not convey the whole picture about the cybersecurity landscape in a country, the report also contains 3 other indices. The other indices are the Global Cybersecurity Index, ICT Development index, and the networked Readiness Index.

Unfortunately, despite the transparency, some aspects are unknown. And that is the reasoning behind the index and the methodology. It is unclear why those specific indicators were chosen to assess the state of cybersecurity in a country or why such a weight was assigned to an indicator. The only explanation given is that this was decided by an expert group. Who the members of the expert panel are is also unknown. It may be assumed that it mainly consists of members of the NCSI project. The About us page shows the names of the core experts, which seems to indicate that there are more besides the core [24].

It also needs to be mentioned that a disadvantage of the index is that it may not show the actual cybersecurity capabilities of a given country. The index assesses if laws, institutions, and crisis managers are in place to respond to cybersecurity threats. But not how effective, successful, or capable they are. While this is only our opinion, it may be hard to argue that for example, Serbia has better cybersecurity capabilities as a country than the United Kingdom and only slightly lower than the United States of America, despite the position in the NCSI of the respective countries [25].

4.4 Comparison

Advantages and disadvantages

The X-Force Threat Intelligence Index offers several advantages for its readers. First off, the report offers aggregated information on cybercrimes conducted globally, identified in the large customer base of IBM. For that reason, it is mostly for large international companies and IT professionals looking to better understand the global cybersecurity landscape. From that also follows the disadvantage that the report is of lesser use for smaller companies and specific countries as the division of regions focus on entire continents and only a few countries of high interest. The largest threats are deeply analyzed, especially the top three attack types and vectors as well as other areas of interest for IBM, such as the financial sector or brand spoofing. This makes the report more focused and points directly to important problem areas. But this also means that other problem areas are mostly left out and, in most cases, it is unknown if more than the areas are researched at all or if there are more that are simply left out on purpose in the report.

This leads us to the biggest issue with the report and that is its lack of transparency. The X-Force report does not openly state any except the most rudimentary methodology. We are therefore unable to confirm, verify, criticize or praise how the research was conducted. It is impossible to say where all the data comes from, from which periods, how long it was processed by IBM, how it was processed or analyzed, and for what reason, if some data was not included, simply left out, or if there is some duplication when an incident has several traits. There is no mention of the limitations, goals, or possible bias that could have made its way into the analysis. Several graphs do not add up to 100% and some even go over. The top ten lists make the report more focused, but it is unknown if there are more items on the list or how many there were in the dataset. Additionally, it is unknown why this specific data is presented or why it was analyzed in that way. It is therefore by the trust in the brand of IBM that the data may be seen as truthful. In our opinion, were it a different and lesser-known organization with the same report, it would not have been acceptable at all.

As mentioned before, the Verizon Data Breach Investigation Report is in stark contrast to the report from IBM as almost all disadvantages listed above for the Threat Intelligence Index are not the case for the Data Breach Investigation Report. This report is very detailed in its structure, researched threats, sources of data, and used methodology. Surprisingly, the report is also freely available to the public.

The base of the report from Verizon is its very large customer base, information gathered by the users of their communication technology [26, 27], and openness to gather information from all trustworthy sources. Another positive is a strong and consistent methodology that is in line with statistical principles. Clear definitions with the use of the VERIS and ATT&CK frameworks. And openness to criticism, admitting mistakes, limitations, and bias.

The information is presented in-depth and since it has consistently been done so since the beginning, it is presented also with a retrospective look at how the threats, vulnerabilities, and patterns of behavior of cybercriminals changed over time. The patterns are a specialty of this report as well as the number of steps an attacker is willing to make, which owes to the large database gathered by Verizon for this report. Although, the information is almost exclusively useful to large multinationals, similarly to the X-Force report, despite a smaller section dedicated to small and medium-sized businesses. The reason is that almost always global data is presented and even when broken down into regions, they are in our opinion far too large, for example, one region being defined as Europe or Africa and the Middle East together. Additionally, the depth, consistency, and large database have the major disadvantage of the report being published half a year later after the last date for the report was gathered. The 2020 report meant that it was published long after the start of the Covid-19 global pandemic, which had a large impact on the whole cybersecurity landscape, which meant the report lost a part of its significance as the data was not as relevant.

The Ponemon Institute Report Separating the Truths from the Myths in Cybersecurity is quite different from the previous two reports but offers its unique advantages. While the other two probably cannot be outdone in terms of their objective nature, factualness, and amount of data gathered on incidents, breaches, and threats, the report from Ponemon help assess the level of understanding of the problem by IT professionals. The drawbacks are of course the reliance on the truthfulness, integrity, and knowledge of the individual respondents but may reveal blind spots of professionals and reveal areas where a report such as from IBM and Verizon should focus on.

Apart from that, the report is simple in its nature with a clear goal, solid make-up of questions, and methodology. The report openly reports on any mistakes and biases. Those being common ones for this type of research, non-response bias, sampling frame bias, and the mentioned self-reported results. What is missing is the explanation as to why these specific questions and areas of interest were chosen.

The National Cybersecurity Index sets itself apart from the other reports because its purpose is not aimed at aiding businesses or IT professionals, but at countries and governments in creating policies, institutions, and legislation that will aid them in dealing with cybersecurity threats. The index is very transparent in its methodology, data gathering, and limitations making it a trusted tool. Although, it is specified what data, how and from where is analyzed, it is not transparent in its reasoning for why the chosen indicators were chosen or why they have such weight in points. The only explanation offered is simply that a group of experts had decided so. It is similarly unclear who all the experts are. This also leads to questions surrounding the results and rankings of countries themselves. We think this is best illustrated by the United States, Serbia, and the United Kingdom occupying the 16th, 17th, and 18th places respectively, seemingly showing that they have similar cybersecurity capabilities, which we would call into question. It may be true that the index only measures if certain legislation and institutions are present in a given country, but if these do not translate to increased cybersecurity capabilities, then it is all the more unclear why these aspects are measured.

Information is published on each of the 160 countries that are listed in the index, which indicators they fulfill, who is the source of the proofs, and the proofs themselves. Any changes may be reported and the changes to a country's report are made immediately after they have been assessed. Apart from information on the index and the different parts that make it up, other indexes are also listed, which help to explain the state of digitalization and cybersecurity of a country's economy, while the NCSI focuses on the governmental aspects, helping to make a better picture of the situation a country is in.

Comparison of CAPEC and ATT&CK

During our analysis of the IBM and Verizon report, we mentioned that they used two different approaches to categorize cyberthreats, ATT&CK and CAPEC respectively. While both share similarities as they both categorize the behavior of malicious actors, their application is different. CAPEC is best used when analyzing application security and describing common attributes and techniques used to exploit their weaknesses. This approach therefore also enumerates these exploits, includes social engineering techniques and how they are used in the context of exploits, and is associated with the Common Weakness Enumeration, CWE [26], which is a community-developed list of hardware weaknesses [29]. This makes CAPEC best suited for application threat modeling, developer training and education, and penetration testing [28].

On the other hand, ATT&CK is focused on network defense. This approach describes the different phases an attack goes through before and after an attack and details specific tactics, techniques, and procedures used to target, compromise, and operate inside a network. This is based on threat intelligence and red team research provides context to understand malicious actors and supports testing and analysis of defense options. This makes ATT&CK best suited for comparing computer network defense capabilities, defending against advanced persistent threats, hunting for new threats, enhancing threat intelligence, and conducting exercises to imitate the actions of attackers [28].

Best-practices

- Transparency—For any quality research transparency is key for its credibility. Any information on what the goals of the research are, who are its intended users, where the data come from and how much, why it was gathered that way, and how it was gathered and used should always be available. Ideally, this should also include information on any bias, study limitations, and why all the areas of research were chosen. Information on the methodology should be complete and the data showed cover 100% of the data used. This has a dual purpose of letting other researchers comment on the methodology and offer ways of improving it as well as increasing credibility as all information may be verified and therefore trusted.
- Data anonymization—Information on incidents and all the more breaches are sensitive for any company to disclose. The reason is that it may harm their reputation and make their clients doubt their ability to keep their data safe. Additionally, laws and regulations in a country may prohibit the disclosure of any data that may point to specific individuals. An automated system for the anonymization of the incident and breach information will therefore increase the willingness to share, improving the quality and accuracy of research.
- Clear frameworks—Great effort should be made to create a clear framework to guide the whole process of research. Clear definitions of attacks, attack vectors, what, how, and who is targeted, and others. And rules for possible situations. For example, if an incident may have several attack types, if yes, then how it is accounted for in research. Otherwise, it may lead to confusing and misleading results. The VERIS, ATT&CK, and CAPEC frameworks are very good examples.
- Questionnaires—Questionnaires such as the one of the Ponemon Institute are a good way to find out the perception of cybersecurity by IT professionals. They may help guide research, point to important topics, or help identify unimportant

and irrelevant ones. Such questionnaires may be a very good addition to a study focusing on incident analysis.

- Timeliness of Information—The cybersecurity landscape is constantly and dynamically changing. New hardware and software, patches, and fixes inevitably create new vulnerabilities, exploits, and ways to breach security. Certain events, such as the Covid-19 pandemic, also have profound effects and may in some cases shift the focus of cybercriminals overnight. For data to be of use it, therefore, needs to be available promptly so that it may warn about those changes and prevent incidents and breaches.
- Focus on important areas—Once a framework was established and it was determined how data should be broken down, for example, based on geography, industry, attack type, and vector, specific areas should be prioritized based on their importance and vulnerability. As the timeliness of the information is crucial and not all areas are equally important it helps to focus the report and save time without missing out on key information. Especially if resources for the compilation of the report are low, prioritization of certain aspects may even be necessary. But based on the previously mentioned best practices, these key issues need to be chosen based on the actual needs of IT professionals and the public, which may both be aided by questionnaires and transparency.
- Consistency—Consistency in data gathering, measured periods and methodology allows for more credibility and trust, but also the ability to compare the data with previous periods and track changes in behavior. This may help show the bigger picture of the problems surrounding cybersecurity and may provide insights that would otherwise be lost.
- Adaptation to current needs—It is obvious that consistency and the dynamic nature of the cybersecurity landscape are conflicting. While a consistent core of the research is important there should always be room to adapt, leave out, add or change the methodology. A sudden change in technology or a global event, which would render a measured item unimportant or another item previously left out into a key concern should make the methodology reflect this.
- Complementing research with other sources—It is clear also from this paper that cybersecurity is a global concern and a globally researched one. Areas that are sufficiently and continually covered by other researchers and institutions should be implemented or expanded based on the current needs and available data to save valuable time and effort and avoid duplicity of work. The issue of cybersecurity is a global one and research should therefore reflect this.
- Encouraging feedback—Humility in research is always good practice. Mistakes are always made, methodologies may always be updated and improved, and data wrongly interpreted. While this should always be avoided, some mistakes and false conclusions inevitably make their way into research. Encouraging feedback may therefore provide important information and help improve research in the future.

4.5 Compatibility with Legislation

General Data Protection Regulation

The General Data Protection Regulation (GDPR) is an EU regulation with its national equivalent in the Slovak collection of laws, specifically law 18/2018 "o ochrane osobných údajov a o zmene a doplnení niektorých zákonov". While this regulation is in itself important for cybersecurity, it mostly does not interfere with research. In §1 section, (a) the law states that it concerns only the protection of personal data of a natural person, therefore not covering legal persons such as companies. But if research would be conducted on natural persons, as would be the case for questionnaire type research among IT professionals for example, then the law may apply. In §2 personal information is defined as any data concerning the identification of a natural person that makes up their physical, physiological, genetic, psychic, mental, economic, cultural, or social identity. In §5, sections (b), (c) and (d) it is more specifically defined what is considered genetic data, biometric data, and data concerning the health of a natural person [30].

This means that data gathered from companies does not present an issue in regards to GDPR. In the case of data gathered from individuals if data does not contain any information which could identify them specifically, there should also be no issue. This could be solved by a randomized and anonymized questionnaire for example. Although, if a questionnaire is conducted with a select group of individuals, such as IT professionals, it should be made untraceable as to who provided the answers. This could be done with a software solution, which automatically deletes or disregards any identifying information.

If for any reason this cannot be done and personal data needs to be gathered, then the law needs to be followed. The base of GDPR is its 7 principles, which are outlined in the paragraphs 6 through 12:

- Principle of lawfulness—Personal data needs to be processed exclusively by lawful means to not breach the rights of a given person.
- Principle of purpose limitation—Personal data may only be used for a specific, clearly stated, and legitimate purpose and nothing else.
- Principle of data minimization—The processed personal data must be adequate, relevant, and limited only to the necessary.
- Principle of accuracy—Processed personal data must be correct, otherwise, it needs to be immediately corrected or deleted.
- Principle of storage limitation—Personal data which may identify a specific person may be stored only as long as is necessary for the stated purposes. It may be stored longer for archiving, scientific or statistical purposes as long as the rights of the person concerned are protected.
- Principle of integrity and confidentiality—Personal data must be processed in such a way that it guarantees adequate security and protects against unlawful processing, loss, deletion, or damage of the data.

- Principle of accountability—The operator is responsible for compliance with these principles and is required to prove it when requested by government agencies. [30]

How these principles are to be kept is outlined in the other paragraphs of this law. §13–25 describe when is the processing of personal data lawful, when and how permissions need to be given from the concerned person, how the different categories may be processed, and other special cases concerning data on crime and granting access to intelligence services. §26 to §28 outlines the rights of individuals to receive a copy of any data they gave to operators, of objecting to the processing of their data, and not be liable for any decisions made based on the automatic processing of personal data, although some exceptions do apply. §29 makes the operator of personal data legally obligated to reply and to implement requests of the concerned persons and makes the operator the responsible party that needs to justify decisions contrary to the wishes of the concerned person. §30 states situations and conditions under which the rights of individuals may be limited, such as security of the country, upholding ethical standards, or protecting the rights of other people. Paragraphs §31–§38 and §39–§41 are probably the most important parts of this legislation, as they define the duties and responsibilities of the operator and any subcontractors in regards to handling personal data as well as how personal data needs to be secured, it also obliges the operator to reports any breach of the protection of personal data to the security agency. §42–§43 obliges the operator to consult any implementation of procedures handling personal information which has a high risk for the rights of concerned persons and to evaluate the risks and to consult this with the security agency if sufficient measures were not or cannot be taken to mitigate the risk. §44–§46 obliges the operator to choose a responsible person if they are a government body or conduct large-scale processing of personal data, to support and keep the independence of the responsible person, and defines the tasks of the responsible person, such as monitoring convergence with the law and working with the security agency. §47–§50 defines how and when data may or may not be transferred to third countries or international organizations. §52–§77 the law defines the rights and obligations of public institutions towards the personal information of concerned persons. These are mostly similar to the ones defined before, with changes made mainly in the domain of transparency based on the nature of a public institution and especially institutions of law and order. §78–§79 defines special cases of how personal data should be treated, for example in the event of the passing of a concerned person, when the concerned person is an employee and others, as well as the obligation to secrecy. §80–§84 define the role of the security agency and its functions, organization, etc. §85–§89 define a code of conduct, certification, monitoring, certified subjects, and the duties of the certifying subject. §90–98 define how an inspection needs to be conducted. §99–§103 defines how the agency should proceed should a personal data protection proceeding be started. §104–106 define penalties for breaking this law. And the final §107–§112 are the common, transitional and final provisions of the legislation [30].

Cybersecurity Regulation

The Cybersecurity Act is an EU regulation with its national equivalent in the Slovak collection of laws in law 69/2019 "o kybernetickej bezpečnosti a o zmene a doplnení niektorých zákonov". Apart from GDPR, which may apply to cybersecurity reports depending on the use of personal data of natural persons, the Cybersecurity Act influences research indirectly. The law establishes the Computer Security Incident Response Team (CSIRT, since 2019 SK-CERT [31]) in §6 as a government cybersecurity body, which is tasked, among others, to gather data on cybersecurity incidents, issue warnings, and act upon them. It also obliges the government to create a national cybersecurity strategy, both of which may be supported by research similar to the ones investigated in this paper [32].

§7 establishes the need to create a national cybersecurity strategy that establishes goals, and priorities for the Slovak republic, identifies measures that need to be taken to increase its resilience, defines security threats, and programs for education, and increases the general knowledge of the populace including professional development in this field, plans for science and research, risks, identifies key partners and others. Meaning the strategy heavily relies on properly defining the security landscape, which in turn relies on gathering the necessary information. §8 creates a unified information system on cybersecurity which crucially has to include alerts, warnings, and other information that may aid in minimizing, preventing, or remedying cybersecurity incidents, which necessitates gathering, concentration, analysis, and evaluation of information on cybersecurity incidents. §15 defines the responsibilities of CSIRT which are preventative—creating awareness, training, monitoring and registering incidents, connecting with the unified information system and managing it—and reactive—sending out warnings, detecting, analyzing, solving, and remedying security incidents as well as support others in it and propose measures. §17 defines when and how it registers a basic service and its provider, §18 defines the identifying characteristics of a provider of a basic service, and §19 the duties of such a provider, which include dealing with, immediately reporting, and gathering information on a security incident as well as cooperating with government bodies. Similarly, these are defined for providers of digital services within §21–§23, which are similar in nature although more tailored towards safeguarding IT infrastructure. §24–§25 are of great interest as they oblige providers of basic and digital services to report cybersecurity incidents, how it is to be done, and when. Reporting is done via the unified information system and is required for providers of basic services in the case of severe incidents and digital service providers, it is for any cybersecurity incident. §26 also allows for voluntary reporting of incidents, although they are processed secondarily if there are sufficient capacities. Lastly, §28–§29 are of interest as they define inspections and audits, both of which may provide valuable information about the actual state of cybersecurity in the inspected and audited institutions [33].

National Strategy on Cybersecurity 2021–2025

The Slovak National Strategy on Cybersecurity is a soon to be ratified document that is the direct result of the Cybersecurity Act and the efforts of the National Security

Authority. The strategy defines Slovakia's overall goals, targets, and priorities in the domain of cybersecurity. The most relevant goals set out by the strategy for conducting research similar to the one analyzed in this paper are 4.1 Trustworthy state ready for threats, 4.3 Resilient private sector, 4.6 Educated specialists and educated public and 4.7 Development of research and development in the areas of cybersecurity [34].

The reason why we consider these goals to be relevant is that they call directly or indirectly for better research into cybersecurity, raising awareness, and providing education for professionals and the public. 4.1 Trustworthy states ready for threats set the goal of working together with citizens by providing information, and recommendations, and taking steps that will increase individual and in turn national security. For effective cooperation with any interesting subjects on all levels of cybersecurity. For improving detection and investigation into incidents and simplifying and improving the way incidents are reported. 4.3 Resilient private sector where the goals call for increasing awareness of operators of basic and digital services about cybersecurity as an important part of their operations and improving the exchange of information on security, experiences, and further development. 4.6 Educated specialists and the educated public set the goal of a systematic, broad-spectrum, and planned dissemination of situational and security awareness based on a good system that responds flexibly to changes in cyberspace. 4.7 Development of research and development in the areas of cybersecurity which sets the goal for good functioning and state-supported science and development in the field of cybersecurity, and good communication between the private, public, and academic sectors [34].

These, we think, support research similar to the ones analyzed within this article. It is needless to say that a good source of information about the current state of the cybersecurity landscape supports all goals set in the national strategy, but the ones mentioned above are the most connected and supportive for such an endeavor.

5 Conclusions

In our assessment of the different methodologies, we can conclude that a positive on its own is the variety of the different approaches. With Verizon's report, we have a very rigorous approach to statistical analysis of incidents and breaches. With IBM's report, we have a more focused, simpler approach that makes it easier to understand and pinpoint key issues. Ponemon's report offers a very different but valuable perspective on the issue. While methodologically well thought out, it also offers insights into the views of IT professionals, who are responsible in their day-to-day for keeping data and IT infrastructure safe within their respective organizations. The National Cybersecurity Index from the e-Governance Academy again fills in the areas of the cybersecurity landscape that are left out by the other reports. How and if a nation as a whole is building up its cybersecurity capabilities based on a set of transparent indicators, but also its digital development makes the NCSI a valuable tool.

Taking all of the approaches and trying to do them at the same time is in our opinion counterproductive. Yet despite this, a certain combination of these approaches to studying the state of cybersecurity may be worth the effort. Understanding the state of the nation and the current views of professionals on this issue first may prove invaluable in developing a report similar to IBM and Verizon. We conclude this as we have found that a common issue among the reports was the missing reasoning behind the research. The best answer we have found was that the topics are researched because an unknown panel of experts saw it to be necessary. Therefore, finding out the current state of the country and the views of professionals may provide the answers as to what areas of cybersecurity and why should be studied and focused on. This may greatly limit the possibility of creating a report which may find little use in practice by IT professionals.

The best practices we identified in the assessed reports were transparency, data anonymization, clear frameworks, the use of questionnaires, timeliness of information, focus on essential areas, consistency, adaptation to current needs, complementing research with other sources, and encouraging feedback. The correct use of statistical methods in the analysis of the data is among the most important.

The legislation was also analyzed in this thesis, specifically, the General Data Protection Regulation, the Cybersecurity Act, and the soon-to-be ratified National Cybersecurity Strategy for the years 2021 to 2025. While studying this legislation we have found that research similar to the ones discussed in this paper is by the law and several cases supported by it. Especially the National Cybersecurity Strategy and the goals that it has set for the period closely align with the creation of a similar report. The Cybersecurity Act in turn provides government agencies with solid sources of data, as it obliges public and private institutions to report incidents or offers a way to do it voluntarily. This data may be directly used for research. GDPR does something similar, but most should be taken as a guide on how to deal with personal data during research. Although it must be noted that an amendment to the Cybersecurity Act is being worked on by the European Union and that an amendment may be possible as well for the National Cybersecurity Strategy before it is ratified. The situation may therefore change shortly.

In conclusion, the digital world offers great benefits for human society, but increasingly great threat. Only by understanding these threats, we may take appropriate measures to keep the nation, its citizens, and businesses safe. And to understand, we require good sources of information, which we can supply with the right methodology.

References

1. IBM: About us. https://www.ibm.com/sk-en (2021). Accessed 6 Apr 2021
2. IBM: Patent Leadership. https://www.research.ibm.com/patents/ (2021). Accessed 6 Apr 2021
3. Fortune 500: IBM. In: Fortune. https://fortune.com/fortune500/2019/ibm/ (2019). Accessed 6 Apr 2021

4. Fortune 500: IBM. In: Fortune. https://fortune.com/company/ibm/fortune500/ (2021). Accessed 6 Apr 2021
5. IBM: X-Force Threat Intelligence. https://www.ibm.com/security/xforce (2021). Accessed 10 Apr 2021
6. Drew, R.: What is X-Force? In: eSecurityPlanet. https://www.esecurityplanet.com/products/ibm-xforce/ (2017). Accessed 10 Apr 2021
7. Newman, D.: IBM X-Force Threat Intelligence Index. In: Converge. https://convergetech media.com/ibm-x-force-threat-intelligence-index-research/ (2017). Accessed 10 Apr 2021
8. IBM Security: X-Force Threat Intelligence Index (2017)
9. IBM Security: (2021 X-Force Threat Intelligence Index (2021)
10. Verizon: Our Company. https://www.verizon.com/about/our-company (2021). Accessed 15 Apr 2021
11. Verizon: History and Timeline. https://www.verizon.com/about/our-company/history-and-tim eline (2021). Accessed 15 Apr 2021
12. Fortune: Verizon Communications. In: Fortune. https://fortune.com/company/verizon/fortun e500/ (2021). Accessed 15 Apr 2021
13. Verizon: Managed Security solutions. https://www.verizon.com/about/our-company/managed-security (2021). Accessed 17 Apr 2021
14. Verizon: Data Breach Investigations Report 2020 (2020).
15. Verizon: The VERIS Framework. http://veriscommunity.net/ (2021). Accessed 18 Apr 2021
16. Verizon: VERIS Overview. http://veriscommunity.net/veris-overview.html (2021). Accessed 18 Apr 2021
17. Ponemon Institute: About. In: Ponemon Institute. https://www.ponemon.org/about/ (2021). Accessed 23 Apr 2021
18. Ponemon Institute: About – Separating the Truths from the Myths in Cybersecurity. In: Ponemon Institute. https://www.ponemon.org/research/ponemon-library/security/separating-the-truths-from-the-myths-in-cybersecurity.html (2018). Accessed 23 Apr 2021
19. Ponemon Institute. Separating the Truths from the Myths in Cybersecurity (2018)
20. eGA: About us. In: e-Governance Academy. https://ega.ee/about-us/ (2021). Accessed 24 Apr 2021
21. E-Estonia: The National Cyber Security Index ranks 160 countries' cyber security status. In: e-Estonia. https://e-estonia.com/the-national-cyber-security-index-ranks-160-countries-cyber-security-status/ (2020). Accessed 24 Apr 2021
22. NCSI: Methodology. https://ncsi.ega.ee/methodology/ (2021). Accessed 6 Apr 2021
23. NCSI: Slovakia. https://ncsi.ega.ee/data-collection/ (2021). Accessed 24 Apr 2021
24. NCSI: About us. https://ncsi.ega.ee/contact/ (2021). Accessed 25 Apr 2021
25. NCSI: Ranking. https://ncsi.ega.ee/ncsi-index/ (2021). Accessed 25 Apr 2021
26. Poniszewska-Maranda, D., Matusiak, R., Kryvinska, N., Yasar, A.-U.-H.: A real-time service system in the cloud. J Ambient Intell Human Comput **11**(3), 961–977 (2020). https://doi.org/10.1007/s12652-019-01203-7
27. Molodetska, K., Veretiuk, S., Gregus, M., Fedushko, S., Syerov, Y.: Simulating the interaction of social internet services actors using irregular attractors. Procedia Comput. Sci. **198**, 688–693 (2021)
28. MITRE: CAPEC–ATT&CK Comparison. https://capec.mitre.org/about/attack_comparison. html (2021). Accessed 24 Nov 2021
29. MITRE: CWE—Common Weakness Enumeration. https://cwe.mitre.org/ (2021). Accessed 27 Nov 2021
30. Zákon, č.: Z.z. o ochrane osobných údajov v znení neskorších predpisov (2018)
31. NBÚ Národná jednotka CSIRT. In: NBU. https://www.nbu.gov.sk/kyberneticka-bezpecnost/sk-csirt/index.html (2019) . Accessed 29 May 2021
32. Zákon, č.: Z.z. o kybernetickej bezpečnosti v znení neskorších predpisov (2018)
33. Chlipala, M., Makatura, I., Pilár, Š.: Zákon o kybernetickej bezpečnosti—Komentár. Eurokódex, Bratislava (2019)
34. NBÚ: Národná stratégia kybernetickej bezpečnosti na roky 2021 až 2025 (2020)

Home Office—Benefit for Employer or Employee?

Phuong Ngoc Nguyen-Hadi Khorsand and Tomáš Peráček

Abstract The main goal is to determine if home offices help businesses or workers. Other research topics include: What is a home office? What do companies gain from a home office? What are the advantages of working from home? Work/life balance is typically promoted via flexible work arrangements. The outcomes vary by worksite. Virtual office employees said telework had no impact on work/life balance. Virtual workers reported poorer work/life balance and personal/family success than conventional or home office workers. A multivariate analysis found that virtual workers damage work/life balance. Working from home affects work/life balance. One aspect may be the absence of external physical constraints. Virtual office workers may struggle to distinguish work from home. The benefits of working from home are many. A happy workforce may contribute to increased profitability and productivity. Let workers work from home has numerous advantages, but this essay argues it helps employers and corporations more than employees. It is easy to communicate: Business communication includes email, meetings, phone calls, video conferencing, and casual conversations. Employers may use several communication alternatives to keep remote staff connected, resulting in a more efficient and collaborative work environment.

1 Introduction

Telecommuting (working from home), is becoming more common in today's industries. The employee benefits from not wasting time stuck in traffic and may instead use that time to do work-related tasks. Hence, this is a perfect setting for staff who do not interact with customers daily; this strategy saves the company money on office

P. N. N.-H. Khorsand · T. Peráček (✉)
Faculty of Management, Comenius University in Bratislava, Odbojárov 10, Bratislava, Slovak Republic
e-mail: peracek2@uniba.sk

P. N. N.-H. Khorsand
e-mail: khorsand2@uniba.sk

space while increasing staff productivity [74]. Public transit may also be credited with reducing traffic congestion and improving the macroenvironment; this award may have certain drawbacks, just as there are to any other employment benefit. Some employees may be distracted or seem bothered by their situations. Furthermore, people who work from home may have a lower feeling of team cohesiveness because they feel less linked to their co-workers, and they may perceive that they have fewer opportunities for promotion [81].

Working from home has evolved over time. For instance, it began when Newton avoided the Great Plague at Cambridge by fleeing to his country home in 1665, where he worked for the next year and a half [1]. His work in these fields would have a huge impact on the development of science over the following several decades, including optics, mathematics, and the law of gravity. Newton himself referred to this era as "the most prolific period of his life" in which he lived and worked. Yes, even for today's perspective, working from home may be the key to increasing productivity. A strategy to eliminate co-worker disruptions while maintaining a work-life balance. Personal and professional aspects that influence productivity while working from home must be considered [27]. Thus, because of the Covid-19 epidemic, researches can thoroughly investigate the consequences of working from home. Working outside the typical workplace is becoming increasingly frequent in today's environment. Some of the terms used to characterize the phenomenon include virtual office, telework, remote work, telecommuting, location independent working, and home office [34]. This study will use the word "working from home" (WFH) to refer to working from a location other than an employee's allotted work area.

Research has been undertaken on WFH and its impact on productivity and well-being. WFH has grown in popularity as internet connections have improved and personal information technology infrastructure has grown. Recently, more than half of millennials and Generation Zers polled stated they would want to be able to work from home more often [31]. According to the Bessa Vilela et al., those who work from home demand more flexibility for various reasons [6]. Many workers, particularly those with children, want a better work-life balance and the flexibility of working from home. Organizational departments report increased productivity due to WFH; however, there are certain downsides. Thorstensson experiment, in which contact center staff were randomly assigned to work from home or in the office for nine months, is an outstanding example of positive outcomes [70]. Those who worked from home increased their overall job performance by 13%; this set of employee were also happier with their occupations. The majority of participants who took part in the experiment said they would rather remain at work since they were lonely at home.

Aside from the obvious advantages of not having to travel, home-based work provides greater control over one's schedule and fewer interruptions, making it a more convenient alternative for many individuals. A home office requires a significant investment in physical and technical resources, although personal comfort is commonly highlighted as an advantage of working from home. People who work from home are more motivated and pleased at work because they have more control over their job and a better work-life balance. Working from home has been related to better levels of satisfaction with leisure time, according to Ta and Doan [69]. Consequently, persons who work from home may be able to devote more time to activities they like.

Feeling isolated from co-workers due to physical and social distance is one of the unpleasant elements of WFH. Another difficulty remote employees had was the inability to shut down their computers and cease working after their regular work hours. WFH was shown to be more challenging for those who had stubborn children, meddling from other family members, neighbors, and acquaintances was also discovered to be a problem. Furthermore, teleworkers may believe that their lack of visibility limits their prospects for advancement, rewards, and positive performance reports. As a consequence of more independence, employees are expected to exert more control over their surroundings. As a consequence of WFH, effective time management skills, self-discipline, and work-life border control measures must be created [70]. As a consequence of increasing flexibility, multitasking and blurring the lines between work and family life are more likely to occur.

In contrast to the commercial sector, the academic community lacks a clear sense of what it's like to work from home and its related advantages and limitations to the employers or employees. Academics for instance have already experienced some productivity drawbacks of working from home since the advent of personal computers in the classroom. According to Stratone et al. research, people who spend long hours at home on their computers report feeling alone and lamenting the lack of support from their co-workers [67].

Participants in online distance education have traditionally been academics with WFH expertise, but this is changing due to Covid-19. Employees reported more autonomy, improved scheduling flexibility, the removal of undesirable distractions, and high levels of job productivity and enjoyment; nevertheless, they also said poor communication and a lack of opportunity for skill advancement [68]. Because of the Covid-19 epidemic, most professionals were forced to work from home for a while, enabling studies to explore the WFH experiences of a broader range of academics at this time. Majority of literatures only know a portion of the factors that contribute to WFH's success right now. Workers and families are all aware that juggling household responsibilities and working from home hinders one's ability to manage their time properly. According to Rafalski and Andrade, women's job is more influenced by family duties than men's work [57]. According to Peráček, women who work remotely are more satisfied with their work-family balance [52]. However, other research has indicated a negative association between work-life balance and well-being. In the case of the COVID-19 epidemic, female workers were more likely to experience a decline in work output and research time owing to childcare duties.

Additionally, during the COVID pandemic, numerous schools, organizations, and government institutions worldwide advocated for a technique known as "smart working" or "artistic working." Therefore, this has had and will continue to harm both workers and enterprises, and the reasons for this must be investigated further. Before the COVID-19 problem, there were many more clever and smarter employees than there are now. The number of persons who died during the pandemic remained extensive, reaching more than 15 million since its official declaration of Covid-19 as pandemic. People who were previously mostly unaware of smart functioning have unexpectedly and excitedly embraced this new mode of operation. Even while it is difficult to forecast the future, smart working, particularly among clerical personnel, may influence how businesses and government agencies conduct their day-to-day operations [47]. More study is required to identify how individuals react to a rapid shift in work mode and how this affects their professional and personal lives.

Because of WFH or working smart, quality of life and productivity have gotten little attention from the scientific community. During the current pandemic lockdown, employers were forced to devise novel ways of working, enabling workers to conduct research on an unprecedented scale from their homes [66]. As a result of the involvement with WFH, workers became vividly aware of both the organization's advantages and disadvantages. WFH agreements have become a legal need for institutions to offer the necessary services, legislation, and infrastructure. However, although some scholars and organizations may have benefited from the new arrangements and want to continue using WFH, others found it more challenging than before. This study meant to provide a complete analysis of home office—benefit for employer or employee.

1.1 Problem Statement

Remote work is the future, and global firms and enterprise employers desire their employees to work from home. Working from home may benefit both employers and employees, including better health and lower costs. Working from home opens up new possibilities for business organizations and management. Due to the coronavirus (COVID-19) epidemic, several firms have been able to sustain operations while prioritizing staff and consumer health [30]. Before the coronavirus outbreak, many companies recognized the benefits of working from home and the improved work-life balance it offered for employees. Workers who have worked for at least 26 weeks have a legal right to request flexible working arrangements, such as home working.

Many people now take advantage of the freedom of working from home. Working from home affords them greater flexibility. The typical picture of a home office is a young individual working from home in casual dress, balancing their personal and professional life. Because that's what "home office" means: flexibility, happiness, and work-life balance go hand in hand [38]. As people work at home, they may care for their children, keep the house clean, or participate in social events for 1–2 h each day. The commute is also unnecessary; this implies that the home office idea is gaining popularity due to its flexibility [46]. Those who work full-time and have a high education degree want a home office. The reasons given are a desire for more temporal autonomy and work-life balance. Everyone wants to be able to work from home as often as possible.

Most homeworkers work longer hours, perform more unpaid overtime, and like their jobs more than their co-workers. Therefore, this is unlike those who wish to work from home but cannot for various reasons. Allowing individuals aged 20 to 35 to work from home may enhance workplace motivation. The study's ultimate conclusion is to explore how working from home impacts present employment circumstances in Germany and then ascertain whether it primarily benefits the employers or the employees. This scientific article will explore the extra benefits of remote working for both employees and employers in the era of Covid-19 globally.

2 Aims and Methodology

The primary aim is to investigate home offices and assert whether it benefits employers or employees. Other areas of concern as per the research questions include:

1. What is a home office?
2. What are the benefits of a home office to employers?
3. What are the benefits of a home office to employees?

Research method (qualitative): A qualitative investigation of the home office benefits to employers and employees was studied. Qualitative data collection and analysis acknowledge the detailed grasp of ideas, reasoning, and reasons that drive a specific situation. The research required a large amount of secondary data for detailed comparisons. Both secondary data on the benefits of home-office to the employers and employees are qualitative.

Using a qualitative method also offers considerable benefits. According to a literature review, the strategy incorporates studies on the study topics. Many researchers use systematic or primary quantitative methods to examine various concepts related to business and office significance [50]. Finally, the qualitative design's open-ended option enabled previously explored themes and challenges to emerge [61]. Also discussed are the advantages of qualitative analysis. A systematic qualitative technique also incorporates outcomes from the company's reports, used where appropriate.

The downsides of systematic qualitative data analysis include the time it takes to make a fast comparison [43]. A researcher cannot simply compare subjective data from human participants since this study did not incorporate human subjects [42]. Instead, the researcher used peer-reviewed articles to complete the study. Finally, either the lack of access to business papers, especially those relating to home-office benefits to employees or employers, prevented the survey from generalizing and making a detailed comparison.

Secondary data: Due to the research's extensive scope, a simple data collection approach requiring minimal technical skills was needed. The articles for the literature review were chosen using non-probability selection. A few criteria must be completed for the study and business publication to be included in the literature review and discussion sections. Articles, for example, must emphasize key terms (home office, employees, or employers). Second, they must be reliable sources from reputable publications sites like Google Scholar, government, or company records.

The studies that met both criteria were used to support the literature review, introduction, conclusion, and discussion sections. The reputable and premium firms' websites were also utilized in this investigation due to their reputation as reliable and trustworthy sources of information.

Aside from that, using secondary data and press releases has several advantages. For example, the researcher saved time by using readily accessible, reliable data [62]. The qualitative data also provided high-quality, comprehensive databases that any researcher could access [39]. On the other hand, the study was very cautious to avoid affecting the overall quality of the research by using secondary data. The researchers found and removed bias by removing secondary data sources. Finding the most critical information took a long time, but predetermined criteria made it faster.

Research methods and strategies: The research employed systematic qualitative data collection methodologies; the technique's data collection approach utilized secondary data [40]. Database engines searched the study's keywords. EBSCO, Google+, Scopus, and Google Scholars gathered secondary data for literature review. The Google search engine was utilized to get reliable luxury brand annual reports. This research would not have been feasible without online resources and secondary data collecting. Hence, the study discovered over a thousand publications using search engines and screened about 20 of the most trustworthy scholarly studies.

Data analysis: The data collection approach uses secondary data; thus, the current data is analyzed using inductive reasoning. Step 1 was to gather data from scholarly articles and journals that investigated and discussed the importance of using home-office on the employees or workers. The second step entailed consolidating the content into the thematic interests of the articles. A range of data sources was investigated to improve data quality, including historical data and data acquired during specific periods. These high-end businesses' reports were also utilized to assess the data's reliability and consistency. This report's outcomes were written in Word Documents, as it did not involve sophisticated analytic software. The researcher ensured the inquiry summary was factual and consistent.

3 Literature Review

3.1 Home Office and Smart Working Concepts

First, let's define some of the terminology employed in this study. Working from home has been referred to in various ways since its inception. Although there is no universally accepted definition of "telework," it is often defined as a method of working "away from the office" using technological connections [16]. Mucha [45] argue that workers may work from any location other than the corporation's traditional workplace, such as their own home or any other site (for example, a satellite office or an external area).

Teleworking has grown into a "smart working art" yet, the word "smart working art" emphasizes what this new mode of operation is projected to give to corporations and individuals alike [53]. Hence, telework that is conducted in an artistic and creative method of smart working is described as telework that people conduct artfully and creatively of smart working [24]. From this vantage point, personnel may favorably embrace the newest information technology environment without being constrained by time or location provided their work duties is achieved [23].

The phrase "smart working art" and "WFH" highlights that teleworking allows employees to "produce maximum value" and "be more productive" while also providing them with greater "spatial and temporal freedom," better working conditions, more creativity, and the ability to influence change [28]. While "teleworking" emphasizes remote work and telecommunications technology, the term "smart working art" focuses on what WFH style can achieve for employees and employers alike and the potential implications and changes [51]. The following elements of smart functioning will be examined in further detail. For instance, workers must be allowed to work mostly from home, and employers must be able to make appropriate modifications to operating procedures and task assignments when and where this is beneficial to the effectiveness of the job obligations [34]. If computer networks and mobile devices are utilized properly, they may benefit both workers and employers [54].

The phrase "smart working" will be used throughout the remainder of the paper, although for the sake of this article, "working from home" will be used as a synonym [55]. This study's primary emphasis is on how smart working influences people and their "concept of work." This problem is important not just for academics but also for businesses, which may better understand the advantages and downsides of adopting smart working practices and therefore focus on the critical elements that must be addressed for a successful implementation [63]. There has already been debate in the literature on the possible negative effects of smart working on workers and the factors that might affect its success (both for people and companies). Hynes [28] identified several critical characteristics, including the nature of the job, clearly stated objectives and standards, the capacity to meet employees' immediate needs, and respect for people's unique preferences. Hynes, who conducted similar research, discovered that there are variables at the business level (for example, innovation atmosphere or style of personnel assessment) and factors at the individual level (for example, work satisfaction and others at the personal level) [28]. Sararu discovered, for example, the capability of self-control) [63].

Potential differences in acceptability and perception among workers have also been widely investigated. The effectiveness of smart working may vary depending on the unique structure of the work environment. Barbieri et al. observes that if too many employees are working from home, it may be difficult to collaborate as a team. Furthermore, each individual's job title and management style may impact the outcome [3]. Moreover, it is important to recognize that smart working may influence the socializing process as well as certain work styles. Again, smart working enhances effectiveness and quality of work—particularly (but not only) for knowledge workers—might suffer significantly. The literature on knowledge management may include useful information on this subject. Dan drew on model of knowledge conversions to support his argument that various information technologies may be more or less successful for interpersonal knowledge sharing based on their ability to transmit tacit or explicit contents effectively [17]. According to Raišiene et al., employees must find a new dynamic balance between the many categories of knowledge (i.e., emotional, intellectual, and spiritual) that characterize people's actions and choices when they use smart working [58].

As Wefersová and Nováčková stated, it is critical to consider whether the term "smart working art" implies that smart working is beneficial to both employers and employees or if this modality may. On the other hand, create implementation barriers or cause additional problems in the workplace and even workers' personal lives [76]. Therefore, this is still a matter of contention in the literature. Smart working has been proven to offer various benefits, including more flexibility and the possibility of improved working condition. In contrast, other studies have shown that it is difficult to implement and has negative consequences. In addition to the necessary investments in information technology, organizational aspects such as difficult employee coordination and cooperation, management's fear of losing control, and anxiety associated with working in isolation may all threaten the successful implementation of a new strategy. The influence on one's personal life should also be considered [48].

Smart working is generally considered as allowing employees more discretion to plan their schedules so that they may accomplish both and establish a better balance between work and personal life. Allowing employees to work from home at least part of the time may provide them with more flexibility and the capacity to combine work and personal or family life better. On the other hand, working from home is often connected with longer hours and worries that flexible working may lead to an increase in work-related stress in personal life. Working from home has long been proposed as a policy measure to reduce road congestion, but it has yet to be implemented [78].

Prior research on telecommuting in the white-collar business identified a loss of social connection, difficulty separating work and home life, and a lack of visibility for promotion opportunities. Barbieri et al. also examined the benefits, such as better time management. People who work from home are also being investigated for the social advantages of this trend, such as better traffic flow, decreased energy consumption, and reduced air pollution and CO^2 emissions. Although numerous kinds of telecommuting have been investigated, one caveat is that home-based businesses and overtime work should not be considered telecommuting due to the impact of smart working on commuting; this entails working at various times throughout the day and from numerous locations [3].

Telecommuting has been shown in studies to influence employee productivity significantly. In a city like Tokyo, where public transportation may cost up to 26% of the yearly budget, telecommuters can save up to 44 h per year in time and congestion and 7–11% on expenditures. Another advantage of working from home is a reduction in work-life conflict, which may help businesses by allowing them to work longer hours. When more flexible work arrangements are offered, there is a clear association between improved job satisfaction and a decreased likelihood of leaving one's employment.

Additionally, regardless of the advantages or the rate of adoption, before COVID19, telecommuting growth remained modest. Consider that not all duties have a telecommuting (work from home) counterpart and that even when telecommuting is possible, it may not be a suitable replacement for those activities [7]. Workers often recognize that trust in management, the ability to preserve the technology involved, and a reasonable workplace culture that places a high value on human resources and member involvement all make telecommuting a more viable alternative [10].

To summarize, the literature addresses a wide variety of critical concerns. A variety of factors, including but not limited to the type of job performed in the smart working modality, organizational characteristics, and working environment, the clarity of goals and policies, as well as the information technologies that a company implements to support smart work, as well as employee attitudinal characteristics, are all critical for smart working success [11]. Second, smart working affects the socialization processes that occur among workers. When information transmission is poor, it is conceivable that the efficiency with which a certain activity may be completed financially diminishes. Some studies emphasize the advantages of smart working, while others warn that businesses and people may face challenges and negative consequences; this sparks debate on the subject.

3.2 Benefits and Challenges of WFH

Because this is such a hot topic, having not as old data is beneficial. During the Covid-19 outbreak, Fadinger and Schymik investigated the trend of people working from home and the consequences. Fadinger and Schymik devised and deployed an online survey to collect the necessary data for investigations [21]. According to Fadinger and Schymik, productivity at work during the Covid-19 pandemic was not significantly different from productivity at home [21]. Working from home has various benefits, including improved health, improved communication with colleagues, and the ease of being at home [22]. According to Dan, the average number of hours worked by respondents who claimed to have school-age children has increased [18]. Working longer hours might be seen as a disadvantage or a drawback. After examining working conditions and surroundings at work and home, researchers discovered the same result: people prefer working from home mainly due to their understanding of the benefits of working from home [77]. In a study conducted by Žofčinová et al., people who worked from home in European countries were questioned and studied [80]. An online survey of European participants revealed the essential pros and downsides of working from home [73]. There is greater flexibility in terms of breaks, meals, and time savings when traveling and minimizing the risk of COVID-19 and infection [72].

Working from home means missing face-to-face interactions with colleagues and having limited access to necessary equipment [33]. According to Ipsen et al., although working from home may be more or less advantageous, it is impossible to determine for sure because of the conflicting sentiments of those who worked from home during Covid-19 [35]. It was mentioned that breaking up long sessions was difficult and that they felt alienated from their colleagues. Working from home has several benefits, like the capacity to focus on work and the flexibility to dine at home. Noise may influence employee satisfaction and productivity if they work from home in the case of a pandemic [79]. According to research on adult participants, workers who work in the same room as their families are more likely to have disturbances in their capacity to focus, pay attention, or relax. Working from home has several disadvantages, one of which increases background noise.

Working from home may be a fulfilling experience, according to Varner & Schmidt research, which utilized the same approach [75]. Not having to commute and being able to focus only on one's work are both highly valued benefits. Moretti et al. highlight many of the same advantages and downsides in assessing the impact of the Covid-19 epidemic on employees [44]. According to many experts and surveys, working from home has several significant drawbacks for employees, including a lack of skills, uncertainty about layoffs, income reductions or insolvency, and so on. They may not be able to oversee or communicate effectively with their subordinates [44]. During the Covid-19 pandemic, Weitzer et al. investigated whether persons who worked from home were as dedicated to their professions and satisfied as before the outbreak [77]. The survey included more workers and employers, and the findings show that the pandemic has a significant impact on employee satisfaction when they work from home. Workers who work from home are more involved in their careers, according to Robelski et al., since they like the independence and flexibility that comes with it [60]. Those who possess these characteristics are more dedicated to their employees and, consequently, happier.

According to Robelski et al., who analyzed data from sampled surveys, employees' attitudes toward working from home are not always positive. A fraction of the severe expense and health consequences of Covid-19 has been alleviated by switching to a home office. Furthermore, it frees up workers' time, reduces commute, improves job security, and lowers the organization's overall operating costs. The bulk of Andrade (2015)'s advantages are acceptable to them [60]. According to Rafalski and Andrade [57], critical disadvantages of working from home include a lack of contact with colleagues, social isolation, organizational challenges, and difficulty finishing tasks. According to a study conducted by Putri and Amran [56], Covid-19 has beneficial and negative impacts on employees who work from home. According to the research results, part-time work from home benefited respondents' mental health and family life, enhancing work productivity. Revutska investigated Czech employees' and managers' mental health and hard work during Covid-19 using questionnaire answers [59]. According to the study results, the mental strain negatively influences the health of employees and supervisors who work from home. According to Putri and Amran, social isolation might be the culprit. During the lockdown, workers who worked from home due to the Covid-19 pandemic had a 20% lower inclination to work and a higher level of social isolation [56].

Working from home during the Covid-19 epidemic may have certain disadvantages; however, Tramontano et al. have provided solutions to overcome these disadvantages using information from prior professional publications and surveys. Establishing clear limits, such as frequent breaks and specified daily routines, may help avoid obesity and a lack of balance between work and home [71]. According to Rafalski and Andrade, working from home offers many downsides to having a proper work-life balance [57]. Many professionals voiced their worries about the strain on families and worked in an online survey after difficult situations like pandemic hits. According to Rafalski and Andrade's [57] conclusions, working from home has a harmful influence on parents with young children. While working from home during the Covid-19 epidemic, some workers may have felt emotional weariness and burnout [56]. Burnout syndrome (emotional tiredness at work) was recognized and examined by Garg and van der Rijst [25], employing an online questionnaire with replies from South American employees. According to comments from participants, Garg and van der Rijst discovered that working from home had a detrimental influence on work-life balance [25].

The coronavirus epidemic endangered Work-from-home prospects for Putri and Amran [56] and their employees, in particular, who work from home at this time? Digitization and innovation have grown due to the convenience and flexibility of working from home. A symptom of increasing digitization, remote labor, according to Rafalski and Andrade [57], aids the financial performance of firms through data and information obtained from other expert surveys and examined afterward. Participants from various nations participated in an online poll undertaken by Putri and Amran [56] to investigate if the advantages and downsides of working from home in the case of a pandemic were different for men and women. According to the findings of a poll, males outweigh females in favor of working from home and saving money by not having to travel. A probable reason for this is that women with children often have to manage their job and family lives, as Dockery and Bawa point out. According to a recent survey, having fewer business meetings was more popular with men than with women [19]. There are no notable variations in the issues or barriers that workers who work from home confront daily, depending on gender.

A home office has been researched by Kumar [41] but from different angles. Using an online poll of employees and employers' respondents, Kumar investigated the productivity of persons who work from home and the aspects that may affect it. According to the findings of a response study, working from home had a higher effect on female workers' productivity in Covid-19 than working in the office [41]. According to respondents, poor communication was commonly identified as a drop in productivity [41]. According to Van Der Lippe and Lippényi [74], who used the same methodology and evaluated respondents' responses, employee productivity rose after moving to a home office. Compared to the last time, their work was better and more efficient. It was shown that homemakers were more prone to acquire weight or suffer from back and neck problems than their male counterparts. Therefore, musculoskeletal discomfort may be regarded as a negative. People who worked from home had inefficient work habits, and Beck and Hensher [4] examined whether the home environment affected this issue. Using an online survey, they gathered information from general workers. This authors found that the impact of remote work on professional isolation and poor work habits and workers' capacity to self-regulate has diminished but is still substantial [4].

4 Discusion

4.1 Home Office

The early commute to work may be difficult and time-consuming due to traffic congestion and other problems. Many individuals now work from home. First, several legal issues must be overcome before evaluating if a home office improves morale. "Home office" refers to "working from home." Phone calls, emails, and internal short messaging systems are examples of this. Employees and independent contractors use home offices as a workspace [58]. Employees may work entirely from home if their employer offers the necessary equipment (such as a laptop computer). Payment may be made hourly or in one lump sum. Employees are usually assigned to achieve within a specified time limit [8]. The structure and division of work may be changed to suit the employees' needs.

Home offices may be designed to match the demands of the individual and the task at hand. Alternative teleworking is the most frequent kind of telework. Teleworkers work from home, eliminating the need for an office. Employees may work from home or in company-provided workplaces. Co-workers who work from home sometimes may share a desk in the main office. In this way, the company saves physical and financial resources. Because the shared workplace cannot accommodate many users at once, all parties must agree on office hours in advance. Mobile teleworking (or mobile working) is an option. Work occurs in several settings, including stations, trains, and cafés. In most cases, an internet connection is needed to work anywhere [26].

Occupational Safety and Health Administration (OSHA) applies to workers who work from home. OSHA discusses how important it is to keep employees and confidential documents safe from outsiders. A good home office must meet the needs of the job contract. The chance to work from home is a good thing; this set includes a computer desk with an adjustable ergonomic chair and an eye-friendly display [65]. By positioning a workstation near the windows, a worker can balance the quantity of light and glare. Employees must be available for contact with their employer within a specific period each day.

4.2 General Benefits

Individuals and small businesses alike have highlighted various advantages to doing business remotely; this study identified a few and will go through some of the benefits in further detail below. For instance, the productivity of workers increases from WFH. An employee's time commuting from home to work is time that may be better spent elsewhere, such as on other activities. According to telework, workers who work from home save 15 days of missed time due to travel (e.g., buses, driving, subways, etc.) [7]. The telework highlights that businesses that demonstrates greater productivity between those who worked from home and those who remained in the office. Best Buy claims that its flexible work program has resulted in a 35% boost in productivity. According to British Telecom, telecommuting has enhanced productivity by 20% while Alpine Access's home-based agents, one of the nation's biggest all-virtual workforces, have increased sales by 30% and reduced customer complaints by 90%. Alpine Access is a major all-virtual enterprise in the United States. American Express teleworkers handled 26% more calls and earned 43% more than their office-based colleagues.

A study conducted by Putri and Amran [56], employees were permitted to work from home while others stayed in the office. Putri and Amran reasoned that allowing employees to work from home would save the company money on office space and equipment. The cost benefits would balance any productivity losses caused by workers no longer having access to an office setting [56]. Kumar affirms that firms got an additional workday every week because those who worked from home completed 13.5 percent more calls than those who stayed in the office [41]. And results of Horváthová and Čajková [27] point to exactly at the level of burden on employers in the form of the obligation to pay the sickness benefit in combination with the insurance company.

Also, sick days are being reduced as stress and personal concerns are responsible for around 78 percent of sick days. People who work from home may need fewer sick days since they may address minor health difficulties without taking a full day off. Employee productivity rises due to improved work-life balance, which raises productivity. Work-from-home employees often have an advantage over their office counterparts in that they may return to work sooner following an illness or accident.

Work from home saves money by not spending on gas, accessories, grooming, clothes, and other expenses due to not having to commute to and from work. Workers also save time since they are not stuck in traffic. Therefore, employees and companies alike benefit from the cost reductions provided by this technique. Using this tool may help save money on overhead and variable expenditures. Some of the funds may be used for growth, while others may be used to increase the company's employee perks. According to Kumar [41], firms saves a huge sum of money per employee every year through WFH. Teleworkers should expect to save between $1,600 and $6,800 per year on their salary. Also, Kumar [41] asserts that improved productivity, less absenteeism, fewer facility expenditures, and reduced turnover are resultant importance of WFH [41].

It is critical to be satisfied with one's job through home office strategy. Employees who are allowed to use this incentive seem to be happier in their employment. When it comes to job happiness, at-home employees outperformed their office colleagues. According to the Kelly et al. research, employees were happier with their occupations. According to a recent poll of this authors of home office employees, working from home decreases stress by 25%, improves nutrition by 73%, enhances loyalty by 76%, and improves work-life balance by 80%. Hence, work-life balance is the essential advantage of working from home [37].

According to a meta-analysis conducted by Kumar, work-family harmony is associated with higher job satisfaction and performance, lower work-family conflict, and a lower likelihood of turnover. The increase in worker perceived autonomy as a result of WFH is proposed to achieve these outcomes. Accordingly, WFH is associated with positive organizational outcomes like turnover intent, commitment, productivity, retention, and performance. According to the findings of Kumar WFH is primarily a "good thing" for individuals and organizations; however, these findings, have certain limitations. WFH intensity or frequency seems to influence outcomes, with high-intensity WFH (defined as more than two or three days per week) associated with negative impacts on work and family connections and a blurring of the borders between the two [41].

Kelly et al. noted that the relationship between WFH and productivity was complex and influenced by social support. In a his study, these researchers discovered that perceived social and organizational support, as well as support for the individual WFH (which included manager trust and support, as well as technical support), were associated with increased job satisfaction and reduced psychological strain (which included manager trust and support, as well as technical support). Social isolation had an impact on these consequences. Although these studies were conducted in a normal operating environment, the results of WFH in times of crisis are less well understood than in normal operational settings. According to this authors analysis, implementing WFH in a crisis differs from implementing WFH in a business-as-usual situation [37].

While the WFH has been regularly recommended in the grey literature as a strategy to continue operations after a disaster, there is no empirical data to back up this claim. Ipsen et al. conducted investigations in the aftermath of the Covid-19 pandemic. Researchers of Ipsen et al. discovered that WFH after a natural disaster helped maintain business continuity and employee well-being. However, this effect was moderated by the personal situation and work role, IT infrastructure and resources, and prior experience with WFH. During the Covid-19 epidemic, economic continuity was shown for those enterprises that employers can adopt WFH, and overall well-being was improved due to WFH's efforts to allow social distancing and viral containment on a larger scale [30]. Working from home has numerous advantages, but it also has many disadvantages, many of which have come to light due to the current economic crisis. A short assessment of the study's concerns has now been given, explaining how these obstacles may vary in the setting of a problem [30].

4.3 General Challenges

Like all others, WFH has a broad range of potential drawbacks while keeping in mind that some of these concerns are based on perceptions or issues not supported by facts. From a managerial standpoint, Ipsen et al. [30] one of the most difficult elements of employees working from home is getting upper-level management's viewpoint. Many supervisors like to see their personnel in action rather than hear about it. Employers worry that employees who work from home and take advantage of the policy are not performing their jobs properly. Therefore, this is due to a lack of self-control. Employers cannot check production in real-time due to a lack of direct monitoring [30]. Supervisors are more likely to trust their workers who work from home if they create clear, quantifiable objectives for everyone [30].

Self-discipline is another problem affecting WFH. Working from a remote location is not ideal for everyone. Working from home requires self-control to avoid getting distracted by domestic tasks or other distractions—employees who abuse the incentive set a terrible example for others. Also, it is vital to communicate in person, which lacks in WFH. A lack of face-to-face interaction, including opportunities for cooperation, maybe a significant disadvantage for workers who work from home. Technology may be able to mitigate some of the negative effects. It may be beneficial for remote employees to visit the workplace for collaboration or networking opportunities. Visibility is another problem associated by WFH. Some data shows that people who work well from home put in more time than their office counterparts; however, establishing this may be challenging. Workers who are more cautious in their communication may ascend to the company's top [26].

A healthy work-life balance is essential, which may be inexistence in WFH. On the other hand, working from home may make it more difficult for certain individuals to create a better work-life balance. Because there is no longer a physical divide between work and home, working from home may make it more difficult for workers to complete their tasks on time. Consequently, folks who work from home may not be as productive as those who work in an office. Changing working methods, whether as part of ordinary operations or in reaction to an emergency, influences the employment and psychological relationships between employees and their employers. WFH's efficacy may be severely hampered if organizations fail to recognize the shift in these dynamics. According to a quick literature surveys, the following are the most significant issues facing WFH. Working from home can cause social isolation, work-home conflict, and an increase in load if these issues are not addressed. As a consequence of these circumstances, a person's overall health (physical and mental) and their capacity to perform at work may deteriorate [26].

Workers may experience social and professional isolation if removed from the workplace's customary social context. People's experience of social isolation was highlighted during the Covid-19 epidemic when the illness cut them off from work, extended family, and recreational activities. Although social involvement was still allowed, social isolation was a harmful side effect of WFH during pandemic. Furthermore, Green et al. [26] observed an increase in isolation due to Covid-19. According to Green et al. [26], organizational social support may help to reduce social isolation. As Green et al. [26] observed during the epidemic, videoconferencing tools like zoom, which is currently commonplace, may play a critical role [26].

WFH, on the other hand, may encourage remote workers to overwork, which may lead to work-family conflict and eventually affect their overall well-being and mental health since work is always 'right there.' As a result of the current pandemic, parents juggling work-family duties while caring for and educating their children face more stress as they attempt to maintain a good work-family balance. Work-family conflict was lowest among distant employees who adopted work-family separation techniques (for example, a dedicated workspace). Even in a business-as-usual context, this was true. Companies must also create a culture and set rules that encourage workers to relax and decompress after a long day. Several countries and businesses have implemented a regulation known as the "right to be unplugged." Hence, working part-time at the office and part-time at home may also assist reduce the harmful effects of WFH on the individual [21].

Working from home may increase the quantity of work that managers must do. Team leaders' experiences and perspectives maybe notably different from team members', with rising expectations on team leaders to govern and coordinate operations and assist workers in difficult conditions, necessitating the usage of improved communication [21]. According to anecdotal evidence, this is also the case for some team leaders in the present pandemic disruption scenario (personal communication).

Companies often avoid WFH since it might hamper communication and cooperation. The use of technology-enabled collaborative tools may improve the availability of online collaboration solutions (such as web-based presentation rooms and workflow management software). According to Fadinger and Schymik [21], enhanced communication tools and high-level ICT assistance have increased production. WFH staff members' management and communication skills also play a role in the organization's success or failure. WFH, unlike other office professions, demands the acquisition of new management and communication skills. With the conventional mindset of wanting to see employees and measuring performance by "time in the seat," management's attitude may be the most difficult to change when it comes to effective work-from-home. WFH managers must prioritize developing and maintaining trust in their working relationships by participating in behaviors that promote open and honest communication [21].

Additionally, communication in a WFH environment must be clearer than in a similar face-to-face contact. A variety of media, including those chosen by employees, must be utilized to convey a company's aims clearly and practically. It has also been shown that remote employees with a feeling of autonomy are more likely to perform well and be pleased. Employee communication is also essential for the flow of information and the coordination of job obligations. Managers may assist home-based employees in developing connections with their co-workers by planning social gatherings and giving informal engagement, learning, and mentorship opportunities [5].

Working from home may negatively influence a worker's physical and mental health since working circumstances are often inferior to those found in an office environment. Employers have health and safety duties to all of their employees, regardless of where they work. It is critical to have the appropriate policies and procedures to satisfy those duties. If there is no private or designated work location in the house, it may affect how remote the job is and blur the lines between work and family life. As a result, it's critical to recognize that, although workers may be OK with WFH moving operational expenses (such as internet, equipment, and heating) to them when WFH is voluntary or temporary during a crisis, the repercussions for the working relationship must be considered [64]. Although the local environment influences operations and IT infrastructure, the surrounding environment also impacts workers' situations and the management under which they work, the voluntary nature of WFH for employees, and policy creation and execution. In Beck and Hensher [4], WFH was not always required owing to other facilities and in-person interaction, which minimized social isolation and communication difficulties. On the other hand, the loss of infrastructure hampered the transition to WFH. WFH, on the other side, was imposed by the government in Covid-19, resulting in infrastructure being spared damage but people being forced to live in isolation [4].

4.4 Benefits for Employees

Workers love the freedom of working from home [13]. Finally, each employee sets their timetable for waking up, routine work, and rest. The employee's total working hours must be adhered to—the flexibility and liberty afforded to make private scheduling sessions between meetings easier. Family care, medical visits, grocery shopping, and childcare are now accessible with work-at-home options. Those who work from the home claim they are more productive since colleagues do not disturb them [49]. Commuters will also be glad that they will not have to cope with daily traffic. The high degree of autonomy, autonomous work scheduling, and absence of colleague interruptions enhanced productivity and motivation. One benefit of remote working is working where and when one chooses. Tips for individuals who wish to work from home yet want to maximize their time: a home, work, and play. Workers may work from anywhere using Wi-Fi as long as they can execute their job well and communicate with their co-workers [12].

Due to variable energy levels, job durations, and personal schedule considerations, such as daycare pick-up or health and wellness days, most professionals cannot work from 9 to 5. Therefore, this flexibility allows employees to prepare for the necessary employment time and meet basic needs while working from home [70]. In 2019, Americans commuted for the first time in 27 min or much longer if they lived or worked in a big city with traffic [29]. Long commutes affect employees, their families, and even their employers; this may save nine months of driving or biking back and forth [32]. Lastly, remote working may help improve people's lives with disabilities and mental health difficulties. They can thrive at work and take care of themselves when needed.

4.5 Employer Advantages

Companies benefit from their workers' increased productivity and sense of well-being, among many more working from home options. Retaining staff is a cost-effective way to reduce turnover. Employers may save money on office space and meals when employees work from home. There would be no additional payments for electricity, beverages, or work utensils, except the rent. Due to the flexibility of self-planning, employees can keep necessary personal appointments without missing time at work. There are various advantages to working from home or utilizing it as a base for part of the workweek for companies. Flexibility and agility in working conditions are enhanced by working from home. Workers who no longer have to be linked to an office may be more suited and ready to work flexible schedules, including weekends. A company's needs may be met by homeworking, such as working with customers in different time zones [1].

Working from home may help retain employees by enabling them to meet child-care commitments, save travel time, and balance work and family life. By allowing their workers to work from home, employers can build a stronger sense of security among their workforce. If companies want to bring in new employees, consider giving home working as an incentive. Allowing the staff to work from home gives them an advantage over competitors. Also, due to fewer interruptions in the workplace, productivity rises. Working from home allows companies to focus on one task at a time. Some employees may want to extend their workdays because they save time driving to and from the office.

Smart working enterprises have access to a broader pool of potential employees instead of those in their immediate area. Firms can recruit the best people since remote work may be done almost anywhere. Having a solid internet connection and computer setup does not need to move an employee. The cost of starting a smart working company is far lower than employing a full-time employee. Businesses have to pay for office supplies, food, and drinks for their employees, so they have to do that [2]. Corporations could save a lot of money if this worked out for them. Even risks such as power outages do not have to be addressed by companies as significantly as possible if they allow workers to work from home, where the employee may be prepared for such a risk [36] or the risk will be spread in this way [36].

While working from home, employees can build trust with their company since they are given more freedom to fulfill their jobs. Staff will be more motivated to produce their best job if they can work from home in a manner that suits them. Also, working from home reduces the stress of commuting, which improves employee health and well-being. Employees may use the time they save to get more sleep, spend more time with their families, exercise, or cook healthier foods. It is possible to save money on office space and supplies to utilities. Her Majesty's Revenue and Customs (HMRC) tax refunds may be available to employees who work from home for UK-based employees [9].

Employees who often visit clients but are not in the office might be convenient. Employees may find it more convenient and cost-effective to work remotely; this may be accomplished by working from home, allowing people to better balance their personal and professional lives. Commuters may now better manage their job and personal lives by making better use of their commute time. Employees may also fit in family responsibilities around their workdays, giving them more time to themselves after hours. Using the internet has made it easier for workers to keep in touch with their employers. It is easier to communicate with colleagues and teams through Skype, which has resulted in more productive meetings [14]. Thus, working from home reduces the risk of immune system damage from burnout, making workers happier and more motivated. Besides, there is less risk of people getting sick in a hospital because they are isolated. Working from home decreases exposure to germs that cause sickness, resulting in fewer sick days. Their colleagues, who feel compelled to show up for work even if unwell, are not helping; this might minimize the number of missed workdays due to illness. An employee's slight cold may also be worked through without spreading the disease to other workforce members.

Work-from-home employees are more productive since they have fewer distractions from their co-workers. Employees might take brief breaks to re-energize and re-motivate as a last resort. Working from home can improve productivity by enabling employees to focus only on their work and set their schedules. Also, Working from home may be an option for specific individuals regarding work-life balance. Employees who have a better work-life balance are more productive and less likely to get burned out [37]. Because people may work from home, they can better organize their schedules. Because there is no commute, going to appointments and family matters is a lot easier.

At ease, employees are another advantage of productive home working. A practical and contented workplace requires careful consideration of ergonomics. If employees can personalize their workstations, they may feel more at ease during the day. Employers may save money on health insurance premiums by hiring healthier workers. Reduced absences due to back and neck pain may be possible with a more pleasant workplace. Working from home offers more freedom to both companies and individuals, and with a good internet connection, long-term travel is doable. Both companies and workers benefit from this for long-term vacation planning. Employers and workers alike must relax and disconnect when they have the opportunity. It is easy to keep up with business even while on holiday in the tropics, where they can work remotely. Having workers in several time zones helps companies to extend their working hours. Employers can assure that someone is working for most of the day since workers begin and conclude their days at various times [15]. Customer service and other round-the-clock services benefit from this.

5 Conclusion

The primary aim is to investigate home offices and assert whether it benefits employers or employees. As per the research questions, other areas of concern include: What is a home office? What are the benefits of a home office to employers?

What are the benefits of a home office to employees? Alternative work environments are commonly employed to promote work/life balance. These results differ by alternative work venue. Telework was perceived as having no influence on work/life balance by virtual office workers. Virtual office workers reported substantially less work/life balance and personal/family success than traditional office or home office employees. According to a multivariate study, virtual employees have a harmful influence on work/life balance. Why is work/life balance a concern for virtual workers? The lack of externally imposed physical restrictions may be one factor. Virtual office professionals may find it hard to discern between work and home.

Giving virtual office employees work-enabling tools may enhance their time density or multitasking ability; this may also impair virtual office workers' work/life balance. The findings demonstrated that working from home was related to more excellent work/life balance for telecommuters. Preceding the daily commute may be one; this saves 3–4 h per week. They also work shorter hours than virtual office workers; hence, this permits them to spend six hours a week more on housekeeping than virtual office workers. Portable computers, pagers, and mobile phones are seldom available to home workers. As a result, they could define boundaries and reduce time constraints.

The freedom and convenience of working from home are only two of the perks that come with it. While employees may enjoy working from home, businesses stand to earn substantially. A work-from-home arrangement may save firms money on overhead costs, boost employee morale, and increase productivity. This article examined the benefits of allowing employees to work from home and its importance to a company's ability to compete following the extensive advantages from both parties. Besides, this may become a need for job searchers as more companies allow remote work environments. Employers may want to introduce remote work possibilities to attract and keep the best employees. Some projects may be completed totally from the comfort of one's own home, while others need travel.

Working from home has several advantages for both employees and employers. A happy workforce may lead a company to be more successful financially and productively. The following are just a few of the many perks that come with letting employees work from home, which this article believes that it benefits employers and corporates more than employees. For example, it is easier to communicate: Email, meetings, phone calls, video conferencing, and informal talks are all forms of business communication; this might make it challenging to maintain track of important tasks or activities. Various communication options are available to employers to keep employees who work remotely in touch; this might lead to a more efficient and collaborative work environment for the company's employees.

At the same time, in the context of public administration, for example at the municipal level, increasing the level of cooperation can enable work from home. The higher level of cooperation enables better interconnection of municipalities and the fact that larger spatial units are expected to be more competitive in a globalized world, will also increase the level of quality of public services [20].

It is possible that working from home will cut down on employee churn. There are fewer motivations for employees to look for a new position. For example, employees may have to leave their jobs to relocate for their spouse's work. This person can keep their job even if they work from home. Reducing employee attrition may be made easier with flexible schedules and high levels of job satisfaction. Employees who work from home have more freedom and control over their plans. A new work environment might be exciting for those who can adjust well. Many advantages might accrue to businesses if their employees are happy in their jobs. A few benefits include increased employee loyalty, productivity, and a better reputation.

For both firms and employees, working from home saves time and money. Many hours are needed to be ready for the day. Additionally, getting to and from work requires a significant amount of time. With this extra time, companies may increase productivity and meet deadlines. Employees may appreciate the spare time for themselves [4barbieri]. When workers work productively, they grow their brand image. Companies that allow employees to work from home may positively affect their public appearance. Employee satisfaction positively affects the public's willingness to support a company.

On the other hand, job searchers may be more excited to apply for a company that gives them more flexibility. In addition, suggestions from the top-down employees who work across the country or worldwide might benefit employers. Employees from various backgrounds might provide fresh perspectives on a company's operations. By moving employees to other marketplaces, employers may learn more about them.

Some employees have difficulty getting to and from work. Employers may increase workplace diversity by enabling employees to work from home. Individuals from a wide range of cultural, regional, and financial backgrounds and those who have difficulty moving about may benefit from working remotely. Thus, employers may lower their environmental footprint by allowing employees to work from home. Smart operating all enterprises may become more environmentally friendly by minimizing energy use and emissions. As a bonus, working from home reduces the amount of petrol and oil purchased and used. Employees may be less likely to use throwaway plates and cups if they bring their own.

Acknowledgements This scientific study was created within the project National Infrastructure Transfer Support Infrastructure in Slovakia—NITT SK II (national project), Operational Program: Integrated Infrastructure, Project co-financed by: European Regional Development Fund.

References

1. Aczel, B., Kovacs, M., Van Der Lippe, T., Szaszi, B.: Researchers working from home: benefits and challenges. PLoS ONE **16**(3), e0249127 (2021). https://doi.org/10.1371/journal.pone.024912
2. Alipour, J.V., Fadinger, H., Schymik, J.: My home is my castle–The benefits of working from home during a pandemic crisis. J. Public Econ. **196**, (2021) 104373
3. Barbieri, B., Balia, S., Sulis, I., Cois, E., Cabras, C., Atzara, S., De Simone, S.: Don't call it smart: working from home during the pandemic crisis. Front. Psychol. **12**, 741585 (2021). https://doi.org/10.3389/fpsyg.2021.741585

4. Beck, M.J., Hensher, D.A.: Working from home in Australia in 2020: positives, negatives and the potential for future benefits to transport and society. Transp. Res. Part A: Policy Pract. **158**, 271–284 (2022)

5. Beňo, M.: The advantages and disadvantages of E-working: an examination using an ALDINE analysis. Emerg. Sci. J. **5**, 11–20 (2021). https://doi.org/10.28991/esj-2021-SPER-02

6. Bessa Vilela, N., Brezovnik, B.: Europe: hell or paradise? An overview of European law and case law. J. Comp. Polit. **11**(2), 65–82 (2018)

7. Blount, Y.: Pondering the fault lines of anywhere working (telework, telecommuting): a literature review. Found. Trends® Inf. Syst. **1**(3), 163–276 (2015). http://dx.doi.org/https://doi.org/10.1561/290000000

8. Bolisani, E., Scarso, E., Ipsen, C., Kirchner, K., Hansen, J.P.: Working from home during COVID-19 pandemic: lessons learned and issues. Manag. Marketing. Chall. Knowl. Soc. **15**(SI), 458–476, (2020). https://doi.org/10.2478/mmcks-2020-0027

9. Brignall, M.: (2022). Working from home: could you be eligible for up to £125 in tax relief? The Guardian. Retrieved 16 February 2022, from https://www.theguardian.com/money/2020/oct/16/working-from-home-tax-relief-hmrc-covid

10. Čajka, P., Olejárová, B., Čajková, A.: Migration as a factor of Germany's security and sustainability. J. Secur. Sustain. Issues **7**(3), 400–408 (2018). https://doi.org/10.9770/jssi.2018.7.3(2)

11. Cajková, A., Jankelová, N., Masár, D.: Knowledge management as a tool for increasing the efficiency of municipality management in Slovakia. Knowl. Manag. Res. Pract., In press, (2021), doi: https://doi.org/10.1080/14778238.2021.1895686

12. Chien., NB, Thanh., NN.: The Impact of Good Governance on the People's Satisfaction with Public Administrative Services in Vietnam. Adm. Sci. **12**(1), 35 (2022). https://doi.org/10.3390/admsci12010035

13. Church, N.F.: Gauging perceived benefits from working from home as a job benefit. Int. J. Bus. Econ. Dev. (IJBED), **3**(3), (2015)

14. Corporate Finance Institute. Home Office. Retrieved 16 February 2022, from https://corporatefinanceinstitute.com/resources/knowledge/other/home-office/ (2022)

15. Courtney, E.: The benefits of working from home beyond the pandemic. FlexJobs. FlexJobs Job Search Tips and Blog. Retrieved 16 February 2022, from https://www.flexjobs.com/blog/post/benefits-of-remote-work/ (2022).

16. Dan, H.: Cultural differences as obstacles in the European economic integration process-a labour market perspective. Online J. Model. New Eur. **15**(1), 76–96 (2015)

17. Dan, H.: The Euro zone—Between fiscal heterogeneity and monetary unity. Transylv. Rev. Adm. Sci. **43E**, 68–84 (2014)

18. Dan, H.: The influence of cultural elements on fiscal behaviour in the European union. Online J. Model. New Eur. **16**(1), 3–19 (2015)

19. Dockery, A.M., Bawa, S.: Is working from home good work or bad work? Evidence from Australian employees. Aust. J. Labour Econ. **17**(2), 163–190 (2014)

20. Dušek, J.: How to measure intermunicipal cooperation in conditions of the Czech Republic. In: Dias, A., Salmelin, B., Pereira, D., Dias, M. (eds.) Modeling innovation sustainability and technologies. Springer Proceedings in Business and Economics. Springer, Cham. https://doi-1org-158yble7e0105.hanproxy.cvtisr.sk/10.1007/978-3-319-67101-7_12

21. Fadinger, H., Schymik, J.: The costs and benefits of home office during the covid-19 pandemic: evidence from infections and an input-output model for Germany. COVID Econom: Vetted and Real-Time Papers **9**, 107–134 (2020)

22. Funta, R., Plavcan, P.: Regulatory concepts for internet platforms. Online J. Model. New Eur. **35**, 44–59 (2021). https://doi.org/10.24193/OJMNE.2021.35.03

23. Funta, R.: Discounts and their effects - economic and legal approach. Danube **5**(4), 277–285 (2014). https://doi.org/10.2478/danb-2014-0015

24. Funta, R.: Economic and legal features of digital markets. Danube **10**(2), 173–183 (2019). https://doi.org/10.2478/danb-2019-0009

25. Garg, A.K., Van der Rijst, J.: The benefits and pitfalls of employees working from home: study of a private company in South Africa. Corp. Board Role Duties Compos. **11**(2), 36–49 (2015). https://doi.org/10.22495/cbv11i2art3
26. Green, N., Tappin, D., Bentley, T.: Working from home before, during and after the Covid-19 pandemic: Implications for workers and organizations. N. Z. J. Employ. Relat. **45**(2), 5–16 (2020). https://doi.org/10.24135/nzjer.v45i2.19
27. Horváthová, Z., Čajková, A.: Framework of the sickness insurance in the Czech Republic and selected countries of the European Union. EJTS Eur. J. Transform. Stud. **7**(1) (2019)
28. Hynes, M.: (2013). What's "Smart" about working from home: telework and the sustainable consumption of distance in Ireland? In C. Fowley, C. English, & S. Thouësny (eds.), Internet Research, Theory, and Practice: Perspectives from Ireland, pp. 225–243
29. Ingraham, C.: Nine days on the road. Average commute time reached a new record last year. https://www.washingtonpost.com. Retrieved 16 February 2022, from https://www.washingtonpost.com/business/2019/10/07/nine-days-road-average-commute-time-reached-new-record-last-year/
30. Ipsen, C., van Veldhoven, M., Kirchner, K., Hansen, J.P.: Six key advantages and disadvantages of working from home in Europe during COVID-19. Int. J. Environ. Res. Public Health **18**(4), 1826 (2021). https://doi.org/10.3390/ijerph18041826
31. Ispas, G.L.: Impact of the covid-19 on the migration in the European Union. Juridical Trib. **11**(1), 30–41 (2021)
32. Jain, T., Currie, G., Aston, L.: COVID and working from home: long-term impacts and psychosocial determinants. Transp. Res. Part A: Policy Pract. **156**, 52–68 (2022)
33. Jančíková, E., Pásztorová, J.: Promoting eu values in international agreements. Juridical Trib. **11**(2), 203–218 (2021). https://doi.org/10.24818/TBJ/2021/11/2.04
34. Jankurová, A., Ljudvigová, I., Gubová, K.: Research of the nature of leadership activities. Econ. Sociol. **10**(1), 135–151 (2017). https://doi.org/10.14254/2071-789X.2017/10-1/10
35. Jayashree, P., El Barachi, M., Hamza, F.: Practice of sustainability leadership: a multi-stakeholder inclusive framework. Sustainability **14**, 6346 (2022). https://doi.org/10.3390/su14106346
36. Kavan Š, Dvořáčková O, Pokorný J, Brumarová L.: Long-term power outage and preparedness of the population of a region in the Czech Republic—a case study. Sustainability **13**(23), 13142 (2021). https://doi-1org-158yble7e0105.hanproxy.cvtisr.sk/https://doi.org/10.3390/su132313142
37. Kelly, T., Poliakoff, J., Mozley, C.: Working from home has benefits—for employers Letters. The Guardian. Retrieved 16 February 2022, from https://www.theguardian.com/money/2021/jun/21/working-from-home-has-its-benefits-for-employers
38. Kłopotek, M.: The advantages and disadvantages of remote working from the perspective of young employees. Organizacja i Zarządzanie: kwartalnik naukowy. **4**, 39–49 (2017). https://doi.org/10.29119/1899-6116.2017.40.3
39. Kryvinska, N., Bickel, L.: Scenario-based analysis of IT enterprises servitization as a part of digital transformation of modern economy. Appl. Sci. **10**(3), 1076 (2020). doi:https://doi.org/10.3390/app10031076
40. Kryvinska, N., Kaczor, S., Strauss, C.: Enterprises' servitization in the first decade - retrospective analysis of back-end and front-end challenges. Appl. Sci. **10**(8), 2957 (2020). https://doi.org/10.3390/app10082957
41. Kumar, D.S.: Employees 'perceived benefits and drawbacks from "work from home" during COVID-19. PalArch's J. Archaeol. Egypt/Egyptol. **17**(6), 2943–2957 (2020)
42. Lacuška, M., Peráček, T.: Trends in global telecommunication fraud and its impact on business. Studies in systems, decision and control **330**, 459–485 (2021). https://doi.org/10.1007/978-3-030-62151-3_12
43. Lawrence, J., Tar, U.: The grounded theory technique is a practical tool for qualitative data collection and analysis. Electron. J. Bus. Res. Methods **11**(1), 29–40 (2013)
44. Moretti, A., Menna, F., Aulicino, M., Paoletta, M., Liguori, S., Iolascon, G.: Characterization of home working population during COVID-19 emergency: a cross-sectional analysis. Int. J. Environ. Res. Public Health **17**(17), 6284 (2020)

45. Mucha, B.: Evaluation of the state of implementation of the European structural and investment funds: case study of the Slovak Republic. Online J. Model. New Eur. **35**, 4–24 (2021). https://doi.org/10.24193/OJMNE.2021.35.0
46. Mustajab, D., Bauw, A., Rasyid, A., Irawan, A., Akbar, M.A., Hamid, M.A.: Working from home phenomenon to prevent COVID-19 attacks and its impacts on work productivity. TIJAB (Int. J. Appl. Bus.) **4**(1), 13–21 (2020)
47. Nakrošienė, A., Bučiūnienė, I., Goštautaitė, B.: Working from home: characteristics and outcomes of telework. Int. J. Manpow. **40**(1), 87–101 (2019). https://doi.org/10.1108/IJM-07-2017-0172(2020)
48. Nováčková, D., Vnuková, J.: Competition issues including in the international agreements of the European union. Juridical Trib. **11**(2), 234–250 (2021). https://doi.org/10.24818/TBJ/2021/11/2.06
49. Oakman, J., Kinsmart working an, N., Stuckey, R., Graham, M., Weale, V.: A rapid review of mental and physical health effects of working at home: how do we optimize health? BMC Public Health **20**(1825), 1–13 (2020). https://doi.org/10.1186/s12889-020-09875-z
50. Palinkas, L.A., Horwitz, S.M., Green, C.A., Wisdom, J.P., Duan, N., Hoagwood, K.: (2015) Purposeful sampling for qualitative data collection and analysis in mixed method implementation research. Adm. Policy Ment. Health Ment. Health Serv. Res. **42**(5), 533–544 (2015)
51. Peráček, T., Mucha, B., Brestovanská, P.: Selected legislative aspects of cybernetic security in the Slovak Republic. Lect. Notes Data Eng. Commun. Technol. **23**, 273–282 (2019). https://doi.org/10.1007/978-3-319-98557-2_25
52. Peráček, T.: Flexibility of creating and changing employment in the options of the Slovak Labor Code. Probl. Perspect. Manag. **19**(3), 373–382 (2021). https://doi.org/10.21511/ppm.19(3).2021.30
53. Perego, F., De Maria, B., Cassetti, G., Parati, M., Bari, V., Cairo, B., Dalla Vecchia, L.A.: Working in the office and smart working differently impact on the cardiac autonomic control. In: 2021 Computing in Cardiology (CinC) 48, 1–4 (2021). doi: https://doi.org/10.23919/cinc53138.2021.9662943
54. Poiană, O., Stretea, A.: Brexitology: A story of renegotiations, referendums and bregrets? Online J. Model. New Eur. **28**, 206–216 (2018). https://doi.org/10.24193/OJMNE.2018.28.11
55. Pokojski, Z., Kister, A., Lipowski, M.: Remote work efficiency from the employers' perspective— What's next? Sustainability **14**, 4220 (2022). https://doi.org/10.3390/su14074220
56. Putri, A., Amran, A.: Employee's work-life balance reviewed from work from home aspect during COVID-19 Pandemic. Int. J. Manag. Sci. Inf. Technol. **1**(1), 30–34 (2021)
57. Rafalski, J.C., Andrade, A.L.D.: Home-office: exploratory aspects of work from home. Temas em Psicologia **23**(2), 431–441 (2015)
58. Raišienė, A.G., Rapuano, V., Varkulevičiūtė, K., Stachová, K.: Working from home—Who is happy? A survey of Lithuania's employees during the COVID-19 quarantine period. Sustainability **12**(13), 5332 (2020). https://doi.org/10.3390/su12135332
59. Revutska, O.: (2021). Human resource management: a shift towards agility due to a pandemic of COVID-19. In Liberec Economic Forum (p. 423)
60. Robelski, S., Keller, H., Harth, V., Mache, S.: Coworking spaces: The better home office? A psycho-social and health-related perspective on an emerging work environment. Int. J. Environ. Res. Public Health **16**(13), 2379 (2019). https://doi.org/10.3390/ijerph16132379
61. Rožman, M., Tominc, P., Crnogaj, K.: Healthy and entrepreneurial work environment for older employees and its impact on work engagement during the COVID-19 pandemic. Sustainability **14**(4545), (2022). https://doi.org/10.3390/su14084545
62. Ruggiano, N., Perry, T.E.: Conducting secondary analysis of qualitative data: Should we, can we, and how? Qual. Soc. Work. **18**(1), 81–97 (2019). https://doi.org/10.1177/1473325017700701
63. Săraru, C.-S.: The European groupings of territorial cooperation developed by administrative structures in Romania and Hungary. Acta Juridica Hungarica **55**(2), 150–162 (2014)

64. Shevchuk, O., Bululukov, O., Lysodyed, O., Mamonova, V., Matat, Y.: Human right to virtual reality in the healthcare: Legal issues and enforcement problems. Juridical Trib. **11**(SI), 302–315 (2021). doi: https://doi.org/10.24818/TBJ/2021/11/SP/03

65. Schall Jr, M. C., Chen, P.: Evidence-based strategies for improving occupational safety and health among teleworkers during and after the coronavirus pandemic. Human factors, 0018720820984583 (2021). doi: https://doi.org/10.1177/0018720820984583

66. Srebalová, M., Vojtech, F.: SME development in the visegrad area. Eurasian Stud. Bus. Econ. **17**, 269–281 (2021). https://doi.org/10.1007/978-3-030-65147-3_19

67. Stratone, M.E., Vataamanescu, E.M. Treapat, L.M.; Rusu, M.; Vidu, C.-M.: Contrasting traditional and virtual teams within the context of COVID-19 pandemic: from team culture towards objectives achievement. Sustainability 14(4558) (2022). https:// doi.org/https://doi.org/10.3390/su14084558

68. Števček, M., Ivančo, M.: The conception and institutional novelties of recodification of private law in the Slovak Republic. The Law of Obligations in Central and Southeast Europe: Recodification and Recent Developments, pp. 33–49 (2021)

69. Ta, T.T., Doan, T.N.: Factors affecting internal audit effectiveness: empirical evidence from Vietnam. Int. J. Financ. Stud. **10**(2), 37 (2022). https://doi.org/10.3390/ijfs10020037

70. Thorstensson, E.: The influence of working from home on employees' productivity. Karlstad Businees School, 1–26 (2020)

71. Tramontano, C., Grant, C., Clarke, C.: Development and validation of the e-Work Self-Efficacy Scale to assess digital competencies in remote working. Comput. Hum. Behav. Rep. **4**, 100129 (2021). https://doi.org/10.1016/j.chbr.2021.100129

72. Triantafillidou, E., Koutroukis, T.: Employee Involvement and Participation as a Function of Labor Relations and Human Resource Management: Evidence from Greek Subsidiaries of Multinational Companies in the Pharmaceutical Industry. Adm. Sci. **12**(1), 41 (2022). https:// doi.org/10.3390/admsci12010041

73. Vacok, J., Srebalova, M., Horvath, M., Vojtech, F., Filip, S.: Legal obstacles to freedom to conduct a business: experience of the Slovak Republic. Entrep. Sustain. Issues 7(4), 3385–3394 (2020). https://doi.org/10.9770/jesi.2020.7.4(53)

74. Van Der Lippe, T., Lippényi, Z.: Co-workers are working from home and individual and team performance. N. Technol. Work. Employ. **35**(1), 60–79 (2020). https://doi.org/10.1111/ntwe.12153

75. Varner, K., Schmidt, K.: Employment-at-will in the United States and the challenges of remote work in the time of COVID-19. Laws **11**, 29 (2022). https://doi.org/10.3390/laws11020029

76. Wefersová, J., Nováčková, D.: Use of digital technologies in business in Slovakia. Stud. Syst., Decis. Control. **376**, 335–355 (2021). https://doi.org/10.1007/978-3-030-76632-0_12

77. Weitzer, J., Papantoniou, K., Seidel, S., Klösch, G., Caniglia, G., Laubichler, M., Schernhammer, E.: Working from home, quality of life, and perceived productivity during the first 50-day COVID-19 mitigation measures in Austria: a cross-sectional study. Int. Arch. Occup. Environ. Health **94**(8), 1823–1837 (2021). https://doi.org/10.1007/s00420-021-01692-0

78. Yang, C., Ha, H.-Y.: Mediator acceptability for sustainable trading management: scale development and validation. Sustainability **14**, 1798 (2022). https://doi.org/10.3390/su1403 1798

79. Žofčinová, V., Čajková, A., Kral, L.: Local leader and labour law position in the context of the concept of 'Smart Cities' through the optics of the European Union. TalTech J. Eur. Stud. **12**(1), (2022)

80. Žofčinová, V., Horváthová, Z., Čajková, A.: Selected social policy instruments in relation to tax policy. Soc. Sci. **7**(11), 241 (2018). https://doi.org/10.3390/socsci7110241

81. Žofčinová, V., Košíková, A.: Selected legislative instruments of family policy supporting work–life balance: a comparison of Italy and the Slovak Republic. Online J. Model. New Eur. **39** (2022)

Knowledge Management of Private Banks as an Asset Improved by Artificial Intelligence Discipline—Applied to Strategic McKinsey Portfolio Concept as Part of the Portfolio Management

Jörg Sträßer and Zuzana Stolicna

Abstract As first part of the scope of this publishing item it is to demonstrate a possibility to support the evolution of the knowledge management discipline for Financial Services industry with the Artificial Intelligence (AI) technology, e.g. Machine Learning (ML). As one of the banking core capabilities the part of strategic Portfolio Management will be focused as use case along the results out of the step mentioned-before. Knowledge Management (KM) is an essential core part of business development and prospering of each individual, group and organization—this discipline is an asset of each organization to do right the right things on a subject matter expert level. It is a major influencer for the setup of many IT systems with its functions [43, 54]. In addition, an existing well-organized Knowledge Management can help to flatten the shortfalls of professionals whilst the knowledge will be saved, improved and up-skilled. KM in a private bank can divide into an external and internal view—to have a look into the external direction, mainly the customer is in the focus in terms of its requests—the market knowledge. Opposite of them it exists bank internal knowledge more or less aligned with the processes as well as workflow chains to satisfy the mentioned customer demands as well as the bank strategy. Moreover, as one approach the knowledge can be separated into the categories explicit and tacit knowledge based on the SECI model [3]. Yet the exploit knowledge visible in documents, data records of knowledge data bases, workflow descriptions, content explanation is easier to retrieve, maintain and use than the tacid knowledge with its thumb rules, heuristics, intuition. The listed items are examples, there are many expressions of these categories existing, also depends on the industry sector [39]. To improve / scale up to a higher level the knowledge management of the private banks the Artificial Intelligence technology can support to generate new knowledge entities, models, scoring points, measurements, probabilities under the usage of various categories of algorithm [17]. As one of the essential strategic parts of a bank there

J. Sträßer (✉) · Z. Stolicna
Comenius University of Bratislava, Bratislava, Slovakia
e-mail: strasser2@uniba.sk

Z. Stolicna
e-mail: zuzana.stolicna@fm.uniba.sk

is to create the own portfolio concept to be ensure the definition of the main indicators, the strategic business units (e.g. Fond business), the appropriate scorings, probabilities drill-downed and related to sub-indicators as well as deviations on the whole business units. The next part of the scope for this article is focused to use the Knowledge Management on the McKinsey "GE-Matrix" or "Nine-box-matrix" adapted for generation of a Strategic Portfolio Management within Private Banking Financial Services. In addition, based on the previous-mentioned aspects this article reflects a short introduction into the possibility for a seamless AI guided workflow circle of strategic portfolio creation based on explicit and tacid knowledge and gives a preview for risk assessment, alternative portfolio model simulations and finally the trade order management related to the regular portfolio revision (re-balancing and upgrading) based on the underlying technology described in this article [24–28].

Keywords Knowledge management · Tacit knowledge · Explicit knowledge · Artificial intelligence · Machine learning · Private banking · Portfolio management

1 Introduction

1.1 Relevance

Data and knowledge are essential for each company, this single items and the derived combination as data constructs has been evolved to a "Quasi" currency [1]. To retrieve, structure, cluster and maintain this collections and classes it needs to spend a significant effort related with investments in time and money. According to a Gartner survey the impacting costs in 2017 for data amendments, caused by poor data quality is $15 million average annual financial cost per banking company [2, 3]. This is one of the challenges which the economy in common encounter, there are even further more like the fact of increasing complexity of data analysis, gathering of current data, the usage of data as baseline for strategic decisions and the convergence of structured and unstructured data—as baseline for the categorization to explicit and tacid knowledge (Fig. 1).

To address this issues some Strategic Planning Assumptions has been defined on behalf of the economy [3]:

- By 2022, 60% of organizations will leverage machine-learning-enabled data quality (DQ) technology for suggestions to reduce manual tasks for data quality improvement.
- Through 2024, 50% of organizations will adopt modern data quality solutions to better support their digital business initiatives.

It shows the willingness of a large part of market player to use AI / ML in terms of automatic straight-through processing and reduction of media breaks.

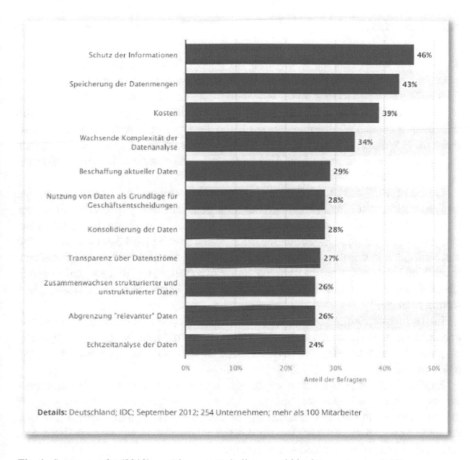

Details: Deutschland; IDC; September 2012; 254 Unternehmen; mehr als 100 Mitarbeiter

Fig. 1 Survey results (2012) most important challenges within data management [4]

1.2 Summary

First in a base explanation, the term Knowledge, Knowledge Management, Artificial Intelligence/Machine Learning will be explained. After that, the preparation steps and preconditions to use KM and AI in Banking sector will be listed, the KM model will be selected and combined with AI/ML techniques. The Strategic Portfolio Management model of McKinsey will be chosen and the process with its Strategic Business Segments will follows and adapted on Private Banking with the described AI approach in particular to increase the tacid knowledge proportion on the evolution of the mentioned strategy discipline. Finally, the conclusion/synopsis and further research opportunities will close this publication item.

1.3   Knowledge

The root or the elementary piece of all knowledge is the raw data item, it can be a number, a sequence of characters or a combination of them. In the next level the data will be changed to an information when it relates to a context, e.g. the sequence of characters "Abb" has got in the context of a german written scientific article the information content of an abbreviation for the german word "Abbildung". Or when the term is written in capital letters ("ABB") in the context of economy it stands for the company short name Asea Brown Boveri—the context makes the difference between two expressions of information.

On top of that—the knowledge will be created by combination, execution, storage and generation of information to new ones. One possible description of the depending items was published in the three-level-triangle [5]: Data could be facts, visualization and sounds, information are formatted, filtered and summarized data and knowledge can be divided into instincts, ideas, rules, procedures steering actions and decisions. In addition, the transformation between them can be formulated as information which are interpretated and context-enriched data and knowledge as an activated and applied information set.

In a further step within the five-tier hierarchy the additional items here are expertise and competence which can be combined with the wisdom of [6] as recognizing of principles and the step internalization of explicit and tacid knowledge of the SECI-model [7].

Explicit knowledge is codified knowledge based in documents, databases, repositories, archives; an important character is that IT services are essential for transfer and storage. Examples are procedures, memos, manuals, guides, articles, notes, databases, checklists, records in general, used as process documentation, hardware inventories, portfolio of traded securities, etc.

Tacit knowledge is intuitive knowledge and know-how, which is rooted in context, experience, practice and values. It is hard to communicate, because it is encapsulated in the mind of experts and the best source of long- term competitive advantage and innovation [8].

Some examples of tacit knowledge are commitment, competence, experience, ideas, intuition, comments, insights, background, skills, observations, expertise, influence, attitude, thinking, deed and related attributes: frequently not identified, unstructured, diffuse, shared with difficulties if at all [5, 9].

Particularly the relationship between these two categories are described by Michael Polanyi, when he mentioned that all knowledge is tacid knowledge or rooted by itself and hence a wholly explicit knowledge—the opposite of tacid—is highly unlikely. A condition is also that explicit knowledge should be tacit understanding in use [10].

Based on this hypothesis and some empirical analysis results it can be stated that tacid knowledge is the category with the most efforts and difficulties to retrieve comparable with the explicit knowledge [11].

Artificial Intelligence with some selected technical segments like Machine Learning, Deep Learning and Neural Networks can support this demanding process in combination with the discipline Knowledge Management [12].

1.4   Knowledge In and For Private Banks

The importance of the knowledge creation process in the banking sector as a strategy for banks to maintain their competitive advantage is essential and comprehensible. That´s why it is beneficial that the banking sector is one of the economies with the most data, information and knowledge feeded and maintained by the various business stakeholders. One of the reasons is that Banking institutes has got a lot of interdependencies between information systems and related processes internal and external. In fact this fundamental network of data with the derived information and knowledge creates a direct relation between growing numbers of data and increasing quality of AI driven results [13–16].

To achieve and to keep the knowledge accuracy in banks on an appropriate level is a real problem to solve, independent of usage of AI or traditional IT. The experience of data clean-up and enrichment projects in the past shows a disproportion of investment of resources like time, budget and human capital to the revenue outcome.

Despite promising approaches like database refactoring the amendment covered only the redesign of technical structures (DB-Scheme, Relations, Data administration, Access possibilities, etc.) while retaining data behavioral and informational semantics—it was not a change from the data quality itself.

The knowledge categories explicit and tacid can also be adapted to the data/information/knowledge soureces like data lakes, warehouses, data bases of the Private Banks like other industries. Unfortunately, it is not possible to show more details according to this segregation, because there is a big gap of studies on the knowledge creation process in the banking sector, most of the limited information are geographically oriented and it exists only a few surveys on empirical basics as well [17].

Nevertheless, some studies has been created in early past years, one from Campanella et al. demonstrates that banks create knowledge by synchronization of explicit and tacit knowledge and that this process generates economic development. In the conclusion it will be stated that "this process should be included in the new best practices of bank management". It is highlighted during that the transformation of tacit knowledge into explicit knowledge and the diffusion of knowledge creates a remarkable challenge for the legacy banking systems [18].

1.5 Knowledge Management (KM)

"KM is as a planned, structured approach to manage the creation, sharing, harvesting and leveraging of knowledge as an organizational asset, to enhance a company's ability, speed and effectiveness in delivering products or services for the benefit of clients, in line with its business strategy. KM takes place on three levels, namely the individual level, team level and organizational level. It is a holistic solution incorporating a variety of perspectives, namely people, process, culture and technology perspectives, all of which carry equal weighting in managing knowledge" [5].

Another incisive definition is "Knowledge management deals with the acquisition, development, transfer, storage, and use of knowledge. Therefore, knowledge management is much more than information management" [19].

That's why is Knowledge Management perhaps best categorized as a science of complexity [20].

The core goal of Knowledge Management is to "provide the necessary knowledge presented in a clear, unambiguous and concise manner, to the extent necessary, to the right person at the right time in order to create (added) value" [5].

One of the benefits to use Knowledge Management is it can help to transform or translate the knowledge outcome from the various special units or topics into the language of the individual recipient.

KM can be part of an common intersection between the pillars people, processes and technology [9].

Moreover, it exists some expressions of KM models like SECI, BI, Capability maturity model, Pyramid to wisdom, Knowledge life cycle and some others [5].

For this article the SECI model will be favorized to gain the tacid knowledge with usage of Artificial Intelligence applied on strategic Portfolio Management for the Private Banking industry.

This model plays an important role in the KM literature and has been widely in use in manufacturing industry and services but has minor applications in the banking sector. In future it could be possible to increase the usage and therefore the expert knowledge as basic for evolution of the Financial Services. It can deliver some new trigger for an ailing industry economy [21].

1.6 Artificial Intelligence

AI as an abbreviation for Artificial Intelligence is the ability of a digital computer or computer-controlled robot to perform tasks commonly associated with humans, such as the ability to reason, discover meaning, generalize or learn.

Some other comparable definitions are: "Artificial Intelligence (AI) is a general-purpose technology that has the potential to improve the welfare and well-being of people, to contribute to positive sustainable global economic activity, to increase innovation and productivity, and to help respond to key global challenges. It is deployed in many sectors ranging from production, finance and transport to healthcare and security" [8, 22].

The WTR (World Trade Organization) has the following defintion for AI: "Artificial intelligence (AI) is the ability of a digital computer or computer-controlled robot to perform tasks commonly associated with humans, such as the ability to reason, discover meaning, generalize or learn from past experience." [23].

Machine Learning (ML) as a sub-method of AI is the ability to adapt to new information and event circumstances and to detect and extrapolate patterns. It delivers decisions and solution constructs after learning trainings sessions [24]. Deep Learning as the conclusion engine of Machine Learning splits the solution generation process into many hierarchical interim layers analogical an artificial neuronal network. After various execution steps including the analysis of the data universe (Big Data Analysis) the solution will be created. The underlying algorithms, functions and procedures generates dedicated conclusions with an interpolation and extrapolation statistical approach.

There are some benefits to use AI/Machine Learning within Knowledge Management, one of them can offers effective ways to replace the possible reluctance of the expert individuals to request for knowledge or opposite of them to reveal knowledge, especially when it follows the 90-9-1 rule of thumb. With the algorithms of ML a part of this gap can be overdriven to retrieve and collect information / knowledge and combine it to new conclusions.

Another advantage to use AI is to handle the huge volume of information on an automatic processing approach, it gives the opportunity to filter the necessary and important facts from the negligible or wrong ones as well as can detect and eliminate some types of bias which distort data relations, classifications and correlations.

With the combination of (un)supervised learning the reinforcement learning algorithm delivers the quality assurance for the conclusion process with its results as well as how big is the deviation of the generated model to the reference one. To shorten time and save performance during the conclusion process the possibility is given to use ADAM optimization algorithm within deep learning for example [25].

It can be stated the purpose of AI for KM is to serve as gatekeeper or watchdog to adjust the knowledge basis and innovations of the dedicated industry or company [8, 12].

As follows some selected Artificial Intelligence, especially Machine Learning methods and algorithms which are relevant for this publication: [26, 27]

- Reinforcement learning: Basics are exploitation or exploration, Markov's decision processes, Policy Learning, Deep Learning and Value learning. An algorithm which learns by trial and error by interacting with the environment. It will be used in case of less of training data, the ideal end state cannot clearly defined or the only way to learn about the environment is to interact with it.
- Supervised learning: Basics are regression and classification, the classification of input data and the type of behavior to be predictable are known, there is a necessity to select the appropriate algorithm to calculate new data, e. g. Bayesian Network

with ideal preconditions (input data accurate and well-structured as well classi-fied). Naive Bayes Classifier, Support Vector Machines and Logistic Regression are also available classification algorithms. In terms of regression problems the Linear Regression, Nonlinear Regression as well as Bayesian Linear Regression can support the solution finding.

- Unsupervised learning: Basics are clustering and associative rule mining prob-lems, the classification of data is not concrete and still undefined, there is a need to find a dedicated algorithm to find patterns and classify the data, e.g. EM-Algorithm with ideal preconditions (data are statistic normal distributed to iden-tify the related classifications and be prepared for pattern matching) or K—Means clustering, Neural Networks as well as Principal Component Analysis.

This snapshot is a high-level view of the main categories of AI—a scope extension or breakdown into detail functions and algorithms can be part for further research activities for application within the Private Banking industry.

2   Theoretical and Conceptual Background

2.1   Artificial Intelligence In and For Private Banks in Context of Data/Knowledge Generation

Based on the findings of an already published article of the author further research outcomes are realized [28].

Data sources as basic for knowledge ingestion

As a result of the AI technical discipline, it offers the generation of conclusions from various data sources—the derived information/knowledge. This data portfolio consists of bank internal sources as well as of bank external static and dynamic oriented data edges—the Big Data pool with various data lakes.

From a business perspective of a bank the underlying data of the daily business has got a widespread of the scope (Risk Management, Trading, Accounting, Supervision, Front Office, Back Office, Corporate Management, etc.). All these are compositions with several weights of real-life, historical, predictive data (models, simulations) from internal and external feed.

- Real-life data
  They deliver one of the most valuable basic source for generation of similar data in combination with data masking.
- Historical data
  Often, they stored as time-series based pillar data; there are single data itself as well as constructs are essential sources too for the extrapolation of future data constructs based on empirical values.
- Data sources bank externally

These are data which the bank not owned, is not able to generate initially; that´s why the organization imports the data as spectrum/cluster/group from external suppliers into the own core banking environment. This could be messages, asset prices, interest rates, lending values, tax data, external security events, etc. from data suppliers, government, public surveys, etc. with an interim step of structuring there are also valuable basics.

Benefits of AI for knowledge creation within Private Banks

- Apparently "senseless" data were be generated, which perhaps would never be created by cognitive manually intervention. But exact those data as knowledge base supports the explorative knowledge creation approach without any reference to the underlying scope.
- The numbers and effort for single use data definition by hand with several characteristics will be reduced, whilst the AI will generate those ones repeatable and similar new ones as proposed by the conclusion algorithm.
- For example: Exactly for the business object security event of interest rate maturity the same one will be generated and another one will be suggested with the same interest rate but another interest period.
- The AI is able to extend the data models universe on horizontal level in general, whilst for semantical similar logical units in comparison with the initial object those new ones will be generated likewise. This is possible conclusion based on existing reference objects stored in historical executed knowledge creation sessions in relation to a strategy paper of the bank management with the hint to publish new products out of the product line.

   For example: The semantical business need for securities with fixed terms will be enlarged with securities with variable terms.
- The data/knowledge object spectrum will be extended on vertical level (Data enrichment): the existing data construct from pre-runs will be extracted out of the data lake, enriched with further attributes and restored again. It delivers the opportunity for example to expand the functional business simulations on a repeatable regeneration sequence, but with a mutated object under changed circumstances.

2.2   Preparation Private Banks for KM/AI

To implement these disciplines with its services, it is necessary to follow some preconditions with related preliminary tasks.

For the operative usage of that combined model in a Private Bank a set of preconditions should be fulfilled:

- Learning Bank
- Bank readiness for KM and AI
- Model level of maturity
- Current data accuracy
- Focus on a topic or investigation area

- Learning Bank passive vs. learning Bank active

  A learning organization is an adaptable and reactive unit depends on impacting internal and external influences—a definition out of organizational development discipline.

  Here we can divide between a learning Bank only to use the knowledge or in addition to create it. Depends on the part of creation the innovation within the bank will increase at scale [8].

  The whole organization should be prepared and open-minded to crossing a well-defined threshold into an area of new findings.

- Bank is ready for KM and AI

  These are the main conditions to prepare a Bank for the mentioned disciplines:

  - The implementation should be defined as strategic project with a preview for the transition into operations.
  - A pre-analysis should be completed to list the first steps, the MVPs (Minimal viable products) and the top scope candidates.
  - It is highly recommended for the initial part to select some simple topics with the chance for lessons-learning and to deliver first results very soon.
  - A dedicated budget is available for that strategic project.
  - All employees on all organizational levels—from top down (Senior Management until assistant)—should be motivated, trained and enabled for the upcoming activities.
  - An overall Change Management should be initialized with a crossover influence to operations tasks as well as project ongoing activities.
  - A public Kick-Off should be communicated with starting of an accompanying communication flow to express current state, success stories as well as improvement areas.
  - The Bank units are informed and committed to start the collaborative mission together.

- Model should be reached a significant level of maturity

  The chosen KM model should be used for a significant number of attempts, preferable for the tacid data ingestion as most demanding step comparable with the explicit knowledge. In a further step the add-on AI functions and algorithms should be included and well-synchronized with the KM model to gain appropriate combination results. Based on the results an iterative refinement of the models prepares for upcoming more demanding requests.

- Current data accuracy is in a well shape

  Data, in particular the belonging quality and accuracy as substantial element of information and knowledge is very crucial with a direct impact on the knowledge accuracy as such.

  That's why the quality and accuracy should be verified in a regular repeatable process with some health checkpoints in use. The accuracy of the data depends on the functional context. For example, is the correct processing information for the expiration of a derivative product more important for the trader or relationship manager than for the risk manager of a bank.

The first one has to ensure that the security related positions either executed by option or closed or expire worthless, the risk-oriented manager has to monitor a spread of risks, e.g., Legacy risks, Credit- and liquidity risks, Market risks, Operational risks.

This point of condition is mostly dedicated to the explicit knowledge and the view for accurate tacid knowledge is mostly neglected. That requested also a clean up for that group of data, information, knowledge, the Best-practice collection and KM database(s)—all these entities, models and tools as a framework are available and well-maintained.

- A focused topic or investigation area is confirmed by the Bank for transformation In some analysis sessions the Bank has to prioritize a collection of "pain points" with a predictable or expected success factor. Moreover, this scope should be open for extension in a later phase of the change process. There are some reasons to select in advance, one of them is to be efficient in the solution finding, another is to mitigate the risk of confusion caused by a mix of various topics in the same time for fixing. The most important aspect is for sure to convince the stakeholders and increase the trust about the new approach and technology.

2.3   Selected KM Model SECI

SECI model as methodical base

Some empirical research shows the relevance of knowledge sharing, knowledge combination and knowledge transfer. Many scholars have highlighted the importance of these factors for the success of KM [7, 12, 29–32]. Numerous variables in our model which measure knowledge sharing, knowledge combination and knowledge transfer in bank organizations positively influence the dependent variables to measure economic value, confirming these KM studies.

With the aid of these processes the core competency of that model is to highlight the retrieval from tacid knowledge, compare and adjust it with other knowledge units, transform it into explicit knowledge and ideally depends on each individual expert an innovation step will be reached/gained.

One of the condition for the success of that never-ending iterative process is that the human-being experts express the tacid knowledge in an open-mind manner and pro-actively. On the other hand some experts can better articulate than another one and the subjective bias can impact the accuracy of the revealed knowledge.

The interaction between tacit and explicit knowledge creates new knowledge [7]. Therefore, by simulating social processes, management must enhance the transition from tacit to explicit knowledge and vice versa by activating the iterative spiral of knowledge [33]. The four modes of the spiral of knowledge are socialization, externalization, combination and internalization [29] (Fig. 2).

In the socialization phase the tacid knowledge will be shared by discoverage with meetings, speeches, observation etc.—from the recipient point of view there are

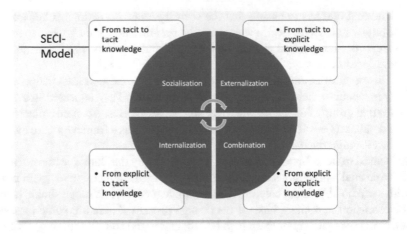

Fig. 2 SECI Model based on [7]

fugitive elements with an appearance in the moment of articulation or cognition. To make it persistent available during the externalization the tacid knowledge will be written down and structured by inductive and deductive inference—it turns to explicit knowledge.

The explicit knowledge items can be combined to new ones—the innovation step(s). With the internalization the knowledge is ready for usage—it will be operational.

As already mentioned, AI as one of the preconditions delivers a layer of technical services to support the portfolio creation in an appropriate manner.

Based on that findings and processes [5, 20, 29, 33] in a wider scope the surrounding aspects of SECI model will be considered in this article as the basic construct and sources for the AI improved KM implementation.

SECI model processes—AI embedded

The four SECI phases will be enriched with AI tasks/activities to initialise the dynamic processing of knowledge generation (Fig. 3).

The entry point to start the knowledge retrieval circle should be a step for setup of the appropriate basic procedures and elements (Fig. 4):

- Combine KM Techniques with AI algorithms
  The existing KM Techniques and Tools can be combined, supported and extended with various AI algorithms related to the category of knowledge item.

  Many of the tacit knowledge capture techniques listed below derived from techniques that were originally used in Artificial Intelligence—more specifically, in the development of expert systems. An expert system incorporates know-how gathered from experts and is designed to perform as experts do.

  But in meantime since this statement was accurate for the current time period the techniques of Artificial Intelligence did an evolution with more retrieval

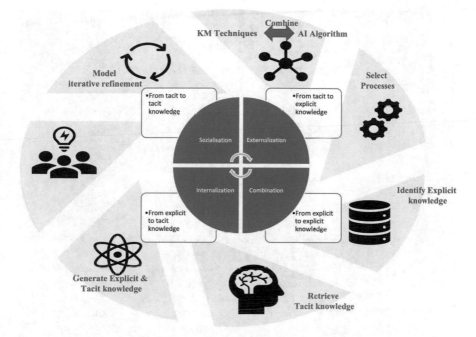

Fig. 3 Surrounding KM/AI activities (Own creation, based on [7, 20])

coverage possibilities and methods to detect knowledge [20], e.g. edge analytics and machine learning algorithm.

- Select Processes (from the process portfolio of the company)
  Identify context sensitive processes or topics for further alignment and focusing and for calibrating of the AI engine. To increase the possibility of a high quality result the processes or topics with its knowledge/information basics should be well-selected in advance, e.g. at least semi-structured, well-categorized, good data and information accuracy.
- Identify Explicit knowledge
  Along the process or topic the data sources should be detected like data lakes, data edges, databases, knowledge bases, model and process archives.
- Retrieve Tacid knowledge
  Based on the KM techniques—AI algorithm mapping the knowledge gathering of the mentioned category can be ongoing.
- Generate Explicit and Tacit knowledge
  With the AI services new or adjusted knowledge from technical aspect will be created.
- Strive for Innovation

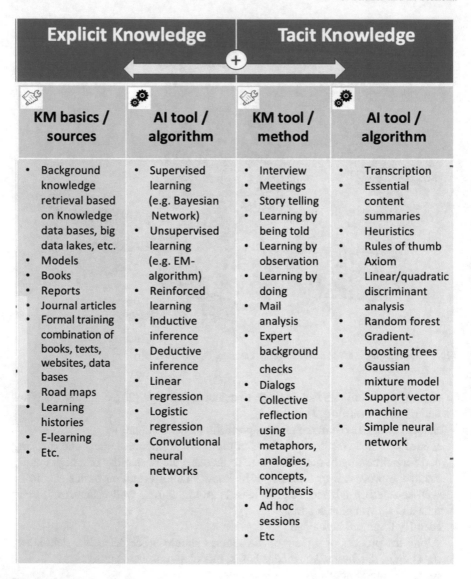

Explicit Knowledge		Tacit Knowledge	
KM basics / sources	**AI tool / algorithm**	**KM tool / method**	**AI tool / algorithm**
• Background knowledge retrieval based on Knowledge data bases, big data lakes, etc. • Models • Books • Reports • Journal articles • Formal training combination of books, texts, websites, data bases • Road maps • Learning histories • E-learning • Etc.	• Supervised learning (e.g. Bayesian Network) • Unsupervised learning (e.g. EM-algorithm) • Reinforced learning • Inductive inference • Deductive inference • Linear regression • Logistic regression • Convolutional neural networks •	• Interview • Meetings • Story telling • Learning by being told • Learning by observation • Learning by doing • Mail analysis • Expert background checks • Dialogs • Collective reflection using metaphors, analogies, concepts, hypothesis • Ad hoc sessions • Etc	• Transcription • Essential content summaries • Heuristics • Rules of thumb • Axiom • Linear/quadratic discriminant analysis • Random forest • Gradient-boosting trees • Gaussian mixture model • Support vector machine • Simple neural network

Fig. 4 Mapping extract (Own creation)

The organization, groups and finally people will be enabled to push the innovation element with the verified KM/AI results and the crossover knowledge symbiosis—it means the combination of explicit with explicit, explicit with tacid and tacid with tacid knowledge.

- Model iterative refinement
  Based on the result quality, adjustment rate and data quality, the described model with its steps will be progressed permenantly.

2.4 Portfolio Management

The Portfolio Management in common reflects the summary of all tasks which are related to decisions for capital investments [34].

Portfolio Management in Private Banks is an important management and analysis instrument to align the strategic business segments to the bank-wide strategy and controlling [35].

Strategic business segments are intrinsic homogeneous and extrinsic heterogeneous units of the general economic scope of a company—in Private Banks there are grouped by various criteria items like sociodemographic, geographic, psychographic variables.

In the figure below some impacting disciplines, processes and knowledge sources are visible—this indicates the complexity of the definition part (Fig. 5).

To have a link onwards to the bank products and services the strategic business segments combine the market segments with them [37].

The Strategic Management delivers the main strategic objectives to define the bank products and services along the sub-objectives. Based on PIMS (Profit Impact of Market Strategies)—a huge collection of data and information as baseline for an empiric definition of strategic business units (volume estimation: more than 25,000 years business experience and greater than 3,000 units delivered by more than 450 companies from various branches and countries). But the Financial services sector is only covered as a small slice of that big data lake [38]—more than 90% of the mentioned data are dedicated to the processing industry [39].

There exists some criticism on that methodology [40]:

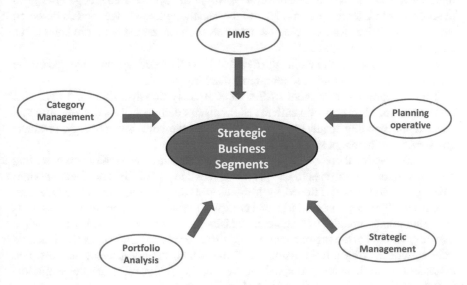

Fig. 5 Strategic business segments (Based on [36])

- Possible spurious statistical correlations: conclusion on causal relations based on correlations
- Neglection of interdependencies between explanatory variables
- Lack of suitability of used multiple regression analysis to verify complex dependency structures
- Number of data aggregation too high triggered by permanent mean calculation.

Moreover, some content-related criticism on the strategy recommendations are listed:

- One-sided focus on the success factor RoI (Return on Investement)
- Poor usage of synergy approach across the strategic business units of the same company
- Missing included branch particularities during the suggestion generation [40].

An additional point of criticism is the fact of not well-balanced industry distribution among the existing information data base—the Financial Services is only coveraged as a niche player.

This reflects the importance of knowledge for the strategy definition, the related sub-strategies, the strategic business units and segments as well as the remaining basics to initialize the portfolio matrix within the Portfolio Management discipline.

Some of the points of criticism can be compensated presumably with a major segment of underlying data and its derived models based on Tacid knowledge. In addition, the added value of these fuzzy information items can be saved after revised analysis of these artifacts as heuristics, rules of thumbs or axioms. Furthermore, in a separate research question it would be useful to verify if this "edge statements" can be linked into a synchronized architectural construct of "Special knowledge laws" and the openness to integrate that model into the remaining knowledge models based on explicit knowledge. An example of a model as such can be the heuristic knowledge cluster #1 "Market influencer".

In the bank operative planning the detail steps will be defined and reflected on the timeline of the short-, mid- and long-term planning.

This supports the alignment of the bank strategy downwards to the operative services and product portfolio with the client market demands. This dynamic process keeps a still pending request to adjust permanent the bank own strategy, products and services to gain a positive revenue.

In bank practice there is one portfolio concept invented by McKinsey consulting company among other ones in use [41]. It is labelled as GE General Electric Nine-Box Matrix and was developed by McKinsey during an engagement for General Electric in 1970 to manage a large and complex portfolio of strategic business units and to support companies take strategic decisions at the corporate level. The strategic planning tool helps the organization to prioritize its investment in different business units by considering three possible scenarios which include protect, harvest, and divest in a basic structure. Adapted on the banking sector there are the scenarios: Invest and Grow, Selectivity earnings, Harvest and Divest.

This matrix can be used by considering of two main success factors of the Financial Services industry—the market attractiveness and strengths of the product and business unit as well as the competitive strength. The market attractiveness can be determined by considering the market (size), profitability, environmental factors, clients, resources, synergy potential, pricing and labor requirements, etc. The market attractiveness is plotted on the vertical axis which includes measures of market attractiveness as high, medium, and low. The higher market attractiveness represents higher profitability which then encourages the other banks to enter this market segment.

After considering the market attractiveness, the competitive strength of the product and business unit is evaluated by analyzing the market share, average profitability, strength of the brand, customer loyalty, etc. This segment is plotted on the horizontal axis of this matrix which is also segmented into three categories i.e. high, medium, and low.

After evaluating matrix attractiveness and competitive strength, each strategic business segment is placed in the matrix which then helps to determine the strategic possibilities for each segment. Thus, the organization can increase or decrease the investment on the basis of the position of each strategic business segment [42, 42].

Comparable with other portfolio matrix concepts, e.g. the four-box matrix invented by Boston Consulting Group, this version involves better estimation of the strategic decision making as much more information is used in this type of decision tool—moreover the range of estimation factors, sub-factors are wider [44].

3 Setup Market Attractiveness/Competitive Strength Matrix for Private Bank

3.1 Example Grouped by Explicit and Tacid Knowledge Enriched with AI Support Discipline (Based on [43])

Baselines:

- Usage of the main success factors Market attractiveness and Competitive strength
- Disaggregation of the both factors into various sub factors
- Each sub factor has its own individual rating
- Consolidation of the rating items to a dedicated summary based on a scoring as well as priorisation metric model
- Usage of the Nine-Box Matrix with the classification shown-above (Fig. 6).

The setup of the Strategic Portfolio Model with its parameters, the retrieval and transition from explicit and tacid knowledge (along the SECI model) related to the Strategic Business Units as well as the placement of the Strategic Business Segments into the various recommendation boxes are summarized into the dashboard below (Fig. 7).

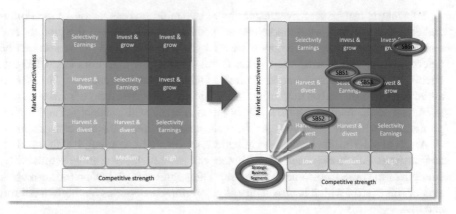

Fig. 6 Market attractiveness-/competitive strength matrix (Basic) in use for SBS placement (Own creation, based on [43])

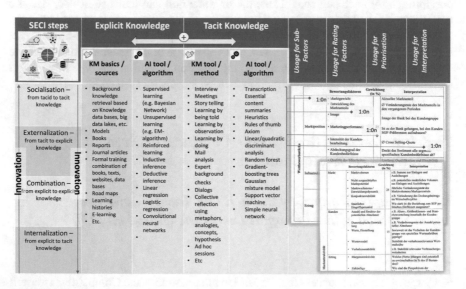

Fig. 7 Strategic portfolio model AI setup (Own creation, based on aspects of [7, 20])

This overview would be applied for a sequence of tasks to create one possible expression of a Strategic Portfolio Model for a Private Bank.

In a first step the criteria catalogue should be identified based on the fix main success factor Market attractiveness and Competitive strength as well as the Strategic Business Segments defined by the bank.

Based on the combination of KM basics and tools as well as AI algorithms for explicit and tacit knowledge as listed in figure xx the sub and rating factors, the interpretation and priorisation will be generated.

It is to emphasize that along that forward conclusion the cardinality 1:0n for the dedicated main factor to the rating factors, interpretations and priorisation exists, e.g. the Market attractiveness relates to none or many of sub factors like Market, Clients, Revenue, etc.; the Market as sub factor relates to many rating factors like Market size, remaining Market potential, Market and Financial growth.

To exploit the power of AI the backward conclusion will also proceed to gain a variety of new elements of the mentioned categories. The start point is here the list of interpretation items and follows the conclusion chain until the main success factor. With that benefit it is possible to compare the results of both approaches and verify these on plausibility and accuracy. In best case it delivers new aspects or maybe even it suggests a new main success factor.

In the next step the Strategic Business Segments will be prioritized based on the knowledge fund. Here is to highlight that particularly the tacid knowledge can deliver a new dimension of scoring. In this nine-box model the usual high score can be 3.00, but with this approach of open scoring based on AI tacid knowledge gathering and evaluation the score number can be extend the mentioned maximum. In this case for the dedicated Strategic Business Segment a concrete action recommendation cannot be given and it results in the "Question mark-box" visible in the matrix with the suggestion to analyze the segment in a deep dive session (Fig. 8).

The critical points of the traditional McKinsey model will be flattened by this approach to combine each knowledge item or its inflexible ingestion process, strong limited generation of new elements with an appropriate AI algorithm (Table 1).

Based on the explained construct in a short preview it will be listed some expansion options.

3.2   Order Trading, Especially High Frequency Trading

The combination of AI core engine functionalities, the knowledge models, data edges, SAA (Strategic Asset Allocation) as well as TAA (Tactical Asset Allocation) offers as construct a proper opportunity for high frequency trading.

According the recommendation results out of the Portfolio Matrix the AI engine can generate via Straight-through-processing a huge amount of trading orders, put it into a dedicated orderbook (e.g. Emerging Markets, Precious Metal Gold funding, etc.) for review by the portfolio or relationship manager. One of the possible algorithms is deep reinforcement learning as special customized function for high frequency trading [46].

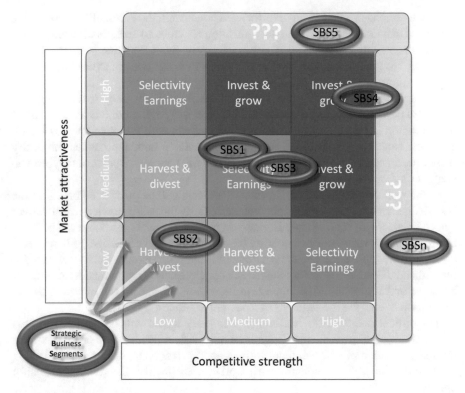

Fig. 8 Market attractiveness-/competitive strength matrix extended (Own creation, based on [43])

3.3 Risk Management Within Banking

This supervision function within Banking should also include into the portfolio management process as described in this publication. It delivers a consolidated solution on a higher level with more strategic data, information and knowledge insights as an add-on for risk aspects like Legacy risks, Credit- and liquidity risks, Market risks, Operational risks, etc.

3.4 Simulation/Portfolio Revision by AI

Portfolio Management is one process like other company business processes with its drill-down sub processes. The AI offers the possibility to simulate the operative (planned) process with another subset of data/information/knowledge as parameter variables with the objective to compare the generated output with the previous one.

Table 1 Mitigation Portfolio weaknesses by AI

Weaknesses of the traditional model [43]	AI remedy tasks
Limited and stiffed selection of the main success factors	Via forward/backward conclusion as well as the listed methods and algorithms the "open-mind" for new generated and recommended factors is given
Limited and stiffed priorisation and rating score	See above, in addition some special functions deliver improved results, e.g. adapted Net Promoter Score, etc
Hard linked combination of the main success factors	See above, in addition a variety of factors can be delivered
Strategic planning tended to become a mechanical character	With the explained approach the possibility for simulations of switched parameters is given, e.g. the main success factors can be replaced by another one or the generated values and numbers triggers the adjustment of the Strategic Business Segments and vice versa
Only focus presence and past oriented	The approach offers some predictive analytics methods and algorithms to include the future perspective
Less flexibility during the process at changes on environment factors	With edge computing it is possible to switch between the process-related data/factor sources
Partly wrong initialisation of causal dependencies of strategic business segments based on unconscious bias [45]	The bias impact factor can be considered and separated especially during the AI training/learning sessions
High uncertainty caused by the big proportion on tacid knowledge	The consideration of tacid knowledge takes a big part and is one of the core competencies of this described approach

One simulation approach, called "Digital twins" or Mirroring can be realized with the usage of AI process and data mining. This technical feature helps to improve the prediction results and reduce the error rate of conclusions [47–49].

4 Conclusion

4.1 Synopsis

The knowledge with its discipline Knowledge Management is fundamental to deliver an added-value for the creation of Private Banking strategies reflected on the placements of the Strategic Business Segments into an appropriate Portfolio model like the Nine-box matrix invented by McKinsey. On top of that a remarkable progress on the evolution of the Bank mission and strategy can be supported by the AI/ML techniques whilst above all the retrieval of Tacid knowledge performs on higher level comparable with the results out of the traditional KM techniques, methods and models. Moreover, it can be stated that after completion of that process many opportunities are given to expand it for a seamless order trading creation, including of risk management as well as simulation similar to a portfolio revision and re-balancing, but with the usage of various models, parameters and strategy figures.

4.2   Further Research

A lot of mentioned and only-touched topics delivers interesting research questions for further creation of publishing items as follows.

- The variation of the Strategic Business items as basic elements for the Strategic Business Segments.
- The Portfolio model can be combined with others with the proceeding of forward or backward conclusion.
- The Strategic model can be multiplied into operative models for product, market, services cluster.
- The simulation of various scenarios can be designed high sophisticated in terms of timestamp focusing—retrospective back to the past, forward thinking with prediction character.
- Generated business recommendations variation based on the shift from explicit knowledge to tacid and vice versa.
- High frequency trading based on data edges of the various sources.
- AI Knowledge Management as a Service for Fintechs as bank-like constructs without banking license.

In near-/midterm future it expands the possibility to gather the knowledge from more external sources too, which opens to participate on the crowdsourcing of a market or industry segment, e.g. to include PIMS as external reference or benchmark it would be necessary to enlarge and improve the part of Financial Services models with the AI technology.

From the audit and security perspective a dedicated artefact should be designed in future to increase the trust on this technology and makes the solution and the generation procedure transparent, in addition it offers the development of a certification branding process.

References

1. Schulz, B.: Mit KI bewältigen Banken das Gold der Neuzeit (2019). Retrieved by https://www.springerprofessional.de/bank-it/kuenstliche-intelligenz/mit-ki-bewaeltigen-banken-das-gold-der-neuzeit/16916438

2. Moore, S.: How to stop data quality undermining your business (2018). Retrieved by https://www.gartner.com/smarterwithgartner/how-to-stop-data-quality-undermining-your-business

3. Chien, M., Jain, A.: Magic quadrant for data quality solutions (2021). Retrieved by https://www.gartner.com/doc/reprints?id=1-27L65HH8&ct=211004&st=sb

4. Statista: Was sind für Ihr Unternehmen die wichtigsten Herausforderungen bei der Datenhaltung und im Daten-Management? (2022); Source: Deutschland; IDC; September 2012; Retrieved by https://de.statista.com/statistik/daten/studie/257898/umfrage/umfrage-bez ueglich-der-wichtigsten-herausforderungen-beim thcma-big-data/

5. Greguš, M., Monika, D.: Slide Deck Knowledge Management. Comenius University Bratislava (2021)

6. Fotache, M.: Knowledge Management between FAD and Relevance, Alexandru Ioan Cuza University of Iaşi, Romania (2006)

7. Nonaka, I., Takeuchi, H.: The Knowledge Creating Company. Oxford University Press (1995)

8. Kabir, N.: Thesis: The Impact of Semantic Knowledge Management System on Firms' Innovation and Competitiveness, Newcastle University Business School (2017)

9. Apurva, A., Singh, M.D.: Understanding Knowledge Management: a literature review. Int. J. Eng. Sci. Technol. (IJEST) 3(2) (2011)

10. Polanyi, M.: The Tacit Dimension. The University of Chicago Press (2009)

11. Maravilhas, S., Martins, J.: Strategic knowledge management in a digital environment: Tacit and explicit knowledge in Fab Labs, in Elsevier. J. Bus. Res. 94 (2019)

12. Goh, S.C.: Managing effective knowledge transfer: an integrative framework and some practice implications. J. Knowl. Manag. (2002)

13. Matt, H.: How machine learning is changing financial services (2020). Retrieved September 21, 2020, https://www.fintechmagazine.com/fintech/how-machine-learning-changing-financial-services?utm_source=ONTRAPORT-email-broadcast&utm_medium=ONTRAPORT-email-broadcast&utm_term=&utm_content=Data+Science+Insider:+April+17th,+2020&utm_cam paign=18042020

14. Litzel, N., Anja, K.: KI Machine Learning = Neue Chancen Für Das Datenmanagement (2020). Retrieved September 21, 2020 https://www.bigdata-insider.de/ki-machine-learning-neue-chancen-fuer-das-datenmanagement-a-907696/?cmp=nl-274&uuid=0218FE69-D4B0-4E0E-A65B-979657C271C3

15. FSB - Finance Stability Board: Artificial Intelligence and Machine Learning in Financial Services - Market Developments and Financial Stability Implications (2017). [Brochure]. Retrieved September 21, 2020 Retrieved from www.fsb.org/wp-content/uploads/P011117

16. BaFin: 'BaFin Perspektiven 01 / 2019' (2019). Retrieved by https://www.bafin.de/SharedDocs/Downloads/DE/BaFinPerspektiven/2019

17. Ali, H.M., Ahmad, N.H.: Knowledge management in Malaysian banks: a new paradigm. J. Knowl. Manag. Pract. 7(3), 1–13 (2006)

18. Campanella, F. et al.: Knowledge management and value creation in the post-crisis banking system. J. Knowl. Manag. 23(2) (2019)

19. Frost, J.: Knowledge Management (2011). Retrieved by https://wirtschaftslexikon.gabler.de/definition/wissensmanagement-47468

20. Dalkir, K.: Knowledge Management in Theory and Practice, Elsevier, Butterworth Heinemann (2005)

21. Knaese, B.: Das Management von Know-how-Risiken: eine Analyse von Wissensverlusten im Investment Banking einer Großbank, Deutscher Universitäts-Verlag (2004)

22. OECD: Recommendation of the Council on Artificial Intelligence (2019). Retrieved from https://legalinstruments.oecd.org/en/instruments/OECD-LEGAL-0449

23. World Trade Organisation: World trade report 2018 [Brochure]. Various, New York, NY (2018)

24. Eberspächer, J., Goetz, T.: Maschinen Entscheiden Vom Cognitive Computing Zu Autonomen Systemen. Knecht-Druck (2014)

25. Brownlee, J.: Gentle introduction to the Adam optimization algorithm for deep learning (2017). Retrieved by https://machinelearningmastery.com/adam-optimization-algorithm-for-deep-lea rning/

26. Kammerdiner, A.R.: Bayesian Networks in Encyclopedia of Optimization, Springer Link (2009)
27. Surbhi, A.: Supervised vs Unsupervised vs Reinforcement (2020). Retrieved by https://www.aitude.com/supervised-vs-unsupervised-vs-reinforcement/
28. Sträßer, J.: Test data generation based on artificial intelligence impacts Banking IT. In: Proceedings of the 36th International Business Information Management Association Conference (IBIMA) (2020)
29. Nonaka, I. et al.: SECI, Ba and Leadership: a Unified Model of Dynamic Knowledge Creation, Elsevier Science (2000)
30. Beijerse, R.P.U.: Questions in Knowledge Management: defining and conceptualizing a phenomenon. J. Knowl. Manag. (1999)
31. Labedz Chester, S. et al.: Interactive knowledge management: putting pragmatic policy planning in place. J. Knowl. Manag. (2011)
32. McGurk, J., Baron, A.: Knowledge management – time to focus on purpose and motivation. Strat. HR Rev. (2012)
33. Nonaka, I.: A Dynamic Theory of Organizational Knowledge Creation, retrieved by https://doi.org/10.1287/orsc.5.1.14(1994)
34. Poddig, T., Brinkmann, U., Seiler, K.: Portfolio Management – Konzept und Strategien, Uhlenbruch Verlag (2004)
35. Newey, L.R., Zahra, S.A.: Portfolio Planning: A Valuable Strategic Tool in The Palgrave Encyclopedia of Strategic Management, Reference Work, Springer Link (2018)
36. Unknown (2018, February 14): Revision von strategisches Geschäftsfeld retrieved by https://wirtschaftslexikon.gabler.de/definition/strategisches-geschaeftsfeld-45861/version-269149
37. Horsch Andreas (2020, April 01): Revision von Strategisches Geschäftsfeld (SGF). Retrieved by https://www.gabler-banklexikon.de/definition/strategisches-geschaeftsfeld-sgf-61667/version-376175
38. McNeely, C., Schintler, L.A.: Big Data Concept in Encyclopedia of Big Data, Reference Work, Springer Link (2022)
39. Pedram, Farschtschian: Private Equity für die Herausforderungen der neuen Zeit: Strategische Innovation für das Funktionieren von Private Equity im 21. Jahrhundert, Campus Verlag (2010)
40. Homburg, C.: Quantitative Betriebswirtschaftslehre: Entscheidungsunterstützung durch Modelle. In: Homburg, Christian. 3. Auflage, Springer Verlag (2020)
41. Tanew-Illitschew, G.: Portfolio-Analyse als Instrument strategischen Bankmanagements, in ÖBA, 30th publishment (1982)
42. Speaking Nerd (2021, November 18): In-depth analysis of usance of GE-McKinsey Matrix, retrieved by https://speakingnerd.com/strategy/strategy-models/ge-mckinsey-matrix
43. Schierenbeck, Henner (2003, September 01): Ertragsorientiertes Bankmanagement, Band 1: Grundlagen, Marktzinsmethode und Rentabilitäts-Controlling, Gabler Verlag
44. Gramlich, S. et al.: Arbeitspapier Nr. 59, Portfolioanalysen im Vergleich, Hochschule Offenburg, Fakultät Medien und Informationswesen (2021)
45. Herm, L.V., Janiesch, C., Fuchs, P.: Der Einfluss von menschlichen Denkmustern auf künstliche Intelligenz – Eine strukturierte Untersuchung von kognitiven Verzerrungen. HMD **59**, 556–571 (2022). https://doi.org/10.1365/s40702-022-00844-1
46. Briola, A. et al.: Deep Reinforcement Learning for Active High Frequency Trading, Cornell University (2021). Retrieved by https://doi.org/10.48550/arXiv.2101.07107
47. Barenkamp, M.: Künstliche Intelligenz als Unterstützungsfunktion der Vorhersage und Prozessexzellenz in Process Mining, Wirtschaftsinformatik & Management 2022/14 (3), Springer Gabler (2022). Retrievable by https://doi.org/10.1365/s35764-022-00404-8
48. Gupta, M.K., Chandra, P.: A Comprehensive Survey of Data Mining. Springer (2020). Retrieved by https://doi.org/10.1007/s41870-020-00427-7
49. Hassani, H. et al.: Digitalisation and Big Data Mining in Banking, MDPI (2018). Retrieved by https://doi.org/10.3390/bdcc2030018
50. Batarseh, F.A.: Big Data Concept in Encyclopedia of Big Data, Reference Work, Springer Link (2022)

51. Gregus, M., Kryvinska, N.: Service Orientation of Enterprises—Aspects, Dimensions, Technologies. Comenius University in Bratislava (2015). ISBN: 9788022339780
52. Kryvinska, N., Gregus, M.: SOA and its Business Value in Requirements, Features, Practices and Methodologies. Comenius University in Bratislava (2014). ISBN: 9788022337649
53. Peracek, T., Srebalova, M., Brestovanska, P., Mucha, B.: Current possibilities of contractual securing the obligations by creating lien on securities. In: 34th International-Business-Information-Management-Association (IBIMA) Conference, pp. 4114–4122 (2019)
54. Bohdal, R., Bohdalová, M., Kohnová, S.: Using genetic algorithms to estimate local shape parameters of RBF RBF. In: WSCG 2017: 25. International Conference in Central Europeon Computer Graphics, Visualization and Computer Vision: Short Papers Proceedings. - Plzeˇn: Vaclav Skala - UNION Agency, S, pp. 59–66 (2017)
55. Vanˇcová, M.H., Ivanochko, I.: Factors behind the long-term success in innovation—focus multinational IT companies. In: Kryvinska, N., Poniszewska-Maraˊnda, A. (eds.) Developments in Information & Knowledge Management for Business Applications. Studies in Systems, Decision and Control, vol. 376. Springer, Cham (2021). https://doi.org/10.1007/978-3-030-76632-0_16
56. Mabey, C., Zhao, S.: Managing five paradoxes of knowledge exchange in networked organizations: new priorities for HRM? Hum. Resour. Manag. J. 27(1) (2017)
57. Al-Emran, M. et al.: The impact of knowledge management processes on information systems: a systematic review. Int. J. Inf. Manag. (2018). Retrieved by https://doi.org/10.1016/j.ijinfo mgt.2018.08.001
58. Kharote, M., Kshirsagar, V.P.: Data mining model for money laundering detection in financial domain. Int. J. Comput. Appl. (2014)

Duties of National Civil Courts and European Commission in Their Cooperation

Rastislav Funta and Matej Šebesta

Abstract The scientific article analyses the degree to which national civil courts and the European Commission must collaborate under primary and secondary legislation. First, it is determined to what extent national civil courts have the authority to request statements and information from the European Commission, whether and to what extent the European Commission is required to respond, and to what degree the national courts are required to consider the response. The role of the European Commission is then considered, insofar as it can appear before national courts as a so-called amicus curiae with self-induced statements; specifically, the right of the European Commission to make such an appearance and the duty of national courts to take such statements into account are examined. Finally, the extent to which national courts are required to submit judgements to the European Commission is explored. The duties (and rights) outlined here are examined in more detail below and placed in the broader context of primary law.

1 Introduction

The European Commission is obligated to cooperate with national courts under the principle of loyalty. Based on the principle of loyalty, the ECJ held in the Delimitis/Henninger Bräu C-234/89) [35] and SFEI C-39/94 [36] rulings that national courts might always question the European Commission about the status of procedures conducted by it [6]. National courts could also approach the European Commission with questions about EU competition law interpretation if "the precise application of Article 85(1) or Article 86 presents particular problems" or to receive

R. Funta (✉)
Janko Jesensky Faculty of Law, Danubius University, Richterova č. 1171, 925 21 Sládkovičovo, Slovak Republic
e-mail: rastislav.funta@vsdanubius.sk

M. Šebesta
Faculty of Social Studies, Danubius University, Richterova č. 1171, 925 21 Sládkovičovo, Slovak Republic

© The Author(s), under exclusive license to Springer Nature Switzerland AG 2023   401
N. Kryvinska et al. (eds.), *Developments in Information and Knowledge Management Systems for Business Applications*, Studies in Systems, Decision and Control 462, https://doi.org/10.1007/978-3-031-25695-0_18

economic or legal information. The EU legislative has basically embraced this understanding of the loyalty obligation in secondary law for antitrust law as per Art. 15 (1) Regulation 1/2003 [41] and state aid law as per Art. 29 (1) Regulation 2015/1589 [43]. Furthermore, for both areas of secondary law, the European Commission was granted the right to appear as an amicus curiae before national civil courts and to request corresponding documents from the courts' files as per Art. 15 (3) sentences 3 and 4 of Regulation 1/2003 and Art. 29 (2) Regulation 2015/1589. The requirement of Member States to provide a copy of judgements on the application of Art. 101, 102 TFEU [17, 18] to the European Commission is only specified for antitrust law [15], not for state aid or merger control law as per Art. 15 (2) Regulation 1/2003. The aim of our scientific study is to analyse the duties of national civil courts and European commission in their cooperation. The duties (and rights) outlined here are examined in more detail below and placed in the broader context of primary law [10].

Our research question and objective are whether national civil courts and the European Commission must cooperate under EU primary and secondary legislation. The data was collected from legislation, respective case law through in-depth document analysis and supported by scientific literature. We want to achieve our objective, particularly through study of legislation, scientific literature, as well as case law. Several scientific methods of knowledge have been used in the exploration and development of our paper. We applied the method of analysis in order to examine the state regarding our research question. The synthesis will allow us to combine partial information into a single unit. By applying a comparative method, we also make different view from EU primary law and secondary law perspective. This allowed us to obtain reliable and valid conclusions and results.

2 Obligation of National Courts and the European Commission

The case law gives the national courts the opportunity to address questions to the European Commission, and the Union legislator has confirmed this in secondary law as per Art. 15 (1) Regulation 1/2003 and Art. 29 (1) Regulation 2015/1589 [37, 39]. The European Commission also encourages national courts to make such requests, also known as "minor referral procedures". A mere right of the national courts to ask questions is not surprising in view of the clearly formulated obligation to cooperate under Art. 4 (3) Treaty on European Union (TEU). Rather, the circumstances under which national courts are allowed to make such a request is decisive. Because Art. 15, 16 Regulation 1/2003 and Art. 28 Regulation 2015/1589 do not answer this matter, it is clarified primarily on the basis of primary law. Another deciding issue is whether the European Commission is required to respond [16]. Finally, there is the issue of national courts' obligation to consider relevant answers from the European Commission when making conclusions.

2.1 Obligation of National Courts to Consult the European Commission

(a) A duty to consult

Consultation with the European Commission is primarily up to the national courts. However, the principle of legal certainty [20] specifies the goal of a uniform assessment of the same facts, because this leads to the stability of legal relationships between economic operators and thus to a stable foundation for the economy in the internal market [19, 22, 24]. Art. 4 (3) of the Treaty on the European Union (TFEU) requires member states to refrain from taking any actions "that could undermine the fulfilment of the Union's goals" and to take all necessary steps to fulfil contractual obligations. Furthermore, the Member States have already delegated to the European Commission a central role in the application of EU competition law [8]. Art. 17 TEU highlights the European Commission's responsibility as guardian of the treaties. However, it must be taken into account that the national courts (as functional Union courts) are not subordinate to the European Commission and, unless the European Commission has sufficient intention to reach a decision, there is no obligation to take them into account in the sense of adopting the opinion of the European Commission. A consultation with the European Commission is also usually a time-consuming process for the national courts, which is not reflected in national procedural law or only to a limited extent and leads to a delay in the court proceedings [12]. If the European Commission already has sufficient intention to make a decision, the national court may not rule contrary decisions. A consultation with the European Commission may be required for the national court in order to fulfill the obligation arising from the loyalty requirement of Art. 4 (3) TEU. As far as a sufficient intention of the European Commission is not (certainly) recognizable, further differentiation is to be made according to the opinion represented here. Obliging national courts to consult the European Commission without any indication that the European Commission is dealing with the same facts would not only contradict the idea of procedural autonomy of the courts, which are called upon to apply Union law independently, but would also be detrimental to effective enforcement of Union competition law and would run counter to the requirement to speed things as stipulated in Art. 47 (2) of the Charter of Fundamental Rights [11, 21, 22, 26, 32]. Finally, there is an obligation to consult if there is already a decision by the European Commission, but there are doubts about the identity of the facts or its scope or correctness, for example, if the court believes that the facts have changed significantly since the European Commission's decision C-574/14, PGE Górnictwo [40]. The same is true if there is a notification from the European Commission, provided that the facts are same or sufficiently similar and the circumstances do not allow for any variation. In these instances, Art. 267 TFEU to the ECJ is open or even essential for disputes about the validity and interpretation of Union law. However, in cases where the court considers a deviation from soft law in the form of a notification, the obligation to consult or refer to the ECJ [28] is not as obvious, because such a notification usually has less authority, as it is often formulated in an abstract manner and thus often does not exactly match the situation under consideration.

(b) Determination of facts and right of application

The above-mentioned obligation of the national courts to take into account the assessment of the European Commission in the case of factual identity or sufficient similarity of facts (depending on the circumstances) results from primary law. It can also be argued that it is in line with the principle of loyalty in conjunction with the idea of legal certainty [21], if facts identified by the European Commission and on which a final decision is based are decisively taken into account by the civil courts, since this corresponds to the conflict avoidance function of the loyalty requirement [5]. In this respect, in the event of ambiguity about the scope of the determination of the facts or the factual identity, a consultation with the European Commission may be obvious or even mandatory. It can be argued that a determination of the facts by a national civil court by means of consultation with the European Commission, particularly in the case of damages proceedings based on EU competition law, can contribute to its effectiveness. Prior to the adoption of Regulation 1/2003 and the implementation of today's Art. 29 Regulation 2015/1589, the ECJ similarly concluded in Delimitis/Henninger Bräu C-234/89 [35] that the right of national courts to ask questions should only apply if it "proves to be compatible with the national procedural provisions." Art. 4 (3) TEU, Art. 15 (1) Regulation 1/2003, and Art. 29 (1) Regulation 2015/1589 may grant courts the authority to request information and opinions from the European Commission. Although the court may receive "legal views" at any moment, it may only request "information about facts, particularly market conditions, from the Commission" to the extent permitted by procedural legislation. Furthermore, neither primary law nor Art. 15 Regulation 1/2003 or Art. 29 Regulation 2015/1589 intervene in the adversarial procedure insofar as the civil courts would be given the opportunity to change the regulatory reality by issuing a specific decision apply to the European Commission. However, EU law does not give the courts any power to do this. The right to ask questions according to Art. 15 Regulation 1/2003 or Art. 29 Regulation 2015/1589 is directed at comments and information, but not to the right to request or even suggest a decision by the European Commission.

A secondary legal innovation results from Art. 5, 6 Directive 2014/104/EU of the European Parliament and of the Council of 26 November 2014 on certain rules governing actions for damages under national law for infringements of the competition law provisions of the Member States and of the European Union, in conjunction with. Recital (15). According to this, the Member States must enable the national courts to order the disclosure of evidence upon (sufficiently substantiated) request of a plaintiff in a damages action on the basis of Union antitrust law and this also vis-à-vis the European Commission. However, it is envisaged that national courts will conduct a strict proportionality test before ordering; the order may be subject to judicial review. Thus, the Directive 2014/104/EU implies that the possibility of a disclosure order by the national court vis-à-vis the European Commission on the basis of national law is condensed into an obligation to do so. A distinction must be established when answering the question of whether national civil courts are required to consult the European Commission. If there are indications of sufficient intention to make a decision, this can trigger an obligation to consult. The same is true if there are questions concerning the basis, scope, or accuracy of a decision. The same is also true if there are informal comfort letters or notices indicating that the European Commission views a certain case differently than the national court.

2.2 Obligation of the European Commission to Respond to a Request from a National Court

It is unclear whether the European Commission feels obliged to respond to inquiries from national courts. In fact, it is hardly conceivable that the European Commission would reject an obligation to provide an answer in principle. At the very least, the responsibility of national courts to pose inquiries to the European Commission must correspond to an obligation of the European Commission to respond. However, this obligation cannot exist indefinitely. In the following, this will first be examined from a purely primary law perspective, followed by a secondary law consideration. The principle of loyalty is formulated in the sense of an equal, mutual obligation to cooperate. Art. 4 (3) TEU obliges the Member States and the Union to support each other, so it also addresses the European Commission in this respect. The European Commission and the national courts are thus part of a "functional unit between the Union and the Member States" for which the reciprocity element [31, 32] is anchored in Art. 4 (3) TEU. Especially when enforcing the Union's competition law [15], the European Commission can dispose of information that only it possesses. Insofar as a national court is obliged to determine whether proceedings before the European Commission have progressed to such an extent that this must be taken into account in the decision-making process, the national court's obligation to make inquiries must go hand in hand with an obligation on the part of the European Commission to respond, since otherwise the national courts will not be able to fulfil their obligation to take any relevant consideration into account. The same applies if a court is obliged to obtain information from the European Commission. The European Commission's obligation to cooperate also includes (to this extent) the transmission of documents and the approval of testimonies.

However, in the First and Franex (C-275/00) [38] decision, the ECJ rightly emphasized that compelling reasons can speak against the transmission of information to the national courts. In particular, aspects of the European Commission's self-organization and the protection of professional and business secrets come into consideration here. The European Commission's right to organize itself is drawn from Art. 13 TEU and Art. 249 (1) TFEU. The ECJ also acknowledged information protection in the case that the operation and independence of the Union, including the execution of the European Commission's (allegedly) functions, were jeopardized. For example, it is not clear if the loyalty requirement in relation to the proper application of Union competition legislation entails a duty to share information, requiring the European Commission to submit internal notes [14]. According to Art. 339 TFEU, the organs of the Union and their Members are obligated not to disclose "information that by its very nature is subject to professional secrecy; this applies in particular to information about companies and their business relationships or cost elements." Even if Art. 339 TFEU does not exclude the publication of material in principle, the provision emphasizes once more that the European Commission's commitment to assist under fundamental law has limits. If the European Commission has a duty to respond to inquiries from the national courts in accordance with the principle of loyalty in primary law [2], this is limited to information that the national civil court really needs in the context of its adversarial proceedings. In this respect, it must also be taken into account that within the scope of Art. 267 TFEU, a referral procedure at the ECJ may be available as an alternative. However, this is only a question of interpretation and validity. In this respect, it seems conceivable that the European Commission will refer the national court to the ECJ. Outside of the situations where a national court is required by Union law to inquire about from the European Commission, the duty of loyalty (only) requires the European Commission to seriously consider an answer, but not to provide an answer in every case. When deciding whether to refuse an answer, reasons relating to the effectiveness of the administrative action of the European Commission within the framework of its right to organize itself can also play a role here, which, however, must not become a blanket declaration for the refusal of information.

At this point we may ask as to whether Art. 4 (3) TEU (or Art. 15 (1) Regulation 1/2003) [4] results in an obligation for the European Commission to identify information that it does not have at the request of the court. The European Commission does not "per se" refuse to "gather information at the request of a national court". However, the European Commission has no duty to act as the investigative body of the national courts. With Art. 15 (1) Regulation 1/2003 and Art. 29 (1) Regulation 2015/1589, the Union legislator did not shift the balance towards an extended duty of the European Commission to reply. It is true that, at least for state aid law, it was the aim of the Union legislator that the courts of the Member States have a right to information from the Commission for the application of Art. 107 (1) and Art. 108 TFEU as well as to opinions from the Commission on questions concerning the application of the state aid rules. While this secondary law does not contain an extended obligation on the part of the European Commission to provide a response, it defines and limits the protection of professional secrets of individuals to whom the information is provided in Art. 28 (2) of Regulation 1/2003 in the event that data are to be assigned. Recital (11) of Regulation 734/2013 (generally) [42] states that the European Commission "shall have due attention for the legitimate interests of enterprises in the safeguarding of their commercial secrets" in state aid proceedings. However, the law on state aid does not include a method for protecting company secrets when information is transmitted to the courts. Insofar as the national courts have an obligation to make inquiries to the European Commission, the European Commission is also obliged to provide an answer, unless compelling reasons of maintaining its functioning or of Art. 339 TFEU stand in the way. In the other cases, the duty of loyalty results in the European Commission's obligation to seriously consider an answer. In this respect, however, it has discretion under primary law to refuse an answer. In particular, it must weigh up whether the forwarding of information and documents or the refusal to do so serves the effective enforcement of Union law [1].

There is a question as to whether Article 4 (3) TEU (or Article 15 (1) Regulation 1/2003) imposes an obligation on the European Commission to identify information not available to it at the request of the court. According to our view, the European Commission does not "per se" refuse to obtain information at the request of a national court. However, the case law on Article 4(3) TEU (or its predecessor provisions) does not indicate that, from the ECJ's perspective, the European Commission would have a duty to act as an investigative body of national courts. For the adversarial proceedings, it should be added that official investigations by the national civil court are hardly relevant anyway and an official investigation in favor of one or the other party does not fit in with this either. It is not evident that the loyalty requirement interferes with the principles of adversarial proceedings in this respect. According to the opinion expressed here, the loyalty requirement does not go so far as to make the European Commission the information-gathering authority for the national civil courts deciding in the adversarial proceedings, because this would not only mean a significant shift in the areas of responsibility, but also a turning away from the adversarial proceedings. Against the background of the primary law requirement of acceleration, however, it must be demanded of the European Commission that it endeavors to provide a timely response so that the court proceedings can also be continued in a timely manner, even if no fixed deadlines apply for this purpose [35].

With Art. 15 (1) Regulation 1/2003 and Art. 29 (1) Regulation 2015/1589, the Union legislator did not shift the balance towards an extended duty of the European Commission to reply. It is true that, at least for state aid law, it was the goal of the Union legislator that the courts of the Member States have a right to information from the Commission for the application of Article 107(1) and Article 108 TFEU as well as to opinions from the Commission on questions concerning the application of the state aid rules. However, it cannot be inferred from this that the obligations of the European Commission should go beyond the scope established here for primary law and that, for example, interested parties should be given a further, albeit indirect, possibility of access to information from the Commission in addition to access to the file, because the actual regulatory text, starting from Regulation 1/2003 up to Regulation 2015/1589, was developed as a right to request information, not as an obligation of the European Commission to actually respond to such requests. While this secondary legislation thus already does not contain an extended obligation of the European Commission to respond, it emphasizes the protection of the professional secrets of those to whom the information is to be attributed in a clarifying and limiting manner in Art. 28(2) Regulation 1/2003 for the case of the intended disclosure of data. Recital (11) Regulation 734/2013 emphasizes (in general terms) that the European Commission should take due account of the legitimate interest of undertakings in the protection of their business secrets in state aid cases. However, a procedure for the protection of business secrets in the transmission of information to the courts is not provided for in state aid law.

The right to effective legal protection, in particular to protect business and professional secrets, requires that the person concerned be able to defend effectively against the disclosure of his secrets to third parties in order to prevent damage from the outset caused by the disclosure of such secrets. The ECJ therefore stated early on, in the AKZO decision [34], that the European Commission must hear the person concerned before disclosing such secrets and give him the opportunity to seek legal protection before the Union courts. This requires the European Commission to issue a decision on disclosure, which the party concerned can then challenge under Article 263 TFEU and, if necessary, thus prevent the disclosure of the trade secrets to a court. However, it should be noted that the AKZO decision did not concern the disclosure of data to a court, but to another company. It is true that the protection of secrets is in principle the responsibility of the European Commission. However, there is nothing to prevent the disclosure to the courts if the national courts protect the secrets. However, the ECJ also assumed that the European Commission may refuse to disclose secrets to a national court in exceptional cases, that the parties concerned must be heard beforehand and that a (contestable) decision must be taken before disclosure. In summary, it can be stated that the data subject, on the basis of the legal framework described above, must have the possibility to take action against the disclosure of (alleged) business and professional secrets by the European Commission. To this end, the person must be heard before information and documents are disclosed and the European Commission must adopt a decision on the disclosure of the data, which can be challenged before the EU courts prior to enforcement. This result, essentially found on the basis of primary law, does not mean, however, that the Union legislator is not free to create secondary law regulations that extend the European Commission's obligation to disclose information, insofar as this remains within the framework of primary law. In this respect, it is also conceivable that effective legal protection with regard to the right of the persons concerned to maintain the confidentiality of their business and professional secrets could be shifted to the national courts [15, 18, 27].

According to the view taken here, the European Commission also has the possibility, within the framework of the primary and secondary legal requirements identified in above, to exempt documents and information from disclosure to national courts if they contain confessed admissions by a cartel offender under a leniency program or were submitted by the cartel offender under that program. The ECJ correctly pointed out that the transmission of such documents to potentially injured parties could impair the effectiveness of these programs. Moreover, the European Commission could be prevented by its self-binding nature from disclosing documents received in the context of a leniency program. To the extent that the ECJ argued for a balancing between the interests justifying the communication of the information and the protection of that information voluntarily provided by the leniency applicant on the issue of the disclosure of such documents, this fits with the result found above, according to which the European Commission is not in principle bound to reply, but considerations relating in particular to the effectiveness of the enforcement of Union law may also speak against a reply.

According to the opinion expressed here, these principles apply in the relationship between the European Commission and national courts in adversarial civil proceedings as follows: The principle of provision in adversarial proceedings enables and obliges the potentially injured party to seek the information itself from the European Commission and the latter must, vis-à-vis this applicant, balance the various interests, i.e. the applicant's interest in asserting its civil law claims based on Union competition law, the public interest of the Union in enforcing Union competition law and the secrecy interests of the undertakings which are the subject of the proceedings before the European Commission. It is not evident that primary law intended to interfere with these principles and to facilitate access to files of the European Commission by obliging the European Commission vis-à-vis the national civil courts to disclose files for the purpose of ascertaining. Therefore, vis-à-vis the national court, the public interest in the effectiveness of the national leniency program identified by the ECJ will already often suffice to refuse the transfer of information and documents. It is not apparent, also with regard to submissions by leniency applicants in cartel proceedings and documents submitted by them, that primary law intended to interfere with the principles of adversarial civil proceedings laid down by national legal systems. Nor does Regulation 1/2003 indicate anything in this regard. Against this background, the hurdles for a refusal to hand over such documents by the European Commission to the civil courts are low under primary law in the sense that it is at the discretion of the European Commission whether the refusal or the handing over of the files is more conducive to the effective enforcement of Union antitrust law. If the European Commission decides in favor of disclosure, the parties concerned must in any case be heard and, if necessary, must be given the opportunity to seek legal protection before the European Union courts; the European Commission must adopt an appealable decision in this regard. However, as already explained above, this does not mean that the Union legislator is not free to create secondary legislation that limits or extends the European Commission's obligation to disclose information, provided this remains within the framework of primary law. The concept of effectiveness prohibits a general exclusion of access to leniency statements. However, the ECJ's view is to be understood in such a way that it did not intend to anticipate secondary law regulations, because the ECJ explicitly refers to a lack of Union law regulation. Since leniency programs are not regulated by primary law, it is up to the Union legislator to decide how to balance effective enforcement of Union (antitrust) law through the effectiveness of these programs (which argues for a high level of leniency protection) and effective enforcement of Union (antitrust) law by way of damages actions by civil courts, which argues for facilitated access to leniency documents held by the European Commission. Particularly in the case of the leniency program practiced by the European Commission with reference to Union antitrust law, the special feature applies that the companies acting as leniency witnesses disclose their information there on the basis of a voluntary corporate decision. Even though such a decision may be made under the pressure that continued silence factually leads to an increased risk of a fine, the latter is only due to the company's (possible) violation of antitrust law and not due to the withholding of information. Due to the voluntary nature of leniency statements and actions, comprehensive protection under Article

339 TFEU cannot apply in this respect and the Union legislator must be free to take secondary regulatory action here; the same applies to settlement statements in settlement proceedings, which are also made voluntarily.

2.3 Obligation of the National Courts to Take into Account Statements by the European Commission in the Context of a Consultation Initiated by the National Court

When the European Commission issues an opinion or transmits information at the request of a national civil court as part of an adversarial procedure, the question arises as to whether the national courts are required to take this into account to a large extent (in the sense of a prohibition on contrary decisions) in their decision-making. However, there may be a question of whether an opinion of the European Commission before a national court according to Art. 15 Regulation 1/2003 or Art. 29 Regulation 2015/1589 is an instrument sui generis, outside of Art. 288 (5) TFEU, could be considered. In the unlikely event that the European Commission reacts by way of a resolution, but also in the case in which the European Commission refers to an existing resolution in its response, the respective resolution is subject to an obligation to take it into account. The same applies if the response from the European Commission indicates sufficient intention to pass a relevant resolution. If, on the other hand, the response of the European Commission does not constitute a decision and does not indicate an intention to do so either, there can be no obligation to give it relevant consideration. The effective legal protection of the party to the civil law dispute potentially burdened by the information and/or opinions of the European Commission would also be impaired if the decision of the national court could be decisively based on this. Ultimately, the opinions of the European Commission requested by the courts, insofar as no sufficient intention to adopt a decision can be inferred from them, only have an orientation effect. However, the national court will have to examine to a particular extent whether the statement of the European Commission indicates an intention to take a decision which would then be subject to increased duties of consideration (in the sense of a prohibition of contrary decisions). An obligation to take into account judicially requested opinions of the European Commission also does not result from secondary law. Art. 16 Regulation 1/2003 only refers to decisions and intended decisions. A corresponding provision for judicially requested opinions of the European Commission is also not contained in Art. 15 Regulation 1/2003 and Art. 29 Regulation 2015/1589. In addition, the recitals of Regulation 1/2003 emphasize that it does not intend to interfere with "national legislation on the standard of proof or the obligation of ... courts of the Member States to clarify legally relevant facts". The exploitation of information received is therefore carried out according to the normal rules of evidence. To the extent that the European Commission answers the courts' questions, there is no obligation on national courts to avoid conflicting decisions. They may also not use the answers of the European Commission as a necessary basis for their decision, but must, in order to enable effective legal protection, come to their own statements that can be reviewed in the course of appeal. However, the answers of the European Commission represent a (non-binding) point of reference, which may make a referral to the ECJ useful or even necessary if the court tends to take a different view. Something else applies only if the response of the European Commission is to be regarded as sufficient evidence of an intention to take a relevant decision or (which would be very unusual) itself constitutes a decision.

3 Cooperation Emanating from the European Commission

Through resolutions and informal comfort letters, the European Commission only indirectly influences the activities of the national courts. Ultimately, the same also applies to notifications, guidelines, etc., with the exception of guidelines that are directly tailored to the cooperation between the European Commission and national courts. In this respect, the European Commission is also understood as an amicus curiae. The term amicus curiae (originating from the English legal area and shaped primarily by US law) is often used to describe a third-party opinion/intervention before a court. The term is not used in EU competition law [29]. Therefore, for the investigation of self-induced statements by the European Commission before the national courts, a focus on the content of the term amicus curiae is not expedient, but an investigation of the concrete mode of operation of this instrument is necessary. In the following, we first examine under primary and then secondary law the extent to which the European Commission is entitled to a self-induced opinion as amicus curiae before the civil courts [9] of the Member States. This is followed by an examination of the extent to which national civil courts must take such opinions into account in their decision-making.

3.1 *Right of the European Commission to Act of Its Own as Amicus Curiae Before National Civil Courts*

(a) Primary Law
   The European Commission's important role in monitoring Member States' application of Union competition law can be found in Art. 17 TEU and Art. 101 TFEU (particularly Art. 105 (1) sentence 1 TFEU) and Art. 107 TFEU. The concept of coherence, which is derived from the principle of loyalty, suggests that the European Commission should be able to intervene in civil law disputes by issuing an opinion in order to work toward the uniform application of EU competition law and to represent the public interest in its effective enforcement [4, 13]. It should be noted, however, that primary legislation does not provide the

European Commission the authority to issue directives to national civil courts [27, 30]. Rather, national courts are expected to apply EU competition law autonomously within the restrictions outlined above. Furthermore, the characteristics of the adversarial procedure under civil law must be considered. When it came to the scope of the European Commission's right to participate, the ECJ relied on secondary law and did not deduce it from the loyalty requirement (but only saw it in its context). Finally, the European Commission recognizes the supplementary legal amicus curiae regulation as a new law. This indicates that the European Commission's right to make amicus curiae statements cannot be derived from core law [7], even though such comments are useful for integrating administrative and civil processes and so appear to make legal policy sense. Because the Union has sole legislative competence for competition law, and in particular for shaping the relationship between the European Commission and national courts, the Union legislature can intervene in a regulatory manner and grant the European Commission a right to comment through secondary law. This result is especially significant in the context of merger control because, unlike in antitrust and state aid law, an amicus curiae opinion by the European Commission is not provided for in secondary legislation.

(b) Secondary Law

For EU antitrust law, Art. 15 (3) sentences 3 to 5 of Regulation 1/2003 and for state aid law, Art. 29 (2) of Regulation 2015/1589, allows the European Commission to participate in national court proceedings as amicus curiae. The European Commission can thus (at least in writing) take proactive action, even in the first instance, and does not have to wait for an invitation from the national court or even the transmission of a judgment, as provided in Art. 15 (2) of Regulation 1/2003. The wording of Art. 15 (3) Regulation 1/2003 ("Assessment of the case") and Art. 29 (2) 2015/1589 ("Assessment of the aid matter") make it clear that the European Commission's Opinion may only be issued with reference to the procedure, while "case-independent circulars" to the courts are prohibited. Likewise, the European Commission is not entitled to introduce new facts into the legal dispute. In fact, however, it cannot be ruled out that the statement by the European Commission will contain additional information, as there is a high probability that the European Commission had dealt with the matter beforehand. As amicus curiae, the European Commission must not act as a supporter of a party, but only in the public interest.

However, the ECJ has convincingly stated that the "grammatical interpretation" already suggests "that the power of the Commission to transmit written opinions of its own accord to the courts of the Member States is solely subject to the provision that the consistent application of the Art. 81 EC or 82 EC so requires" [39]. This makes it likely that the Community/Union legislator intended this characteristic to have an independent meaning, even if nothing is apparent in this regard from the legislative materials of the Regulation 1/2003. Thus, if the national court is allowed to take a contrary decision anyway, it could be argued that the assessment of the European Commission on the coherence requirement (which can be reviewed by the national court only to a limited extent) would not be relevant. Nevertheless, statements of the European Commission on the coherence objective can help the national court to appropriately take into account the opinion of the European Commission, even if it would only have an orientation effect. As amicus curiae, the European Commission has the power to present its own statements before national courts, but only if it can demonstrate why this is necessary for the consistent application of relevant Union law. In this regard, the European Commission acts only in the public interest and is not authorized to add additional factual submissions to the legal dispute. Because the right to amicus curiae statements cannot be derived from primary law and secondary law only provides for regulations for Union antitrust and state aid law, the European Commission has no ability to make statements of its own accord in proceedings before national civil courts concerning EU merger control issues [25].

In the Inspecteur van de Belastingdienst [39], the ECJ addressed the question "whether the Commission is empowered under Article 15 (3) of Regulation No. 1/2003 to submit, on its own initiative, a written opinion to a national court in proceedings in which the question is whether a fine imposed by the Commission for an infringement of Article 81 EC or Article 82 EC may be deducted in whole or in part from the taxable profits", which it answered in the affirmative. The condition of the European Commission's involvement, "the consistent application of Articles 81 EC or 82 EC," could be met "even if the proceedings in question do not concern the application of Articles 81 EC or 82." Sentence four of recital (21) does not contradict this, "according to which the Commission may submit written comments on the application of Art. 81 EC or 82", each of which does not take precedence over the regulation itself. Therefore, "the third sentence of the first subparagraph of Article 15(3) of Regulation No 1/2003 must be interpreted as meaning that, under that provision, the Commission is empowered to send, on its own initiative, a written opinion to a national court in proceedings concerning whether all or part of a fine imposed by the Commission for infringement of Article 81 EC or Article 82 EC can be taxable profits may be deducted." This may be questionable. Art. 15 (2) and (3) sentence 1 Regulation 1/2003 speak of the "application of Articles 81 and 82 of the Treaty" and precisely not of other regulations that become relevant as a consequence of the application of Art. 101, 102 TFEU. However, we are of the opinion that the broad interpretation of the ECJ contradicts the genesis of the provision. The ECJ's teleological reasoning that a right of the European Commission to submit observations serves the effective enforcement of EU antitrust law is not sustainable, since it is questionable whether this really corresponded to the legislative intent of the EU. It is correct that the function of the European Commission as primus inter pares must end where its competence ends. Art. 15 of Regulation 1/2003 and Art. 29 of Regulation 2015/1589 cannot be understood otherwise. However, to the extent that national regulations or practices would not satisfy the principles of equivalence and effectiveness with regard to Union competition law, it can be argued with the ECJ that this should be covered by the European Commission's secondary law power to issue self-induced amicus curiae opinions. This is expressed in the coherence requirement related to Art. 101, 102 or Art. 107 (1), Art. 108 TFEU.

However, both Art. 15(3) Regulation 1/2003 and Art. 29 Regulation 2015/1589 do not provide for any fundamental interference with the civil procedural law of the Member States. In particular, the negotiation maxim should remain untouched. Therefore, the self-evident principle formulated for state aid law in the recitals that the European Commission must comply with national procedural rules and practices, including those concerning the safeguarding of the rights of the parties also applies to Regulation 1/2003. The European Commission does not become a party, nor can it interfere with the disposition maxim, i.e. the sovereignty of the parties over the subject matter of the dispute, nor does it have the right to have the oral proceedings reopened specifically for it or to have its own date set for oral statements. In such cases, as far as the national rules of procedure provide for it, the court may in principle also reject a statement of the European Commission, but will as a rule in view of the principle of the duty of loyalty under Community law from Art. 4 (3) TEU only reject such a request in exceptional cases. According to Art. 15 (3) Regulation 1/2003, the European Commission may, for the sole purpose of preparing its opinions … request the court or tribunal of the Member State concerned to forward to it, or arrange for the forwarding of, any document necessary for the assessment of the case. Art. 29 (2) Regulation 2015/1589 contains a similar provision. Art. 15 (3) Regulation 1/2003 and Art. 29 (2) Regulation 2015/1589 also contain restrictions on use which mean that the documents transmitted to the European Commission may not be used directly in its administrative proceedings, but may only be used for the exclusive purpose of amicus curiae participation, which is to be understood as a prohibition on the use of evidence.

In practice, in the event of a request from the European Commission, the courts will transmit the entire procedural file to the latter. However, the obligation cannot be unlimited. With regard to documents whose content has been or will be the subject of public proceedings, there should not be any reservations as a rule. If, on the other hand, documents or information are subject to rules on the protection of secrets in civil proceedings, these must be observed—as is the case with other national procedural law: despite the above-mentioned prohibition of the European Commission to use evidence, there is a risk that the European Commission will use findings from such documents—as a reason for its own investigations and thus to the detriment of the party concerned. Even if, based on the clear wording of Art. (3) Regulation 1/2003 and Art. 29 (2) Regulation 2015/1589 ("exclusively for") and in comparison with Art. 12 (2) Regulation 1/2003, it could be argued that the documents received in the context of the amicus curiae opinion may not be further used by the European Commission in this manner either, a factual effect on the officials of the European Commission cannot be excluded to initiate further investigations (e.g. a sector inquiry) on the occasion of the information received in order to be able to investigate and ultimately punish (allegedly) unlawful conduct on the basis of consideration in this way. Insofar as civil procedural law permits the non-public or restricted public submission of documents and information in adversarial proceedings before the civil court, this follows a system that has been balanced out over decades of practice with the aim of balancing the interests of the opposing parties. Now, if a party fears that certain sensitive information—which civil procedure law would actually protect—may be disclosed to the European Commission, this may contradict the principle of equality of arms, which requires that each party be reasonably enabled to present its point of view as well as its evidence under conditions which do not place it in a distinctly disadvantageous position vis-à-vis its opponent. Here, in deciding which documents or information to send to the European Commission, the court must balance the obligation under the Regulations and Article 47 of the Charter of Fundamental Rights, and interpret the Regulations restrictively if necessary. In view of the clear wording of the regulations, however, only clarification by the ECJ is likely to provide clarity.

3.2 Duty of National Courts to Take into Account Self-induced Amicus Curiae Statements by the European Commission

(a) Primary Law

The considerations regarding opinions and information of the European Commission, which the latter submits upon request of a national court, also apply if the European Commission submits an opinion on its own initiative. Insofar as the European Commission submits a relevant decision or can refer to such a decision, decisive consideration obligation emanates from this decision. Here, too, it is true that the parties generally could not sufficiently present themselves to the European Commission before it issued its opinion and that no legal protection against the opinion is possible. From the perspective of primary law, such self-induced opinions can therefore only have an orientation function, which is not connected with a duty to take into account in the sense of a prohibition of contrary decisions. Nor can the court refer to the opinion as a necessary justification for its decision but must come to its own decision that can be reviewed in the court of appeal.

(b) Secondary Law

It is questionable whether an obligation—going beyond primary law—to take into account self-induced amicus curiae statements by the European Commission (in the sense of a ban on contrary decisions) results from secondary law. However, in contrast to judicially requested opinions and information, the normative text on self-induced amicus curiae opinions of the European Commission in Art. 15 (3) Regulation 1/2003 and Art. 29 (2) Regulation 2015/1589 emphasizes the objective of a "coherent application" of substantive law. It is questionable whether this is to be understood as an indication by the legislator that the national courts must take the amicus curiae opinions of the European

Commission into decisive consideration. In this respect, it must first be examined whether the objective of a "uniform application of Community competition law" ("uniform application") of Art. 16 Regulation 1/2003 is congruent with the objective of a "consistent application" in Art. 15 (3) sentence 3 Regulation 1/2003 and Art. 29 (2) Regulation 2015/1589.

Advocate General Mengozzi wishes to understand the concepts of "uniformity" and "consistency" as synonymous [35]. According to the case law of the ECJ "a recital in a regulation may help" to "shed light on the interpretation of a legal provision" but does not itself constitute "such a provision". It is true that the Advocate General is right that in the recitals the terms "coherence", "uniformity" and (in English) "consistency" are used in different equivalents. However, this does not necessarily lead to the conclusion that they should be used synonymously. Rather, the opposite seems to be the case: Art. 15 of Regulation 1/2003 and Art. 29 of Regulation 2015/1589 speak of "coherence" and Art. 16 of Regulation 1/2003 of "uniform application", i.e. a different terminology was chosen. With regard to the self-induced opinions of the European Commission, the result is the same as with regard to the court-induced opinions, i.e. these essentially have a non-binding orientation effect, which in individual cases may result in an obligation to refer the matter to the ECJ if the national court takes a different view.

3.3 Obligation of the National Civil Courts to Transmit Judgments to the European Commission According to Art. 15 Regulation 1/2003

In the context of the European Commission's activity as amicus curiae, which is regulated by secondary law, the national courts are obliged to submit the necessary file contents to the European Commission. In addition, although not primary law, secondary law, albeit only for Union antitrust law, provides for the obligation of the Member States to submit a copy of any written judgment of a national court on the application of Art. 101, 102 TFEU. According to the wording of Art. 15 (2) Regulation 1/2003 it only applies to completed proceedings. Thus, the obligation to refer is not suitable for the European Commission to intervene in national civil court proceedings; at most, an amicus curiae opinion in the next instance may be considered, provided that the parties continue the proceedings. In the light of Art. 15 (1) Regulation 1/2003, a restrictive understanding is necessary. In fact, Art. 15 (1) Regulation 1/2003 is to be interpreted restrictively. The starting point, according to the view expressed here, is the wording in Art. 15 (1) Regulation 1/2003 ("judgment … on the application of Article 81 or 82 of the Treaty"). This can at least be understood as an indication that only decisions in which Art. 101 or Art. 102 TFEU play a decisive role are to be relevant. If the Union legislator had wanted the cartel authorities to be informed of the initiation of proceedings already before a final first instance decision, it would have regulated this.

4 Conclusion

Primary and secondary law provides the European Commission and the national courts with instruments by means of which the European Commission can fulfill its role as primus inter pares and the national courts can contribute to a coherent application of Union competition law as functional Union courts. Thus, national courts can request information from the European Commission and ask for opinions. Under certain circumstances, they are even obliged to do so under European Union law, in particular if there is a sufficient intention on the part of the European Commission to take a decision and consultation promises clarification in this respect. There is also a duty to consult if there are doubts about the basis, scope or correctness of a decision (in which case the national court can alternatively take the route to the ECJ for questions of interpretation and validity of the decision). An obligation of the national civil court in adversarial proceedings to turn to the European Commission for fact-finding cannot be justified under European Union law. The duty of national courts to address questions to the European Commission corresponds to an obligation of the European Commission to reply, unless compelling reasons for the preservation of its functioning or overriding interests of protection of Art. 339 TFEU stand in the way. Within the framework of the loyalty requirement, the European Commission must seriously consider a response in all other cases but has a discretion under primary law to reject a response, weighs up whether a response to the court or its rejection serves the effective enforcement of Union law more and can additionally take into account the protection of business secrets. Specifically with regard to leniency submissions in antitrust proceedings, it can (in addition to the aspect of self-binding through the guidelines of its Leniency Program) include in the consideration whether the effective enforcement of Union antitrust law is preferable through the protection of the leniency program or through the facilitation of the civil enforcement of damage claims. The possibility of the national courts to ask the European Commission for opinions and information on their own initiative and its duty to reply corresponds to the right of the European Commission as amicus curiae to submit opinions to the national courts on its own initiative, but only if it justifies why this is necessary for a coherent application of the relevant Union law. In this respect, the European Commission acts exclusively in the public interest and is not authorized to introduce new factual arguments into the legal dispute. Since the right to amicus curiae opinions cannot be derived from primary law and secondary law only provides for regulations for Union antitrust and state aid law, there is no possibility for the European Commission to issue self-induced opinions in proceedings on EU merger control issues before national civil courts. As far as the European Commission provides opinions or information or self-induced opinions to the national civil courts upon their requests, there is no obligation of the national courts to take them into account (in the sense of a prohibition of contrary decisions) and no right to use them as a necessary basis for a decision, but they only have an orientation function. Union competition law, antitrust law, state aid law and merger control law, is enforced administratively by the European Commission, but from civil law perspective by national courts. This parallel application leads to a tension and the danger that the European Commission and national civil courts contradict each other in their decisions. A uniform application of EU competition law, on the other hand, avoids zones of divergent application of the law as well as inappropriate forum shopping and thus contributes to a uniform internal market [23]. A further focus of the public discussion on the cooperation between the European Commission and national civil courts will remain useful. This is true not only because the tension between official and (increasing) private enforcement is still underestimated in some cases [30], but also because (also due to Brexit) the competition between the court systems of the Member States is likely to intensify. It therefore seems correct that the European Commission wants to increase the transparency of its activities with a homepage offering an overview of the status of the proceedings conducted before it and even informing about specific questions of cooperation.

References

1. Bakkar, Y., Ögcem, A.R.: Democracy and economic development: disentangling the effect of elections and rule of law. TalTech J. Eur. Stud. 4 (2019). https://doi.org/10.1515/bjes-2019-0042
2. De Baere, G., Roes, T.: EU loyalty as good faith. Int. Comp. Law Q. **64**, 829–874 (2020). https://doi.org/10.1017/S0020589315000421
3. Friedery, R., Horváthy, B., Ziegler, T.: Európai Unió alapjogvédelmi rendszere [European Union system for the protection of fundamental rights]. [in:] Vanda, L. (edt) Emberi jogi enciklopédia Budapest (2018)
4. Funta, R.: Social networks and potential competition issues. Krytyka Prawa **12**, 193–205 (2020). https://doi.org/10.7206/kp.2080-1084.369
5. Funta, R., Golovko, L., Juriš, F.: Európa a Európske právo [Europe and European Law]. 2. doplnené a rozšírené vydanie, Brno: MSD (2020)
6. Horváth, M.: Princípy v občianskom súdnom konaní [Principles in civil proceedings]. Konsenzus v práve. Banská Bystrica: Belianum (2013)
7. Izarova, I., Szolc-Nartowski, B., Kovtun, A.: Amicus curiae: origin, worldwide experience and suggestions for East European countries. Hung. J. Leg. Stud. **60**(1), 18–39 (2019). https://doi.org/10.1556/2052.2019.60103
8. Jones, A., Sufrin, B., Dne, N.: EU Competition Law: Text, Cases, and Materials. Oxford University Press, Oxford (2019)
9. Király, L.: Gyorsabb, Egyszerűbb, olcsóbb, hatékonyabb? az új magyar polgári perrendtartás általános rész osztott perszerkezetének hatékonysági elemzése [Faster, simpler, cheaper, more efficient? efficiency analysis of the shared structure of the general part of the new Hungarian civil procedure]. Akadémiai Kiadó, Budapest (2019)
10. Kindl, J., Kupčík, J., Mikeš, S., Svoboda, P.: Soutěžní právo. C.H.Beck, Praha (2021)
11. Králik, J., Králiková, K.: Základná inštitucionálna báza ochrany ľudských práv [The basic institutional basis for the protection of human rights]. Masarykova univerzita, Právnická fakulta, Days of Public Law. Brno (2007)
12. Lazar, J., Dulak, A., Dulaková-Jakúbeková, D., Jurčová, M., Kirstová, K., Novotná, M., Muríň, P.: Občianske právo hmotné 2. Záväzkové právo: Právo duševného vlastníctva [Substantive civil law 2. Law of obligations: Intellectual property law]. Bratislava: Iuris (2018)
13. Mihálik, J., Šramel, B.: Supervision of public prosecution service over public administration: the case study of Slovakia. Public Policy Adm. **2**, 192–202 (2018)

14. Miskolczi-Bodnár, P., Szuchy, R.: Joint and several liability of competition law infringers in the legislation of central and Eastern European member states. Yearb. Antitrust Regul. Stud. **10**(15), 85–109 (2017). https://doi.org/10.7172/1689-9024.YARS.2017.10.15.5

15. Miskolczi-Bodnár, P.: Indemnification and harm caused by infringement of Antitrust rules from Private Law point of view. [In.] Osztovits, A. (ed.) Recent Developments in European and Hungarian Competition Law, Budapest: Károli Gáspár Református Egyetem, Állam- és Jogtudományi Kar (2012)

16. Niels, G., Ralston, H.: Two-sided market definition: some common misunderstandings. Eur. Compet. J. 118–133 (2021). https://doi.org/10.1080/17441056.2020.1851478

17. Nováčková, D., Peráček, T.: The common European investment policy and its perspectives in the context of the achmea case law. TalTech J. Eur. Stud. **11**(1), 153–169 (2021). https://doi.org/10.2478/bjes-2021-0010

18. Osztovits, A.: Quantifying harm in action for damages based on breaches of article 101 or 102 of the treaty on the functioning of the European union – some remarks on the draft guidance paper of the European commission. [In:] Osztovits, A. (ed.) Recent Developments in European and Hungarian Competition Law, Károli Gáspár Református Egyetem Állam- és Jogtudományi Kar (2012)

19. Peráček, T.: Impact of European legislation and case-law on elimination of discrimination in employment relations in EU. Online J. Model. New Eur. **36**, 65–82 (2021). https://doi.org/10.24193/OJMNE.2021.36.04

20. Portuese, A., Gough, O., Tanega, J.: The principle of legal certainty as a principle of economic efficiency. Eur. J. Leg. Stud. Westminst. Sch. Law Res. Pap. **13–13**, 1–35 (2013). https://doi.org/10.2139/ssrn.2332016

21. Siman, M.: Zásady právnej istoty a legitímnej dôvery v judikatúre Súdneho Dvora Európskej únie [Principles of legal certainty and legitimate expectations in the case law of the Court of Justice of the European Union]. Acta Iuridica Olomucensia (AIO), roč. 7, č. **2**, 91–98 (2012)

22. Sitek, M.: The human rights to communicate in the light of the development of IT technology at the turn of the XX and XXI centuries. [In.] Sitek, M., Uricchio, A.F., Florek, I. (eds.) Human rights between needs and possibilities. Józefów: WSGE (2017)

23. Stehlík, V., Hamuľák, O.: Legal issues of EU Internal Market: Understanding the Four Freedoms. Univerzita Palackého v Olomouci, Olomouc (2013)

24. Stehlík, V., Petr, M., Hamuľák, O.: Handbook on EU Internal Market. Univerzita Palackého v Olomouci, Olomouc (2016)

25. Smulders, B., Gippini-Fournier, E.: Some Critical Comments on the Report of the Global Competition Law Centre on the Directly Applicable Exception System and the Direct Applicability of Article 81 (3) EC: Positive Enforcement and Legal Certainty, pp. 1–12 (2009). https://doi.org/10.2139/ssrn.1476962

26. Svák, J., Grŭnwald, T.: Nadnárodné systémy ochrany ľudských práv [Transnational human rights systems]. Wolters Kluwer, Bratislava (2019)

27. Szegedi, A.: Several thoughts on the coming into force of the new hungarian civil code, with respect to company law. [In.] Czudek, D., Kozieł, M. (szerk.) Legal and Economic Aspects of the Business in V4 Countries. Brno, Csehország: Centrum Prawa Polskiego (2014)

28. Šmejkal, V., Svobodová, M.: ECJ´s new role – guardian of open but not socially inclusive Europe? East. Eur. J. Transnatl. Relations **2**, 11–32 (2018). https://doi.org/10.15290/eejtr.2018.02.02.01

29. Šmejkal, V., Dufková, B.: Soutěžní právo EU Casebook - rozsudky SDEU 2006–2020. Wolters Kluwer ČR, Praha (2021)

30. Šramel, B., Horváth, P., Machyniak, J.: Peculiarities of prosecution and indictment of the president of the Slovak Republic: Is current legal regulation really sufficient? Soc. Sci. **3**, 1–20 (2019)

31. Šramel, B., Machyniak, J., Gutan, D.: Slovak criminal justice and the philosophy of its privatization: an appropriate solution of problems of Slovak justice in the 21st century? Soc. Sci. **2**, 1–14 (2020)

32. Tridimas, T.: Fundamental rights, general principles of EU law, and the charter. Camb. Yearb. Eur. Leg. Stud. **16**(2014), 361 392 (2014)
33. Whish, R., Bailey, D.: Competition Law. Oxford University Press, Oxford (2021)

EU Case Law

34. C-53/85, AKZO Chemie BV and AKZO Chemie UK Ltd v Commission of the European Communities, ECLI:EU:C:1986:256
35. C-234/89, *Stergios Delimitis v Henninger Bräu AG*, ECLI:EU:C:1991:91
36. C-39/94, *Syndicat français de l'Express international (SFEI) and others v La Poste and others*, ECLI:EU:C:1996:285
37. C-339/00, *Ireland v Commission of the European Communities*, ECLI:EU:C:2003:545
38. C-275/00, *European Community, represented by the Commission of the European Communities v First NV and Franex NV*, ECLI:EU:C:2002:711
39. C-429/07, *Inspecteur van de Belastingdienst v X BV*, ECLI:EU:C:2009:359
40. C-574/14, *PGE Górnictwo i Energetyka Konwencjonalna SA v Prezes Urzędu Regulacji Energetyki*, ECLI:EU:C:2016:686

EU Secondary Legislation

41. Council Regulation (EC) No 1/2003 of 16 December 2002 on the implementation of the rules on competition laid down in Articles 81 and 82 of the Treaty
42. Council Regulation (EU) No 734/2013 of 22 July 2013 amending Regulation (EC) No 659/1999 laying down detailed rules for the application of Article 93 of the EC Treaty
43. Council Regulation (EU) 2015/1589 of 13 July 2015 laying down detailed rules for the application of Article 108 of the Treaty on the Functioning of the European Union

Significance of an Ethical Culture for Young Employees

Nadja Pade and Rozália Sulíková

Abstract Fast changing businesses and developments such as the shortage of skilled workers, the Corona pandemic and the consequences of political instability are putting the needs of employees once again in the spotlight. And here also—and especially—the needs of the youngest generation, which is already conquering the labour market. The Organizational Culture makes a major contribution to motivating and retaining employees and the consideration of an Ethical Organizational Culture is becoming more and more interesting as the return to values and meaning is gaining relevance in these turbulent times. The importance of an Ethical Organizational Culture for young employees is in the focus of attention as this cohort seems to have special requirements to their employers.

1 Introduction

1.1 Relevance

Entrepreneurial activity is increasingly observed and critically judged by the public for already many years [1]. But the demand for the assumption of moral responsibility has become louder and louder, not just in the last two and a half years under the pandemic influence. The young generations influence dealing with the topic of ethics in organizations. Expectations of Generation Z include not only meaningful work but also fair treatment, Ethical leadership, and social responsibility of the organizations. Especially when choosing the next employer, the ethics of the organization are important for these young new employees [2]. Facing the war for talents, it is once again worthwhile to think about what future working generations need and expect from their employer in terms of ethics. This is where an Ethical Culture comes into

N. Pade · R. Sulíková (✉)
Comenius University, Odbojárov 10, Bratislava 820 05, Slovakia
e-mail: rozalia.sulikova@fm.uniba.sk

© The Author(s), under exclusive license to Springer Nature Switzerland AG 2023   421
N. Kryvinska et al. (eds.), *Developments in Information and Knowledge Management Systems for Business Applications*, Studies in Systems, Decision and Control 462,
https://doi.org/10.1007/978-3-031-25695-0_19

focus as well as the role of leadership to establish and develop an Ethical Culture in organizations.

The fact, that influencing factors are difficult to prove directly in organizational cultures, since culture develops historically in the learning process and interactively via the socialization development, [3] is complicating an ad-hoc investigation.

Therefore, it is an appropriate approach to base this work on a literature review to explore this topic.

1.2   Goals and Objectives

This work aims to contribute to a better understanding of the significance of an Ethical Organizational Culture for young employees. It wants to provide with relevant definitions to create a common understanding among the stakeholders concerned and will pay particular attention to the influence of leadership on an Ethical Organizational Culture and thus indirectly on the younger generations.

The chapter also addresses the ethical behavioural requirements of the young generation and works on the question of why ethics in organizations are particularly relevant for young people. It also clarifies the question of whether there are ethical principles that are violated. Of particular importance will also be the topic of what influence organisations can have on the Ethical Organizational Culture and what special position leader and leadership takes in this context.

The focus of this chapter can help researchers in the disciplines of Human Resources, Generation Management and Ethical Management as well as managers and other actors in organizations to reach a better understanding of this specific subject and thus give concrete implications for practice.

This work is structured as followed. First, we provide with an extensive theoretical background of these areas. After introducing the definitions of Generation Management, Generation Z, Ethical Organizational Culture and Ethical Leadership, the scientific methodology of this chapter is presented, followed by the discussion and the results. The work is closing with the conclusion including a summary and an outlook on further research work.

2   Theoretical Background

The following section provides an overview of the state of the art on the topic of this chapter. It gives an overview of relevant terms and deals with the integration of business ethics in organizations, the specifics of the young generation and their ethical aspirations as well as the possibility of organizations influencing ethics in the company.

2.1 Classification of Relevant Terms

2.1.1 Generation Management

Retain employees to ensure the company's success: this applies not only in the war for skilled workers as well as in the war for junior talents. Successful Generation Management is one of the keys. The change in the age structure and the resulting consequences must be adequately countered, also to keep performance at a competitive level. When Generation Management is successful, it leads to a positive Organizational Culture and increases employee engagement and motivation [4].

Generation Management also pays special attention to the coexistence of generations. Here, among other things, the type of communication, for example, the promoted exchange, has overriding importance. But it is also significant to consider the differences between the generations themselves and how the organization and the managers deal with them.

Holz is postulating, that various performance parameters can be distinguished in which younger people are superior to older people; likewise, however, older people are attributed to certain characteristics in which they have an advantage. Knowing the strengths of one's employees not only helps to deploy them optimally for the company but also to create positive experiences for the employees. Following Holz, the strengths of the younger employees include spontaneity and activity, power, speed (for example in reaction), flexibility, willingness to take risks, openness as well as more up-to-date training, career orientation, and enthusiasm to continue training. Strengths of older employees are serenity and overview, sense of responsibility, quality awareness (in the sense of accuracy), judgment, conflict and cooperation skills, experience knowledge, expertise and company-specific understanding as well as communication skills and reliability [5].

2.1.2 Business Ethics

Business Ethics has got two main disciplines: business and ethics. The combination, Business Ethics, is concerned with the norms and values that underlie business decisions and the consequences that result from them. Ethics in general is meant to be the doctrine of right action [6]. Specifically, Business Ethics deals with which entrepreneurial decisions are to be judged as good or are to be judged as bad from an ethical point of view. Corporate ethical decisions justify corporate actions and define the moral and social responsibility of a company [7].

There was not always an understanding of the need for this justification, since the belief prevailed that business studies, and thus also the teaching of entrepreneurial decision-making, was in itself an ethical discipline since the rational principle was equivalent to the rational ethical imperative [8]. Wien and Franzke are postulating that the most important approaches to Ethics Management have been developed in the USA already since the 1930s [9].

With the example of an entrepreneur accepting additional costs to follow his Business Ethics, it becomes clear, that entrepreneurial actions are not automatically ethically underpinned [10]. In the context of this work, Business Ethics is often related to the values of the organization.

2.1.3 Organizational Culture

Hofstede is defining culture as the applications and principles of an organization. He postulated that these principles and applications differentiate from others of other companies and verify however staff carries themselves with one another and with external parties. This is how they choose a way to become involved in productive assignments [11].

Hofstede also stated that domestic and regional factors are influencing the culture of any organization and eventually impact the actions of organizational staff [11]. Organizational Cultures include standards and orientation patterns in different mediation mechanisms and forms of expression. They are complex, elusive phenomena that develop alongside the formal order in companies [3]. Keller sees a central role for managers in shaping the Organizational Culture [12].

The culture of a company creates, among other things, internal identification, coordinates the actions of individual company actors and shapes the image of the company. It has, among other things, functions such as the stabilization function, the function of conveying meaning and the motivational function [13]. All sources are almost unanimous in concluding that the Organizational Culture also contributes to the success of the company. For this work Organizational Culture and Company Culture are used as synonyms.

2.1.4 Young Generation

This work is designating the young generation, to Generation Z. The name Generation Z was formed by marketing experts and trend scouts for young people with birth years from 1996. It is so far the youngest generation to join organizational structures and the labour market [14]. Generation Z is also known as Digital Natives [15]. Figure 1, Generation Z, is showing some main attributes of this young generation.

Taking a closer look at other classifications of generations of importance, literature is naming the oldest generation, to be taken into account in the world of work, the silent generation—born 1925–1945 [17]. As these people have gradually disappeared from the labour market and are only active in organizations in exceptional cases, this work summarizes the baby boomer generation as the oldest active generation. It is the high-birth cohort, approximately born between 1946 and 1962 [18]. This generation is of main significance as they provide the very many executives in companies just like Generation X, born in the years 1963 to 1980. The last to mention active generation is Generation Y, born between 1981 and 1995 [18].

> First true digital and mobile native generation from birth

> Diversity is an expectation

> Driven by traditional opportunities for advancement and development and improved economic security and better benefits

> Entrepreneurial tendencies

> Wants frequent in-person feedback

Fig. 1 Generation Z (adapted from [14, 16])

Generation Z is the most diverse generation regarding ethnicity, religion and the structure of family [19]. This generation is growing up in a time when gender equality has become the norm [20]. For this work young employees are to be equated with the young generation—meaning Generation Z, even if the literature does not always explicitly refer to Generation Z when referring to 'young employees' and 'young generation'.

2.1.5 Leadership

Leadership is present where there is a goal-related exertion of influence [21]. Through leadership, employees are to be made to achieve certain goals, which are usually derived from the company's goals [22]. Hölzerkopf is defining leadership as a process of steering in a certain direction [23]. Leadership can influence in different ways, generally speaking, a distinction can be made between leadership through structures and leadership through people.

Leadership through structures always occurs when an employee is influenced in a goal-related way without a leader directly exerting this influence. The origin can be manifold and could, for example, take place through organizational charts, regulations, incentives or the Organizational Culture [24].

Leadership by people refers to the lived reality of the above-mentioned regulations and rules, for example, how flexibly and creatively the rules are worked with. Here, the behaviour of the supervisor, for instance, the way of conducting conversations or the way of clarifying goals, becomes an essential part of leadership [7].

Looking at leadership and leading, the role of power plays an important role in the relationship between leaders and employees. In the original definition, Max Weber postulates: "Power means any chance to assert one's own will within a social relationship, even against opposition, regardless of what this chance is based on." [25].

In organizations, power is mostly based on position, i.e., job descriptions and employment contracts. However, social psychology distinguishes between different

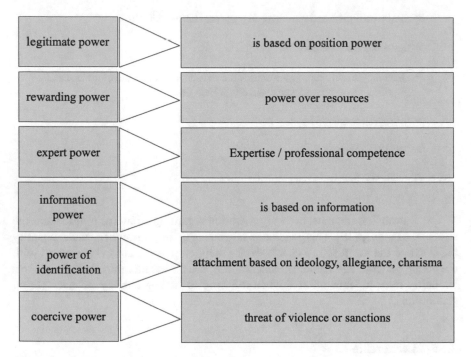

Fig. 2 Types of Power (adapted from [26])

types of power. Figure 2 provides an overview of these. It is important to know, that every leader has got different sources of power on to base leadership behaviour.

Power can have a strong impact on ethics in organizations and vice versa. Again, here comes the Organizational Culture in focus.

And finally, looking at leadership and the young generation, Agarwal and Vaghela found out in their study, that Generation Z likes to work independently and does not think, that there is a need to be supervised [27].

2.2 Integration of Business Ethics in Organizations

Internal and external demands are possible reasons why organizations try to anchor attitudes and values in their Organizational Culture. They usually find their way into, for example, the corporate philosophy, the corporate principles, or corresponding guidelines. In general, they summarise the behaviour that employees are expected to show with statements that provide orientation. In addition to the co-determination enshrined in the law in Germany, different instruments are regulating ethics in organizations.

These aspects play an important role for young generations regarding employer branding in the process of recruitment as well in the topic of employee retention and loyalty. Some of the expressions of ethics in organizations are briefly presented below.

2.2.1 Corporate Guidelines: Codes of Ethics and Codes of Conduct

Ethical principles are often anchored in the Corporate Guidelines. These are understood as framework conditions and include, among other things, the company's treatment of its employees, which is also important from a moral point of view [26]. Guidelines might include as well equal rights, safe and healthy working conditions, development of employee skills, right to form employee representation, open communication as well as trust and respect. They can also provide with concrete quantities, as an example with a mandatory share of women in the management team.

Basic ethical principles in organizations include the values of an organization—communicated internally to the employees as well to the outside, such as customer and suppliers [7]. These principles contain, among another things, do's and do not's, guidelines, and instructions for daily work, such as dealing with gifts, invitations and rules for dealing with information.

More and more companies passing Codes of Conduct or Ethics. These are mainly ethical principles of action in a written form for executives and employees. In addition to antitrust and competition law requirements, Codes of Conduct and Ethics also include global guidelines, such as on human rights, child and forced labour, and so forth. Although the work with a Code of Conduct is mostly voluntary, the motivation to let executives and employees sign this paper could be the fact, that business partners expect other companies, they have got business with, to have a Code of Conduct [26].

2.2.2 Corporate Social Responsibility

A voluntary commitment of an organization in a social, environmental, or institutional direction which exceeds the actual central business is Corporate Social Responsibility [28].

The concept of Corporate Social Responsibility requires company management to take social concerns into account in addition to profit orientation [29]. Corporate Social Responsibility can be as well a competitive advantage: a research by Marakova et al. showed, that the realisation of perceptions of Corporate Social Responsibility has statistical results in the importance of competitive benefit [30].

2.2.3 Strategies Supporting Compliance with Ethical Principles

Strategies supporting compliance with ethical principles in organizations are various. Following Beeri et al. promoting ethics in business starts with having a Code of Ethics and being sure that it is effectively transmitted to all employees and other relevant parties [31].

Vance and Harris are postulating that there are organizational processes, which can be introduced and manifested within the organisation and can protect against unethical behaviour. In their view, further strategies and processes—in addition to a Code of Conduct or Ethic—could be: change in leadership, installation of Ethic Officers and a re-training of the employee's set of moral reasoning, as a help to become sensible to ethical issues [32]. Besides that, their literature review has led to organizational factors, which can solidify ethics (Fig. 3):

Wien and Franzke also give special importance to Organizational Culture and the norms it sets for Business Ethics and their implementation, they also see Ethics Management as the result of strategic and operational leadership [7].

Fig. 3 Organizational components facilitating ethics (adapted from [31])

2.2.4 Importance of Organizational Culture for Implementing and Living Ethics

The employee's behaviour is determined by individual ethics and institutional norms and values, which form a written and unwritten ethical framework. This framework is not only grounded in the above-mentioned Corporate Principles but as well in the Organizational Culture [26]. This shows that it needs an Organizational Culture as a fertile ground for the implementation of Business Ethics.

Ethics should always be considered in conjunction with culture, among other things, ethical considerations and attitudes should not be perceived as disruptive. A company's goals should not conflict with ethics [29]. In principle, Business Ethics can only be effective if the Organizational Culture is in line with the respective convictions and values [1].

2.3 Organizational Ethics and the Young Generation

2.3.1 Specifics of the Young Generation

For understanding the special relationship between ethics and the young generation it is important to know, what distinguishes the young generation. From main importance is to identify what motivates them. The content theories of motivation theories state that an employee's motivation is based on the fact, that needs and motives are recognized and might be used as an incentive in organizations [33].

During the stages of working life, the needs of employees change. Organizations—and leaders—should react flexibly to these changes. Most people start their professional life with highly motivated [5]. Oertel postulates that the needs of those starting in the job tend to be characterized by material modesty, a low need for security and self-actualization, but high social needs and esteem [18]. Generation Z is intrinsically motivated and wants to make the world a better place [34].

If we look at the characteristics and requirements of Generation Z it seems that Oertel's general approach is fitting to the needs of the youngest generation at first. Since the generation is commonly not missing anything and they usually grow up in conditions characterized by abundance, the need for security and material is high, but mainly still satisfied by their family of origin. But the need for safety is growing fast among Generation Z in working life and goes hand in hand with the dwindling commitment to relationships and a non-binding attitude [35]. However, it is important to the generation that they have a job that they enjoy. As Generation Z is challenged to deal with the increasing limitlessness of options as it grows up, this generation accordingly seeks orientation [36]. After deciding for a new employer they need an effective onboarding [2]. Porter et al. are supposing that younger employees may place a high value on social justice or volunteering [37].

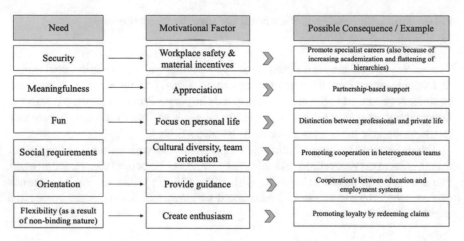

Fig. 4 Young generation: overview of needs and motivational factors (adapted from [36] and [38])

To meet this generation's needs, employers must respond to these diverse requirements. Figure 4 provides an overview of the various requirements, the motivational factors, and the possibilities for the organizations to react.

2.3.2 Ethical Behavioural Requirements of the Young Generation

Generation X experienced a high divorce rate within the family, which led them to be independent parents due to the unpredictability of family life and therefore encourage their offspring, Generation Z, to think independently [39]. Therefore, Generation Z values connections to family, order, structure, work ethic and a sense of predictability [38].

Generation Z wants a good relationship with their staff members and teamwork is important, a good work environment is motivating them. Moreover, this generation is sensitive to society and would therefore appreciate working for companies that are committed to society [27]. Generation Z has different norms than the generation before [40]. Ethics at workplace are very central for this generation.

For them, work-life balance means more than a career and they appreciate meaningful work more than a high salary [41]. They want to use their life for something meaningful [42].

Generation Z favours sincerity, autonomy, flexibility and personal freedom, absence can lead to frustration among colleagues, lower productivity, low morale and lack of employee engagement [27].

According to Oertel, children in the late 1990s were very interested in environmental issues. Today's young adults consider ethical goals, such as solving environmental problems or securing peace, more important than material things and are willing to get involved [18].

Knowledge of ethical needs is important in that otherwise companies may not recruit the most suitable employees due to false expectations about the ethical behaviour of workers. For example, the hypothesis is that young employees may have a tendency not to behave in a very responsible or ethical way [43].

In terms of leadership, the young generation wants to be taught the meaning and importance of the task, as well as the ethical value of their work [44]. In addition, it can be said that the values, social influence and ethics of a company play a significant role for this generation, especially when choosing a job [2].

Regarding external perception, there are always complaints about the younger generations in the context of generational conflicts in the workplace, for example, their lack of work ethic or lack respect for the authority of their elders [18]. Interesting enough as studies have shown that older employees had shown less ethical behaviour than younger employees [43].

2.4   Influence of Organizations on Ethics

2.4.1   Ethical Organizational Culture

In an Ethical Organizational Culture values and norms have a strong impact on the organization. There is an acknowledgement of the overall responsibility of all stakeholders and ethical values, and action is taken morally—based on the needs and interests of all stakeholders and not just one's own. Frey is postulating that establishing an Ethical Organizational Culture means changing norms, structures, and routines. But it is not enough to teach ethical values, they must also be lived: it is important to find multipliers from the top management who can act as role models for these values [45].

Nahar and Nigar are proposing that an Ethical Organizational Culture can be understood as a collection of values, norms and techniques which are playing a significant role in the made performance of corporations. According to this, Organizational Culture is one of the biggest issues affecting the standards of organizational work, and only one of the positive effects of an Ethical Organizational Culture is that employees are more innovative and drive the organization forward when the Ethical Organizational Culture is accepted as helpful [46]. They suggest from their research the following elements of an Ethical Culture (Fig. 5):

Existence of
- ethical policies, which are easily accessible and communicated
- ethic assessment, that can provide information on the orientation and attitude of employees towards ethics
- advisory and educational sessions for staff facing ethical dilemmas
- ethical discourses about "what is right" and "what is wrong" in the context of a particular situations

Fig. 5 Elements of an ethical organizational culture (adapted from [46])

According to Kotras, Organizational Culture is linked to the ethical environment of the company [47].

Concerning the current crisis, the author would also like to endorse French and Holden who are defining an Ethical Organizational Culture as a corporate environment, which is moderating bad effects during change processes [48].

In the authors understanding an Ethical Organizational Culture is the target to be aimed at being a culture that combines actual values with perspectives, which probably have been handed down from generations, and which give all members an orientation in a fast-changing world.

2.4.2   Ethical Leadership and Ethical Decision-Making

In the last years, Ethical Leadership has begun to be considered as its own leadership style as it was previously usual to focus on research only on ethical parts of leadership [49]. This work focuses on an Ethic Leadership style as an own style within Responsible Leadership.

Today's leader's challenge is to show an attractive future and support to focus in a complex and uncertain world. Leaders need to have purposefulness and a vision. They need to create and run a business environment, which is characterized by trust. In this context Responsible Leadership can be understood as a social-related and ethical phenomenon between leaders and followers [50]. Coming back to the role of power within the relation between leader and employee, this means, that it is important for the responsible leader to evaluate the sources of power ethically. Responsibility starts with a legitimate leadership relationship. Leaders have to keep in mind that, however, if a subordinate's distress or dilemma is abused, for example, to obtain an employee's consent to worsening working conditions, or if a poor information situation is used to deceive the employee, then from an ethical point of view it may be the case that—despite an employment contract—there is no legitimate relationship [51].

The transition from a generally responsible leadership to Ethical Leadership is fluid. Following Maak and Pless, a responsible leader, inter alia, needs emotional and ethical qualifications [50].

Based on a qualitative study Brown et al. defined Ethical Leadership as "… the demonstration of normatively appropriate conduct through personal actions and interpersonal relationships, and the promotion of such conduct to followers through two-way communication, reinforcement, and decision-making." [49].

Kalshoven et al. distinguished Ethical Leadership in the following dimensions: fairness, integrity, people-orientation, role clarification, ethical guidance and power-sharing [52]. Returning to the issue of power, Bellingham postulates, that ethical leaders want to empower their staff through development and support, this gives them freedom and the opportunity to show initiative [53].

Martindale is stating that responsible leaders can motivate their employees by providing with norms and values of the Organization's Culture [16]. In addition, leadership style seems to be the biggest influencer in the tension between employee values and expectations and market values imposed on the sector from the outside [54].

Ethical leaders base their work on reliability, they are—among other things—honest and fair. Brown et al. are stating that a leader's personality characteristic are honesty, integrity and trustworthiness—their traits are responding to the 'moral person' as one pillar of Ethical Leadership [49]. They stand for ethical standards and also work against unethical behaviour [55].

To ensure an ethical understanding of leadership, leadership guidelines are often derived in addition to the corporate principles to provide ethical direction.

Decision-making is of particular importance in the context of Ethical Leadership. In the end, an organization and its actions are the results of decisions, decisions whose responsibility usually lies with the leaders [56]. Many, if not all decisions, also include ethical aspects, so we talk of Ethical Decision Making, which includes confidence and fairness, responsibility and looking after others. Acknowledging these preconditions, the Ethical Decision-Making process involves balancing all available options, excluding unethical views, and selecting the best ethical alternative.

In the end, however, there is still room for manoeuvres that each manager has—depending on the size of the scope for decision-making. But there is supervision as a special reference group, managers are observed, reflected upon and assessed by employees [26].

It can be stated that leaders spending effort in the Organizational Culture, can inspire all employees [57]. When a leader is following an ethical standard and is recognized as honest and trustworthy this is followed by a healthy working relationship [49]. Nahar and Nigah have been showing with their study that the effects of Ethical Leadership are trust in the leader and a strong work engagement [46].

2.4.3 Violation of Ethics in Organizations

Blickle and Nerdinger are defining possible challenges related to ethics in organizations. They talk, for example, about gender-based discrimination, ethical violations in personnel selection procedures, when deciding on personnel development measures, when implementing organizational changes and in exit processes [58]. Also, micropolitics and leadership are possible fields of violation of ethics in organizations.

Beeri et al. describe the negative impact of unethical behaviour in public organizations as obvious. For them unethical behaviour is the greatest danger to a democracy [31]. Dempsey defines moral responsibility in companies as individuals taking responsibility for the wrongdoing of others by actively sharing values with them [59]. When we talk about the violation of ethics in organizations, the courage of the employees also plays a major role. Carsten and Uhl even assume that employees must not only be trained to show moral behaviour but also to confront with unethical behaviour [60].

Panigrahi and Al Nashash are postulating that participation and engagement at work lead to a decrease in unethical behaviour such as idleness or wasting time [61].

Kozika makes a similar observation: as loyalty, commitment and motivation increase, the probability of actively damaging the company decreases [62].

The literature research for this chapter has not given any implication in the context of the young generation and the infringement of ethics, but following Kotra's research, the tendency to accept non-ethical performance is stronger among younger employees [47]. In general, it can be said that individuals who have high levels of narcissism, Machiavellianism and psychopathy—the so-called dark triad associated with unethical behaviour—embedded in their personality can be massively damaging to their organization [63].

3   Method

The research of this work is based on a literature review. An essential part of the analyses was the method of qualitative content analyses by Mayring and Fenzel. To interpret the findings the coding system MAXQDA developed by Kuckartz was used for the structured analysis [64]. The intention was to answer the following research question: To what extent is an Ethical Culture important for young employees?

The following databases and library catalogues were used for literature searches: SpringerLink, SCOPUS, ECONSTOR, Researchgate, ProQuest, EBSCO Host, ZBW, Google Scholar, Google Search, and Harvard Business Review.

The author has identified search terms as a part of the targeted literature search such as: Generation Management, Generation Z, ethics, Ethical Culture, young employees, and needs of Generation Z. The most important professional area of analysis is human resource management.

All sources that meet the applicable scientific requirements for the level of detail and quality of elaboration were classified as relevant. The research yielded 139 potential sources, of which 64 were classified as relevant and included as references. All relevant data were compiled, then categorised and finally analyzed. For a general overview of the Structured Content Analysis approach see Fig. 6.

Using MAXQDA, four first-level subcodes were derived from one main code representing the topic "Generation Management": "Organizational Culture and ethics", "Corporate Culture", "Young generation", "Leadership" and "Ethics in organizations". From there 16-s level subcodes have been developed inductively and deductively. Three third-level codes have also been emerged inductively and deductively subsidized to the following topics: "Specifics of Generation Z".

Literature was encoded using the structured content analysis method. This involved coding the passages into 456 codes and then analysing them concerning the research question. The synthesis of all data led to the following findings and implications in this chapter. For an overview of the codes and subcodes created and their hierarchy, see Fig. 7 (Fig. 8).

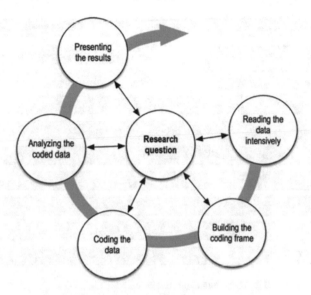

Fig. 6 Five phases of a qualitative content analysis [64]

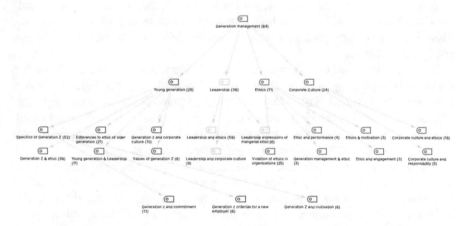

Fig. 7 Hierarchical code-subcodes model by MAXQDA (*Source* Authors own creation)

4 Discussion

An important challenge that organizations are facing today is the retention of their most valuable benefit: people [4].

Due to demographic change, the age structure of organizations has changed and is still transforming. The examination of the needs of the different generations can deal with the fact of shortage of skilled workers, maintains a positive employer branding and secures the competitiveness of a company.

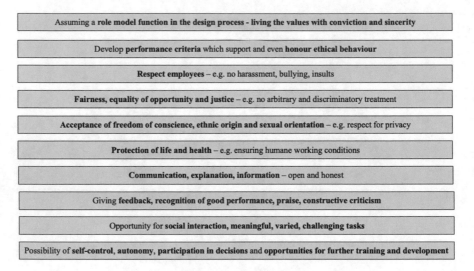

Fig. 8 Manual of ethical organizational cultures (adapted from [67])

It is worth putting the youngest generation into focus, as they are an already present cohort and will be a growing future employee community for organizations. This chapter is working on the question of how important an Ethical Culture for this young generation is.

To shed light on the relevance of an Ethical Culture, it is important to know that literature is supporting the interconnection and mutual influence between Ethical Leadership Style and an Ethical Organizational Culture. The positive consequences of Ethical Leadership can be equated with the characteristics of an Ethical Organizational Culture and lead to the deduction that it takes an Ethical Leadership Style to implement and develop an Ethical Organizational Culture.

Practically this interconnection has been as well substantiated by various empirical studies and research, which have focused, for example, on the context between Ethical Leadership and employee engagement [46] or the connection of Ethical Leadership and factors like improvement of cooperation, increased motivation and optimism among employees [65].

This could also indicate that the lack of an Ethical Leadership Style could, conversely, mean that there is no influence on the establishment and development of an Ethical Organizational Culture. It can be assumed that a leader, who does not pursue an Ethical Leadership style, cannot establish an Ethical Organizational Culture. As mentioned before Schiemann and Jonas postulate that people, who combine pronounced traits of narcissism, Machiavellianism, and psychopathy, have disastrous consequences for an organization [63]. But, interestingly enough, if those leaders enter an Organizational Culture, where they feel valued and have got the impression that the organization is concerned about their well-being, research shows that they show less counterproductive work behaviour [66].

If we take a look at the specifics and behavioural requirements of Generation Z it seems that they want to be flexible and independent on the one hand but, in terms of leadership, they want to be taught the meaning and importance of their tasks, as well as the ethical value of their work [44]. As stressed before, the values, social influence and ethics of a company play an important role for them [2].

Knowing that leading through structures as well takes place through the Organizational Culture this seems to be an adequate possibility to work with Generation Z.

5 Results

Derived from the last years and current developments the interest in Ethical Organizational Culture will probably even increase in importance. The acute situations of the worldwide pandemic and the war site in Ukraine have shown that questions such as: what is right and what is wrong? will continue to be of great concern in the future—in the corporate context and for Generation Z.

The role and the influence that a leader takes in the Ethical Culture in organizations are of crucial importance. As well as the fact that it will help employees to have confidence in their leaders. Trust in leaders is strengthened by establishing ethical approaches. Employees are experiencing this by knowing that their interests are perceived and considered, even further, that decisions are made fairly and ethically. Organizations should therefore take full responsibility for ensuring that leaders promote these approaches [66].

Literature highlights the importance of ethics for Generation Z but is hardly giving any implications on the special meaning of an Ethical Organizational Culture for these young people. But it seems that an Ethical Culture, as a derivation of ethics in companies, and with its connection to Ethical Leadership is very important for them.

Besides general ethical integrations, like implementing an Ethical Code of Conduct and other regulations, the author has compiled in the following overview (Fig. 6), some general approaches discussed in this paper and further advice from the sources used.

Successful Generation Management is based, among other things, on knowing where the company stands in terms of age structure, knowing the peculiarities of the different generations and employees and on the approach of sensitizing managers accordingly [68].

On the other hand, many different needs do not require a completely different strategy. Especially in leadership and support through Human Resources Management, factors such as flexibility and appreciation can be used to create motivation and understanding for all target groups.

However, every company has its uniqueness, so the management of different generations must also be adapted to the respective circumstances of an organization. Only those who know the needs and wishes of their employees can operate successful

Generation Management and can thus respond to the needs of the young generation in terms of ethics and Ethical Organizational Culture [69].

6 Conclusion

6.1 Synopsis

If a company wants to attract high potential today, it must take its perspective and rethink, not on the surface, but deep in the core of its business model, in its values [41]. Moreover, a value-based organisation, in general, has long ceased to be a question of philosophy but must be seen as a prerequisite [70].

The results of the literature analyses have shown the importance of ethics for the younger generations in the business context. To attract this generation, companies need to emphasise their efforts to be good global market players. This generation needs to do a job with ethical values [31].

In literature, there has no direct link between the younger generation and the concrete need for an Ethical Culture be found, but an indirect connection and the significance of an Ethical Culture can be recognized, as ethics in organizations is reflected through an Ethical Organizational Culture. Also, the anchoring of work-life balance and the enablement to live teamwork in heterogeneous groups in organizational and Human Resources policies will lead to motivation and loyalty of the young generation. Leaders and Human Resources Management can meet these expectations not only by establishing an Ethical Culture but also, among others, with the offer of flexible work and working time arrangements, open-minded recruitment, and diversity as the main columns of Human Resources work.

To promote intergenerational cooperation and understanding as well as to fulfil the expectations of the generations, there should be a constant exchange between leaders and employees as well as between the generations.

Successful intergenerational activities could e.g., include workshops, round tables or joint project work sponsored by Human Resources. But also, feedback-based methods and approaches, like mentoring and coaching, could help to ensure that the intergenerational dialogue is profitable for both sides [36].

To promote cooperation and understanding, special mentoring programs could be implemented. Approaches to coaching "old for young" are interesting and could be investigated deeper in the future. Here, older employees are placed at the side of younger colleagues and accompany them through their daily work. On the one hand, this relieves the older employees, who no longer necessarily want to expose themselves to the demands of a management function.

But not only Human Resources Development is required—Human Resources in its integrated function must ensure that ethical principles are consolidated in the management guidelines and the learning culture of organizations. Especially the previously mentioned ageing process must be met by an effective Human Resources

Diversity Management System [71]. To keep the balance between new and already-proven, among others, communication and participation, transparency and interaction as well as openness should rely upon [69]. And not only to take seriously the importance of an Ethical Culture for young employees.

6.2 Further Research

There is a comprehensive study about Generation Management and the consideration of the individual generations with their requirements and needs.

There is quite a lot of literature available on Generation Z, but there are not yet many studies on this generation in organizations, as these employees are still very young and many of them are still in education and only now or in the future joining a company. Further research could be done on this topic—also in an empirical framework to have more practical implications for organisations.

It would be interesting to examine the changes in the significance of Ethical Leadership and Ethical Organizational Culture after the global crises in general and for young employees.

On the other hand, the literature does not yet consider the significance of the violation of ethical principles in companies and its importance for the young generation. Here, too, further research would be interesting.

For sure, there is a lot of effort for organizations now and in the future—and here especially for leaders and Human Resources Management—to stay adaptable to all employees' needs and motivational factors.

References

1. Grabner-Kräuter, S.: Zum Verhältnis von Unternehmensethik und Unternehmenskultur. Zeitschrift für Wirtschafts- und Unternehmensethik **1**, 290–312 (2000)
2. Expectations of Generation Z: Effective onboarding. Human Resource Management International Digest **29**(6), 33–35 (2021)
3. Schreyögg, G., Geiger, D.: Organisation: Grundlagen moderner Organisationsgestaltung. Mit Fallstudien. Springer Fachmedien Wiesbaden, Wiesbaden (2021)
4. Mahmoud, A.B., Fuxman, L., Mohr, I., et al.: "We aren't your reincarnation!" workplace motivation across X, Y and Z generations. Int. J. Manpow. **42**(1), 193–209 (2021)
5. Holz, M.: Demografischer Wandel in Unternehmen: Herausforderung für die strategische Personalplanung, 1st edn. Gabler, Wiesbaden (2007)
6. Petersen, T., Quandt, J.H., Schmidt, M.: Führung in Verantwortung. Springer Fachmedien Wiesbaden, Wiesbaden (2017)
7. Rosenstiel, R., Regnet, E., Domsch, M.E.: Führung von Mitarbeitern: Handbuch für erfolgreiches Personalmanagement. Schäffer-Poeschel, Stuttgart (2020)
8. Albach, H.: Betriebswirtschaftslehre ohne Unternehmensethik!: Zeitschrift für Betriebswirtschaft **75**, 809–831 (2005)
9. Wien, A., Franzke, N.: Unternehmenskultur. Springer Fachmedien Wiesbaden, Wiesbaden (2014)

10. Scherm, E., Julmi, C.: Strategisches Management: Theorie, Entscheidung, Reflexion. De Gruyter, Berlin, Boston (2019)
11. Hofstede, G., Hofstede, G.J., Minkov, M.: Cultures and Organizations: Software of the Mind; Intercultural Cooperation and Its Importance for Survival. McGraw-Hill, New York (2010)
12. Keller, K.: Nachhaltige Personal- und Organisationsentwicklung. Springer Fachmedien Wiesbaden, Wiesbaden (2018)
13. Sackmann, S.: Unternehmenskultur: Erkennen – Entwickeln – Verändern. Springer Fachmedien Wiesbaden, Wiesbaden (2017)
14. Lanier, K.: 5 things HR professionals need to know about Generation Z: thought leaders share their views on the HR profession and its direction for the future. Strateg. HR Rev. **16**, 288–290 (2017)
15. Bata, T., Bejtkovsky, J.: The employees of baby boomers generation, generation X, generation Y and generation Z in selected Czech. Corporations as conceivers of development and competitiveness in their corporation. J. Compet. **8**(4), 105–123 (2016)
16. Martindale, N.: What does a motivation strategy for a future workforce look like? Employer Benefits Online Retrieved from: https://employeebenefits.co.uk/issues/january-2016-2/what-does-a-motivation-strategy-for-a-future-workforce-look-like/. (accessed: Juli 10, 2022). (2016)
17. Twenge, J.M., Campbell, S.M., Hoffman, B.J., Lance, C.E.: Generational differences in work values: leisure and extrinsic values increasing, social and intrinsic values decreasing. J. Manag. **36**(5), 1117–1142 (2010)
18. Oertel, J.: Generationenmanagement in Unternehmen. Deutscher Universitäts-Verlag, Wiesbaden (2007)
19. Wiley, J., Grubb, V.M.: Clash of the Generations: Managing the New Workplace Reality, John Wiley & Sons, Incorporated. ProQuest Ebook Central (2016). http://ebookcentral.proquest.com/lib/zbw/detail.action?docID=4714037. Accessed 30 June 2022
20. Eilers, S.: Generation Z in Deutschland. In: Scholz, C.: Generation Z im Vier-Länder-Vergleich, pp. 57–118. München: Rainer Hampp Verlag (2019)
21. von Rosenstiel, L., Molt, W., Rüttinger, B.: Organisationspsychologie. Kohlhammer, Stuttgart (2005)
22. Comelli, G., von Rosenstiel, L.: Führung durch Motivation: Mitarbeiter für Unternehmensziele gewinnen. Vahlen, München (2011)
23. Hölzerkopf, G.: Führung auf den Punkt gebracht. Gabler Verlag, Wiesbaden (2005)
24. Wunderer, R.: Führung und Zusammenarbeit: eine unternehmerische Führungslehre, 8., aktualisierte und erw. Aufl. Luchterhand, Köln (2009)
25. Weber, M.: Wirtschaft und Gesellschaft. Jazzybee Verlag, Altenmünster (2012)
26. Franken, S.: Verhaltensorientierte Führung: Handeln, Lernen und Diversity in Unternehmen. Springer Fachmedien Wiesbaden, Wiesbaden (2019)
27. Agarwal, H., Vaghela, P.S.: Work values of Gen Z: bridging the gap to the next generation. In: NC-2018 - National Conference on Innovative Business Management Practices in 21st Century, Faculty of Management Studies, Parul University, Gujarat, India (2018)
28. Homann, K.: Sollten Unternehmen neben einer ökonomischen auch eine soziale Verantwortung haben? ifo Schnelldienst **24**, 3–6 (2018)
29. Schreyögg, G., Koch, J.: Management: Grundlagen der Unternehmensführung. Springer Fachmedien Wiesbaden, Wiesbaden (2020)
30. Marakova, V., Wolak-, A., Tuckova, Z.: Corporate social responsibility as a source of competitive advantage in large enterprises. J. Compet. **13**(1), 113–128 (2021)
31. Beeri, I., Dayan, R., Vigoda-Gadot, E., Werner, S.B.: Advancing ethics in public organizations: the impact of an ethics program on employees' perceptions and behaviors in a regional council. J. Bus. Ethics **112**, 59–78 (2013)
32. Vance, N.R., Harris, A.S.: Ethics as management principles. J. Leadersh. Account. Ethics **8**(3), 11–21 (2011)
33. Stock-Homburg, R., Groß, M.: Personalmanagement: Theorien – Konzepte – Instrumente. Springer Fachmedien Wiesbaden, Wiesbaden (2019)

34. Leslie, B., Anderson, C., Bickham, C., et al.: Generation Z perceptions of a positive workplace environment. Empl. Responsib. Rights J. **33**(3), 171–187 (2021)
35. Mangelsdorf, M.: Von Babyboomer bis Generation Z: Der richtige Umgang mit unterschiedlichen Generationen im Unternehmen. GABAL Verlag, Offenbach am Main (2015)
36. Klaffke, M.: Millennials und Generation Z – Charakteristika der nachrückenden Arbeitnehmer-Generationen. In: Klaffke, M.: Generationen-Management. Wiesbaden: Springer Fachmedien Wiesbaden, pp. 57–82 (2014)
37. Porter, T.H., Riesenmy, K.D., Fields, D.: Work environment and employee motivation to lead: moderating effects of personal characteristics. Am. J. Bus. **31**(2) (2016)
38. Ivanovska, L.P., Kiril, P., Iliev, J.Λ., Magdincheva, S.M.: Establishing balance between professional and private life of generation Z. Res. Phys. Educ. Sport. Health **6**(1), 3–9 (2017)
39. Tapscott, D.: Grown Up Digital: How the Net Generation Is Changing Your World. McGraw-Hill, New York city (2008)
40. Bencsik, A., Juhász, T., Horváth-Csikós, G.: Y and Z generations at workplaces. J. Compet. **6**(3), 90–106 (2016)
41. Schlaepfer, K., Welz, M.: Das dynamische Unternehmen: Wie Wertewandel, Innovation und Digitalisierung zum Erfolg führen. Schäffer-Poeschel, Stuttgart (2017)
42. Herget, J.: Unternehmenskultur gestalten: Systematisch zum nachhaltigen Unternehmenserfolg. Springer, Berlin (2020)
43. Pilukienė, L.: Importance of motivation and work pay of young employees in the value creation chain of a business company: assessment of changes in and formation of expectations. Bus. Manag. Educ. **15**(2), 211–226 (2017)
44. Machwürth, S.: Generation Y und Z führen. Arbeit und Arbeitsrecht **7**, 416–417 (2016)
45. Frey, D., Nikitopoulus, A., Peus, C., et al.: Unternehmenserfolg durch ethikorientierte Unternehmens- und Mitarbeiterführung (2010)
46. Nahar, A., Nigah, R.K.: Ethical organizational culture - a way to employee engagement. Glob. J. Enterp. Inf. Syst. **10**(4), 18–29 (2018)
47. Kotras, M.: Corporate culture and its connection with external and internal public comparative economic research. Central East. Eur. **13**(5), 27–47 (2011)
48. French, S., Holden, T.: Positive Organizational behavior: a buffer for bad news. Bus. Commun. Q. **75**, 208–220 (2012)
49. Brown, M.E., Treviño, L.K., Harrison, D.A.: Ethical leadership: a social learning perspective for construct development and testing. Organ. Behav. Hum. Decis. Process. **97**(2), 117–134 (2005)
50. Maak, T., Pless, N.M.: Responsible leadership in a stakeholder society – a relational perspective. J. Bus. Ethics **66**(1), 99–115 (2006)
51. Göbel, E.: Unternehmensethik. Grundlagen und praktische Umsetzung., 3., überarb. und aktualisierte Aufl. UVK-Verlags-Gesellschaft, Konstanz (2013)
52. Kalshoven, K., Den Hartog, D.N., De Hoogh, A.H.B.: Ethical leadership at work questionnaire (ELW): development and validation of a multidimensional measure. Leadersh. Q. **22**(1), 51–69 (2011)
53. Bellingham, R.: Ethical Leadership: Rebuilding Trust in Corporations. HRD Press, Amherst, Massachusetts (2003)
54. Townsend, K., McDonald, P., Cathcart, A.: Managing flexible work arrangements in small not-for-profit firms: the influence of organisational size, financial constraints and workforce characteristics. Int. J. Hum. Resour. Manag. **28**(14), 2085–2107 (2017)
55. Engelbrecht, A.S., Heine, G., Mahembe, B.: Integrity, ethical leadership, trust and work engagement. Leadersh. Org. Dev. J. **38**(3), 368–379 (2017)
56. Stainer, L.: Ethical dimensions of management decision-making. Strateg. Chang. **13**(6), 333–342 (2004)
57. Leigh, A.: Ethical Leadership: Creating and Sustaining an Ethical Business Culture. Kogan Page, London (2013)
58. Blickle, G., Nerdinger, F. W.: Ethik und kontraproduktive Prozesse in Organisationen. In: Schuler, H., Moser, K. (eds.) Lehrbuch Organisationspsychologie, 6th edn, pp. 639–661. Hogrefe, Göttingen (2019)

59. Dempsey, J.: Moral responsibility, shared values, and corporate culture. Bus. Ethics Q. **25**(3), 319–340 (2015)
60. Carsten, M.K., Uhl, M.: Ethical followership: an examination of followership beliefs and crimes of obedience. J. Leadership Organ. Stud. **20**(1), 49–61 (2013)
61. Panigrahi, S.K., Al-Nashash, H.M.: Quality work ethics and job satisfaction: an empirical analysis. Organ. Mark.: Policies Process. eJournal (2019)
62. Kozica, A.M.F.: Personalethik: die ethische Dimension personalwissenschaftlicher Forschung. Lang, Frankfurt am Main, New York (2011)
63. Schiemann, S.J., Jonas, E.: Streben nach Macht fern von Ethik: Die "dunkle Triade" bei Führungskräften und die Folgen für Organisationen. Organ. Superv. Coach. **27**(2), 251–263 (2020)
64. Kuckartz, U.: Qualitative text analysis: a systematic approach. In: Kaiser, G., Presmeg, N. (eds.) Compendium for Early Career Researchers in Mathematics Education, pp. 181–197. New York City, Springer International Publishing (2019)
65. Resick, C.J., Martin, G.S., Keating, M.A., et al.: What ethical leadership means to me: Asian, American, and European perspectives. J. Bus. Ethics **101**(3), 435–457 (2011)
66. Palmer, J.C., Komarraju, M., Carter, M.Z., Karau, S.J.: Angel on one shoulder: can perceived organizational support moderate the relationship between the Dark Triad traits and counterproductive work behavior? Pers. Individ. Differ. **110**, 31–37 (2017)
67. Göbel, E.: Unternehmensethik. Grundlagen und praktische Umsetzung. UVK Verlagsgesellschaft mbH, Konstanz, mit UVK/Lucius, München (2017)
68. Schuett, S.: Demografie-Management in der Praxis. Springer, Berlin (2014)
69. Riedel, P.: So gelingt das Generationenmanagement (2020). Retrieved from: Personalwirtschaft https://www.personalwirtschaft.de/news/hr-organisation/so-gelingt-das-generatio nenmanagement-98788/. Accessed June 16 2022
70. Blanchard, K.H., O'Connor, M.J., Ballard, J.: Managing by values. Berrett-Koehler Publishers, Oakland (1997)
71. Thom, N., Zaugg, R.J.: Moderne Personalentwicklung: Mitarbeiterpotenziale erkennen, entwickeln und fördern, 3, aktualisierte Gabler, Wiesbaden (2008)

European Digital Strategy and Its Impact on the Conclusion of Selected Types of Business Contracts

Tomáš Peráček, Boris Mucha, Štefan Palatický, Konstantin Keller, and Andreas Mußmann

Abstract At the moment of the information age, the knowledge and digital economy has become part of the pan-European economy as a result of the fourth industrial revolution. Despite this fact, a number of unanswered questions about the correct use of digital technologies, in particular when concluding contracts, have long arisen in business practice. In our view, this is a serious problem which, due to the lack of interest of legal theorists in particular, is an unexplored area. For this reason, answers to the ambiguities and problems that have arisen are relatively difficult to find. The aim of this study is, in particular, to examine the current legislative options and the associated problems of electronic conclusion of selected types of contracts. The setting of this objective is based directly on current needs and emerging practical problems in business practice. In the processing of the issue, we applied the primary analysis of legislation in view of the nature of the topic examined. However, we also use scientific literature, case law and analogy of law. This scientific study provides qualified answers to serious problems of business practice. The added value of our research is in a broader context to critically examine selected application problems and propose appropriate ways of improvement.

T. Peráček (✉)
Faculty of Management, Comenius University, Odbojárov 10, Bratislava, Slovak Republic
e-mail: peracek2@uniba.sk

B. Mucha
Ministry of the Interior of the Slovak Republic, Pribinova 2, Bratislava, Slovak Republic

Š. Palatický · K. Keller · A. Mußmann
Faculty of Public Policy and Public Administration, Danubius University, Richterova, 1171
Sládkovičovo, Slovak Republic
e-mail: konstantin.keller@web.de

A. Mußmann
e-mail: andreas@musfam.de

© The Author(s), under exclusive license to Springer Nature Switzerland AG 2023   443
N. Kryvinska et al. (eds.), *Developments in Information and Knowledge Management
Systems for Business Applications*, Studies in Systems, Decision and Control 462,
https://doi.org/10.1007/978-3-031-25695-0_20

1 Introduction

Trading is one of the cornerstones of every economy and has found its application in the current fourth industrial revolution. The current possibilities of digital interconnection not only improve the overall efficiency of trading, but also accelerate innovation by introducing new business models that can be realised much faster. As a result of this activity, electronic trade provides several advantages. These, on the one hand, allow entrepreneurs to move faster at a lower cost to a certain market, on the other hand, it allows customers to buy goods and services in a "non-stone" store from the comfort of not only their offices but also households. The change in social set-up in 1989, the transition to a market economy and the gradual, albeit very slow pace of digitisation of society, made it possible to start distance trading [41].

The European Union is also aware of this situation, which, through the European Union's Digital Strategy, invests in the digital skills of all Europeans, protects people from cyber threats (attacks by hackers, ransomware, identity theft), ensures that AI develops in a way that respects people's rights and trusts, accelerates the deployment of ultra-fast broadband for households, schools and hospitals across the EU, expands Europe's supercomputing capacity to develop innovative solutions in medicine, transport and the environment [8].

As this document shows, the main benefits of e-commerce for society are lower costs, increased interaction, great opportunities in every area of life, as well as infrastructure support and individual benefits for organizations, individuals and society. E-commerce according to Peráček (2022) expands the market to national and international markets [34]. With minimal costs, an entrepreneur can easily and quickly find the necessary business partners. He agrees with this view Oláh et al. (2020) and adds that it also facilitates administration and reduces the need for paper-based information storage [30]. It is also an improvement of reputation, flexible improvement of customer service or increased productivity. However, the main advantage of e-commerce is the possibility to purchase at any time from any place. Scientific literature in the field of personnel management such as Joamets and Chochia (2021) considers the possibility of working from home to work as a fundamental advantage, i.e. without the need to move to work. This results in reduced traffic on the roads and reduces air pollution [19].

2 Objective and Methodology

The aim of the scientific study is to examine selected aspects of the electronic conclusion of selected types of commercial contracts and their subsequent analysis. The establishment of this rather demanding goal is based directly on the current long-term unsolved needs and emerging practical problems in Slovak business practice. In addition to the main objective, we have also chosen two sub-objectives, which are:

- identify the biggest legislative drawbacks of e-contracting;
- critically assess the current e-commerce legislation and, if necessary, provide "de lege ferenda" proposals, i.e. legislative proposals for improvement.

In particular, we want to achieve the objectives set by means of a thorough study of relevant legislation, professional as well as scientific literature. Due to the nature of the scientific study, we use several scientific methods of knowledge suitable for knowing the law. This concerns, in particular, the use of a critical analysis method to examine the legal situation and regulation as well as abstraction. Using the comparative method, we make different opinions available not only to economists, but especially to lawyers, not only on the appropriateness of legislation but also on the interpretation of individual legal institutes. In this way, we seek a multidisciplinary view of the issue under investigation. Due to our many years of experience in the field of commercial law, we also use doctrinal interpretation. Selected legislation of European and national law is an essential source of knowledge needed to achieve the main objective [36].

Our intention is also consistent with the structure of this scientific study, which, in addition to the introduction and combined discussion with the conclusion, is divided into three core chapters. Each of these chapters comprehensively examines a single contractual type. At the end of the study we will provide answers to the objectives set.

3 Results

There are several definitions for e-commerce. Most people think of buying and selling goods and services over the internet. It follows at first sight that e-commerce takes place only in a virtual-internet environment. However, as it states Funta and Králiková (2022), this is not entirely true, it points to the European initiative on e-commerce [15]. In its view, electronic commerce is understood to mean commerce carried out by electronic means, as well as the carrying out of marketing or the sale and purchase of goods or services and the transmission of relevant information over electronic networks. E-commerce is thus a traditional commerce, but carried out by electronic means. Some economists Jančíková and Pásztorová (2021) and Troitino et al. (2020) however, they have a different view on this concept and point to the OECD definition. E-commerce can be characterized as all forms of electronic transactions between organizations and individuals based on the creation and transmission of electronic data, including text, sound and visual presentations. It identifies or is associated with the effect that the exchange of electronic data may cause in the area of institutions and processes that relate to such activities [17, 44].

4 Contract of Sale

In practice, the contract of sale as a relative commercial obligation relationship, governed by Sections 409 to 475 of the Commercial Code, is the most commonly used type of commercial contract between entrepreneurs in their business activities. With reference to the principle of disposition, the legislature designated as mandatory only the provisions of Sections 444, 458 and 459 of the Commercial Code [9]. By way of purchase contract, the seller undertakes to deliver to the buyer movable goods (goods) determined individually or in quantity and type and to transfer ownership of this item to the buyer and the buyer undertakes to pay the purchase price. The essential parts of the contract are:

1. identification of the parties, i.e. who is the buyer and who the seller is;
2. the seller's obligation to deliver the goods to the buyer individually or in terms of quantity and type,
3. the seller's obligation to transfer ownership of the goods to the buyer,
4. definition of the subject matter of the contract,
5. the buyer's obligation to pay the purchase price.

From a formal point of view, the contract does not require a written form; it can be concluded not only orally but tacitly.

As is apparent from Section 409 of the Commercial Code, only movable goods may be the subject of a purchase contract. This term is not regulated by the Commercial Code. According to Horvath et al. (2017) the Civil Code contains only a legal definition of real estate, which is land and buildings connected to the ground by a solid foundation [16]. It follows from the above that the subject of the purchase contract may be various objects, e.g. cars, fuels, energy, etc. The purchase price must be expressed and paid in money, otherwise it would be a contract of exchange which is governed by Section 611 of the Civil Code. The purchase price can be determined in three ways, namely:

1. the law leaves its amount to the agreement of the parties;
2. if the parties have not agreed on the amount of the purchase price, the purchase contract must include a method of its subsequent determination, unless the parties express their intention to conclude it even without fixing the purchase price,
3. the buyer is obliged to pay the purchase price at which such or comparable goods were sold at the time of conclusion of the contract under contractual conditions similar to the content of this contract.

As Funta (2014) if the purchase price is determined by the weight of the goods, its net weight is decisive in doubt. A special rule applies in the case of a contract for the supply of goods which are yet to be produced. Such a contract shall be deemed to be a contract of sale, unless the party to whom the goods are to be delivered undertakes to hand over to the other party a substantial part of the items necessary for the manufacture of the goods. However, a contract of sale shall not be deemed to be a contract under which the bulk of the obligation of the party to supply the goods consists of carrying out an activity or the obligation of that party includes the assembly of the goods [13].

The seller is obliged to deliver the goods to the buyer, hand over the documents that apply to the goods and allow the buyer to acquire ownership of the goods in accordance with the contract and the Commercial Code. The goods are handed over by the seller to the buyer at a certain agreed place. If the seller is not obliged under the contract to deliver the goods at a specific place, the delivery of the goods is effected by handing over to the first carrier for transport to the buyer. The exception is the case where the contract determines the dispatch of the goods by the seller. The seller is obliged to allow the buyer to exercise the rights under the transport contract against the carrier until the buyer has these rights under the transport contract [22].

The Commercial Code also deals with a situation where the parties have not agreed to dispatch the goods by the seller and the goods are individually specified or determined in the contract by type, but they are to be delivered from certain stocks or to be produced, and the parties knew at the time of conclusion of the contract where it is located or where it is to be produced. In such a case, the delivery takes place when the buyer is allowed to dispose of the goods at that place. Another possibility of fulfilling the seller's obligation to deliver the goods is to allow the buyer to dispose of the goods at the place where the seller has its registered office or place of business, or domicile or organizational branch, provided that the seller notifies the place to the buyer in due time [2].

Since the buyer may not always have information about the dispatch of the goods, the statutory regulation protects his rights in a special way, so that if the delivery of the goods is carried out by sending them and the goods handed over to the carrier are not clearly and sufficiently marked as a shipment for the buyer, the effects of the delivery will occur only if the seller notifies the buyer without undue delay of the dispatch of the goods and specifies the dispatch of the goods in the notification. If the seller fails to do so, the delivery of the goods is carried out through the carrier who handed over the goods to the buyer [25].

The time of delivery of the goods is determined as follows, so that the seller is obliged to deliver the goods:

1. on the date specified or specified in the contract in the manner specified in the contract,
2. at any time during the period specified or specified in the contract in the manner specified in the contract, unless it is apparent from the contract or from the purpose of the contract known to the seller at the conclusion of the contract that the period of delivery within that period is determined by the buyer.

Unless otherwise provided by the contract, the time limit for delivery of the goods shall begin to run from the date of conclusion of the contract. However, if under the contract the buyer has to fulfil certain obligations prior to delivery of the goods, such as, for example, to provide drawings necessary for the production of the goods, to pay the purchase price or part thereof, or to arrange for its payment, this period begins to run only from the date of fulfilment of this obligation. In the event that the seller fulfils his obligation earlier and delivers the goods before the specified time, the law allows the buyer to take over or refuse the goods [3].

Section 416 of the Commercial Code resolves a situation where there is no agreed delivery period and thus modifies Section 340(2) of the Commercial Code to the effect that if the delivery period of the goods is not agreed, the seller is obliged to deliver the goods within a reasonable period of time, taking into account the nature of the goods and the place of delivery.

Together with the goods, the seller is obliged to hand over to the buyer the documents necessary for the takeover and use of the goods, as well as other documents specified in the purchase contract, while the handing over of the documents takes place at the time and place specified in the contract or when the goods are delivered at the place of delivery. In the event that the seller has handed over the documents before the specified time and the documents contain incorrect data or other defects, the seller may, until the specified time of delivery of the goods, remove their defects, but only on condition that it does not cause undue difficulties or expenses to the buyer. Any claim for damages of the Buyer shall be retained [7].

A certain exception for the handling of documents of goods is the provision of Section 419 of the Commercial Code, which stipulates that the documents necessary for the receipt of the transported goods or for the free handling of the goods or when importing them for their clearance, the seller is obliged to hand over to the buyer:

- at the place of payment of the purchase price, if the handover is to take place at the time of payment, otherwise at the registered office or place of business or at the place of residence of the buyer
- in time so that the goods can be freely disposed of or taken over by the buyer at the time of their arrival at their destination and clear the imported goods without undue delay.

Quantity, quality, manufacture and packaging of goods

According to Kyncl (2012) other obligations of the seller include delivery of the goods in the quantity, quality and design determined by the contract and must pack them or equip them for transport in the manner specified in the contract [26]. If the contract does not specify the quality or the completion of the goods, the seller is obliged to deliver the goods in quality and in a copy that is suitable for the purpose specified in the contract. If this purpose is not specified in the contract, then for the purpose for which such goods are normally used. In certain cases, the characteristics of the goods may also be determined by the sample or masterpiece delivered to the buyer prior to the conclusion of the contract, resulting in the seller's obligation to deliver the goods with the characteristics of the sample or the model submitted to the buyer.

In the event of a discrepancy between the determination of the quality or the making-up of the goods according to this sample or model and the destination of the goods described in the contract, the determination described in the contract is directly decisive by law and, if there is no contradiction in those determinations, the goods must have characteristics both as described in the contract and according to the sample supplied.

The packaging of goods is taken for granted today. The purpose of the packaging is to protect the goods from damage. However, if the contract does not specify how the goods are to be packed or prepared for transport, the seller shall pack or prepare the goods for transport in a manner which is customary for the goods transported in the course of trade. If this method cannot be determined in the manner necessary for the preservation and protection of the goods [27].

Quantity of goods. The purchase contract considers the determination of the quantity of goods as one of the basic requirements. However, the subject of trade is also things for which it is not possible, for objective reasons, to deliver exactly the required quantity e.g. iron ore, etc. In approximate determination of the quantity of goods, the seller is entitled to determine the exact quantity of goods to be delivered. However, this right may be granted to the buyer by the contract and, unless the parties agree otherwise, the deviation may not exceed 5% of the quantity specified in the contract. Such a possibility shall also apply where the nature of the goods indicates that their quantity, as determined in the contract, is only approximate. The difference between the quantity of goods specified in the contract and the quantity of goods actually delivered may be no more than 5% of the quantity specified in the contract, unless the contract or previous practice between the parties or commercial practices indicates otherwise [1]. Such a derogation means that the seller is entitled to payment of the purchase price only for the goods actually delivered [18].

Defects of goods. In the event that the seller breaches his obligations to deliver the goods in quality, quantity and execution as determined by the contract, the goods have factual defects. The delivery of goods other than those determined by the contract and defects in the documents necessary for the use of the goods are also considered to be defects. If it is apparent from the transport document or from the proof of delivery of the goods or from the declaration of the seller that the goods are delivered in a smaller quantity or only part of the goods, the goods are not defective.

The subject of the purchase contract may also be goods for the production of which the material was provided by the buyer, which later proved unsuitable. In the event that, according to the contract, goods were used in the manufacture of goods delivered by the buyer, the seller shall not be liable for defects of the goods caused by the use of these items in two cases:

1. if, having exercised professional diligence, he could not reveal the inadequacy of those items for the manufacture of the goods; or
2. if the unsuitability of the provided items for the production of the goods was brought to the attention of the buyer, but the buyer insisted on their use.

The seller may also be released from liability for defects of goods of which the buyer was aware at the time of conclusion of the contract or having regard to the circumstances in which the contract must have been known, except where the defects relate to the characteristics of the goods which the goods should have had under the contract. The seller is also liable for a defect that the goods have at the moment when the risk of damage to the goods passes to the buyer, even if it becomes apparent only after that time. The seller's obligation to deliver quality goods is multiplied by his liability for any defect that arises after the delivery of the item to the buyer, if it was caused by a breach of his obligations. According to Troitino et al. (2020) § 426 of the Slovak Commercial Code allows the seller to deliver the goods with the consent of the buyer before the time of destination and until the expiry of this period he has the right to:

1. deliver the missing part or quantity of the goods delivered; or
2. deliver replacement goods for defective goods delivered; or
3. defects of delivered goods repaired

Provided that the exercise of this right does not cause disproportionate inconvenience or disproportionate expense to the buyer. However, the Buyer retains the right to compensation for any damages [44].

According to Kelblová (2013) the buyer's obligation is to inspect the goods as soon as possible after the risk of damage to the goods has passed, taking into account their nature. If under the contract it is necessary to send the goods by the seller, the search may be postponed until the goods have been delivered to their destination [21]. A different situation occurs if two conditions are met:

1. the goods are routed during transport to another destination or the buyer resells them without the buyer having the opportunity to inspect them appropriately to the nature of the goods
2. at the time of conclusion of the contract, the seller knew, or at least had to know, of the possibility of such a change of destination or such re-dispatch.

As a result, the inspection of the goods can be postponed until the goods are transported to a new destination. Failure to comply with the obligation to inspect the goods properly and in due time by the buyer has fatal consequences for him, which consist of the forfeiture of the claim for defects detectable during the inspection. Unless he proves that these defects already had the goods at the time of passing the risk of damage to the goods.

The buyer's right for defects of the goods expires, i.e. it cannot be granted in court proceedings if the buyer does not report to the seller about defects of the goods without undue delay after:

- the buyer found the defect, or
- the buyer, when taking professional care, should have detected the defects during the inspection, which he is obliged to carry out, or
- the defects could be detected at a later time with professional care, but no later than two years from the time of delivery of the goods or from the arrival of the goods to the destination specified in the contract.

The Act according to Chochia & Nassi (2021) places particular emphasis on the fact that the loss of the rights of the buyer from defects of the goods will be taken into account by the court only if the seller contests in court proceedings that the buyer has not fulfilled his obligation to inform him in due time of the defects of the goods.

These effects do not occur if the defects of the goods are due to facts about which the seller knew or had to know already at the time of delivery of the goods [10].

Quality guarantee

Starting with Section 429 of the Commercial Code, the Commercial Code provides for a quality guarantee, which is understood to mean a legal act whereby the seller assumes in writing an obligation that the delivered goods will, for a certain period of time, be capable of being used for an agreed (otherwise usual) purpose, or that it retains agreed (otherwise customary) properties. The assumption of the guarantee obligation itself may arise either from the contract or from the statement of the seller, in particular in the form of a guarantee card. The effects of the acceptance of this obligation also have the indication of the length of the warranty period or the durability or usefulness of the delivered goods on its packaging. Not all types of goods are subject to a general two-year warranty period [11]. If a different warranty period is specified in the contract or in the warranty statement of the seller, this period applies. The beginning of the warranty period may be linked to:

(a) the day of delivery of the goods,
(b) a specific date, if it results from the content of the contract or warranty declaration
(c) from the date of arrival of the goods to the destination, if the seller is obliged to send the goods.

The warranty period does not run for the period for which the buyer cannot use the goods for its defects, for which the seller is responsible. Section 431 of the Commercial Code exempts the seller from liability for defects covered by the quality guarantee in cases where these were caused after the risk of damage to the goods was passed by external events and were not caused by the seller or persons with whose help the seller fulfilled his obligation (e.g. carrier).

Legal defects of goods

In addition to factual defects, which are generally visible to the naked eye, the goods may suffer from legal defects, meaning that the goods sold are encumbered by the right of a third party if the buyer has given his consent to this restriction. The goods also have legal defects if they are encumbered by the right of a third party arising from industrial or other intellectual property and:

(a) if this right enjoys legal protection under the law of the State in whose territory the seller has its registered office or place of business or domicile; or
(b) if, at the time of the conclusion of the contract, the seller knew or ought to have known that this right enjoys legal protection under the law of the State in whose territory the buyer has its registered office or place of business or, where applicable, his domicile, or under the law of the State where the goods were to be resold or used, and the seller was aware of that sale or place of use at the time of conclusion of the contract.

According to Čajková and Gogová (2022), the dispositive provision of Section 434 of the Commercial Code, the claim does not arise from legal defects [5]. These are situations in which the buyer was aware of the third party's right at the time of the conclusion of the contract or the seller under the contract was obliged to follow the documents submitted by the buyer in fulfilling his obligations. The Buyer is obliged to notify the Seller of the exercise of the right of a third party to the goods, stating their nature, without undue delay after becoming aware of it under threat of loss of their judicial enforceability. There is also an exception to that rule, according to which those effects do not occur if the seller was aware of the third party's assertion of the right at the time when the buyer became aware of it. For the buyer's claims of legal defects of goods, the provisions of Sections 436 to 441 of the Commercial Code concerning buyer's claims for defects of goods shall apply [2].

Claims for defects of goods

Delivery of goods with defects means a breach of contract. If the contract is breached in a material manner, as referred to in Section 345(2) of the Commercial Code, the buyer is allowed to:

(a) require the removal of defects by delivering replacement goods for defective goods,
(b) delivery of missing goods and require the removal of legal defects,
(c) require the removal of defects by repairing the goods if the defects are repairable,
(d) require a reasonable discount on the purchase price; or
(e) with draw from the contract.

The buyer's right to choose between the above options is subject to timely fulfilment of the aforementioned notification obligation, together with the notification of the possibility of resolving his claim. The buyer may not change this claim without the seller's consent. The exception is the situation where it turns out that defects of the goods are irreparable, or their repair would involve disproportionate costs, which means that the buyer may require the delivery of replacement goods if the seller so requests without undue delay after the seller has notified him of this fact.

If the seller does not remedy the defects of the goods within a reasonable additional period or if he or she informs before its expiry that the defects will not be removed, the buyer may withdraw from the contract or demand a reasonable discount from the purchase price, especially in situations where the defective item can be used. Even in this case, the timely failure to notify the choice of the claim for defects of the goods has significant consequences, namely the buyer's right to claims for defects of the goods in such a way as in the case of insignificant breach of contract. However, the buyer retains a claim for damages as well as a contractual penalty if it has been negotiated [23].

In the event of a non-essential breach of the contract, the buyer has only the right to choose between two options, namely:

1. delivery of missing goods and elimination of other defects of goods,
2. discount on the purchase price.

If the buyer does not claim a discount from the purchase price or does not withdraw from the contract as permitted by law, the seller is obliged to deliver the missing goods and remove legal defects of the goods. The remaining defects must be rectified at his choice either by repairing the goods or by delivering equivalent goods. However, it must ensure that, in the chosen manner, the buyer does not incur unreasonable costs. If the buyer requests the removal of defects of the goods, at the same time before the expiry of the additional reasonable period, which he is obliged to provide to the seller for this purpose, he may no longer make other claims for defects of the goods, except for the claim for damages and the contractual penalty. The exception is the situation where the seller has informed him that he will not comply with his obligations even within a specified period of time [40].

The discount on the purchase price is another option to resolve the buyer's claim for defective performance of the contract. The actual entitlement to a discount on the purchase price is the difference between the value the goods would have had without defects and the value of the goods delivered with defects, the time at which the proper performance was to be carried out is decisive for determining the values. The buyer is allowed to reduce the purchase price paid to the seller when paying for the goods, or if the purchase price has already been paid, the buyer may request its return up to the amount of the discount together with the interest agreed in the contract, otherwise at the maximum permissible amount laid down by law or on the basis of the law. If the interest is not so determined, the debtor is obliged to pay the usual interest required for loans granted by the banks at the place where the debtor has its registered office at the time of conclusion of the contract [45].

The late notification of the defect of the goods to the seller has the consequence that the buyer can already claim a discount on the purchase price or use the right to a discount on set-off with the seller's claim for payment of the purchase price only with the seller's consent, unless the seller knew about the defects at the time of delivery of the goods, whereas in the case of legal defects, the period of exercise of the right by a third party is decisive. The buyer's protection is strengthened by his possibility not to pay part of the purchase price until the defects are removed, which would correspond to his entitlement to a discount if the defects were not removed.

Furthermore, it is necessary to draw attention to claims for defects of the goods, as these do not affect the claim for damages or contractual penalties. The buyer, who is entitled to a discount on the purchase price, is not entitled to claim compensation for lost profits due to the lack of quality of the goods to which the discount applies.

Withdrawal from the contract is the most extreme way of dealing with a situation where a contract is terminated. In general, the law allows the buyer to withdraw from the contract only if the defects were notified to the seller in time and if he can return the goods in the condition in which he received it. There are also according to Žofčinová & Košíková (2022) two exceptions to this rule:

1. if the impossibility of returning the goods in the condition stated therein is not due to the act or omission of the buyer, or
2. if there is a change in the condition of the goods as a result of a properly carried out inspection in order to detect defects of the goods [47].

The restriction of the buyer's ability to withdraw from the contract does not apply even if, prior to the discovery of the defect, the buyer sold the goods or a part of it to a third party, or the goods were consumed or altered in full or in part during their normal use. In this case, he is obliged to return the unsold or unconsumed goods or the forged goods and provide the seller with compensation up to the amount in which he benefited from the said use of the goods.

Delivery of more goods

The seller is generally obliged to perform properly and on time, i.e. deliver only the ordered quantity of goods. However, if the seller delivers a larger quantity of goods than specified in the contract, the buyer may accept the delivery or may refuse to accept the excess quantity of the goods. If the buyer accepts the delivery of all or part of the excess goods, he is obliged by law to pay for it the purchase price corresponding to the purchase price specified in the contract [39].

Acquisition of title

As is apparent from the diction of Section 409 of the Commercial Code, only movable property may be the subject of a purchase contract, where, unlike immovable property, the intabulation principle of the acquisition of property rights does not apply. For this reason, the Commercial Code governs in a special way, starting with Section 443, the method of acquiring ownership in such a way that 'the purchaser acquires ownership of the goods as soon as the delivered goods have been handed over to him'. Before handing over, the buyer acquires ownership of the goods transported only if he is authorized to dispose of the consignment.

Respecting the principle of dispositives on which the Commercial Code is based, the parties are allowed to agree in writing on the acquisition of ownership by the buyer before it is put into disposition, but subject to the fulfilment of two conditions:

1. the subject of purchase is goods determined individually or by type and at the time of transfer of ownership will be sufficiently marked to distinguish from other goods,
2. the method of acquisition of ownership is agreed between the parties, otherwise notified to the buyer without undue delay.

The parties may also agree in writing that the buyer acquires ownership of the goods later than provided for in the Commercial Code. If the content of this reservation of ownership does not indicate otherwise, it is assumed that the buyer has to acquire ownership only by paying the purchase price in full. According to Troitino et al. (2017), the legal protection of the buyer's good faith also consists of the fact that the buyer acquires ownership of the goods even if the seller is not the owner of the goods sold, unless at the time when the buyer had the ownership right to acquire, he knew that the seller is not the owner and that he is not entitled to dispose of the goods for the purpose of selling them [43].

Obligations of the Buyer

A contract of sale is a synalagmatic (two-sided) obligation relationship; each Party has both rights and obligations. The buyer's main obligations are to pay the purchase price for the goods and to take over the delivered goods in accordance with the contract. The purchase price can be determined by:

1. by agreement, i.e. the buyer is obliged to pay the agreed purchase price,
2. the method of determining e.g. according to the budget,
3. if the price is not agreed in the contract and the manner of its determination is not determined, and if the contract is valid in accordance with Section 409(2) of the Commercial Code, the seller may demand payment of the purchase price at which such or comparable goods were sold at the time of conclusion of the contract and under contractual conditions similar to the content of this contract.

If the purchase price is determined by the weight of the goods, its net weight is decisive in doubt [9].

The method, place and time of payment of the purchase price may be agreed by the parties in any way. In the event that this does not happen, there is a statutory provision according to which if the purchase price is to be paid when the goods or documents are handed over, the buyer is obliged to pay the purchase price at the place of delivery and if the contract does not indicate otherwise, the buyer is obliged to pay the purchase price when the seller in accordance with the contract and this Act allows the buyer to dispose of the goods or with documents enabling the buyer to dispose of the goods. The seller is allowed to make the delivery of goods or documents subject to payment of the purchase price.

In the course of trade, the goods sold are transported very often and this is remembered by the law in connection with the payment of the purchase price. If the seller has to send the goods according to the contract, he may do so on condition that the goods or documents enabling the handling of the goods are handed over to the buyer only upon payment of the purchase price. This applies only if this condition does not conflict with the agreed method of payment of the purchase price. The proverb "Don't buy a cat in a bag" also applies among entrepreneurs, from which it can be inferred that the buyer is not obliged to pay the purchase price until he has the opportunity to inspect the goods, unless the agreed method of delivery of the goods or payment of the purchase price would be contrary to this.

The buyer is according to Čajková et al. (2021) obliged to provide the seller with cooperation and to do all the actions necessary under the contract and the Commercial Code to enable the seller to deliver the goods. The buyer is obliged to take over the delivered goods if the contract itself or the law does not provide for the possibility to refuse to accept them [6].

In special cases, according to the contract, the buyer is obliged to determine the form, size or characteristics of the goods within the agreed period. If the deadline is not agreed, then within a reasonable period of time after the seller's request has been reached. Alternatively, the seller can determine it himself taking into account the needs of the buyer, as long as they are known to him, but this is without prejudice to other claims of the seller. In the event that the seller has made this determination himself, he must communicate details of this to the buyer and set a reasonable time limit. Within this period of time, the buyer may notify the seller of a derogating determination and if the buyer does not do so after such notification has been made and within a specified time, the determination notified by the seller is binding.

As stated above, the most important obligations of the buyer include payment of the purchase price and, as a result, the seller is entitled to require the buyer to pay the purchase price, to take over the goods and to fulfil its other obligations until the seller has exercised the right of breach of the contract which is incompatible with that requirement [24].

In accordance with the provisions of Section 454 of the Commercial Code, when securing the obligation to pay the purchase price was agreed, the buyer is obliged to hand over to the seller all necessary documents proving that payment of the purchase price was secured in accordance with the contract in time before the time agreed for delivery of the goods. Non-fulfilment of this obligation allows the seller to refuse delivery of the goods until the delivery of these documents. If the buyer does not ensure payment of the purchase price or retrospectively within a reasonable period specified by the seller, the seller may withdraw from the purchase contract under the law.

Risk of damage to goods

According to Funta and Plavčan (2021) very sensitive issue is the transfer of the risk of damage to the goods transferred to the buyer according to the law, either at the time when he takes over the goods from the seller or, if he does not do so in time, at a time when the seller allows him to dispose of the goods. The buyer violates the contract by not taking over the goods [12]. In the event that the buyer has to take over the goods from a person other than the seller, the risk of damage to the goods passes to the buyer within the time specified for the delivery of the goods, if he was allowed to dispose of the goods at that time and the buyer was aware of this possibility. If the buyer is allowed to dispose of the goods or if he becomes aware of this option only later, the danger passes only from the time when he had the opportunity to learn about it.

There is a different situation in the transport of goods. If the seller is obliged under the contract to hand over the goods to the carrier in a specific place for the transport of the goods to the buyer, the risk of damage to the goods shall pass to the buyer by handing it over to the carrier in that place. In the event that the purchase contract also includes the obligation of the seller to send the goods, but the seller is not obliged to hand over the goods to the carrier in a certain place, the law stipulates that the risk of damage to the goods passes to the buyer when the goods are handed over to the first carrier for transport to the destination. The mere fact that the seller handles the documents relating to the transported goods no longer affects the passing of the risk of damage to the goods. In this regard, we refer to the dictation of Section 458 of the Commercial Code, according to which the risk of damage to goods, determined by type and not taken over by the buyer, does not pass to the buyer until the goods are clearly marked for the purpose of the contract by marking on the goods or transport documents or specified in the message sent to the buyer or otherwise defined [14].

Even in the event of danger of damage to the goods, the contracting parties are allowed by way of derogation to regulate the issue of this liability before the period referred to in Sections 455 to 458 of the Commercial Code, but only for goods individually determined or for goods determined by type, if, at the time of the passing of the risk of damage, it is sufficiently separated and distinguished from other goods of the same kind [9].

In particular in international trade, goods are often purchased during transport; since it is relatively impossible to ascertain his condition, the legislator shifted the risk of damage to the goods before the conclusion of the contract of sale to the moment when the goods were handed over to the first carrier. The exception is when it is established that the seller knew or should have known, taking into account all the circumstances, that damage to the goods had already occurred, which means that the damage is borne by the seller and not by the carrier. The occurrence of damage to the goods, which occurred after the transfer of its danger to the buyer, does not affect his obligation to pay the purchase price except where the damage to the goods occurred as a result of a breach of the seller's obligation.

Preservation of goods

The Commercial Code also addresses the position of the parties in the event of a delay. If the buyer is in delay with taking over the goods or paying the purchase price in cases where the delivery of the goods and payment of the purchase price is to take place at the same time, and the seller has the goods with himself or may otherwise dispose of it, it is on the seller's shoulders to take appropriate measures to preserve the goods. The law grants him a retention right to the goods until the buyer pays the reasonable costs incurred by the retention of the goods [9].

Paragraph 463 of the Commercial Code obliges the buyer who has taken over the goods and intends to refuse them to take measures appropriate to the circumstances of the preservation of the goods and is also entitled to reimbursement of these costs together with the retention right to the goods [9]. If the buyer has the possibility to dispose of the goods after transporting the goods to their destination and at the same time exercises the right to refuse them, he is obliged to take over the goods and have them with him on behalf of the seller, if he can do so:

1. without payment of the purchase price
2. without undue inconvenience and expense.

However, this obligation does not arise if the seller or the person appointed by the seller to care for the goods is present at the place of destination. The obligation to preserve the goods pursuant to Sections 462 to 464 of the Commercial Code may also be fulfilled by depositing the goods in the warehouse of a third party on behalf of the other party and may demand payment of reasonable costs incurred by virtue of the concluded storage contract within the meaning of Sections 527 to 534 of the Commercial Code. The party who is in delay with the takeover or take-back of the goods or the payment of the purchase price to be made on receipt of the goods, or the payment of the costs associated with the fulfilment of the obligations under Sections 462 to 464 of the Commercial Code, may be called upon to comply with this obligation.

According to Kryvinska and Bickel (2020), the other party is entitled in the invitation to take over the goods, the form of which is not laid down by law, to determine a reasonable period for this and, after its vain expiry, to sell the goods in an appropriate manner [24]. However, prior to that sale, it is obliged to notify the late party of the intention to sell the goods, which may also be notified when the takeover period is fixed.

Special rules shall apply to perishable goods or where their preservation entails disproportionate costs. The obliged party must take reasonable steps to sell it in order not to depreciate it and, as far as possible, to notify the other party of the intended sale. After its sale, the party who sold the goods has the right to retain from the proceeds an amount corresponding to the reasonable costs associated with fulfilling the obligations under Sections 462 to 464 of the Commercial Code, i.e. the preservation of the goods and their actual sale. The remainder of the proceeds obtained shall be returned to the other party without delay [9].

Termination of the contract

The Commercial Code does not contain special provisions dealing with the issue of the termination of the purchase contract or the obligation from it. As a general rule, the obligation of the parties to the contract is extinguished by performance. Special methods of termination can also be included in the case of recurring performances, agreement to terminate the contract, termination of the contract, withdrawal from the contract, frustration of the purpose of the contract and others.

5   Commercial Agency Contract

A commercial agency contract is a special contractual type, falling within the category of relative commercial obligation relationship. The provisions of Sections 652 to 672a of the Commercial Code are relatively detailed, of which Sections 655(1), 655a, 660(2) to (4), 668(3), 668a, 669, 669a, 672a are mandatory [9]. The exceptionality of this named contract lies in the fact that only an entrepreneur can be a commercial agent and principal. By an agency contract, the commercial agent, as an entrepreneur, undertakes for the principal to pursue an activity aimed at concluding a certain type of contract ('the shops') or to negotiate and conclude transactions in the principal's name and on his behalf, and the principal undertakes to pay a commission to the agent. The mandatory requirements of the contract are:

- the designation of the parties, i.e. the commercial agent and the principal;
- an obligation on the agent to engage in an activity for the principal in order to conclude a certain type of contract or to negotiate and conclude transactions in his name and on his behalf;

- the principal's obligation to pay a commission to the agent.

According to Tokareva et al. (2021), the Act then negatively defines the person of the commercial agent, i.e. who cannot be the agent:

- a person who, as an authority, may bind the principal;
- a member or member of a cooperative, legally empowered to bind other members or members of the cooperative, if the cooperative is represented; or
- liquidator, receiver for administration, receiver of bankruptcy estate, special administrator or arrangement trustee represented [42].

The commercial agency contract must be in writing under the penalty of nullity and, unless otherwise provided for in the provisions of Sections 652 to 672a of the Commercial Code, this type of contract shall also apply mutatis mutandis to the mediation contract. The subject of the commercial agent's obligation is to seek out candidates for the conclusion of trades as defined in the contract, which, if it determines that the commercial agent performs legal acts on behalf of the principal, these rights and obligations are governed by the provisions of the Commercial Code on the Mandate Contract.

The possibility of a commercial agent to perform legal acts on behalf of the principal is subject to the granting of a written power of attorney containing the particulars set out in the Civil Code, which, as a "lex generalis" of private law, allows a natural or legal person to be represented by another natural person or a legal person, provided that they grant powers of attorney to perform legal acts for them [28]. In the interests of legal certainty, the scope of the representative's authorization must always be indicated in the power of attorney. In general, the theory of civil law defines power of attorney as a unilateral legal act addressed to third parties who certify the existence and scope of the representative's authority to act in the name and on behalf of the principal. In terms of the number of legal acts covered by the power of attorney, we distinguish:

- a fully-fledged (general) authorising representative for all legal acts;
- power of attorney special (specially) authorising a representative to do either one act or only certain types of legal acts.

Without a power of attorney, the commercial agent is not entitled to enter into trades on behalf of the principal, to accept anything for him or to do other legal acts. The duties of the commercial agent shall include in particular:

- in a specified territorial area to carry out with professional care an activity which is the subject of his obligation and, if this territorial area is not specified in the contract, there is a legally rebuttable presumption that the commercial agent is to operate in the territory of the Slovak Republic,
- carry out the agreed activity with due diligence and in good faith;
- must observe the interests of the principal, act in accordance with the principal's mandate and reasonable instructions and provide the principal with the necessary and accessible information;

- report to the principal on the evolution of the market and all circumstances relevant to the interests of the principal, in particular for its decision-making in connection with the conclusion of trades;
- if the contract also includes the conclusion of trades by an agent on behalf of the principal, enter into these transactions only on the terms and conditions specified by the principal, unless the principal has given his consent to the other procedure;
- if it is unable to carry out its activities, inform the principal without undue delay;
- to act as part of their commitment to carry out concluded transactions on the instructions of the principal and in the interest of the principal, which are or must be known to the agent, in particular in resolving irregularities arising from closed transactions.
- for the principal's need to keep the documents he has acquired in connection with his activity for as long as such documents may be relevant for the protection of the principal's interests [9].

In mutual relations with the commercial agent, the mandatory provision of Section 655a of the Commercial Code obliges the principal to act honestly and in good faith and in particular is obliged to:

- provide the commercial agent with the necessary data relating to the subject matter of the business;
- to provide the commercial agent with the information necessary for the performance of the obligations arising from the contract, in particular to inform the commercial agent within a reasonable period of time that it presupposes a substantial reduction in the scope of the activity compared to what the commercial agent could normally expect;
- inform the agent within a reasonable period of time that he has accepted, refused or has not carried out the transaction procured by the agent;
- hand over to the agent all the documents and aids necessary for the performance of his obligation, which remain the property of the principal and the commercial agent is obliged to return them to the agent after the termination of the contract, unless, by their nature, the commercial agent has used them in the performance of his business [9].

A commercial agent shall, in the course of his duties, propose to enter into dealings only with persons who are expected to fulfil their obligations, provided that he does not guarantee the fulfilment of the obligations by a third party with whom he has proposed to the principal to enter into a business with or with whom he has entered into business on behalf of the principal, unless:

1. to do so in writing, and
2. on condition that he receives a special remuneration for taking over the guarantee.

The commission

According to Pavelek and Zajíčková (2021) the agent is entitled to commission:

1. agreed in a commercial agency contract, or

2. corresponding to practices in the field of activity according to its place of performance, taking into account the type of business which is the subject of the contract
   [31].

Commission also means the remuneration of a commercial agent for an activity
carried out depending on the number or level of trades executed. In that regard, as in
the case of an intermediation contract, a commercial agent is entitled, in addition to
commission, to reimbursement of the costs associated with his activity only if it has
been negotiated and if the contract does not provide otherwise unless he is entitled
to commission from the transaction to which the costs relate.

In order to prevent speculative conduct of the commercial agent, the law does not
grant him the right to commission and to the agreed reimbursement of costs in cases
where he was acting as a commercial agent or intermediary for the person with whom
the principal entered into the business at the time of the conclusion of the trade [32].
A commercial agent shall be entitled to commission for trades carried out during the
term of the contractual relationship if:

1. the trade is concluded as a result of its activities;
2. the transaction is concluded with a third party which it acquired prior to the
   conclusion of the transaction for the purpose of carrying out transactions of that
   kind.

However, the agent is not entitled to commission even if the above conditions are
met, where the right to that commission belongs to the previous agent itself and, in the
circumstances of the case, it would not be fair to distribute the commission between
the two agents. The supplementary provision of Section 660(1) of the Commercial
Code provides for a right to commission if no agreement has been concluded pursuant
to Section 661 of the Commercial Code which arises at the time when:

- the principal has fulfilled an obligation arising from the trade; or
- the principal is obliged to fulfil an obligation arising out of a trade with a third
  party; or
- the third party has fulfilled an obligation arising from the trade.

The following mandatory regulation protects the commercial agent from
attempting by the principal to delay or otherwise make the payment of his commission conditional. In that regard, the right to commission arises at the latest when the
third party has fulfilled its part of the obligation or was obliged to fulfil it if the principal has fulfilled its part. However, if a third party is required to fulfil his obligation
only after more than six months after the conclusion of the trade, the agent shall be
entitled to a post-trade commission [37]. The commission shall be payable at the
latest on the last day of the month following the end of the quarter in which it was
due.

The principal is required by law to issue a written confirmation to the agent with a
breakdown of the main items necessary for the calculation of the commission, even
without requesting it, at the latest on the last day of the month following the end of
the quarter in which the commission became due. This is without prejudice to the

commercial agent's entitlement to request the information available to the principal and necessary to check the calculation of the commission.

The basis for determining the commission is the extent of the obligation fulfilled by the third party. In this case, the non-performance due to the reason for which the principal is responsible shall also be included in the basis thus determined. The activity of a commercial agent may be agreed only to ensure that the principal is procured an opportunity to enter into a business with a third party with certain content. In such a case, the agent is entitled to a commission as soon as the opportunity is acquired, regardless of the outcome [39].

The right to commission shall cease if it is apparent from the circumstances of the case that the trade between the principal and the third party will not take place and if the failure to do so is not the result of the circumstances for which the principal is responsible, unless the business provides otherwise. In so doing, the commission already paid must be refunded if the right to it has ceased for that reason. Due to the hardness of the law in this section, the parties are allowed to derogate from this statutory provision, but always for the benefit of the commercial agent.

Non-exclusive commercial representation

As Sararu (2008) the Slovak Commercial Code allows the parties to agree on a non-exclusive agency. This means that the principal may also entrust other persons with a commercial agency which he has negotiated with the agent. A commercial agent may carry out an activity to which he or she has committed himself to the principal also for other persons or to enter into trades which are the subject of a commercial agency, on his own account or on behalf of another person. In the case of non-exclusive commercial representation, the law allows the principal to have several commercial agents [35].

Exclusive Sales Representation

In the case of an agreement between the parties on an exclusive agency the principal is not authorised to use another commercial agent in a specified territorial area and for a specified range of shops. Likewise, the commercial agent may not, to this extent, perform a commercial agency for other persons or enter into trades on his own account or on behalf of another person. A breach of this obligation shall give rise to a right for the party concerned to claim damages within the meaning of Sections 373 et seq. The Commercial Code as well as the right of withdrawal, as referred to in Section 672 of the Commercial Code [33].

The principal is entitled to enter into transactions which are covered by the exclusive agency even without the cooperation of the commercial agent, however, unless the contract stipulates otherwise, it is obliged to pay the agent a commission on these transactions as if these transactions had been concluded with his cooperation. However, this is a supplementary provision which can be excluded by agreement between the parties.

Dismantling of commercial representation

A commercial agency contract can be terminated in several ways. In the first place, the agent's obligation expires at the end of the period for which the contract was concluded. If, after its expiry, the parties continue to abide by the contract, the law presupposes that the contract has been extended without restriction.

Furthermore, the contract is agreed for an indefinite period if:

- this is determined by the contract, or
- the contract does not contain a provision on the time for which it is to be concluded; or
- there is no limitation in time from the purpose of the contract.

A contract agreed for an indefinite period may be terminated by either party by denunciation. The following mandatory provision of Section 668(3) of the Commercial Code governs the length of the notice period, which is:

- one month on termination given during the first year of the contract;
- two months at notice given during the second year,
- three months if the contract lasts three years or more.

The Contracting Parties are allowed under the judgment of the Supreme Court of the Slovak Republic (Supreme Court of the Slovak Republic). According to Skóra et al. (2022) in the period of 33/2006, agree only on an extension of the notice period and, if they so agree, the time limit by which he is bound by the principal may not be shorter than the period to be respected by the commercial agent [38]. As a general rule, the period of notice expires at the end of the calendar month, which may, however, be regulated by the contract by way of derogation. Such rules also apply in the case of a fixed-term contract which has been converted into a contract concluded for an indefinite period in accordance with Section 667 of the Commercial Code, provided that the notice period is calculated taking into account the duration of the contractual relationship which preceded the conversion into a contract of indefinite duration.

We also draw attention to the view of the court, according to which an agency contract agreed for an indefinite period can be terminated by either party by notice. If one party has expressed in writing the intention to terminate the agency contract in a certain and comprehensible manner towards the other party, the nullity of the termination clause as an ancillary element of the contract shall not invalidate the termination [43].

The commercial agent shall be guaranteed in a mandatory manner the right to compensation for damage suffered as a result of the termination of contractual relations with the principal, provided that:

1. he has not been paid the commission due to the performance of the commercial agency in accordance with Paragraph 655(1) of the Commercial Code, even though he has benefited substantially from his activity;
2. he has not been reimbursed for the costs incurred by him in connection with his obligation and the agency contract confers on him the right to reimbursement.

In the event of termination of the contract, the commercial agent shall be entitled to severance payments subject to two conditions, which are:

- for the principal has acquired new customers or significantly developed business with existing customers and the principal has substantial benefits arising from their dealings with them
- the payment of severance payments, taking into account all the circumstances, in particular the commission forfeited by the agent and resulting from the transactions carried out with those customers, is fair; those circumstances also include the use or non-use of the so-called 'competition clause' under Paragraph 672a.

The commercial agent's right to severance payments thus granted shall cease if, within one year of the termination of the contract, he fails to notify the principal that he is exercising his rights to payment [44].

The legislature limits the amount of the severance allowance by the supplementary provision of Section 669(2) of the Commercial Code, according to which its amount may not exceed the average annual commission calculated on the basis of the average commission received by the agent during the last five years. However, the grant of severance pay does not mean the termination of the commercial agent's right to compensation under Section 668a of the Commercial Code. We draw particular attention to the existence of a right to severance payments even if the termination of the contract occurs as a result of the death of the commercial agent [46]. A right to severance shall not arise if:

- the principal has withdrawn from the contract for breach of a contractual obligation by the agent, which gives rise to a right of withdrawal;
- the agent has terminated the contract and such termination is not justified by the circumstances on the part of the principal or the age, invalidity or illness of the commercial agent and if he cannot reasonably be required to continue his business;
- under the agreement with the principal, the commercial agent transfers the rights and obligations from the contract to a third party.

As Poiana (2015) in the event that the exclusive representation has been agreed for a fixed term, either party may terminate the contract lawfully if the volume of the trade has not reached the volume of trades specified in the contract within the last 12 months, otherwise appropriate to the sales opportunities [33]. The right to severance payments also belongs to the commercial agent when the contract is terminated by the principal. After the termination of his obligation under Section 668 or Section 670 of the Commercial Code, the commercial agent shall be entitled to commission even if:

1. the trade was carried out mainly as a result of the activities of the agent and within a reasonable period of time after the end of the contract;
2. in accordance with the conditions laid down in Section 659a, the order of the third party was received by the principal or commercial agent before the termination of the contract,
3. the third party's obligation was fulfilled only after the end of the contract.

According to Števček and Ivančo (2021) Section 672, which is closely linked to the provision of Section 655 of the Commercial Code governing exclusive (exclusive) commercial representation, establishes the right of the parties to withdraw from the contract, which is not only the commercial agent but also the principal 40. This means that if, at the time of the arrangement of the exclusive agency:

- uses the represented other commercial agent, *is the authorised agent to withdraw from the contract,*
- the commercial agent performs an activity which is the subject of his obligation towards the principal and for other persons, the *principal may withdraw from the contract.*

The Competitive Clause

Nováčková and Vnuková (2021) claims that, the commercial agency contract may agree in writing that the agent may not, for a maximum period of two years after the termination of the contract in the designated territory or towards a specified group of customers in that territory, carry out, on his own account or on behalf of a foreign country, an activity which has been the subject of a commercial agency or any other activity which would be competitive with the principal's business [29]. That clause restricts in some way the commercial agent's right to conduct business freely upon termination of the contract, and therefore the mandatory provision of Paragraph 672a of the Commercial Code lays down three conditions for the validity of that clause:

1. it must be agreed in writing,
2. is limited in time to a maximum of two years,
3. it applies only to a certain territorial area, certain customers and certain shops.

In the event of undue restriction of the rights of the commercial agent, the court may restrict or invalidate such a competitive clause.

6 Conclusion

In view of the stated objective of our scientific study, we concluded by analysing scientific and professional literature, legislation and selected case law of the Supreme Court of the Slovak Republic, using the Institute of Legal Logic, that the legal order of the Slovak Republic also contains in the context of the European Digital Strategy relatively complex e-commerce legislation and allows for the electronic conclusion of not all but only selected types of commercial contracts. Quite extensive legislation fulfils its purpose and regulates mainly commercial representation from both the substantive and procedural aspects. In doing so, however, it provides the commercial agent, as a general rule, the weaker party in the contractual relationship with the necessary and effective protection. At the same time, in the context of the principle of availability and contractual freedom on which the Commercial Code is based, in the case of a purchase contract, the agreement allows both the buyer and the seller to grant more rights than the Commercial Code guarantees them.

However, the added value of our investigation is a number of important findings. We critically assess the long-term lack of interest of the legislator to codify and thus unify the regulation of e-commerce, which is literally atomised. In a number of laws applicable to e-commerce, such as the Civil Code, the Commercial Code, the Electronic Commerce Act, the Consumer Protection Act, the Electronic Signature Act, the Consumer Credit Act and others, an experienced lawyer working in this field also has a problem. In the case of a layman, many often incomprehensible legislation literally has to be horrifying for him. The pitfalls are also apparent in the duplication of legislation of institutes such as the sales contract or the terms as a consumer contained in several regulations with different meanings. In today's information age, codification of e-commerce into a separate code would remove many barriers, barriers and, in particular, the concerns of potential entrepreneurs. In our view, such a positive intervention by the legislator would clearly have a positive impact on job creation in today's post-Covid-19 era.

Acknowledgements This article was created thanks to support under the Operational Program Integrated Infrastructure for the project: National infrastructure for supporting technology transfer in Slovakia II – NITT SK II, co-financed by the European Regional Development Fund.

References

1. Aksamovic, D., Simunovic, L.: The EU scheme for the state aid rules in the air transport sector during the Covid-19 crisis. Juridical Trib. **12**(1), 51–67 (2022). https://doi.org/10.24818/TBJ/2022/12/1.04
2. Androniceanu, A., Popescu, C.R.: An inclusive model for an effective development of the renewable energies public sector. Adm. Manag. Public **2017**(28), 81–96 (2017)
3. Čajka, P., Abrhám, J.: Regional aspects of V4 countries' economic development over a membership period of 15 years in the european union. Slovak J. Polit. Sci. **19**(1), 89–105 (2019). https://doi.org/10.34135/sjps.190105
4. Čajková, A., Čajka, P., Elfimova, O.: Personality and charisma as prerequisites for a leading position in public administration. In: Springer Proceedings in Business and Economics, pp. 199–211 (2018). https://doi.org/10.1007/978-3-319-74216-8_21
5. Čajková, A., Gogová, A.: Case study of the knowledge management process in selected department of state administration in Slovakia. Stud. Syst. Decis. Control **420**, 533–545 (2022). https://doi.org/10.1007/978-3-030-95813-8_21
6. Cajková, A., Jankelová, N., Masár, D.: Knowledge management as a tool for increasing the efficiency of municipality management in Slovakia. In: Knowledge Management Research and Practice, In press (2021). https://doi.org/10.1080/14778238.2021.1895686
7. Čajková, A., Šindleryová, I.B., Garaj, M.: The covid-19 pandemic and budget shortfalls in the local governments in Slovakia. Sci. Pap. Univ. Pardubic. Ser. D: Fac. Econ. Adm. **29**(1), 1243 (2021). https://doi.org/10.46585/sp29011243
8. European Commission: Shaping Europe's Digita Future. [online] [cit. 1.6.2022]: https://ec.europa.eu/info/sites/default/files/communication-shaping-europes-digital-future-feb2020_en_4.pdf

9. Federal Assembly of the Czech and Slovak Federal Republic: Act no. 513/1991 Coll. Commercial Code as amendet. [online] [cit. 1.6.2022]: https://www.slov-lex.sk/pravne-predpisy/SK/ZZ/1991/513/20211228

10. Chochia, A., Nässi, T.: Ethics and emerging technologies – facial recognition. Revista de Internet, Derecho y Politica 34 (2021). https://doi.org/10.7238/idp.v0i34.387466

11. Dănişor, D.C., Dănişor, M.C.: Modern solidarity and administrative repression. Juridical Trib. **11**(3), 472–496 (2021). https://doi.org/10.24818/TBJ/2021/11/3.04

12. Funta, R., Plavčan, P.: Regulatory concepts for internet platforms. Online J. Model. New Eur. **35**, 44–59 (2021). https://doi.org/10.24193/OJMNE.2021.35.03

13. Funta, R.: Discounts and their effects - economic and legal approach. Danube **5**(4), 277–285 (2014). https://doi.org/10.2478/danb-2014-0015

14. Funta, R.: Economic and legal features of digital markets. Danube **10**(2), 173–183 (2019). https://doi.org/10.2478/danb-2019-0009

15. Funta, R., Králiková, K.: Obligation of the European Commission to review national civil court judgements? Juridical Trib. **12**(2), 215–226. https://doi.org/10.24818/TBJ/2022/12/2.04

16. Horvat, M., Magurová, H., Srebalová, M.: Protection of consumers' rights in railway in the Slovak Republic. Yearb. Antitrust Regul. Stud. **10**(16), 177–190 (2017). https://doi.org/10.7172/1689-9024.YARS.2017.10.16.9

17. Jančíková, E., Pásztorová, J.: Promoting eu values in international agreements. Juridical Trib. **11**(2), 203–218 (2021). https://doi.org/10.24818/TBJ/2021/11/2.04

18. Jankurová, A., Ljudvigová, I., Gubová, K.: Research of the nature of leadership activities. Econ. Sociol. **10**(1), 135–151 (2017). https://doi.org/10.14254/2071-789X.2017/10-1/10

19. Joamets, K., Chochia, A.: Access to artificial intelligence for persons with disabilities: Legal and ethical questions concerning the application of trustworthy AI. Acta Balt. Hist. Philos. Sci. **9**(1), 51–66 (2021). https://doi.org/10.11590/ABHPS.2021.1.04

20. Joamets, K., Chochia, A.: Artificial intelligence and its impact on labour relations in Estonia. Slovak J. Polit. Sci. **20**(2), 255–277 (2020). https://doi.org/10.34135/sjps.200204

21. Kelblová, H.: Right to privacy and some methods of direct marketing. Acta Univ. Agric. Silv.Ulturae Mendel. Brun. **61**(7), 2277–2283 (2013). https://doi.org/10.11118/actaun201361072277

22. Kelblová, H.: Legal aspects of some internet marketing instruments. Acta Univ. Agric. Silviculturae Ulturae Mendel. Brun. **60**(2), 117–124 (2012). https://doi.org/10.11118/actaun201260020117

23. Krejčová, A., Rašticová, M.: Retire – why yes, why not? Results of a study on the Czech seniors. J. East. Eur. Central Asian Res. **7**(3), 316–328 (2020). https://doi.org/10.15549/jeecar.v7i3.413

24. Kryvinska, N., Bickel, L.: Scenario-based analysis of IT enterprises servitization as a part of digital transformation of modern economy. Appl. Sci. **10**(3), 1076 (2020). https://doi.org/10.3390/app10031076

25. Kryvinska, N., Kaczor, S., Strauss, C.: Enterprises' servitization in the first decade - retrospective analysis of back-end and front-end challenges. Appl. Sci. **10**(8), 2957 (2020). https://doi.org/10.3390/app10082957

26. Kyncl, L.: Four authorization protocols for an electronic payment system. Lecture Notes in Computer Science (including subseries Lecture Notes in Artificial Intelligence and Lecture Notes in Bioinformatics), pp. 205–214 (2012). https://doi.org/10.1007/978-3-642-25929-6_19

27. Matějková, J., Pavelek, O.: The protective purpose of the contract and the liability of an expert towards a third party in Czech, Austrian, and German private law. Balt. J. Law Polit. **12**(2), 163–185 (2019). https://doi.org/10.2478/bjlp-2019-0016

28. Matúšová, S., Nováček, P.: New generation of investment agreements in the regime of the European Union. Juridical Trib. **12**(1), 21–34 (2022). https://doi.org/10.24818/TBJ/2022/12/1.02

29. Nováčková, D., Vnuková, J.: Competition issues including in the international agreements of the eropean union. Juridical Trib. **11**(2), 234–250 (2021). https://doi.org/10.24818/TBJ/2021/11/2.06

30. Oláh, J., Aburumman, N., Popp, J., Khan, M. A., Haddad, H., Kitukutha, N.: Impact of industry 4.0 on environmental sustainability. Sustainability (Switzerland) 12(11), 4674 (2020). https://doi.org/10.3390/su12114674

31. Pavelek, O., Zajíčková, D.: Personal data protection in the decision-making of the CJEU before and after the lisbon treaty. TalTech J. Eur. Stud. 11(2), 167–188 (2021). https://doi.org/10.2478/bjes-2021-0020

32. Poiană, O.: An overview of the European energy policy evolution: from the European energy community to the European energy union. Online J. Model. New Eur. 22(1), 175–189 (2017). https://doi.org/10.24193/ojmne.2017.22.09

33. Poiană, O.: Regional cooperation and national preferences in the Black Sea region: a zero-sum game perpetuated by energy insecurity? Secur. Democr. Dev. South. Cauc. Black Sea Region 14, 307–332 (2015). https://doi.org/10.3726/978-3-0351-0836-1

34. Peráček, T.: E-commerce and its limits in the context of the consumer protection: the case of the Slovak Republic. Juridical Trib. 12(1), 35–50 (2022). https://doi.org/10.24818/TBJ/2022/12/1.03

35. Săraru, C.S.: Considera ţii cu privire la limitele libertătii contractuale în dreptul public impuse de integrarea în uniunea europeană. Transylv. Rev. Adm. Sci. 1, 131–140 (2008)

36. Săraru, C.S.: The European groupings of territorial cooperation developed by administrative structures in Romania and Hungary. Acta Juridica Hung. 55(2), 150–162 (2014)

37. Săraru, C.S.: Public domain and private domain. In: Administratve law in Romania, pp. 84–100 (2019)

38. Skóra, A., Srebalová, M., Papáčová, I.: Administrative judiciary is looking for a balance in a crisis. Juridical Trib. 12(1), 5–20 (2022). https://doi.org/10.24818/TBJ/2022/12/1.01

39. Suchomelová, A., Procházka, J., Ďuriník, M.: Personal interest branding: source of price premium. J. Int. Consum. Mark. 29(1), 27–34 (2017). https://doi.org/10.1080/08961530.2016.1236309

40. Števček, M., Ivančo, M.: The conception and institutional novelties of recodification of private law in the Slovak Republic. In: The Law of Obligations in Central and Southeast Europe: Recodification and Recent Developments, pp. 33–49 (2021)

41. Tkachenko, R., Izonin, I., Kryvinska, N., Dronyuk, I., Zub, K.: An approach towards increasing prediction accuracy for the recovery of missing IoT data based on the GRNN-SGTM ensemble. Sensors 20(9), 2625 (2020). https://doi.org/10.3390/s20092625

42. Tokareva, V., Davydova, I., Adamova, E.: Legal problems of the use of orphan works in digital age. Juridical Trib. 11(3), 452–471 (2021). https://doi.org/10.24818/TBJ/2021/11/3.03

43. Troitino, D.R., Chochia, A., Kerikmäe, T.: The incapacity of the union to act as a reliable actor in the international arena. In: European Union: Political, Economic and Social Issues, pp. 11–32 (2017)

44. Troitiño, D.R., Kerikmäe, T., Chochia, A.: Foreign affairs of the European Union: how to become an independent and dominant power in the international arena. In: The EU in the 21st Century: Challenges and Opportunities for the European Integration Process, pp. 209–230 (2020). https://doi.org/10.1007/978-3-030-38399-2_12

45. Žofčinová, V., Čajková, A., Král, R.: Local leader and the labour law position in the context of the smart city concept through the optics of the EU. TalTech J. Eur. Stud. 11(2), 3–26 (2022). https://doi.org/10.2478/bjes-2022-0001

46. Žofčinová, V., Horváthová, Z., Čajková, A.: Selected social policy instruments in relation to tax policy. Soc. Sci. 7(11), 241 (2018). https://doi.org/10.3390/socsci7110241

47. Žofčinová, V., Košíková, A.: Selected legislative instruments of family policy supporting work–life balance: a comparison of Italy and the Slovak Republic. Online J. Model. New Eur. 39 (2022)

AI in Customer Relationship Management

Gutola Bokayo Roba and Petra Maric

Abstract By the impact of globalization on market business conditions, companies must adapt their business more and more quickly to new situations. New strategic directions, especially Artificial Intelligence (AI), have become imperative for the successful business of today's companies. An intelligent company makes better decisions faster and outperforms its competitors. Consumer Relationship Management (CRM) serves as a business strategy for choosing customer management to optimize long-term value. CRM also requires a business philosophy focusing on consumers and a business culture that supports effective marketing, sales, and service processes. CRM integrates people, processes, and technology to maximize relationships with all consumers. Through the collaborative CRM, all communication towards clients is carried out, while their responses to the information system arrive via the operating part of the CRM. Analytical CRM, through a detailed analysis of a multitude of data based on expert knowledge, creates an image of each client, his needs, and desires to develop more vital interconnections.

1 Introduction

1.1 Relevance

Customer Relationship Management (CRM) aims to improve commercial relationships between businesses and their clients through apt sales management, productivity enhancement, and adequate contact management [1, 2, 3]. Generally, CRM reduces operational costs for businesses improves customer experiences, interactions, and procedural efficiency [4, 3]. There exist three components of CRM. They

G. B. Roba (✉) · P. Maric
Faculty of Management, Comenius University Bratislava, Odbojárov 10, P.O. Box 95, 820 05 Bratislava 25, Slovak Republic
e-mail: gutola1@uniba.sk

include operational CRM, Collaborative CRM, and Analytical CRM [5]. Each category targets different aspects of Customer Relationship Management. However, the most significant difference between Operational and the two remaining components, collaborative and analytical CRM, is that operational CRM dwells on varying official front-end entities [5–9].

Modern times seem to change the dynamics in business operations. It provides the need to change with time and enhance customer relationship management techniques [7]. Change gets characterized by varying cultural improvements and developments, life, fashion, behavior, and preferable lifestyles [6–8]. They also formulate the main factors enabling the integration of Artificial Intelligence in CRM. Therefore, it is essential to integrate AI to develop a nexus between the changing modern and apt customer relationship models [9]. It is possible for incorporations, both big establishments with gigantic market shares and small business entrepreneurs, to develop marketing strategies based on the clients' orientations [9, 10]. It can get done using modern technology and Information and multi-agent systems technologies. In effect, subsystems intellectualization is influential in bridging the gaps between real-life merits for improved CRM and the theoretical aspects [11]. They are responsible for effective interactions with customers, meeting specific company requirements, and training for solutions to uncertainties.

Companies get faced with many challenges using standard CRM systems. A good example is how salespeople focus on profitable business items and criteria. However, most of their task force needs to concentrate on data recording [11, 12]. In this light, handling gathered customer related information becomes relevant to fulfill their demands. In this instance, AI integration in CRM plays an essential role since data can be recorded automatically, and patterns get observed [12]. The goals of introducing Artificial Intelligence and CRM components relate to maximizing profits as soon as possible, standardizing, centralizing, expanding sales in the global market, and simplifying delivering insurance to customers.

The research aims to provide general Information on artificial Intelligence and to point out that artificial Intelligence influences the effectiveness of the insurance sector to a great extent. In the scientific sense of the lower rank, scientific descriptors as a goal in this field have great significance in terms of actuality, the attractiveness of the topic, and potentially practical importance. Expanding knowledge about AI, its advantages and disadvantages, and its impact on the insurance industry, albeit descriptively, enriches knowledge (quantitative and qualitative) about a familiar segment of this global problem.

The topic is broad and challenging to present all these aspects precisely. For this purpose, the following pages are focused on two main themes, AI and CRM. Due to the actuality of the topic, only secondary literature such as books and online sources was used during this work.

1.2   Research Objectives and Goals

Two questions arise from the above background information. They include:

- What is the scope of Artificial Intelligence in Customer Relationship Management?
- And How do companies increase sales and customer experience with AI driven CRM?

Therefore, for these questions to get answered the research entails several objectives. First, the paper shall comprise of a literature review for how AI has got used in relations to CRM, a definition of the important terms and their explanation.

In the process the author aims to explain the concept of Artificial Intelligence in businesses and how enterprises utilize AI and CRM. It shall require a close analysis of the relationship between AI and CRM, sales, marketing, and customer services. More so, the papers also aim at displaying a definitive future of AI and Customer Relationship Management, and the ethical concerns of Artificial Intelligence.

2   Theoretical and Conceptual Background

2.1   Customer Relationship Management

Several authors attempt to define CRM. According to [13], CRM is a philosophy and business strategy that, with the help of technology and business systems, aims to improve human interactions in the business environment.

According to the definition offered by Ronald Swift in his *book Accelerating Customer Relationships*, Customer Relationship Management or CRM, entails: "Striving the whole company to understand customer behavior better and gain the ability to influence them through a variety of forms of meaningful communication; to continuously improve the ability to attract new and retain existing customers and raise the level of their loyalty and usefulness."

CRM is a business and customer communication strategy that collects and uses Information about clients to increase their satisfaction, loyalty, and profitability. It is often used to synthesize CRM to combine client acquisition, customer retention, service quality, and profitability [14, 15]. CRM is a key component of business organizations. CRM is a process of data collection, analysis, and appropriate actions and measurements [15].

According to [16], there are three key elements for a successful CRM initiative:

- People,
- Processes,
- Technology.

Fig. 1 Factors enhancing
successful CRM [15]

In addition, some factors influence CRM. Both cases mention technology, processes, and people or their interaction. As can be seen from Fig. 1, several factors influence CRM [15].

Organizations, technology, interaction, customer relationships are vital in creating factors that affect CRMs [15–16]. All these factors are interlinked and interconnected and, as such, encourage regulation of CRM. The CRM strategy is essential for the needs and behavior of clients to develop stronger relationships with them. A more helpful way of thinking about CRM relates to a process that will help gather a lot of Information about customers, sales, marketing, which will create efficiency, responsibility, and market trends. Reference [16] states that this knowledge enables better customer service, increases call center efficiency, simplifies sales and marketing processes.

2.2   CRM Components

Automated complete CRM systems consist of three components interlinked and upgraded, and often CRM solutions offer only one of these elements. Components are operational, analytical, and collaborative CRM (Fig. 2) [16].

Operative CRM

According to [16–19], Operative CRM deals with creating Information by entering data into the information system through client monitoring applications. For example,

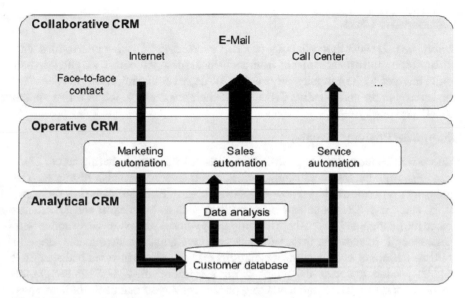

Fig. 2 CRM components [16]

when a bank client calls the bank's call center and requests insight into the account's balance. In such an instance, the screen operator already has all Information about previous queries, remarks, suggestions, and client account information. These applications also have a function to enter new notes based on established contacts with the client [18, 19]. New and previously entered data get stored in a database without any analysis, for which the analytical part of the CRM is in charge.

Marketing Automation

By analyzing customer data and using knowledge about them, businesses can monitor the campaign's success [20]. The results of past campaigns are used as a knowledge base for future campaigns (close-loop campaign management). The richer database of businesses increases, the greater the power of the company to understand consumers and introduce real marketing one by one [19, 20]. CRM products offer campaigns to save time, workforce, and cost savings campaigns and make incomes from investments that can be quantified. CRM marketing initiatives—essentially should serve: "Cross-selling" and "Up-selling"; Consumer retention; Predicting consumer behavior; Consumer profitability and value modeling; Channel optimization; Personalization and Marketing based events [20].

Sales Automation

Salesforce Automation (SFA) is originally intended to improve the sales force's productivity and encourage sellers to document and discuss activities in their field [21]. SFA products are becoming more and more focused on cultivating consumer relationships and increasing consumer satisfaction.

Collaborative CRM

Reference [22] states that collaborative CRM oversees establishing interactions with clients through all available media, from traditional personal contact, via phone, email to the Internet, to the currently very current WAP. All communications to clients are made through the collaborative CRM, while their responses to the IS arrive via the CRM operating system.

Enterprise Resource Planning

Enterprise Resource Planning (ERP) offers easier Information sharing across organizations, from purchasing, manufacturing, and financing to human resources. The company's procedures address how they work [22–24]. ERP automated key functions in the enterprise. The prominent ERP vendors, such as PeopleSoft and SAP, automated the features and linked them to companies whose system was never connected. Reference [25] states that these products require significant investments, often in millions of dollars, but ERP implementations often duplicate invested budgets. ERP enables vendors to access a single checking system, buying agents can look at the history of supplier prices, and marketing product managers can track shortcomings.

Supply Chain Management

Supply chain management (SCM) involves the automation of a more significant part of the supply chain and the pinnacle of the thinking of many executives [25–28]. According to [28], SCM is an everyday use of technology to improve B2B (business to business marketing) processes and increase speed, agility, control of the right time, and customer satisfaction. References [28, 29] state that the success of SCM depends on two main factors. The first is that all companies in the supply chain have to accept the partner's cooperation as a strategic goal and tasks that must be operational priorities. The second is that the availability of Information through the supply chain must be guided by strict policies, discipline, and daily monitoring.

Supplier Relationship Management

Supplier Relationship Management (SRM) helps enterprises to analyze suppliers, whether they consider strategic suppliers, those with whom the company has established a collaborative common benefit strategy or suppliers of consumer goods that the company has selected based on the price [30].

Partner Relationship Management

References [30, 31] state that Partner Relationship Management (PRM) is a subset of CRM, allowing companies to ensure their partners' satisfaction. PRM tools provide partner profiles, which enable the company to understand the characteristics of a business partner and monitor its success and contribution [31, 32]. Enterprises can use this information to improve their relationships with partners through additional training or joint marketing activities.

Analytical CRM

Analytical CRM, which is the most complex and expensive segment of the entire CRM system, through a detailed analysis of data-based and expert knowledge, creates an image of each client, their needs and desires, intending to develop more robust interconnections [33].

Data Warehouse

References [32, 33] state that Data Warehouse (DW) is a central element of the CRM system. Data sources, both internal and external, provide data that describe consumers. The data collection system converts input data into an electronic medium if it has not already been done [33]. The Data Warehouse system prepares storage information, deposits them, describes them so that they can be retrieved, and performs management and control functions.

Data Mining

Data Mining describes how a user receives previously unknown Information from a large DW. The process is similar to the mining business, so it got such a name [34]. It entails selecting, exploring, and modeling large amounts of data to detect previously unknown business improvement tricks [33, 34]. Data Mining uses techniques and algorithms in statistics, Artificial Intelligence, and other fields to uncover significant "hidden" stacks in large datasets [35].

Online Analytical Processing—OLAP

Reference [36] states that in Online analytical processing (OLAP), as the name itself speaks, the user within the system uses the client workstation to enter instructions and receive answers to the queries. OLAP was conceived by E. F. Codd, who, together with C. J. Date, was responsible for the idea of relational databases [36, 37]. In 1993, he wrote an example of identifying 12 requests for the digestion of multidimensional data [37]. One of the requirements was that users could perform the processing process using the mouse to point and click rather than to search the menu. The code considered that the system could manipulate 20 or more dimensions [37, 38]. The OLAP from a relational database takes a snapshot and transforms it into a dimensional data structure, which allows you to meet the set queries. Such a structure gets formed from a star schema in the center of the table of facts with which several tables of dimensions are connected [39]. OLAP analysis results are queries and reports.

Customer data includes address, customer history, hobbies, assets, education, health, debts, etc. Companies use these data to identify features such as loyalty, preferences, and salaries [33, 35, 38]. CRM is not just a technology or a system that is in the function of establishing and developing customer relationships. It is much more; moreover, nowadays, CRM is a key component of the survival of modern business entities [39]. Therefore, current business entities need to recognize the importance of CRM and implement it in their business.

3 Artificial Intelligence

3.1 *Artificial General Intelligence*

Artificial Intelligence (AI) is the ability of a digital computer or computer-controlled robot to perform tasks commonly associated with intelligent beings [40]. The artificial intelligence field, shortened by AI, is trying to understand intelligent entities. Thus, one of the reasons for studying this area is to understand ourselves better managing [40, 41–45]. However, unlike philosophy or psychology, which also explores Intelligence, the field of Artificial Intelligence seeks to build artificial entities and their understanding.

Reference [41] states that during exploring artificial Intelligence, researchers start from two primary goals. The first and most crucial goal is to build intelligent machines. The second is to understand the nature of Intelligence or what psychologists call g-factor in intelligence tests [42]. Both plans have, in their essence, the need to define the notion of Intelligence.

According to [43], AGI—Artificial General Intelligence describes research focused on creating machines capable of generally intelligent actions. General Artificial Intelligence must be able to master different areas and learn new areas with which it has not faced before. It is not necessary to have the same capabilities in all domains (Fig. 3) [22].

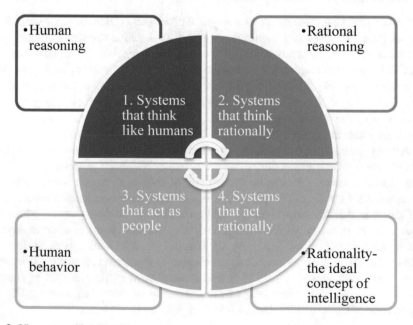

Fig. 3 Views on artificial intelligence systems [22]

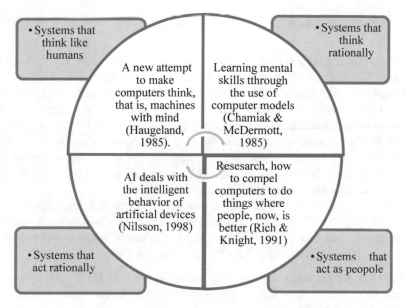

Fig. 4 Definitions of artificial Intelligence are organized into four categories [22]

The above categories deal with thought processes and reasoning, while the two below deal with the behavior. Categories on the left measure performance in matching with human performance. On the right, they measure success compared to the ideal concepts of Intelligence we call rationality (Fig. 4) [22].

Scientists researching AI often talk about intelligent machines, but there is very little agreement among the essential Intelligence [43, 44]. They are complex because machines want to give away attributes that they cannot define. Artificial Intelligence, therefore, suffers from the lack of definition of its field of activity, and artificial intelligence research focuses mainly on intelligence components such as learning, locking, problem-solving, perception, and language use [44]. Intelligence features do not have to be present at all, but they are an integral part of the intelligent entity [46]. They include internal knowledge, knowledge of the world, goals, plans, and creativity.

Communication is only possible between intelligent subjects, regardless of Intelligence's different types. Internal knowledge is the expectation that the intelligent entity has some knowledge of itself [39, 43, 46]. He should know when he needs something, think about something, and know-how to think about it. Knowledge about the world assumes that Intelligence involves awareness of the outside world and the ability to find and use Information that someone has about the outside world [43]. It also implies the possession of memory in which experience has been coded and can be used as a guide to processing new experiences [43, 44]. Goals and plans relate to goal-oriented behaviors in terms of knowing the goal and understanding the plan for achieving the goal.

3.2 Approach Based on Turing's Test

Alan Turing based his theory of classic computer science in 1936. In 1950, entitled Computing Machinery and Intelligence, he asked: can machines think? Not only did he defend the claim that it was possible, but he proposed a test to determine if the program had achieved it [46]. The test is now known as the Turing test. It is based on the fact that a suitable judge cannot recognize whether he is talking to another person or with a computer program or whether the program is a human or not [46, 45]. Turing also proposed protocols for conducting his tests, such as the program and the judge with the judge interacting separately and only in textual form, via a medium such as a teleprinter [45].

It would only check the ability of the candidate to think, not their appearance, in the test. Turing's test and the arguments he cited have led many scientists to believe that he is right and how to pass the test [45, 47]. A machine that would eventually pass the Turing test should have the following capabilities:

- Natural language processing to communicate successfully in the natural language.
- Representation of knowledge—automatically memorizing what it knows and receives based on the sensor.
- Reasoning—to use learned information to answer questions and make new conclusions.
- Machine learning—to adapt to new circumstances.

Scientists have begun to write programs to research ways to pass the test. In 1964, Joseph Weizenbaum wrote Eliza's computer program designed to imitate a psychotherapist [45, 47]. He assessed that a psychotherapist is a particularly simple imitation person who gives him indefinable answers about himself and asks questions based on user questions and answers. Typically, such programs are based on two strategies [47]. First, the input words entered by the user are scanned, followed by specific keywords and grammatical forms [45, 48]. To pass the complete Turing test, the computer must master:

- Computer vision
- Robotics (Fig. 5) [45].

3.3 Types of Artificial Intelligence

(1) According to [49], there are four types of artificial Intelligence (AI):
(2) Non-psychological AI—this type of research includes the construction and programming of computers to perform tasks that, according to Marvin Minsky, require the Intelligence of human beings [49]. Researchers of non-psychological AI do not have any pretensions to the psychological realism of the programs or devices they build, whether they will perform them as a human.

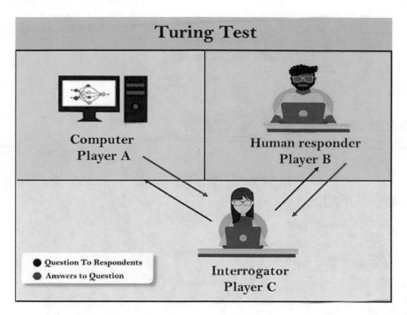

Fig. 5 Turing test [45]

(3) Weak Psychological AI—this research type gets driven by the view that a computer is a valuable tool in studying the brain (consciousness). Practically, we can write computer programs or make devices that simulate human psychological processes and then test our predictions as the alleged processes work [49]. We can tune these programs and devices together with other programs and devices that simulate various alleged mental processes and thus test the degree to which the AI system can fully simulate human mental activity.

(4) Strong Psychological AI—according to this view, our brains (minds) are computers, so duplicates can be made using other computers. Chen [49] wrote that the true ambition of mythical propositions is to make the general purpose of intelligence-reason. According to [49], the ultimate goal is to make a person, or, far more accessible, an animal.

(5) Parapsychological AI—research of this type, such as a strong psychological AI, takes a seriously functionalistic view according to which mental activities can be realized on many different types of physical devices [49]. Paedopsychological AI blames the strong psychological AI for being chauvinistic because it is only interested in human Intelligence. Paedopsychological AI claims to be interested in all intelligible ways of Intelligence that can be achieved (Fig. 6) [39].

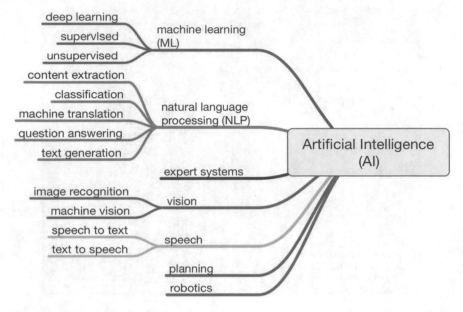

Fig. 6 Artificial intelligence [39]

4   Artificial Intelligence and CRM

4.1   AI and CRM

Every organization comes with different challenges, and one of the main is effective data management to comply with the user's need [39, 44]. Companies are looking for a large quantity of Information generating interaction with users. There is a better method for solving these challenges, and one of these is the integration of AI to CRM. AI and CRM can improve relations with clients [43, 45]. Often the question arises what is the main focus of CRM, and does it need to be combined with artificial Intelligence? Reference [39] states that the primary goal of CRM is understanding customers and then offering them customized and valuable products— always adapted to their needs. CRM has a database based on which realizes its goals and relates to storing and collecting data from clients [50, 51]. Due to the large amount of data that organizations must solve; CRM needs to adapt to the era of digital transformation.

4.2   The Advantages of AI and CRM

Reference [52] states that there are five basic categories of potential benefits from applying artificial Intelligence in the function of CRM. They include Increase in revenue, increased profit, enhanced customer satisfaction, savings, and increased market share.

- **More efficient data management**

CRM stores all data from social networks, redirects them, and organizes them to provide the company's real value [51, 52]. As a result, the company's marketing departments can implement resources that will enable them to react faster and more precisely at the user's request.

- **Optimized sales strategy**

Thanks to automatic data entry and intelligent analysis, more precise profiles of those customers are created that will most likely purchase their products. References [48, 51, 52] state that companies based on customer foresight can take personalized actions to avoid invasive advertising and provide material of real interest for every opportunity.

- **Better development of application applications**

Intelligent data analysis is performed by integrating AI and CRM [48, 51]. Applications will be created with easy-to-use interfaces that increase conversion rates and guarantee the customer unprecedented experience and quality.

4.3   Examples of CRM and AI

Based on AI, CRM achieves its efficiency that contributes and saves time and cost and the supply of valuable products. Reference [53] noticed that Artificial Intelligence can optimize our campaign based on-site visitors' analysis in real-time. In this way, AI can engage a client with the right content [52, 53]. Thanks to artificial Intelligence, a huge amount of data about clients available in real-time social channels are analyzed. Based on this, you can predict the success of marketing initiatives and campaigns and learn how to improve them [51]. CRM and AI can anticipate the needs of the client. CRM and AI operate to actively and continuously monitor client status, automatically activate events, and inform the right person based on this status, effectively eliminating any chances of a client seeking help through breakouts [48]. The task of CRM and AI is to strengthen customer service representatives by providing them with the right Information at the right time to respond to customer inquiries in a more informed way.

One of the examples of CRM and AI links is Salesforce Einstein software. Salesforce Einstein has been designed to help any company improve its relationship with its customers [48, 51–53]. Thanks to machine learning, deep learning, predictive analysis, natural language processing, and data collection, Einstein is referred to the entire Salesforce Customer Success Platform. Salesforce is used in sales, marketing, business analysis, customer service, and support [53].

Siri (Apple), Cortana (Microsoft), Google Assistant, Alexa (Amazon), Watson (IBM), Bixby (Samsung) are some of the most famous virtual personal assistants who can help us in everyday activities [51]. In Europe, artificial Intelligence already makes us healthier, gives us cleaner air and energy, keeps us (our data) on the Internet, in traffic, and improves the quality of our work [55, 53]. Like any other tool, artificial Intelligence is there to help us with people, and it is precisely this focus on AI that supports EU access. The development of Artificial Intelligence is not an end in itself; it must serve the greater good of the whole society.

Artificial Intelligence has been penetrating the devices we have been surrounded by for years. The progression of machine learning to automate processes and help in a range of applications does not occur only in the consumer world but also in the business environment [47, 48]. Business analytics is increasingly using hardware learning to track business processes in real-time. In Florida, SAP has presented the world with a new set of its SAP C/4HANA software packages that use machine learning and are aimed at maintaining a better customer relationship or Customer Relationship Management (CRM). SAP C/4HANA is a step-in database compression and analysis, translating into higher revenues [48, 55].

5 Adoption and Implementation of AI in Customer Relationship Management

Until 2022, the economic impact of AI on CRM is expected to generate approximately USD 1.1 trillion in increased gross domestic product and 800,000 net job creation globally. Artificial Intelligence is projected to revolutionize the way people work due to the rise of massive data volumes, processing capacity, and machine learning technological leaps [49]. AI is expected to improve employee performance, particularly in areas involving customer relationship management [49]. Companies can assess the effectiveness of AI technologies based on factors such as enhanced productivity, expanded market prospects, cost savings, and return on investment (ROI) [48, 49]. As mentioned in the previous chapter, integrating AI into CRM is critical for creating intelligent, impressive, and impactful customer experiences in a timely manner [47–49].

Customer Relationship Management (CRM) is viewed as a strategic undertaking for a company. A large number of customer data is processed to build CRM activities, which helps firms discover customers' purchase intentions [48, 49]. However, humans struggle to analyze such large volumes of consumer data in an accurate and timely manner, and firms confront multiple obstacles and, at times, crisis circumstances in the absence of an automated solution [49]. Thus, AI technology linked with CRM could be a method to overcome such issues and provide a solution for overcoming such crises, frequently without the need for human interaction. AI-CRM has the potential to keep businesses afloat during decision-making.

6 Discussion

Users understand that new technologies must be transformed and adapted to improve their users' experience and get insights that they have not been able to. The influence of data on today's business is great but drawing useful and concrete insights from them is a real challenge. It is very important to connect data inside and outside the organization, and this is only possible by using the latest technologies. Today's industry leaders recognize limitations in their digital capabilities—60% believe they cannot create a personalized experience, 53% lack confidence in their strategic decisions, and 31% believe that their data's quality, accuracy, and completeness are not satisfactory. Tomorrow's disruptors will be organizations that can converge digital business with a new digital intelligence level. The new CRM systems and AI will improve user experience with faster and simpler processes to anticipate user needs better and address their demands.

In recent years, it's no longer enough to have an IT strategy. Businesses must also have a data strategy. These are, for example, Information about your users, partners, and the entire ecosystem and industry in which you operate. Among all these data, some are structured and part of the unstructured data; the biggest challenge is how to achieve unstructured data "talking" and get some business benefit from them. Today, such data can be much better-used thanks to technology (AI).

AI requires the creation of new forms of work, skills, new jobs, business models, and entire industries. New jobs are created in cybersecurity, data science, health, finance, and other sectors. Artificial Intelligence means helping human Intelligence and abilities rather than replacing people with machines.

Artificial Intelligence aims to increase human Intelligence; cognitive systems will be embedded in the business and social processes but will remain under human control. It is very important to get to know the public about the purpose and significance of AI. There is much debate about whether artificial Intelligence reduces the number of jobs, and fears are understandable, but while one job disappears, others arise.

It is defiantly that artificial Intelligence has become mainstream and that all industries have fully redefined those cognitive technologies. I believe that in five to ten years, no one will even notice that cognitive technology is behind everything that will have a great impact on our lives and make them more convenient, easier, and more enjoyable. It will be experienced as a completely normal and common thing, and it will be around us everywhere, from offices, doctors, restaurants, etc.

7 Conclusion

7.1 Synopsis

Companies must adjust their business more and more swiftly to new scenarios as a result of the impact of globalization on market business conditions. New strategic orientations, including Artificial Intelligence (AI), have become critical for today's businesses to succeed. A smart business makes better judgments faster and outperforms its rivals. Consumer Relationship Management (CRM) is a business technique for deciding how to handle customers in order to maximize long-term value. CRM also necessitates a customer-centric corporate philosophy and a company culture that promotes efficient marketing, sales, and service procedures. CRM combines people, procedures, and technology to improve customer relationships. All customer communication is handled by the collaborative CRM, while their reactions to the information system are handled by the CRM's operating component.

Newer artificial intelligence methods seek to create intelligent, creative agents who learn in interaction with the environment. Evolutionary computer science uses the principles of evolution and natural selection in seeking solutions using genetic algorithms. Created intelligent agents interact with the environment by changing their behavior and adapting to changes without human guidance. There are many examples of intelligent systems applied in today's affiliated society. The amount of data to be classified, processed, and returned to users becomes immense. The predictions and recommendations made by smart software are gaining or losing significant amounts of money. Human brain research and neuromorphic computing architecture are expected to bring a breakthrough in artificial intelligence construction over the next several years. Artificial Intelligence is nevertheless a relatively young scientific discipline whose development in recent years has been caused by an increase in the investment of significant money resources in research and the development and expansion of the Internet. In this way, they help companies in making business decisions.

7.2 Conclusion

This sudden development has also raised several questions about the consequences of using artificial Intelligence; on privacy, security of the individual, and economy. The business entities currently on the market are in an environment of robust competition and developed distribution channels, and goods and services are more than necessary. It is not enough to be average; one should strive for the very top and business excellence. Of course, it is essential to "know" how to survive in such an environment, prepare plans, and make the right decisions. To do this, companies need to have the correct Information. Having the right and required Information timely can make a big difference in the company's business, while possessing bad Information leads to wrong business decisions and even revenue loss.

Using information technology and it's tools makes it easy to make the right decisions and enables survival on the market. Artificial Intelligence is now very present in its weak form in various technologies and areas of human activity. Knowing that Artificial Intelligence does not need to be intelligent in general terms but intelligent enough to solve the problem, a great shift from academic research to the application of techniques developed in Artificial Intelligence has been made. Today, there are many examples of the use of intelligent systems. The commercial application of artificial Intelligence can be monitored from expert systems and machine learning. The first steps are made to the present-day development of artificial neural networks of large scale that allow computers to learn, recognize objects from the environment, use natural language.

Several factors affect CRM. Organization, technology, interaction, customer relationships are key in creating factors that affect CRMs. All these factors are interlinked and interconnected and, as such, encourage regulation of CRM. In combination with AI, CRM becomes the main factor in the success of companies. But not everything is on the system itself. For users to use it at all, they should educate themselves and know how to "cooperate" with them when deciding. All companies are involved in the digitization process because the labor market and competitiveness require it.

Based on all the above facts and theoretical analysis, we can conclude that AI in combination with CRM is one of great importance for companies and greatly affects the company's performance. [41]

References

1. Al-Zahrani, A.: How Artificial Intelligent Transform Business, p. 13 (2018)
2. Agent-based online analytical mining in CRM system. IEEE, Tucson, AZ, USA (2005). https://doi.org/10.1109/ICNSC.2005.1461181
3. Kracklauer, A.H., Quinn Mills, D., Seifert, D.: Collaborative Customer Relationship Management—Taking CRM to the Next Level. Springer (2004)
4. Safer, A.M.: Data Mining Technology Across Disciplines (2009)
5. Ali, M., Lee, Y.: CRM sales prediction using continuous time-evolving classification. Proceedings of the AAAI Conference on Artificial Intelligence 32(1) (2018)
6. P. Valter, P. Lindgren and R. Prasad.: Artificial intelligence and deep learning in a world of humans and persuasive business models. IEEE, Global Wireless Summit, Cape Town, South Africa (2017)
7. Goldenberg, B.J.: CRM Automation. Prentice Hall PTR, New Jersey, USA (2002)
8. Chagas, B.N., Viana, J.A., Reinhold, O., Lobato, F.M., Jacob, A.F., & Alt, R.: Current applications of machine learning techniques in CRM: a literature review and practical implications. IEEE/WIC/ACM International Conference on Web Intelligence (WI), 452–458, Santiago, Chile (2018). https://doi.org/10.1109/WI.2018.00-53
9. Bergmann, R.: Experience Management: Foundations, Development Methodology, and Internet-Based Applications. Springer-Verlag Berlin Heidelberg New York (2002)
10. Bergeron, B.: Essentials of CRM: A Guide to Customer Relationship Management (Essential Series) (2002)

11. Rygielski, C., Wang, J.-C., Yen, D.C.: Data mining techniques for customer relationship management. 483–502 (2002)
12. Todman, C.: Designing a Data Warehouse. Hewlett-Packard Company, Prentice Hall (2001)
13. Marinchak, C.M., Forrest, E., Hoanca, B.: Artificial intelligence: redefining marketing management and the customer experience (2018)
14. Dirican, C.: The impacts of robotics, artificial intelligence on business and economics (2015)
15. Partridge, D., Hussain, K.M.: Artificial Intelligence and Business Management (1992)
16. Mandel, E., Georgescu, V.: Methods of AI based CRM (2017)
17. Bagheri, E., Cheung, J.C.K.: Advances in Artificial Intelligence. 31st Canadian Conference on Artificial Intelligence. Canadian AI 2018, Toronto, ON, Canada (2018)
18. Empirical comparison and evaluation of classifier performance for data mining in customer relationship management (n.d.)
19. Tsui, E., Garner, B.J., Staab, S.: The role of artificial intelligence in knowledge management (2000)
20. Ngai, E.W.T., Xiu, L., Chau, D.C.K.: Application of data mining techniques in customer relationship management. 2592–2602 (n.d.)
21. Phillips-Wren, G., Ichalkaranje, N., Jain, L.C.: Intelligent Decision Making: An AI-Based Approach (2008)
22. Goertzel, B., Wang, P.: Advances in Artificial General Intelligence: Concepts, Architectures and Algorithms (2007)
23. Norris, G., Hurley, J.R., Hartley, K.M., Dunleavy, J.R., Balls, J.D.: E-Business and ERP—Transforming the Enterprise. Wiley (2000)
24. Kimiloglu, H.: What signifies success in e-CRM? 246–267 (2009)
25. Gacanin, H., Wagner, M.: Artificial intelligence paradigm for customer experience management in next-generation networks: challenges and perspectives. 188–194 (2019). https://doi.org/10.1109/MNET.2019.1800015
26. Min, H.: Artificial Intelligence in supply chain management: theory and applications. 13–39 (2009)
27. Huang, M.-H., Rust Roland, T.: Artificial intelligence in service (2018)
28. Sterne, J.: Customer Service on the Internet: Building (2000)
29. Reynolds, J.: A Practical Guide to CRM. CPM Books (2002)
30. Ranjan, J.: Role of knowledge management and analytical CRM in business: data mining-based framework. Emerald Group Publishing Limited (2011)
31. Dyche, J.: The CRM Handbook: A Business Guide to Customer Relationship Management (2001)
32. Freeland, J.: The Ultimate CRM Handbook: Strategies and Concepts for Building Enduring Customer Loyalty and Profitability. McGraw-Hill (2002). Greenberg, P.: CRM at the Speed of Light: Capturing and Keeping Customers in Internet Real Time. McGraw-Hill (2004)
33. Choy, K.L., Lee, W.B., Lo, V.: Design of a case based intelligent supplier relationship management system—the integration of supplier rating system and product coding system (2003)
34. Kumar, C.: Artificial Intelligence: Definition, Types, Examples, Technologies. https://medium.com/chethankumargn/artificial-intelligence-definition-types-examples-technologies-962ea75c7b9b (2018)
35. Lee-Kelley, L.: How e-CRM can enhance customer loyalty (n.d.). Nadimpalli, M.: Artificial intelligence—consumers and industry impact. OMICS International (2017). https://doi.org/10.4172/2162-6359.1000429
36. Metaxiotis, K., Terzakis, K., Moulids, E., Psara's, J.: Decision support through knowledge management: the role of the artificial Intelligence. Inf. Manag. Comput. Secur. **11**(5), 216–221 (2003)
37. Missbach, M., Stark, T., Gardiner, C., McCloud, J., Madly, R., Tempe's, M., Anderson, G.: SAP on the Cloud, pp. 7–8. Springer, Heidelberg (2016)
38. Berry, M.J.A., Linoiff, G.S.: Data Mining Techniques: For Marketing, Sales, and Customer Relationship Management (2004)

39. Pandey, P.K.: Role of artificial intelligence in business (2018)
40. Gentsch, P.: AI in Marketing, Sales and Service: How Marketers Without a Data Science Degree Can Use AI, Big Data and Bots (2018)
41. Akerkar, R.: Artificial Intelligence for Business (n.d.)
42. Ransbotham, S., Kiron, D., Gerbert, P.: Reshaping Business With Artificial Intelligence: Closing the Gap Between Ambition and Action. Cambridge 59 (n.d.)
43. Ranjan, J., Bhatnagar, V.: Role of knowledge management and analytical CRM in business: data mining-based framework. Learn. Organ. **18**(2), 131–148 (2011)
44. Russell, S.J., Norvig, P.: Artificial Intelligence: A Modern Approach. Malaysia (2016)
45. Deb, S.K., Jain, R.: Artificial Intelligence—creating automated insights for customer relationship management (2018)
46. Siddike, M., Kalam, A., Spohrer, J., Demirkan, H., Kohda, Y.: People's interactions with cognitive assistants for enhanced performances. In: Proceedings of the 51st Hawaii International Conference on System Sciences, January (2018)
47. Swift, R.S.: Accelerating Customer Relationships: Using CRM and Relationship Technologies. Prentice Hall PTR, New York (n.d.)
48. Femina Bahari, T.: An efficient CRM-data mining framework for the prediction of customer behaviour (2014)
49. Chen, T.F.: Applying artificial intelligence in CRM: case studies of intelligent virtual agents and pegasystems. 878–882 (2012)
50. Davenport, T.H., Klahr, P.: Managing customer support knowledge. **40** (1998)
51. Poniszewska-Maranda, Kaczmarek, D., Kryvinska, N., Xhafa, F.: Studying usability of AI in the IoT systems/paradigm through embedding NN techniques into mobile smart service system. Computing **101**(11), 1661–1685 (2019). https://doi.org/10.1007/s00607-018-0680-z
52. Thomspon, B.: What is CRM. The customer relationship management primer. What you need to know to get started (2002)
53. Fedushko, S., Ustyianovych, T., Gregus, M.: Real-time high-load infrastructure transaction status output prediction using operational intelligence and big data technologies. Electronics **9**(4), 668 (2020). https://doi.org/10.3390/electronics9040668
54. Ertel, W.: Introduction to Artificial Intelligence (2011)
55. Akerkar, R.: Introduction to Artificial Intelligence. Springer, Cham (2018)
56. Jadhav, T.G., Patel, K.: Methods and apparatus for using artificial intelligence entities to provide information to an end user (2018)

Competitiveness of International IT Companies—Comparison of Strategies, Their Strengths, and Weaknesses

Ester Federlova⬤, Ondrej Čupka⬤, and Peter Vesely⬤

Abstract The article deals with the competitiveness of international companies. The article aims to guide managers and companies to properly determine the strategy for the implementation of cloud solutions by identifying the position of international IT companies in a competitive environment. The subject of the research is four selected IT companies, providers of cloud solutions. The relationships between individual providers of cloud solutions are examined by analysis. The starting point for the elaboration of the article is annual reports, professional articles, and publications of domestic and foreign authors. In the first section, we deal with theoretical knowledge in the field of competitiveness, strategies, and analysis of the external and internal environment of the company. The second section deals with the main and partial goals. In the third section, we deal with research methods and work procedures. We characterize selected IT companies. The last section summarizes the achieved results of the work and discussion.

1 Introduction

Today we live in a "world of information", where information technology has become part of our daily lives. Students are more open to learning using modern technology and focus more on online learning, especially in this situation. Thanks to information technology, we were able to reorient ourselves to this type of teaching and thus maintain the process of distance learning. Information technology can be seen in almost all areas, including work, learning, leisure, and health. New technologies

E. Federlova (✉) · O. Čupka · P. Vesely
Faculty of Management, Comenius University in Bratislava, Odbojarov 10, 831 04 Bratislava, Slovakia
e-mail: federlova2@uniba.sk

O. Čupka
e-mail: cupka9@uniba.sk

P. Vesely
e-mail: peter.vesely@fm.uniba.sk

© The Author(s), under exclusive license to Springer Nature Switzerland AG 2023
N. Kryvinska et al. (eds.), *Developments in Information and Knowledge Management Systems for Business Applications*, Studies in Systems, Decision and Control 462,
https://doi.org/10.1007/978-3-031-25695-0_22

are constantly coming. Cloud computing is a major technological innovation of the twenty-first century, enabling companies to store and access content without forced installation. This technology has received the fastest adoption than any other. Companies appreciate their flexibility, flexibility, affordability, automatic updates, and speed. At the beginning of writing the article, we set the main goal and smaller sub-goals. The main goal of the article is to guide managers and companies to properly determine the strategy for the implementation of cloud solutions by identifying the position of international IT companies in a competitive environment. The article is divided into the theoretical and practical parts. The structure of the article consists of four sections. In the first section, we focus mainly on the theoretical basis, where we focus on identifying the main concepts such as competition, competitiveness, and strategy. We analyzed the company's external and internal competitive environment, where we examined the factors that have an immediate impact on the effectiveness of the set goals. At the same time, it affects the success of the company and the performance of its activities. In the second section, we defined and determined the main and partial goals. In the third section, we characterized the objects of research— four selected the largest technology companies and providers of cloud solutions for companies. In the final section, we analyzed the four largest providers, focusing on their strategies, strengths, and weaknesses. These companies have been leading providers for several years. From the information obtained, we have set a strategy for implementing cloud services in smaller companies, but it can also serve as a guide for the manager.

2 Conceptual Background

The term competition means competition or competition. It is characterized by a large economic, ethical, cultural, social, or political scope. It is also a relationship between two or more companies that can compete with each other [1]. One of the basic features of a market economy is the customer's ability to freely decide where and from whom to buy. This fact leads potential competitors to compete for the consumer, who is important to make sure that the supplier is the right one. It is important to show him that it is his product that best meets customer needs [2]. In the market, we may encounter competition between supply and demand, competition between consumers, and competition between manufacturers. There may be price or non-price competition between manufacturers. In terms of market conditions, we distinguish between perfect and imperfect competition [3]. There are many different definitions for perfect competition. For the perfect competition, we divide the conditions into necessary and optional, which have a secondary character. Competition in perfect competition is atomic between entities. The competition is offered by many smaller producers and bought by many small consumers, with neither participant affecting the price directly. The products on the market are identical, i., the products of different companies are homogeneous in a given market. The participant may enter and exit the market freely without restrictions. Buyers and sellers alike have complete information about products, prices, price developments, and quality. The factors that a manufacturer needs for production can be transferred at any time to the manufacturers, who also know how to make the most efficient use of them. The places where manufacturers produce are close to each other, so there are no differences in shipping costs. Agreements between consumers and the manufacturer (manufacturer) are strictly prohibited [4]. There is almost no perfect competition in the real economy. It is based on a level playing field for each market participant. Companies accept prices as a given parameter, the product is homogeneous, the entry of a new manufacturer into the market is free and information about the company's activities is perfect. The way companies are to maximize profits and reduce costs. In a market economy, most companies move in imperfectly competitive markets. In imperfect competition, firms are considered to be price makers, which means that the company has the freedom to set product prices. If a company can significantly influence its price on the market, we can call it an imperfect competitor. There are three basic forms of imperfect competition—full monopoly, oligopoly, and monopolistic competition [3].

2.1 Competitiveness

We can perceive competitiveness from different angles according to individual theories. We perceive them, for example, from the point of view of management, economics, or marketing. The concept of competitiveness has a preserved basis in the word competitive and competitive. The meaning of this word presents the ability of the entity to compete or compete in the market. Competitiveness is one of the characteristics that allow an entity to win in competition with other entities and its evaluation is also related to the nature and conditions of the competition. The winning entity can be a business entity that can claim a certain competitive advantage in the competition and gain an advantage over the rivals. Being competitive permanently means creating a competitive advantage for the company in the future faster than a competitor copying today's ones. Competitiveness also represents the optimal use of capital and human resources. Organization, innovation, motivation, and employee leadership are also key to success. The concept of competitiveness also includes its ability to generate profit. It is equally important for companies to stay in the domestic market but also to expand into foreign markets [5].

It is also the most appropriate use of human capital resources. Various comparisons show that international differences in company productivity are, in most cases, conditioned by the situation within the company [6].

The latest view of competitiveness points to the fact that competitiveness is made up of four dimensions: identity, integrity, mobility, and sovereignty. We can define identity as our personality, which defines the environment. Mobility is defined as the ability of a company to respond to changes in the external and internal environment. Sovereignty determines whether a company can make and make decisions about the development of the company. Finally, integrity defines corporate cohesion [7].

The importance of these factors changes from year to year with the development of new strategies and development production. These factors are defined by knowledge, technology, and external factors environment. The company's costs are related to the inputs to the production process. Quality products later apply to outputs. Usually, these methods of improvement are aimed at improving only one factor. The goal of businesses is usually achieving the highest quality at the lowest cost, low price, and in the shortest time [8]. Competitiveness is made up of 4 basic dimensions. The first is identity, which is defined as the very peculiarity of society, which defines the surrounding environment. Integrity is the second dimension of the unity of the company, which is given by the identification of employees with the company as a whole. Mobility is given as the ability of a company to respond to change in the internal and external environment of the company. The last dimension is the sovereignty of the company, which determines whether the company can actually make decisions about its development and whether they are also feasible. These four dimensions (especially the identity of the company) have a significant impact on the competitive potential of the company. When evaluating the company's competitiveness, it is not only financial indicators or evaluation of the achievement of important marketing activities but also other factors, such as employee relations, corporate culture, company image, external behavior, and others [7].

2.2   Factors of Company Competitiveness

The basic factors of competitiveness and their products are quality, cost, and time. The importance of these factors changes from year to year with the development of new strategies and development production. These factors are defined by knowledge, technology, and external factors environment. The company's costs are related to the inputs to the production process. Quality products later apply to outputs. Usually, these methods of improvement are aimed at improving only one factor. The goal of businesses is usually achieving the highest quality at the lowest cost, low price, and in the shortest time [8].

2.3   Dimensions of Company Competitiveness

Competitiveness is made up of 4 basic dimensions. The first is identity, which is defined as the very peculiarity of society, which defines the surrounding environment. Integrity is the second dimension of the unity of the company, which is given by the identification of employees with the company as a whole. Mobility is given as the ability of a company to respond to change in the internal and external environment of the company. The last dimension is the sovereignty of the company, which determines whether the company can actually make decisions about its development and whether they are also feasible. These four dimensions (especially the identity of the company) have a significant impact on the competitive potential of the company. When evaluating the company's competitiveness, it is not only financial indicators or evaluation of the achievement of important marketing activities, but also other factors, such as employee relations, corporate culture, company image, external behavior, and others [7].

2.4 Analysis of the External Environment

The analysis of the external environment focuses on the incoming factors that have an impact on the company's strategic position, and we analyze the various opportunities and threats that could affect this activity. It is important to observe and evaluate the individual factors of the macroenvironment in the environment in which the company operates. Today, companies are moving more in an open economy [9].

Environmental analysis includes PESTLE analysis. PESTLE analysis sometimes referred to as PEST analysis, is one of the concepts used by managers in marketing principles. In addition, this concept is used by companies as a tool to monitor the environment in which they operate or plan to launch a new project/product/service or other. PESTLE analysis is a mnemonic tool that designates its parts as P for political, E for Economic, S for social, T for technological, L for Legal, and E for environmental factors. It provides a view of the whole environment from a bird's eye view and many different angles that one wants to observe and explore, at a time when thinking about a certain new project, product development, service, or plan [10].

Porter's model of five competing forces is considered a credible and practical alternative to the widely used SWOT analysis. The five key factors that the model uses to identify and assess potential opportunities and risks are competitive rivalry, the threat of new entrants, the threat of substitutes, the bargaining power of suppliers, and the bargaining power of customers. One of the keys to an organization's success is its ability to understand the company's activities and the marketing strategies of its competitors. The extent to which there is the rivalry between competitors varies between industries and market sectors. Regardless of the number of key competitors the company faces, it is essential for its longevity that you understand the differences between the rivals. The number of potential entrants varies considerably and is a key factor that you need to quantify. Industries that require a high level of investment and expertise are much more difficult for new organizations to infiltrate and challenge existing providers, which protects the profit levels of existing players. The threat of substitutes is the availability of a product that the consumer can buy instead of the product in the industry. The availability of near-substitute products can increase the industry's competitiveness and reduce potential profits for companies in the industry. It shapes the competitive structure of the industry and affects the organization's ability to achieve profitability. Every organization needs to create relationships between buyer and seller and between market and suppliers. The distribution of energy within these relationships varies, but if it is up to the supplier, he can use this influence to determine prices and availability. When using the Porter model, you need to assess the balance of power in your market. Customers also have significant bargaining power in markets where there is an easy transition between different products without incurring conversion costs.

When using the model, keep in mind: the pace of change and market structure, Porter's model is based on the idea of competition [11].

In the analysis of the competitive environment, we examine the factors that have an immediate impact on the effectiveness of the set goals. At the same time, it affects the success of the company and the performance of its activities. Important factors include competitive market position, customer profiles, reputation among suppliers and creditors, and the ability to attract suitable employees. By developing a detailed analysis of competitors, we can predict their growth or profit in the future. When preparing analyzes, we look not only at potential and existing competitors but also at our customers, so that we can properly plan strategic steps, and adapt to customer demand and needs. Companies are sometimes forced to call on suppliers or creditors due to special conditions such as deferred payments or extended invoices, so it is important to maintain fair and good relations with them. Attracting and retaining suitable employees is also a key task for the company's success [12].

2.5   Internal Environment Analysis

The strategic analysis also deals with the internal environment of the company. There are many factors to which the company is exposed to the external environment, which sometimes we cannot influence and we have to adapt to them. In the internal environment, the company can influence its activities. We divide the resources that the company must use effectively to achieve and maintain a competitive advantage into the physical, human, financial, and intangible resources of the company. For a company to have a successful result, it must correctly and realistically evaluate the above-mentioned internal resources. The goal of the company's internal analysis is to identify the company's strengths, which we must maintain and build on. We have to manage the weaknesses of the company and eliminate them [13].

SWOT analysis is an important part of the strategic management planning process. SWOT (strengths, weaknesses, opportunities, threats) are proposed by many as an analytical tool that should be used to categorize significant environmental factors, both internal and external. The SWOT analysis was praised for its simplicity and practicality. The analysis was widely accepted as a framework. SWOT analysis should not be considered as a static and analytical tool with an emphasis only on its output. It should be used as a dynamic part of the business management and development process. The use of SWOT analysis is usually closely linked to the "mechanical" approach. SWOT analysis does not have to be mechanical. It is recommended to adopt an approach with an emphasis on process values and company outputs [14].

For example, McKinsey's model 7S can help us improve the performance of our organization or determine the best way to implement the proposed strategy. This analysis will provide a framework that can be used to examine the likely effects of future changes in the organization or to reconcile departments and processes during a merger or acquisition. We divide it into seven parts:

1. Strategy: the strategy is the company's plan to build and maintain a competitive advantage over the competition;
2. Structure: by this point, we mean structured departments and teams, including who is accountable to whom;
3. Systems: daily activities and procedures that employees use to perform work;
4. Shared values: are the core values of an organization, as shown by its corporate culture and general work ethic. When the model was first developed, they were called "overarching goals" and placing shared values at the center of the model emphasizes that these values are central to the development of all other critical elements;
5. Management style: adopted management style;
6. Collaborators: this category includes employees and their abilities;
7. Skills: the real skills and competencies of the organization's employees [15].

Companies are affected by various events that shape the financial structure and at the same time determine the company's financial situation. The financial situation has a significant impact on the company's image. The financial situation is influenced by internal as well as external factors. It represents the relationship between sources of capital and needs. To assess the financial situation, companies use financial analysis, considered the most effective and fastest method. Using the internal financial situation, we assess the company's liquidity and look at its past results. We use external analysis to assess its investment potential [16].

A good and regular assessment of financial health is important for the fulfillment of strategic goals. The person performing the financial analysis in the company should be sufficiently authorized. In small companies, financial analysis can also be performed by the owner of the company. In larger companies, she is in charge of the financial manager or CFO [17].

2.6 Company Strategy

The strategy has long been considered an important part of the future. It is an important prerequisite for success in many areas. Strategy is a kind of manifestation of how we can achieve our goals. We determine the future goals of our organization, we determine the priorities that will lead us to the expected success. By defining long-term goals, we create space for taking strategic positions in the market [18].

When formulating and evaluating strategies, we must be aware that we do not divide strategies into right and wrong. All strategies are created based on an analysis of the internal and external environment. If a company has more than one strategy to choose from, it should choose the most effective one. When choosing, we should also take into account the suitability, reachability, and acceptability of the strategy. The company should set goals and ask itself whether the strategy is in line with the organization's values. The company must have enough money to implement the chosen strategy. The risks and willingness of employees to adopt a strategy must also be assessed. The company's role is to communicate the strategy correctly through the available platforms [13].

Today, we encounter a large number of strategies and their division. The corporate or corporate strategy is at the top management level. This strategy defines the basic framework for business strategy. At this stage, strategic business units are singled out, the basic strategic goals are determined and the basic directions for individual business units are defined.

At the level of business units, there is a business or business strategy.

At the third level, there are functional strategies that maintain and create a competitive advantage.

We divide the basic types of functional strategy into [18]:

1. Marketing strategy: marketing plays an important role in strategic management. It connects society and the external environment. At this stage, there are important strategic verdicts in the area of market position selection or the idea of a marketing mix;
2. Production strategy: in this strategy, it is very important to deliver services and products at competitive costs and of high quality;
3. Financial strategy: the company, through its strong financial position, is easier to compete and the company is thus more flexible. Important tasks include: efficient use capital, securing resources, and setting a reasonable price for capital;
4. Personnel strategy: proper management, retention, and development of employees is a crucial and essential part of any company. The return on invest-ment and investment in employees and human resources is in terms of return long-term;
5. Scientific and technical strategy: this type of strategy requires an advanced level of used technologies. In applying this strategy, services are created and products are produced at a high level. This activity is influenced by the life cycle of the company.

3 Methodology

The main goal of the article is to guide managers and companies to properly determine the strategy for the implementation of cloud solutions by identifying the position of international IT companies in a competitive environment.

We divided the main goal into smaller sub-goals, which we addressed individually.

1. Summarization and elaboration of the theoretical part with clarification of the main concepts in the field of competitiveness, strategy, and approximation of factors that affect the internal and external environment of the company.
2. Characteristics of individual international IT companies;
3. Determination of working procedures;

4. Product comparison—the cloud of selected IT companies;
5. Comparison of strategies of individual IT companies—cloud services;
6. Analysis and evaluation of strengths and weaknesses of individual IT companies;
7. Creating a draft strategy for managers and smaller companies at deploying cloud solutions.

3.1 Work Methodology and Research Methods

In the article, we used several research methods. In the theoretical part, publications, books, and internet resources for the given helped us to deepen information about our data available for analysis. We used several research methods in the article. In the theoretical part, we analyzed the available scientific publications, books, and Internet resources the given helped us to deepen information about our available data for analysis. In the practical part of the article, we obtained and analyzed information on individual IT companies from the official websites of companies and publicly available reports and analyzes.

3.2 Characteristics of the Research Object

Today, many international companies come up with cloud services and achieve better results every day. These cloud service providers provide SaaS, PaaS, and IaaS. The data was previously stored on hard drives, which were not reliable and secure because anyone could access the drive. Today, cloud computing services have replaced hard drive search technology and come up with a new concept called cloud technology, in which data is stored in the cloud.

Cloud service providers offer software as a service (SaaS), platform as a service (PaaS), and infrastructure as a service (IaaS).

Software as a service is a cloud service provided by a cloud company. In SaaS, the customer provides software that can be either for a specific time or life. SaaS uses the Internet and delivers the application to customers. Most SaaS applications do not require any downloads because they can be used directly through a web browser.

Platform as a Service is a platform for developers where they can create an application to customize a previously created application. This service is also provided via the Internet and all administration is performed by the company or any third-party provider.

Infrastructure as a Service is a form of a cloud computing system that provides basic computing, networking, and storage resources to consumers on-demand, over the Internet.

Many companies provide cloud computing services and are very reliable. For the article we have chosen four selected international companies and their products, which we will analyze in more depth:

1. Amazon—Amazon Web Services (AWS)—2006
2. Microsoft—Microsoft Azure—2010
3. IBM—IBM Cloud—2011
4. Google—Google Cloud 2008

Amazon Web Services began offering IT infrastructure services to businesses in the form of web services—commonly known as cloud computing in 2006. Today, Amazon Web Services provides a highly reliable, scalable, and low-cost cloud platform that manages hundreds of thousands of companies in 190 countries around the world. With data center locations in the US, Europe, Brazil, Singapore, Japan, and Australia, customers in all industries enjoy a variety of benefits [19].

Microsoft Corporation is an American multinational technology company based in Redmond, Washington. Develops, manufactures, licenses supports, and sells computer software, consumer electronics, personal computers, and related services. Its best-known software products include Microsoft Windows operating systems, Microsoft Office, Internet Explorer, and Edge web browsers. Its flagships are hardware products for the Xbox and Microsoft Surface game consoles for touchscreen personal computers.

IBM is a leader in cloud platforms and cognitive solutions. The company dates back to 1911 and is one of the largest employers in technology and consulting in the world. It has more than 350,000 employees and serves clients in 170 countries. With Watson, a data-based AI trading platform, they build industry solutions to real-world problems. For more than seven decades IBM Research defines the future of information technology with more than 3000 researchers in 12 laboratories on six continents.

Google is an American search engine company founded in 1998 by Sergey Brin and Larry Page, a subsidiary of the holding company Alphabet Inc. More than 70% of the world's online search requests are handled by Google, which is at the heart of most Internet user experiences. Its headquarters are located in Mountain View, California. Google started as an online search engine, but now offers more than 50 Internet services and products, from e-mail and online document creation to software for mobile phones and tablets.

While writing the article, we chose to proceed systematically. In the beginning, we determined the main goal that we want to fulfill in the article. Before writing the theoretical part, we searched for and collected a large number of domestic and foreign sources of information, which we then incorporated into the article. The topic of the article is the competitiveness of international IT companies—a comparison of strategies, their strengths, and weaknesses, and that is why we focused on the current view of the fastest growing and most popular service in the IT industry. As objects of research, we chose cloud service providers—the four largest players in the market. Through thorough analysis, we obtained information about their strategies, strengths, and weaknesses.

During the writing of the article, our knowledge was based on available and relevant sources, which dealt with competitiveness, analysis of the internal and external environment of the company, and strategies and tools for assessing strengths and weaknesses. By comparing several IT companies, we used a comparison based on available and published information. We focused mainly on similar and different features of individual companies and on comparing individual, available data.

We used the analysis to analyze the individual elements, in the case of examining strategies, strengths, and weaknesses. The opposite of analysis is synthesis. Through synthesis, we combined individual knowledge about individual IT companies and their products, which led us to gain new connections.

In the work, we analyzed and analyzed a large amount of data. To select the right information, we used selection as a method that helped us write the theoretical and practical parts. Using deduction, we progressed from general knowledge to individual knowledge and statements.

4 Results

In part, the results of the work and discussion focused on the comparison of the largest international IT companies, and providers of cloud solutions. For comparison, we chose the four largest providers based on market share and achieved sales. These companies target corporate and mid-range customers, generally offering high-quality services with excellent availability, good performance, high security, and good customer support. Four of the largest cloud service providers are part of the research, as part of the largest IT companies in the world. In this chapter, we will focus on the cloud, which allows users to access the same files and applications from almost any device, because calculations and data storage is performed on servers in the data center, not locally on the user's device.

In the beginning, we will describe the history of the cloud and compare what advantages and disadvantages a cloud can bring to a company. We will compare the strategies of the Amazon Web Services, Microsoft Azure, Google Cloud, and IBM Cloud platforms. We will evaluate the strengths of the platform and at the same time describe the weaknesses that give companies room for improvement.

4.1 History of the Cloud

The emergence of a modern cloud began with the launch of the public cloud Amazon Web Services (AWS) in 2002. At that time, there were virtually no competitors. There were known benefits to use the cloud, such as flexibility and scalability. Actual use cases to convince potential users have not yet been available. Cloud has indeed offered the dream solution to the technical and managerial problems that many small and medium-sized businesses and organizations have suffered. The cloud has eased the burden of server maintenance, the initial investment in computing resources, and demand-driven web services scaling. As more websites and workflows began to look for a place in the cloud, in the next decade the cloud evolved over two different generations. The traditional definition of what a cloud is was realized during the first generation—a centralized infrastructure in data centers that host a lot of computing and storage resources. The momentum gained during this time allowed application owners to typically take advantage of the traditional two-tier architecture in which cloud providers hosted the back-end, while users sent all their requests from the web and mobile applications to the cloud. The OpenNebula research project was launched to develop a fully cloud-based software package in 2005. AWS's Elastic Compute Cloud (EC2) was made available to the general public in 2006. During this time, the technology was further developed and solutions aimed at strengthening data centers. Thanks to them, applications relied on them and at the same time reduced risks for users. Microsoft has entered the market with the introduction of Azure services.

Based on the first-generation cloud, the services provided have been dramatically enriched and competition has increased. Because it was possible to monitor the resources used in the cloud, the concept of the cloud became more credible. Two significant changes during the implementation of the first and second-generation cloud initially went unnoticed. The first concerned off-cloud processing and the second concerned heterogeneous cloud resources. While centralized clouds (which were remote from end-users) became popular, research began on whether some of the user's requirements could be processed outside the cloud [20].

4.2 Advantages and Disadvantages of Cloud Computing

Many companies are getting to cloud solutions to streamline their processes, store data, and maximize their processes.

Advantages of cloud computing [21]:

1. Easy to implement: Cloud solutions allow companies to maintain the same applications and business processes without having to deal with technical service requirements. Companies have easy and fast access to a cloud infrastructure that is easy to manage from the Internet;
2. Accessibility: companies can access their data anywhere, anytime. The cloud infrastructure maximizes business productivity and efficiency by ensuring that the application is always accessible. This allows for easy collaboration and sharing between users in multiple locations;
3. No hardware is needed: companies using cloud solutions do not need physical storage because they have everything they need in the cloud. It is also important to have data backed up in the event of a disaster;
4. Costs: Overhead costs for cloud services are kept to a minimum, allowing businesses to use extra time and resources to improve enterprise infrastructure;
5. Flexibility for growth: the cloud is easily scalable, so companies can add or remove as needed. As companies grow, so will their system;

6. Fast loading: cloud computing provides faster and more accurate loading of applications and data. With less downtime, this is the most effective recovery plan.

Disadvantages of cloud computing [22]:

1. Loss of control: When transferring data and services to the cloud, the company passes on its data and information and transmits minimal control to the customer. For companies that have their own IT staff, they will not be able to handle problems on their own;
2. Possible outages: cloud computing systems are based on the Internet, service outages are always a possibility, and they can occur for any reason;
3. Security and privacy: companics implement the best security standards and certifications, but there is always a discussion about possible risks;
4. Limited functionality: cloud providers offer cloud services at different levcls. Some cloud service providers tend to offer limited versions and allow only the most popular features. Before you sign up, you need to make sure you know what your cloud service provider is offering;
5. Bandwidth problems. For ideal performance, clients must properly plan and not pack large numbers of servers and storage devices into a small set of data centers.

4.3   Competitiveness of Individual Cloud Providers

In the case of the competitiveness of individual cloud providers, the company forms and develops individual factors. They are influenced by various political laws and the economic systems of the country in which they operate. It is important how companies are technologically equipped and their financial strength. An equally important factor is the human capital, as they organize almost all activities in society. Staff is needed also to train and qualify to maintain the latest trends. Marketing is also a condition, through which the company creates relationships between the company and customers. From the point of view of market factors, customers are affected by price and quality of service, and strategy settings of individual companies. From the external environment, companies are conditioned by different standards and conditions. These statistics show us the market share of leading cloud vendors' computers worldwide in 2015, 2016, and June 2019. In the years to June 2019, Microsoft generated the largest share of the revenue from cloud computing and controlled 11.7% of the market. In total, the cloud computing market was valued at 246 billion that year's US dollars. Gartner has developed the Magic Quadrant, which reflects the changing dynamics offered by cloud services and how the company's customers adopt them. Ultimately, cloud providers with a high scale and a wide range of services beyond infrastructure as a service (IaaS) have found strategic importance in Gartner and Magic Quadrant corporate clients and must develop to reflect in the future them as much as possible. As the market changes and evolves, the company reviews the criteria for change providers in this magic quadrant. As a result of the evaluation, the quadrant may change [23].

The basic inclusion and exclusion criteria for the Magic Quadrant include [23]:

1. Market participation: companies must sell public cloud IaaS as a stand-alone service, without the use of any managed services;
2. Market and market power: companies must be among the best global providers for the relevant segments must be audited ISO 27001 data centers on at least three continents;
3. Business skills relevant: companies must offer public IaaS cloud service globally, be able to offer consolidated invoicing, provide ongoing customer support and be willing to negotiate tailor-made contracts;
4. Technical possibilities: companies must have public cloud IaaS services and PaaS, which are suitable to support critical, production, and labor loads.

IBM Cloud

The IBM cloud platform combines platform as a service (PaaS) with infrastructure as service (IaaS) to provide an integrated experience. The platform extends and supports small development teams and organizations as well as large corporate companies. Deployed globally in data centers around the world, the solution is based on IBM. The cloud works reliably in a tested and supported environment that you can trust. You can deploy native cloud applications while ensuring portability workload. One of IBM's responses to declining revenue is restructuring some of its business activities, moving to areas like cloud, blockchain, and artificial intelligence. The company is trying to address its underinvestment in cloud development software and platform products. For example, its IBM Cloud platform aims to help customers move their operations to cloud environments. At the beginning of 2019, IBM acquired open-source technology company Red Hat, which offers a hybrid cloud platform. As a major and innovative player in the IBM industry constantly spends about $ 5 billion on research and development, such as advances in quantum computer technology and artificial intelligence [24] (Fig. 1).

This chart shows IBM's worldwide revenue in billions of US dollars. In 2020, it generated a global segment of technology services IBM made $ 25.81 billion in company revenue for 2020 accounting for about $ 73.62 billion. In 2016, IBM for the first time, changed its segment reports to reflect the company's shift from companies dealing with hardware, software, and services towards having become a company in the field of cognitive solutions and cloud platforms. In 2019, IBM's cloud and cognitive software revenues reached more than $ 23 billion, of which $ 9.5 billion was generated subsegment of cloud and data platforms. This is an increase of 8.6 billion US dollars in revenue generated by this subsegment in 2018 [26].

Whether you need to migrate applications to the cloud, upgrade your existing applications using cloud services, ensure data resilience against regional failure or le, or average new deployments to innovate and develop your cloud applications, open the platform architecture is built to suit the use case. IBM Cloud offers services as a platform as a service and infrastructure as a service. This cloud organization can deploy and access your resources such as network storage and computing performance, using the internet. Several tools help customers draw from deep industry

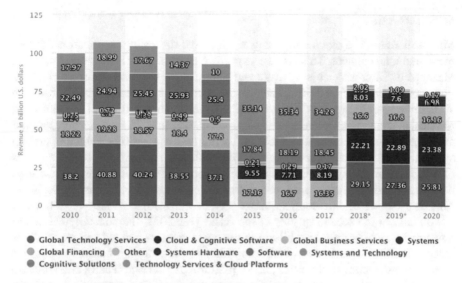

Fig. 1 IBM sales worldwide from 2010 to 2020 by segment [25]

expertise. The speed and agility of the cloud meet customer requirements and make them feel satisfied. An IBM customer can easily find opportunities to grow, create new revenue schemes and improve operational efficiency. IBM cloud user does not have many obstacles compared to traditional technologies. The IBM cloud eliminates the complex problem and challenges facing large companies. IBM cloud computing services also help home manufacturing businesses appliances, retailers, and medical supplies. He uses it because offers the best services at the lowest price [27].

Amazon Web Services (AWS)

Amazon Web Services (AWS) is the most comprehensive and widely accepted cloud platform in the world, which offers more than 175 full-fledged data center services worldwide the world. Millions of customers—including the fastest-growing fastest-growing startups, the largest companies, and leading government agencies—use AWS to reduce costs, agility, and faster innovation. AWS has the largest global cloud infrastructure. No other provider cloud does not offer so many regions with multiple low-availability availability zones latency, high throughput, and highly redundant networks. AWS has 77 zones of accessibility in 24 geographical regions around the world and announced plans for 18 additional accessibility zones and 6 additional AWS areas in Australia, India, Indonesia, Japan, Spain, and Switzerland. Amazon Web Services is a cloud computing platform that provides services like computing power, database storage, content delivery, and many other features that help with business integration. Amazon Web Services is flexible, scalable, and reliable and as a result, many companies implement it in their work. They do not exist no initial costs and the customer only has to pay for what he used. He is one of the leading cloud service providers [28].

Microsoft Azure

Microsoft Azure is a cloud computing service used for testing, application deployment, and management. This process is performed on a global data center network managed by Microsoft. It is a private and public cloud platform. It uses virtualization, which distinguishes the link between the operating system and the CPU using an abstraction layer known as a hypervisor. The hypervisor transforms all functions of a physical machine, such as hardware and server, into virtual. There are a large amount available virtual machines and a lot can be run on each virtual machine operating system. There are many servers in the Microsoft data center, each server consists of a hypervisor through which multiple virtual machines can operate. With Azure, developers and IT professionals can easily manage and deploy their applications. This chart shows us Microsoft's quarterly cloud computing revenue in billions of US dollars. In the second quarter of the fiscal year 2021, the company Microsoft generated approximately 15.6 billion in revenue from smart cloud services dollars. Microsoft offers its customers cloud computing services through Azure. Its solutions include infrastructure as a service (IaaS), platform as a service (PaaS), and software as a service (SaaS) that can be used, for example for virtual computing, analysis, networking, and data storage [19].

Google Cloud Platform

Google's cloud platform is one of the leading cloud computing services that is provided by Google and runs on the same infrastructure as Google for its end-user products. The Google cloud platform is used for Google and YouTube searches. Google Cloud offers various services such as data analysis, machine learning, and data storage.

The data stored in Google Cloud is secure and easy to access. Offer various services from infrastructure to platform. Google also provides a firm commitment to security and stability. With the help of the Google cloud platform, the user can freely think about the code and features that need to be developed without worrying about operating pages. Here, most services are fully managed and customers can easily focus on your work. In this, machine learning and using the API are very simple. API also very easily helps in translating speech detection languages. He, therefore, prefers among customers [29] (Fig. 2).

This graph shows us the global revenue from Google Cloud for the years 2017–2021 in billions of US dollars.

Selected Financial Indicators

Interest in cloud computing is still growing and thanks to that we have a lot of information about it. On the one hand, it saves money for companies, and on the other hand cloud providers still profit. The cloud services market reached—$ 37 billion in the fourth quarter of 2020 dollars, an increase of $ 33 billion in the third quarter of 2019. Pandemic COVID-19 accelerated the digital transformation and included an accelerated transition to the cloud (Fig. 3).

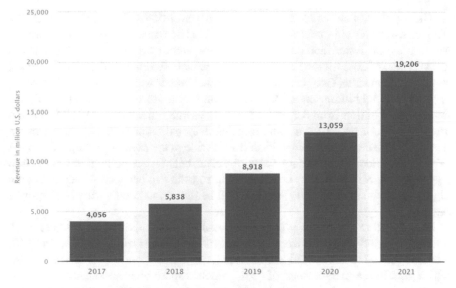

Fig. 2 Global revenue from Google Cloud 2017–2021 [30]

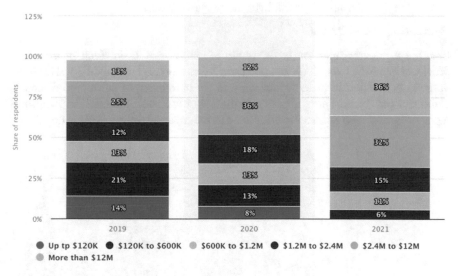

Fig. 3 Annual spending on public cloud companies

In 2021, 36% of enterprise respondents plan to spend more on the public cloud $12 million. Public cloud spending remains high as organizations from around the world are constantly migrating to the cloud as part of their digital transformation plans. Organizations are investing more in collaboration, remote work technology, and infrastructure [31].

The big three were as usual in 2020—Amazon, Microsoft, and Google, taking Alibaba was firmly entrenched in fourth place and IBM came in fifth place. However, Microsoft grew faster than competing Amazon and at the end of 2020 reached a market share of 20% for the first time. Software giant Redmond since 2017 doubled its share. Meanwhile, Google and Alibaba took home 9%, respectively 6%. Google brought in $ 3.3 billion, out of $ 2.98 billion in the 3rd quarter of 2020 Alibaba raised $ 2.22 billion from $ 1.65 billion in the same time frame. AWS has been a great success for over 10 years and remains in an extremely strong position in the market despite increasing competition from a wide range of strong companies in the IT industry. Microsoft considers it a worthy rival, but at some point, it has to run into a growth wall. Microsoft may continue to narrow the gap between Amazon, but the bigger Microsoft Azure is, the harder it will be for maintaining a truly high growth rate [32] (Fig. 4).

Statistics show the returns of the largest vendors in the public cloud market Infrastructure as a Service (IaaS) from 2015 to 2019 in billions of US dollars. In 2019, Amazon generated about $ 20 billion in sales of infrastructure as a service dollar, which gave him a market share of 45%, which is more than the other four largest sellers in the market. It is expected to be among the three main segments of public cloud services, software, infrastructure, and platforms, the infrastructure will be a service the fastest level of growth over the next five years.

Comparing Cloud Prices

Creating a price comparison in the cloud is not an easy task. Each of the cloud providers regularly lower its prices and each offers unique price models and discount options. In addition to price reductions and the possibility of additional discounts cloud service providers also offer new availability zones, new products, and improvements. Amazon Web Services is often seen as the cheapest option. This is not always

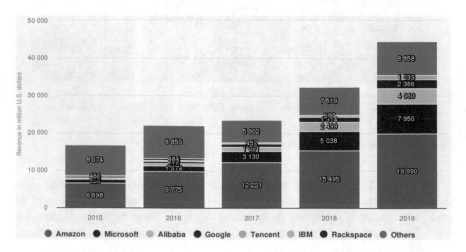

Fig. 4 Revenues from the public cloud market in 2015–2019

the case. AWS is at the forefront of cloud computing due to its large turnover and market share. The reason why Amazon Web Services is considered the cheapest is a concept in which they present the so-called Virtuous cycle. AWS was also the first to be launched. Selected cloud service providers continue to reduce the prices of cloud instances, add new discount options, add instances, and abandon settlement increments. By observing many changes in the prices for cloud instances, they are visible everyone year. Cloud providers add new instances each year, and in some cases and entire families of instances. The possibilities of discounts increase from year to year. Microsoft Azure has introduced dedicated instances that will save up to 72%. AWS presents one year of convertible instances, and Google has introduced one-year and three-year tied use discounts. As for changes in invoicing, AWS has made a big change. Billing per hour shifted from hours to seconds for EC2 and several other services. Google always provided billing per second but reduced the minimum time from 10 to 1 min. Cloud providers turn users away from relying on local disks and pushing them toward the attached storage [33].

Discounts on Cloud Services

With the recent availability of Azure Reserved Instances, they also offer AWS and Google publicly available discounts of up to 75% in some cases in exchange for a commitment to companies that will use the cloud service provider for 1 or up to 3 years. IBM offers public discounts only for monthly use, which can save companies about 10%. Google also offers a discount on usage and does not require any commitment. It offers an automatic discount for each type of instance that has been active in the region for more than 25% of the month. If the service runs all month, the company can provide a discount of up to 30% [34].

Cloud Prices of Individual Platforms

Amazon Web Services offers pricing models for more than 160 cloud services. The companies only pay for the services they have used. As soon as customers stop service they do not incur any additional costs or termination fees. Three basic factors cause the main costs: calculations, storage, and data transmission. Prices of AWS services are established independently, and transparently and are available on request. Customers can choose and pay for exactly what they need. Customers can get started with AWS Free Tier, which allows them to gain free experience in various programs. For example, customers can use the trial version. Companies can also use AWS Pricing Calculator where they can calculate their costs. Microsoft Azure provides cloud payments via credit card or by sending monthly invoices. However, the customer must decide at the beginning. The customer can use a price calculator for a combination of required products. Since the development of testing to production workloads realize additional savings that match development and deployment needs. They combine discounts and quotes to significantly reduce your cloud costs. They state that AWS is five times more expensive than Windows and SQL Server.

Google Cloud Pricing is also not published on the official website, because each company needs a different product combination. They present themselves transparent and innovative approach to prices. By creating an account, they can companies start running their tasks for free. New customers receive 300 credits USD for the operation, testing, and deployment of workloads free of charge. All customers can use more than 20 products for free, up to monthly limit use. Thanks to the "pay as you go" pricing structure in Google Cloud, companies pay only for the services they use. They don't have to pay any initial fees or none termination fees. The IBM Cloud platform also offers its customers the opportunity to start using their services for free. Lite accounts provide free access to more than 40 services, including IBM Watson API. You will receive a \$ 200 credit when migrating from Lite to Pay-as-you-go. The credit is valid for 30 days. With a service of 350 or more, companies will gain access to more than 350 products at discounted prices using PayGo with binding use or subscription upgrade. IBM Cloud reports the lowest prices in 67 scenarios to beat Microsoft, Google, and AWS [33].

4.4 Cloud Platform Strategy

The cloud strategy is an outline of the cloud's role in the organization, not a plan to move everything to the cloud. Today, most organizations do not have a formal cloud strategy, although, within a year By 2022, 70% of organizations will have it. Businesses with such a plan have a more coherent approach to using the cloud and optimizing resources and costs. Cloud strategy should not only be for the IT team but also for him. The summary is important to summarize the document for leaders in all departments, including top management. This section should also emphasize people in the company's cloud line to demonstrate its importance beyond the IT department [35].

IBM Cloud Platform Strategy

IBM's cloud strategy has evolved as customer expectations for the cloud. IBM has invested heavily in hybrid and multi-cloud basis, based on industry standards and open source. The core of the cloud IBM's strategy is to understand that customers need a unified and efficient way to create modular application services, transform and modernization of legacy solutions, and administer all services, regardless of where these workloads are located. All of these services must be cataloged, secured, administered, and governed by rules. The only way to achieve a unified approach to cloud computing is to combine strategy based on open source and standards. Therefore, it is easy to understand why IBM created a platform focused on Kubernetes. The IBM cloud model allows the customer to have the same services available in IBM's private and public cloud, as well as in any public cloud and services third parties on the margin. Performance and versatility of open source, including Linux, open API, and Kubernetes, helped IBM transform its software to become modular and unified services that are easier to manage. IBM has an advantage in providing computing platforms to support some of the largest and most comprehensive jobs in the world. IBM has many years of experience implementing domain applications based on sophisticated business processes. These best practices are based on the expertise of Global Business Services and Global Technology Services. These organizations provide business and technical advice to customers worldwide in vertical markets from financial services to retail and transportation. In planning for dramatic business and technical change, all of these industries depend on hybrid and multi-cloud environments. Business executives understand inherent the risk of relying on one supplier for all situations. Managers in smart business, therefore, require an environment that combines different public and private clouds, hosted services, edge computing, and data center options from many different suppliers. This hybrid environment must be safe as well as manageable regardless of where the services and workload are deployed. IBM's customer base is complex, so there are different requirements to move to the cloud. IBM has defined three entry points or customer support models. Migration of existing services to the cloud. Most large companies have to deal with large numbers of obsolete legacy applications and infrastructure. These businesses need a path to modernization and reorienting of their applications to gain the flexibility and agility needed for business transformation. These businesses can start by lifting existing applications to could be hosted in the cloud using virtualization through VMware services. This entry point is designed to help businesses that want to make the most of their infrastructure expenditure. Modernization of existing applications and services [24]. Organizations transform their applications using a container platform to upgrade and migrate existing applications. The computing environment is analyzed to determine which applications can be reconstructed as micro services and which applications will be needed rework. After containerization, it is possible to connect these microservices as part of the cloud basis using open standards. After refinancing these existing ones application developers can quickly create new applications. New cloud-native cloud-native innovative applications. Using native cloud services such as Knative and Serverless, they can developers create next-generation applications that can be deployed in an open environment source Kubernetes. Business executives understand the inherent risks of relying on one supplier for all situations. Leaders in smart business, therefore, require an environment that combines various public and private cloudy, hosted services, edge computing, and data center capabilities from many different vendors [25].

Amazon Web Services Platform Strategy

If we focus on Amazon Web Services' strategy, many years since its the launches focused only on the cloud, which seemed great for new startups and existing companies such as Netflix, which has been able to build its entire service on the platform Amazon Web Services. From the point of view of cloud competitiveness, this situation is also always the case of leaving Microsoft in a better position to provide hybrid cloud solutions, because they were already in the data center, which AWS left behind many large potential corporate accounts. This situation is changing rapidly and AWS is making 3 major efforts. They are the first already the aforementioned outposts. Customers can connect to their premises and use the same APIs, integration with other AWS services, and central AWS console management. The second effort is AWS Local Zones, which is a newly deployed service, which means that you can deploy to specific locations. The third effort is AWS WaveLength—which is AWS push integration with the core of the 5G mobile network. The mobile application may have the network itself as an advantage with ultra-low ultra-low latency [36].

Microsoft Azure Platform Strategy

Microsoft CEO Satya Nadella immediately after joining sent the first e-mail to his employees, where she wrote about the future importance of the cloud. She stressed that their industry does not respect traditions—it only respects innovation. She started with a big transformation of the company's strategy. An important step was to focus on a smooth architecture throughout digital reality. They focused on addressing real customer needs using a differentiated approach to the cloud—architecture for hybrid consistency, developer productivity, artificial intelligence capabilities, trusted security; and compliance with regulations. In addition, customers choose Microsoft Cloud for their operations, because productivity and security cover the entire digital headquarters, including the cloud, Windows 10, Dynamics 365, Enterprise Mobility & Security, and Azure. Azure revenue grew 90% in four quarters as it provides what the business Customers want: consistency. The provision of colocation services themselves or connections between local data centers and the public cloud is not enough to penetrate customer needs. They need consistency between development environments, operating models, and technology stacks. Azure provides this consistency across the entire stack, including identity, data, application platform, security, and news—everything on the edge and in the cloud. Their hybrid cloud is an important part of the strategy. Azure is gaining huge market share among global companies for its large strategic deployment. The change has also occurred in the financial services industry, which is now downloaded to the cloud. Not only do simple relationships with suppliers work, but they also try to build your software capacity, you need a trusted partner to they are more interested in whether they can build their technological capabilities. Microsoft focuses on strategic traders and long–term business—based relationships success and corporate innovation in the cloud. The hybrid is not just a transition point for them, it is a strategic goal. CFO Microsoft claims their unique ability to provide a distributed hybrid model for Microsoft Office smart cloud continues to attract customers to the company. Large banks and financial service providers are using Azure and its data services to improve customer experience. In their technologies, they try to focus on ensuring that has helped you become able to create your software. AI is the new heart competitive advantage. The main currency of any future business will be the ability to convert their data into artificial intelligence, which increases competitive advantage. They are trying to continue to improve their cognitive services to provide each business with powerful elements for creating artificial intelligence applications [29].

Google Cloud Platform Strategy

Google Cloud's main strategy is to continue the trend of hybrid and multi-cloud computing. In both cases, they provide cloud-era technology customers the opportunity to make better use of existing assets and take advantage of newer methods of calculation, storage, and analysis of data. The multi-cloud strategy gives companies the freedom to use the best possible cloud for every workload. Using a single cloud there are increased costs. We are committed to Google Cloud customer needs by providing choice, flexibility, and openness. Google Cloud CEO Thomas Kurian argues that extensive organizational changes and investments for the last year are returning. The main strategy was also to launch an aggressive one drive for enterprise customers, developed an open-source game plan, hybrid and multi-cloud plan with the new Anthos platform, and introduced the vertical solutions sector-specific. Google Cloud has also introduced more than 280 new products and improved its affiliate program. Kurian said Google Cloud is moving at 100% strategy to tie partners to business with customers because—not even financially by strengthening Alphabet, its parent company worth almost 162 billion dollars could not match the power that this channel brings. "Our cloud strategy is quite simple, Kurian said." We see customers in every industry who want to adopt digital technology [37], either to reduce costs, the growth of their highest revenues, or changes [31].

4.5 Strengths and Weaknesses of Individual Cloud Services

Companies should know their strengths and weaknesses, which gives them baa better understanding of how society works. Understanding the strengths will keep them ahead in many things. It also helps them grow more. Knowing their weaknesses gives them a clearer understanding of the thing that can hold them back.

Strengths of the IBM Cloud Platform

IBM Cloud sets itself apart from other providers by leveraging its extensive history with the computing infrastructure of the Power Systems family. This supports the workload as well diverse such as SAP HANA, Oracle ERP, and newer styles, such as those focused on deep learning. In addition, IBM Cloud has significantly improved overall VM delivery times and enhancements control plane. IBM Services acts as an embedded channel and SME for IBM businesses Cloud, which sets it apart from other cloud providers that rely on third parties. IBM as one company can provide companies with transformation services, to become more agile and a cloud infrastructure in which they can deploy their agile workload. IBM switched to hybrid and multi-cloud multi-cloud messaging calculations. It offers businesses options and tools that allow them to choose the cloud environment best for their specific application requirements while taking IBM Cloud during places the selection process in place of a specialized or special offer [23].

Weaknesses of the IBM Cloud Platform

IBM Cloud remains a comprehensive platform resulting from older offerings and uneven product development profits. Although IBM has listed the infrastructure Gen 2 in 2019, the new IaaS offering includes a subset of SoftLayer's capabilities and is k available in only five countries. It is noteworthy that IBM was 10 years ago by launching cloud-based software-defined networking capabilities. All the time there is a lack of broad platform support for identity and access management, despite promoting innovative efforts in the financial services sector. IBM supports hybrid cloud stories using OpenShift, which enables portability between cloud providers. However, this approach will result in limited business benefits because you require general use of cloud provider features. IBM Cloud has an eligible world market share in the PaaS application category. Some of IBM's PaaS offerings for example; IBM Cloud Functions have little acceptance and are perceived by IBM as an older technology provider [23].

Strengths of the Amazon Web Services Platform

Amazon Web Services (AWS) continues to be a leader in many critical dimensions of the computer integrated process systems market, including the overall market share and bidding capabilities. AWS has the largest share of the world market on offer IaaS and PaaS databases. AWS has the capabilities, resources, and motivation for vertical integration and providing end-to-end solutions to customers. The company can suggest everything from the beginning on their servers to the embedded operating systems in the peripherals devices and a complete software package between them. AWS is a financially extremely well-managed company and generates more than 50% of operating revenues for the whole of Amazon. Unlike some other providers in this magic quadrant, AWS's healthy margins mean an improvement in its parent company [23].

Weaknesses of the AWS Platform

Growing concerns about Amazon's size and reach, along with AWS's leading position, give a break to some of its partners and customers. In addition, AWS risks that as a result. The company struggles with open source communities, alienating software developers who are its basic circuit. Open Source benefited mainly from a wide range of offers AWS without AWS providing perfect service. AWS has poor coherence in its ever-expanding offers. The company's leading position in IaaS and dbPaaS creates a misleading effect for other offers. AWS organizational design that allows developers to work as semi-autonomous units, rather creates inconsistencies between products than a coherent whole, especially for new services. The effective use of AWS also requires the thinking of application developers and combined with poor coherence can be for many businesses a real challenge. Customers continue to believe that AWS is declining prices significantly; however, declines are often not universally applied to all services. For example, AWS has not recorded the most frequently provided repository for the AWS computing service since the 2014 price reduction, which is almost half the life of the company, despite the dramatic falling market prices [23].

Strengths of the Google Cloud Platform

The strengths of Google Cloud are Kubernetes and TensorFlow, which have been innovations that have moved the market and changed the direction of enterprise IT. Such innovations have served to enable other cloud service providers, but they have also brought the developer's mind share to the Google Cloud Platform. Google's long-term strategy is brought to the Google Cloud Platform as managed services of other partners focused on an open-source. The Google Cloud Platform has seen significant growth over the past year year-on-year market share in terms of IaaS and dbPaaS, albeit from a lower base, compared to other providers. Google has also made significant profits by closing several critical capabilities gaps between the Google Cloud, platform and Microsoft Azure, its closest competitors in terms of market share and options [23].

Weaknesses of the Google Cloud Platform

Some clients remain cautious about Google's commitment to serving the g needs of corporate clients when they are placed in the context of SAP's preferences for Microsoft Azure and the slowness of GCP in the implementation of some highly offered partnerships. GCP lacks the capabilities of aPaaS focused on enterprise and support for Oracle and continues to struggle with corporate thinking in this area. Financially, GCP revenue is a small fraction of Google's total revenue. The success of GCP can further undermine the overall health gross margins of the company. Google's much-vaunted networking opportunities have been in over the past year have been the source of many GCP outages, which has been devastating to customers. One outage was multiregional in scope and affected GCP customers and Google consumer services like G Suite and YouTube. This resulted in a complete GCP network unavailability for some customers [23].

Strengths of the Microsoft Azure Platform

Microsoft Azure offers a complete set of solutions for a wide range of applications workloads and applications. This is evident from Microsoft's Azure partnerships Oracle, SAP, and VMware, continue Azure's capabilities in the field containers and without servers and end with compelling solutions for a peripheral and hybrid environment. Microsoft is working together to provide better service to software developers, in particular through their efforts with open-source systems. Microsoft is at the forefront of cloud providers with hyper scales when it comes to marketing a stake in the PaaS application developer segment, thanks to their suite of tools that include Azure DevOps and Github. Microsoft Visual Studio Codespaces, which is currently in beta version, is the first impressive development environment for hosted cloud applications that bridge the use of public cloud and ultra-popular development tools, for example, Visual Studio Code. Ordinary businesses often have a strategic connection to Microsoft, which gives Azure significant sales advantages in this market segment. Microsoft Azure has in addition a particularly strong share in mind with Gartner's corporate clients in Europe. Microsoft Azure Cloud Services provides a comprehensive set of solutions for companies to accelerate their digital transformation initiatives and take advantage of new technologies for the benefit of the company. One of the best benefits of Microsoft Azure Cloud Services is the scalability that the platform provides to businesses [23].

Weaknesses of the Microsoft Azure Platform

The weakest aspect of Microsoft may be the lowest ratio of availability zones to areas of any vendor and the availability zone model is supported by a limited set of services. Concerns about the overall architecture and implementation of Azure, despite efforts to be resilient and improved service availability metrics during the past year. Microsoft does not provide customers any form of guaranteed capacity; nor are prepaid agreements and reserved instances capacity guarantees. When in a few weeks it occurred in several European regions to customer-related capacity shortfalls related to COVID-19, a small number of customers has not been able to secure reserved cases or capacity for which already they paid. Microsoft Unified Support can be very expensive, especially for those customers who did not have support services covering their entire portfolio in the past Microsoft. Although the company's professional competence has improved over the last year Microsoft in the field sales and Microsoft is improving Azure technical support, Gartner clients continue to report concerns about the quality of this experience [38].

4.6   Designing a Cloud Strategy Solution for Managers and Smaller Companies

Information technology and cloud computing technology is evolving very fast last few years. Before cloud services, people used applications and programs stored on a physical server. Companies often have their data centers for which they must have often many IT professionals. This is a difficult enough situation for a company to sustain productivity. This makes companies more focused on cloud solutions. Cloud computing is a technology that provides virtualization and provides on-demand storage. Cloud is an opportunity for a new business model and applies to every type of company. It has an integrated backup and recovery process data that is always secure. It can bring improvements to companies' competitiveness, agility, and profitability. The cloud is also flexible for employees who can also be flexible thanks to it. They have access to all data and documents from any environment and equipment. For many companies accustomed to operating for local data centers, the adoption of the cloud can also mean a major shift. For everyone, the company needs to prepare a comprehensive and flexible cloud strategy. All right the strategy developed will guide every step. One of the best ways to get started creating a cloud plan is to find out the current state of people, processes, and technologies. Adopting the cloud will change the way an organization works. The first important step is to understand the importance of the cloud in society. Some important decisions regarding IT matters are left to business owners, directors, and managers on their IT team. Sometimes it is necessary to look at it from a business perspective as well as benefits and opportunities. If companies don't use a cloud solution today, it could be a competitive disadvantage for them. The cloud is not just a way of storing data, much more. It can offer us faster, cheaper, safer, and more scalable ways to develop new products and services. It is a framework for the future success of companies. Without changing the view of cloud technology, we cannot develop a business strategy. The second step is to understand the benefits of the cloud. Cloud solutions will help the company achieve business goals, reduce security risks, increase team productivity and reduce duplication of data. The company will save money on the security and maintenance of in-house systems. Cloud data centers are too expensive to access allowing many medium-sized businesses. They are difficult to maintain and staff. The third step is to focus on the company's corporate goals, systems, technologies, and the processes that the company currently uses. You need to gather information about company goals, current needs, and key stakeholders.

One of the suggested steps may be to carry out a brainstorming where they are documented goals of individual departments, such as the IT department, HR department, finance department, and the Executive Director. The next step is to assess the current state of IT in the company. In the IT industry, it is routine to assess readiness for the introduction of new technologies. The fifth step is to identify and select a cloud model. The choice is public, private, hybrid cloud, or multi-cloud. The public cloud offers its services to the public. The largest providers are Microsoft, Amazon, IBM, and Google, which are mentioned even at work. They offer internet access and service charges are based on models pay as you go. It is not necessary to have your own IT for public cloud infrastructure. The end-user also has no control and reach over where IT is infrastructure saved. The public cloud has limited server configuration options. Is designed for a large number of cloud tenants, which allows data isolation, access, and security for each customer? A private cloud offers the same features as well as benefits as a public cloud. It works on dedicated infrastructure in the data center. Information no they are shared with other participants and organizations and provide the highest form of security company data. The downside is the return on investment in the private cloud. An organization implementing a private cloud is responsible for running and managing IT resources instead of transferring this responsibility to the cloud service provider. The hybrid cloud is a combination of public and private clouds. Some of the important data is located in a private cloud. Other data is accessible and located in the public cloud. This option is also investment intensive due to the purchase and maintenance of hardware. Multi-cloud is the connection of several public clouds in one architecture. They distribute applications and data on various specialized platforms. It focuses on several of the same matters as the public cloud can mitigate the strengths and weaknesses of multiple platforms identified where critical process hosting can take place. The sixth step is to choose the right type of cloud service. If we have a question cloud model solved, we will focus on choosing the right type. In this step, he has a company to choose between infrastructure as a service (IaaS), software as a service (SaaS), and Platform as a Service (PaaS). With IaaS, companies will have direct control over application configuration. With SaaS, the software that cloud providers deliver updates automatically for all users. PaaS serves developers in application development and testing.

The next step is to select and prioritize the cloud initiative. You should have the company identify the target departments where the cloud services will be located. In each department, the company chooses the main IT services that support business activities. The company will determine which IT services have the highest priority— activities that are key to cloud selection. If we already know which services we would use the cloud for, the company can ask questions about the impact of cloud adoption on staffing, the benefits of migration of services to the cloud, or whether there will be cost savings from providing the service through the cloud.

To transfer services to the cloud, it is also necessary to evaluate the readiness of the current cloud services. This evaluation concerns data security and security. The company should be able to control and protect access to sensitive data. Data must also be protected against unauthorized manipulation, they must be confidential but at the same time always available to the company. The company should focus on these points before the introduction of services. In step ten, we will focus on the risks we might face. Is it important to develop a small risk identification plan and possible risk reduction steps and potential problems we might encounter?. There are key areas on which we should focus. If in the company we do not yet have a team specialized in IT, it is important to find an external partner. The company should provide appropriate technology so that employees can access the cloud even in times of outage. If we have a cloud strategy in place, it is important to communicate your plan properly to employees. It is essential to inform them about the importance and benefits the cloud company will bring. The company can inform employees via e-mail or through an internal site. The report should be simple and concise, for everyone to understand. The last step is to implement a cloud strategy. Every company's process implementation takes different lengths because it is a unique process. For the strategy to be successful, the company must have a clearly defined plan, employees must be well trained. During the implementation process and even after its completion, the cloud must still be supported.

5 Discussions

In the last part of the article, entitled Discussion, we evaluate the results, compare them with other research on a similar topic, and focus on the benefits and use of the article. We will formulate basic recommendations for the needs of the practice. The article aimed to provide a guide and companies to correctly determine the strategy for the implementation of cloud solutions by identifying the position of international IT companies in a competitive environment. We managed to develop a guide for designing a strategy for implementing cloud solutions in the company in twelve steps, from finding out the importance of the company's cloud consumption to the implementation and maintenance of the cloud in the company. I think this proposal may be a small guide for managers who would be thinking about implementing the cloud in their companies or in the individual departments that need it. In this work, we also identified the position of individual IT companies. From these providers, companies can choose their cloud provider providers. We have identified four selected cloud providers that are considered the largest in the world based on market share and revenue in recent years. We described their strategies, strengths, and weaknesses. Based on these indicators, companies can choose just that. Of course, they do not have to choose from the companies listed in our work. However, they have sophisticated service offerings, offering various packages for free and a high level of company data security. While writing my dissertation, I encountered several papers that dealt with the competitiveness of IT companies. The authors did not deal with the design of a cloud strategy in any work. I met with work that analyzed the competitiveness of IT companies based on the analysis of the pillars of the global competitiveness index and based on strategies and comparisons. He thus proposed an alternative strategy for international IT companies in a competitive environment. Further work focused on the competitiveness of international IT companies based on the development of SWOT analysis and SPACE matrix. Data for work was available and easy to find, as the issue progresses very quickly and many companies are trying to move to the cloud as quickly as possible. I would identify the search for information on cloud pricing as a weakness. Large international companies cater to companies individually and send them request prices. They thus create additional discounts for them based on commitment. The COVID-19 pandemic has accelerated this decision and more and more companies need to secure their data to enable their employees to work from home. I am satisfied with the choice of topic and the elaboration of the manual, which can be an aid for someone who has not yet encountered the issue. The student can work on the topic of the implementation of a new cloud strategy for a company that did not have it before.

6 Conclusions

The article dealt with the competitiveness of international IT companies—comparing their strategies, strengths, and weaknesses. The article aimed to provide guidance/information for managers and companies to properly determine the strategy for the implementation of cloud solutions by identifying the position of international IT companies in a competitive environment.

In the first section, we dealt with theoretical knowledge. Domestic and foreign sources have helped us to clarify the main concepts of the competitive environment, strategies, and analysis of the competitive environment that affect companies in achieving their goals.

In the second section, we defined the main goal and the partial goals that we met.

In the third section, we described international IT companies with a focus on cloud solutions. We have determined the evaluation methods and their sources. The key evaluation method for us was the analysis of individual documents related to the activities of individual companies. At the same time, in the third section, we described the working procedures, thanks to which we reached the main goal.

In the fourth section, we focused on the results of the work. In the results section, we defined the cloud, described the emergence of this innovative technology, and presented the main advantages and disadvantages of cloud solutions. To bring competitiveness and choice to cloud companies closer together, we have focused on the four largest cloud solution providers. We described Amazon Web Services, Microsoft Azure, Google Cloud, and IBM cloud in more detail. We compared their strategies, strengths, and weaknesses. The strengths of a company can be real strengths that need to be guarded and strengthened, and the weaknesses can be weaknesses that need to be worked on.

We see the Amazon Web Services platform as a global leader in providing the world's largest cloud infrastructure. The main strategy and strong point are to focus on the customer. Every step taken is designed to serve customers. The front page is also an extensive set of services. One of the weaknesses is setting limits and the additional cost of technical support. Its weaknesses include hybrid solutions and older systems.

The second platform examined is Microsoft Azure, developed by the well-known company Microsoft. Unlike the Amazon Web Services platform, it came on the market four years later in 2010. The main strategy of the platform is the hybrid cloud, which they take not only as a breakthrough point but as the main strategic goal. It has stronger PaaS functions and is designed based on a state-of-the-art security process. Microsoft Azure is primarily a solution for larger businesses. It may seem too complicated and expensive for small and medium-sized companies.

Google offers its public cloud computing solutions called the Google Cloud Platform. Google Cloud's main strategy is to continue the trend of hybrid and multi-cloud computing. At the forefront are their Google Cloud data centers, which run on half the power of a typical data center and 100% renewable energy. It offers fewer features but saves on storage costs and ease of use.

The last platform evaluated was the IBM cloud platform, which came on the market last in 2011. The main strategy is to focus on the hybrid cloud and provide services for large companies to which they can guarantee the security of their data. IBM Cloud Services are designed to suit a specific use case. This is one of the reasons why the IBM cloud is not as popular compared to other companies. But it is a highly secure and reliable platform. There is efficiency as well as flexibility in data security.

In the last section, we designed a cloud strategy for smaller companies or managers who would like to implement cloud solutions in the company. The company must be prepared for such a change and develop a detailed and strategic plan. Companies need to evaluate which departments need to move to the cloud and which can use it most effectively. Such innovations can bring huge benefits to companies in storing, backing up data, and simplifying information sharing. Employees can access data when they need it.

References

1. Linhart, J.: Slovník cizích slov pro nové století: základní měnové jednotky, abecední seznam chemických prvk°u, jazykovědné pojmy, 30 000 hesel. Dialog, Litvínov (2005)
2. Borovsky, J.: Manazment zmien - cesta k rastu konkurencieschopnosti. Eurounion, Bratislava (2005)
3. Lisý, J.: Ekonómia (2016)
4. Tokarova, M.: Protimonopolna politika: teoreticke aspekty vyvoja konkurencie, sutazivosti a protimonopolnej politiky. Sprint, Bratislava (2002)

5. Jáč, I., Rydvalová, P., Zizka, M.: Inovace v malem a strednim podnikani. Computer Press, Brno (2005)
6. Šrédl, K.: Vyjadřování podnikové konkurenceschopnosti pomocí souhrnné analýzy trhu a jeho efektivnostiSPU, 2001, s. 487–490 (2001). ISBN 80-7137-867-4. Zborník vedeckých prác z medzinárodných vedeckých dní 2001, Nitra
7. Mikoláš, Z.: Jak zvýšit konkurenceschopnost podniku: konkurenční potenciál a dynamika podnikání (2005)
8. Kádár, G., Kádárová, J.: Hodnotenie faktorov konkurencieschopnosti podnikov. p. 6
9. Sedlackova, H., Buchta, K.: Strategicka analyza. C.H. Beck, V Praze (2006)
10. What is PESTLE Analysis? An Important Business Analysis Tool. https://pestleanalysis.com/what-is-pestle-analysis/ (2011). Accessed 21 May 2022
11. Porter's Five Forces—Free Management eBooks/porter-s-five-forces-free-management-ebooks.pdf/PDF4PRO. PDF4PRO. https://pdf4pro.com/view/porter-s-five-forces-free-man agement-ebooks-1525ee.html (2018). Accessed 22 May 2022
12. Pearce, J.A., Robinson, R.B.: Strategic Management: Formulation, Implementation, and Control, 10th edn. McGraw-Hill Irwin, Boston, Mass., London (2007)
13. Thompson, J., Scott, J.M., Martin, F.: Strategic Management: Awareness & Change, 8th edn. Cengage Learning, Hampshire (2017)
14. Pickton, D., Wright, S.: What's SWOT in strategic analysis? Strateg. Chang. 7, 101–109 (1998). https://doi.org/10.1002/(SICI)1099-1697(199803/04)7:23.0.CO;2-6
15. The McKinsey 7S Framework—Strategy Skills From MindTools.com. https://www.mindtools. com/pages/article/newSTR_91.htm. Accessed 22 May 2022
16. Štangová, N.: Finančná analýza firmy. pp. 91–96 (1996)
17. Gundová, P., Šrámeková, S.: Finančná Analýza Ako Nástroj Riadenia Podniku. Financial Analysis as a Tool of Corporate Governance (2019)
18. Papula, J., Papulová, Z.: Strategické myslenie manažérov, 1. vydanie. Kartprint
19. About AWS. Amazon Web Services, Inc. https://aws.amazon.com/about-aws/. Accessed 22 May 2022
20. History of the Cloud | BCS. https://www.bcs.org/articles-opinion-and-research/history-of-the-cloud/. Accessed 21 May 2022
21. 12 Benefits of Cloud Computing and Its Advantages. Salesforce.com. https://www.salesforce. com/products/platform/best-practices/benefits-of-cloud-computing/. Accessed 21 May 2022
22. Disadvantages of Cloud Computing. Cloud Academy. https://cloudacademy.com/blog/disadv antages-of-cloud-computing/ (2019). Accessed 21 May 2022
23. Magic Quadrant for Cloud Infrastructure and Platform Services. Gartner. https://www.gartner. com/en/documents/3989743. Accessed 22 May 2022
24. Poniszewska-Maranda, D., Matusiak, R., Kryvinska, N., Yasar, A.-U.-H.: A real-time service system in the cloud. J. Ambient Intell. Hum. Comput. 11(3), 961–977 (2020). https://doi.org/ 10.1007/s12652-019-01203-7
25. IBM Cloud. https://www.ibm.com/cloud (2022). Accessed 22 May 2022
26. Statista—The Statistics Portal—IBM Sales Worldwide from 2010 to 2020 by Segment. Statista. https://www.statista.com/. Accessed 30 May 2022
27. IBM Cloud & Cognitive Software Revenue by Subsegment 2020. Statista. https://www.sta tista.com/statistics/1118015/ibm-cloud-and-cognitive-software-revenue-by-subsegment-wor ldwide/. Accessed 22 May 2022
28. What is Cloud Computing? Everything You Need to Know. SearchCloudComputing. https:// www.techtarget.com/searchcloudcomputing/definition/cloud-computing. Accessed 22 May 2022
29. Cloud Computing Services | Microsoft Azure. http://azure.microsoft.com/en-us/. Accessed 22 May 2022
30. Saha, D.: Google Cloud BrandVoice: How Computing Has Evolved, And Why Multi-Cloud Is The Future. Forbes. https://www.forbes.com/sites/googlecloud/2020/07/16/how-computing-has-evolved-and-why-multi-cloud-is-the-future/. Accessed 22 May 2022

31. Google Cloud Revenue 2021. Statista. https://www.statista.com/statistics/478176/google-pub
    lic-cloud-revenue/. Accessed 30 May 2022
32. Fedushko, S., Ustyianovych, T., Gregus, M.: Real-time high-load infrastructure transaction
    status output prediction using operational intelligence and big data technologies. Electronics
    9(4), 668 (2020). https://doi.org/10.3390/electronics9040668
33. Haranas, M.: Enterprises Spend $178 Billion On Cloud Services, Doubling Data Center Market.
    CRN. https://www.crn.com/news/cloud/enterprises-spend-178-billion-on-cloud-services-dou
    bling-data-center-market (2022). Accessed 22 May 2022
34. Pricing—Cloud Services I Microsoft Azure. https://azure.microsoft.com/en-us/pricing/details/
    cloud-services/. Accessed 21 May 2022
35. Gartner: A 10-Step Cookbook for Building a Cloud Strategy I CIO Dive. https://www.cio
    dive.com/news/gartner-a-10-step-cookbook-for-building-a-cloud-strategy/561315/. Accessed
    21 May 2022
36. Levitzky, R.: Amazon's 2020 Cloud Strategy and My 5 Takeaways from AWS Re:Invent
    2019. Medium. https://medium.com/ranlevitzky/amazons-2020-cloud-strategy-and-my-5-tak
    eaways-from-aws-re-invent-2019-bbb9fb196fef (2019). Accessed 20 May 2022
37. Cycles T Text Provides General Information S Assumes No Liability for the Information
    Given Being Complete or Correct D to Varying Update, Text SCDM Up-to-DDTR in the Topic:
    Cloud Computing. Statista. https://www.statista.com/topics/1695/cloud-computing/. Accessed
    22 May 2022
38. InfoWorld—Technology Insight for the Enterprise. InfoWorld. https://www.infoworld.com/art
    icle/3237566/cloud-pricing-comparison-aws-vs-azure-vs-%20google-vs-ibm.html. Accessed
    21 May 2022

Definition of Terminology in Security Sciences

Ondrej Čupka⬤, Ester Federlova⬤, and Peter Vesely⬤

Abstract Cybersecurity has only recently become a major topic in the Slovak republic, mainly because of the introduction of new legislation, such as the General Data Protection Regulation and the Cybersecurity Act as well as the impact of a pandemic that moved many human activities into the digital world. This legislation makes public and private institutions legally responsible for keeping the nation's IT infrastructure and data safe under the threat of legal consequences. This sudden shift saw many companies and public institutions scrambling to meet the new legal demands, but have often found a lack of professionals, institutions, systems, but mainly information. Cybersecurity is a young science with its terminology still in debate. Within this paper we will be defining key terms necessary for discussions on this topic.

Keywords Cybersecurity · Terminology · Security sciences · IT infrastructure · Legislation

1 Introduction

The digital landscape is constantly evolving and dynamic. New software and hardware, patches, updates, and fixes as well as new tactics and exploits both make up and change this landscape. Cybersecurity always needs to adjust and adapt to these changes to remain effective and relevant in protecting data and IT infrastructure. As a very young scientific field, it was not at the forefront of public and private

O. Čupka (✉) · E. Federlova · P. Vesely
Faculty of Management, Comenius University in Bratislava, Odbojarov 10, 831 04 Bratislava, Slovakia
e-mail: cupka9@uniba.sk

E. Federlova
e-mail: federlova2@uniba.sk

P. Vesely
e-mail: peter.vesely@fm.uniba.sk

© The Author(s), under exclusive license to Springer Nature Switzerland AG 2023   521
N. Kryvinska et al. (eds.), *Developments in Information and Knowledge Management Systems for Business Applications*, Studies in Systems, Decision and Control 462,
https://doi.org/10.1007/978-3-031-25695-0_23

policy. But as more and more public, private and commercial life move to the digital world, which was rapidly accelerated by the Covid-19 pandemic, the potential to cause damage increases. With the stakes so high that a successful malware attack may disrupt a vital service, such as an oil pipeline or powerplant, cause a standstill in a factory or violate the rights of millions, the field is equally gaining in importance. This was further aided by the introduction of legislation such as the General Data Protection Regulation or the Cybersecurity Act, which oblige businesses and public institutions alike to keep their data and IT infrastructure safe under the threat of legal consequences. It is therefore in the interests of all to make the field of cybersecurity as effective as possible. To do this first and foremost requires a good definition of the vulnerabilities, risks and threats, motivations and behavior of cybercriminals, the common attack types, the source of them, the measures that can be taken against them, the state of awareness among professionals and the public and much more. Unfortunately, this is the type of information that the Slovak republic is currently lacking. In the article, we are going to start with the terminology of security management and information security management to create a baseline for discussing this topic.

2   Conceptual Background

The emergence of modern society, based on technological progress and the discovery of new raw materials and energy sources, has brought with it unprecedented prosperity for mankind, but also unprecedented risks. Each system naturally carries instruments for its prosperity as well as for its demise. Improved means of travel have brought the whole world closer together with all the economic and social benefits, but it also allows for global pandemics and international crime. The Internet has enabled the emergence of the world's richest businesses, and made enormous amounts of information available to the world, but has also enabled the emergence of cybercrime, hackers, and constant monitoring and analysis of digital users. Ulrich Beck, therefore, called the current society a society of risk. Security sciences, security management, and particularly cybersecurity are therefore becoming crucial for correctly identifying and preventing these risks [1].

3   Terminology

In this chapter, we will look at security management terminology and cyber security terminology. The use of terminology starts from very simple occasions in our ordinary life and develops to higher communicative levels. However, if terminology for non-professionals is an option, for specialists it is a necessity. Terminology plays an important role in the understanding of contexts and specialized texts. Understanding the intricate terminological details of the technical and scientific contexts helps us comprehend what the main message of the document is, and it helps specialists to transmit the content more effectively. Terminology helps individuals realize the interaction between the units of specialized texts and the whole context which is often a subconscious mechanism of knowledge acquisition. It also develops an interest in the formation of new words and terms.

3.1 Terminology of Security Management

Risk

"Risk is a potential threat to security... that is objectively conditional and cannot be excluded without changing these objective conditions." Risk can be considered as the lowest level and the first symptom of the negative development of the security situation. The various degrees of escalation of the situation are security risk, security threat, crisis, unarmed or armed conflict, and finally war. Geert Hofstede considers risk to be a certain probability that an unwanted event will occur. In doing so, it highlights the difference between uncertainty and risk. It makes the risk subject to direct it to a specific person, object, or phenomenon. Uncertainty is a vague feeling that has no clear source or consequence [1].

A security risk can be classified according to the origin, scale and resources and by the use of armed forces. According to the place of origin, we distinguish between external security risks, which include uncontrolled arms trade or the spread of international terrorism, and internal security risks, such as the risk of organized crime or the rise of extremism. Depending on the extent of the risk we divide it into global risks, such as deterioration in the quality of the environment, declining sources of mineral resources or the emergence of new superpowers, and regional risks, such as the risk of the spread of the conflict between Ukraine and Russia or the risk of conflicts due to uneven economic growth, which are a risk in the context of Europe. And depending on the use of armed forces in military and non-military risks [1].

Security

Security is the condition of a particular system or object where the probability of damage or loss is at an acceptable level. In terms of human security, we are talking about a state without fear and flaws, by which we understand protection from violence and aggression, job security, income, moral and physiological security, family and health security, etc. Porada admits that it is a very complex multidimensional social phenomenon. Mareš defines it as a situation where the threat is eliminated to the lowest possible extent, in particular for the state and an international organization. He also defines, together with Zeman, security as objective and subjective, which emphasizes the frequent mismatch between perceived security and reality. Volner defines security as a state of absence of dangers or a state in which mankind is not aware of possible dangers. Hofreiter draws attention to changes in its definition and priorities which always happen in the light of the needs of the present. He cites the example of the end of the bipolar world with the break-up of the Soviet Union when the emphasis on security shifted from national defense issues to issues of international relations [1].

Security is therefore an objective state in which there is no danger, but is perceived subjectively. In our opinion, this reflects the multidisciplinary and interdisciplinary nature of security sciences, which require input from all areas of human activity and strong cooperation between its actors. Only under these conditions will it be possible to ensure objectivity to the highest extent, to correctly identify factors influencing security, and determine their priority and relevance. We therefore also understand that security is a dynamic field, where priorities change based on current needs and available information. And so, for example, securing national borders, that is to prevent unauthorized persons from entering the sovereign territory of the state, may, under certain conditions, be more or less important than safe healthcare, which does not register high numbers of preventable deaths or safe education, which does not educate future citizens in a way that could cause the decline of society and the state [1].

Danger

The danger is a state of a human system, an object, or an area of a protected interest when there is a high probability of loss or damage. To some extent, we can talk about danger as the opposite of security. These security objects are more or less likely to turn a threat into real damage or harm. This susceptibility is influenced by the object itself, its surroundings, or its relationships with it. The concept of danger can be considered synonymous with the concept of threat. However, some authors distinguish these concepts. A danger is a hidden feature of the system causing negative security events that threaten the system itself or its surroundings. We can divide danger into different groups according to the origin of the threats, the characteristics of the target objects, how the danger is realized, and others. In the event of danger, we also distinguish absolute danger, when it can cause an adverse event for all persons equally, and, much more often, relative danger, where the adverse situation affects only a part of the persons in a given population [1].

Threat and Challenge

The threat is a term often used in military environments. As well as a security challenge, which is an increasingly widely used term that expresses some optimism and the ability to face and overcome these challenges. It puts security threats in a different light, where they are overcome and can even represent an opportunity. The term has come into mainstream use through NATO and U.S. security concepts, stressing that "facing security challenges means actively addressing human rights issues and the moral aspects of maintaining security." Several authors consider the concepts of challenge, danger, and threat as degrees of risk and escalation of tension. With the lowest risk posed by a challenge, medium risk posed by danger and the highest by threat [1].

Hybrid Warfare

The Security Strategy of the Slovak Republic defines hybrid threats as a set of coercive and subversive activities, conventional and unconventional, military and non-military methods, and tools used in a coordinated manner to achieve specific political objectives without a formal declaration of war and below the threshold of the usual response. They may include centrally managed intelligence and information activities, including cyberattacks on key institutions, aimed at influencing, the use of non-state actors, including paramilitary groups, or the deployment of unmarked state actor forces. Such hybrid activities may commence before openly declared military operations take place. They polarize society, and bring uncertainty, thereby undermining the legitimacy and credibility of actions of state institutions and the democratic constitutional order, thus hurting the realization of the security interests of the states. Hybrid activities may also aim to weaken public support for the fulfillment of international obligations or paralyze the international community's response [1].

Reference Object

Reference objects are entities that are existentially threatened and can legitimately claim the right of survival [1].

3.2 Cybersecurity Terminology

Incident Versus Breach

An incident in the context of cybersecurity is a security event that compromises the integrity, confidentiality, or availability of an information asset. A breach on the other hand is an incident that results in the confirmed disclosure, not just potential exposure, of data to an unauthorized party [2].

Hacking

The terms hacking or the person hacker, have mostly negative connotations in popular culture, where hacking is mostly viewed as an exclusively illegal act. This is not always the case and therefore a broader definition should be applied. Hacking is applying technology or technical knowledge to overcome some sort of problem or obstacle. From this definition, hacking is not only criminal, which leads to the different types of hackers. Black hat hackers or cybercriminals are using security vulnerabilities to gain unauthorized access or control over data, devices, or systems, mostly for financial gains. White hat hackers or ethical hackers test systems for vulnerabilities with the permission of their owners to make them more secure. In between are grey hat hackers, which do not ask for permission to test systems or may even act as mercenaries trading knowledge of vulnerabilities to the highest bidder [3].

Malware

Malware or malicious software refers to any intrusive software developed by cyber-criminals to steal data and damage or destroy electronic devices and computer systems. Examples of common malware include viruses, worms, Trojan viruses, spyware, adware, and ransomware [4].

Viruses

A computer virus is a type of malicious code or program written to alter the way a computer operates and is designed to spread from one computer to another, similar to a real virus, hence the name. A virus operates by inserting or attaching itself to a legitimate program or document that supports macros to execute its code. In the process, a virus has the potential to cause unexpected or damaging effects, such as harming the system software by corrupting or destroying data [5].

Trojan Virus

A trojan horse or trojan is a type of malware that is often disguised as legitimate software. Trojans can be employed by cybercriminals to gain access to a system. Users are typically tricked by some form of social engineering into downloading and executing Trojans on their system. Once activated, the trojan can enable cyber-criminals to spy on the user, steal sensitive data, and gain backdoor access to the system [6].

Worms

A computer worm is a type of malware that spreads copies of itself from computer to computer, similar to a virus. But the worm can replicate itself without any human interaction, and it does not need to attach itself to a software program to cause damage [7].

Spyware

Spyware is loosely defined as malicious software designed to enter an electronic device, gather data about its user, and forward it to a third party without permission or prior notice. Spyware can also refer to legitimate software that monitors data for commercial purposes such as advertising. However, malicious spyware is explicitly used to profit from stolen data. In both cases, spyware's surveillance activity leaves the device open to data breaches and misuse of private data. Spyware also affects network and device performance [8].

Adware

Adware, also known as advertisement-supported software, generates revenue for its developers by automatically generating adverts on a user's screen, usually within a web browser. Adware is typically created for computers but can also be found on mobile devices. Some forms of adware are highly manipulative and act as an opening to the system for malicious programs [4].

Ransomware

Ransomware is malware that employs encryption to hold a victim's information at ransom. A user's or organization's critical data is encrypted so that files, databases, or applications cannot be accessed. A ransom is then demanded to provide access. Ransomware is often designed to spread across a network and target database and file servers, and can thus quickly paralyze an entire organization. It is one of the largest threats in cybersecurity, generating billions of dollars in payments to cybercriminals and inflicting significant damage and expenses for businesses and governmental organizations [9].

Social Engineering

Social engineering refers to all techniques aimed at getting a target to reveal specific information or perform a specific action for illegitimate reasons. Either by using psychological manipulation to get further access to an IT system where the actual objective of the scammer resides, e.g. impersonating an important client via a phone call to lure the target into browsing a malicious website to infect the target's workstation or impersonating a bank manager to extract knowledge about the victim's bank account information [10].

Phishing

Phishing or brand spoofing is the practice of sending fraudulent communications that appear to come from a reputable source. It is usually performed through email. The goal is to steal sensitive data like credit card and login information or to install malware on the victim's machine [11].

Denial of Service

A denial-of-service attacks flood systems, servers, or networks with traffic to exhaust resources and bandwidth. As a result, the system is unable to fulfill legitimate requests. Attackers can also use multiple compromised devices to launch this attack. This is known as a distributed-denial-of-service (DDoS) attack [12].

Man-in-the-Middle Attack

Man-in-the-middle (MitM) attacks, also known as eavesdropping attacks, occur when attackers insert themselves into a two-party transaction. The data then flows directly through their device making it possible to interrupt the traffic, filter, and steal data [12].

Scan-and-Exploit

Scan-and-exploit is a practice of cybercriminals where before an attack a system is scanned for vulnerabilities, which are then exploited to increase the success chance of an attack. This practice uses the same technologies used by cybersecurity professionals to fix and patch identified security holes. Scan-and-exploit has become one of the most popular attack vectors of cybercriminals as it can be used on many devices at the same time and help the attacker identify the easiest target [13].

Brute Force Attack

A brute force attack is a relatively old attack method used by hackers, but still popular and effective. It uses trial-and-error to guess login information, encryption keys, or find a hidden web page. Hackers work through all possible combinations hoping to guess correctly. These attacks are done by brute force meaning they use excessive forceful attempts to try and force their way into private accounts [14].

Cryptojacking

Cryptojacking is a practice of using malicious code designed to hijack the idle processing power of a victim's device and use it to mine cryptocurrency. Victims are not asked to consent to such activity and even may be unaware that it is happening in the background. While in the past this was possible only when victims downloaded and ran an executable on their devices, today it is often done in a browser with JavaScript without the need to install any software [15].

Nation-State Attackers

Nation-state attackers operate similarly to cybercriminals with the exception that the goals are not primarily motivated by financial gains, but by the interests of national governments. These governments exploit the anonymity of the internet to pursue policy goals with plausible deniability. The goals of such an attack are as various as the interests of a nation. The goal may be to steal trade, military or medical secrets. To gain leverage in negotiations, information on persons of interest, such as spies, key figures in government, or dissidents, may be done with the purpose of blackmail. In preparation for war, to destabilize a country, influence its elections, erode confidence in government, sow dissent among the populace, or directly before an attack disable military capabilities by targeting information and defense systems and key infrastructure, such as energy supply, and for retaliation [16].

3.3 Security Science

Security science deals with fulfilling the value of security and its need, analyzing the feeling of security, and trying to get it as far as possible into the objective plane. In this sense, it analyses and defines risks, threats, and dangers, but also the manager and his role in the process of creating a safe environment. The beginnings of this science were in 1989 when sudden changes in the world's geopolitical situation associated with the disintegration of the USSR conditioned its emergence. Security science is necessarily a multidisciplinary science because the state of security is constantly influenced by several factors of an objective and subjective nature [1]. For this reason, it relies on the knowledge of natural and social sciences to be as effective as possible in maximizing security, minimizing risk, eliminating threats, and overcoming dangers [17].

As a young science, security science is not uniform and opinions differ on its nature and field of action. Tadeusz Hanousek considers it necessary to develop this science to offer security management the necessary theoretical basis to minimize or eliminate threats. The University of Žilina speaks directly about security management as an activity that aims to minimize the risk to social values. It uses elements of risk and crisis management for this purpose. Ulrich Beck, describes our society as a society of risk, one that is threatened by the side effects of scientific, technological, and therefore social progress, where the state of emergency becomes the norm. It describes a certain domino effect of these side effects, which ultimately affect policy, thus also creating room for political catastrophe. Side effects may include threats to health, economic or social changes, which are dependent on the growth and decline of markets, crises, the opening up of new markets, legislative and bureaucratic pressures, and others [1].

The Objective of Security Science

The determination of the objective of security efforts at the level of science is still an object of lively debate. The basis of science is objectivity and therefore requires a quantifiable and measurable goal, but security is more of a qualitative phenomenon. Buzan identifies 4 key issues in which the authors of the security literature differ: "whether the state is the reference object exclusively, whether to include internal as well as external threats, whether to extend security beyond the military sector and the use of force, and whether to perceive security as a phenomenon inseparable from a set of changing threats, dangers and emergencies" [1].

In our opinion, security, as a universal value shared to varying degrees throughout humanity and as a factor that is often a condition for the success of human activities, should not be limited to external security threats or threats perceived only through the military sector. We agree with the viewpoint of doc. Pawera, who emphasizes the influence of non-military factors in creating a secure environment. Factors such as economy, politics, culture, the state of science, public activity, and others directly influence the process of creating a security situation. From food security to social stability, security is a pervasive factor that security science must properly reflect. Sak stresses that the state of security can only be defined in the sense of a specific entity and based on its target status. Relating to mankind the state of security is one in which progress and growth can occur, it is not an unchanging state. Sak, therefore, defines static and dynamic security for these purposes. Static security occurs when an entity maintains a certain state, function, and structure that has been defined for it, i.e., it does not change or develop but is stable. Dynamic security ensures positive development and change of status and function of a given entity, i.e., its development as well as its existence, function, and structure are ensured. These two types of security can be clarified by the example of a child. Their static security lies in maintaining their current and unchanged state, as in their mental and physical health. Dynamic security also speaks of its growth and development and therefore of qualitative change into a fully developed adult individual [1].

Security should also take into account factors outside the set of changing threats, hazards, and emergencies. Security can in many cases be seen as a set of human decisions and therefore activities that improve or worsen a security situation. The quality of decisions and how safe they are is closely linked to external and internal processes within and around a person. From this, there is a clear need for security sciences to include economics, psychology, and sociology, which reveal the inner life of humans and society. It is often because of economic, sociological, and psychological factors and their impact on all human activities that cause the emergence of risks, dangers, and threats [1].

Homeland Security—Police Science Theory

According to data from the Ministry of Interior of Slovakia, 61.392 offenses were reported for the year 2018. On average, every hour there are seven offenses, 1,3 robberies every day, and one murder every 5,4 days. Of these crimes, 61% have been solved. However we might perceive these figures, police forces are key to maintaining internal security. The basic task of the police is to maintain public order and enforce the law. These tasks are performed by the police with the possibility of using legitimate force, even physical, with the possibility of using violence in critical situations. This also leads to the attributes of police activity. The attribute of force, with the possibility of restricting human rights and freedoms and applying violence. The attribute of security emphasizes the protection of objects, persons, life, health, and property, but also public order, the state's interest, and prevention of any threats to them. Finally, the attribute of non-punishment, makes it a condition for the police not to use the attribute of force in their activities for punishment, which can only be given by criminal law institutions [1].

Distinctive Features of Police Security Activities

The police are an armed, uniformed, and state-established security force. Members of the police force are therefore authorized to use weapons in critical situations. Its unity is typical of continental Europe, where it is divided into more economically independent units, into regional departments. These are hierarchically directly subordinate to the Police Presidium. This differs from the Anglo-Saxon system, where police forces are more decentralized and more tied to the local or regional area, which is due to different historical developments [1].

The basis for the functioning of the police is the Law on the Police Force, which declares that the police is a security force that "performs tasks in matters of internal order, security, the fight against crime, including its organized forms and international forms, and the role that arises for the Police Force from the international", which "in its activities is governed by the Constitution, constitutional laws, laws and other generally binding legal regulations and international treaties" [18]. It follows then from the above that police-security activities:

- are aimed at protecting social values. This characteristic is paramount to police activity and is linked to the principle of offensiveness. The police have and must have a constant effort not only to act when an unlawful act occurs but also to prevent it. Preventative measures require constant vigilance and threats of inevitable sanctions in the case that an unlawful act occurs. The principle of non-discrimination also becomes important, since all citizens and groups of people defined by law and legal obligations under international treaties and their property, regardless of the form of ownership, have the right to this protection [1].

- are apolitical in nature. Even though being apolitical can become an unfulfilled ideal in practice, it remains an important ideal, the fulfillment of which in many respects is the basis for the functioning of a democratic state. The fulfillment of this ideal is directly linked to the essence of the rule of law. Violation of this characteristic is often associated with serious consequences for the security and stability of the country [1].
- are legal. All activities must necessarily be based on the law, and so must be legal. This need for legality stems from the right of the police to interfere with citizens' rights and freedoms. Such intervention must therefore only be carried out by the law. The law therefore clearly identifies all aspects of police activity, from the organization of the police force to the powers and responsibilities, to the structure, content, methods, and means applicable to the actual performance of police activity [1].
- it has a specific and creative character. The fight against illegal activity is not final. Its infinity causes individual players to constantly change and it is impossible to hope for a final and total victory. Over time the rules also change. Crime comes from unknown places, legislation changes, as well as the form and methods of crime. This requires the police to constantly adapt to all sorts of situations and conditions. But in its capacity to adapt it is limited by the principle of legality, which allows it to act only as permitted by law. To some extent, this puts the police at a disadvantage against criminals, but there often are other values that must be respected, the protection of which requires the limitation of the power of the police force. These may include values such as liberty, the presumption of innocence, or respect for individuals' privacy [1].
- it is systematic. The police, by their nature, content, and form, make up a system. Under this system, we understand measures and actions aimed at combating criminal, and anti-social activities, and protecting public order, health, and property [1].

Principles of Police-Security Activities

As we mentioned in the previous section, police activity is governed by certain principles and certain basic rules of action. Contrary to the principles that we form ourselves from our beliefs and experience, the principles governing police activity are based on objective social reality, take into account theories of the rule of law, and are based on the legal status of the police, and practical police activity. The basic principles according to prof. Porada includes democracy and humanism, legalism, the scientific process, adequateness, subsidiarity, unity of prevention and repression, and secrecy [1].

Concept, Content, and Functions of Police-Security Activities

Police security activity is a concept expressing a system of measures and actions that defines the fight against antisocial activity, and the protection of public order, life, health, and property. These measures and acts are defined by laws, regulations, international treaties, ethical principles, and scientific knowledge. The content of police security activities is subject to the security strategy of the sovereign country managing the territory. According to the principle of subsidiarity and systematism, it is hierarchical. Therefore, the police and security authorities further elaborate its valuable content, starting with the police presidium and the police president and going all the way to the local police station and individual police officers [1].

The basic functions of police-security activities are the protective, coercive and educational functions. Police action aims to prevent illegal activity. In this sense, the police can be understood as a guardian of the law. For example, protection of the right to property against theft, the right to life, etc. If prevention fails, the protection of these rights and values has been unsuccessful, so it is its task to ensure that punishment is unavoidable for those who caused it. In this sense, it serves a coercive function, which should be balanced with the educational function, to not only ensure punishments but also to strengthen the prevention of future illegal activities. These include other functions indirectly related to the fight against crime and the protection of public order. Such as public administration through services to citizens, such as issuing documents, information security, e.g., by providing information about their activities, or by public relations, e.g., through spokespersons and the media. As we have indicated, police-security action is not only aimed at prevention but also at repression. The main elements of repression or coercion and enforcement are activities of revealing, clarifying, and searching [1].

Police and Cybersecurity

The role of the police in cybercrime is in principle the same as in regular crime. To prevent crimes or if they do happen, then to investigate and capture potential criminals. Although the realities of cybercrime are very different and pose a great challenge many conventional police activities are ineffective. Despite Slovakia having legislation for cybersecurity, nowhere in any Slovak legislative texts is cybercrime defined. This points to the first challenges for law enforcement in cyberspace, a lack of clear definitions, ways of measuring, and processes for addressing cybercrime. From this comes a lack of professionals in both law enforcement and justice institutions [19]. If this can be observed with professionals then all the more with the vast number of common users that use the internet every day for both private and work-related activities. The spread of these technologies has far outpaced the level of understanding of how they work and what risks are associated with their use. The interconnectedness, openness, and lack of control mechanisms of the internet also make sharing of information on vulnerabilities, tools to exploit them, and misuse of stolen data much faster than they could be stopped or prevented. The internet also does not have any borders which makes tracing offenders as well as persecuting them very difficult if not impossible. Furthermore, only a fraction of crimes committed in cyberspace are reported to the police. Another challenge is the question of values and where to find the right balance between surveillance and privacy, security and liberty, or justice and the cost of enforcing it [20].

Global data underlines the magnitude of the challenges posed by cybercrime. In the year 2018, there were 812 million recorded attacks on systems with the use of malware of which more than 90% were carried out via email [21]. In the same year, another 204 million attacks happened with the use of ransomware [22]. The average cost of such an attack is estimated at 133 thousand USD. It is expected that globally by 2021 an organization will be under attack every 11 s [21] and that the economic damage caused by cybercrime will reach 6 trillion USD [23]. All the while about 5% of cybercriminals are caught by law enforcement [24] and from them, only about 1% are convicted [25].

External Security, Military Science

Two world wars have forever marked human history and show the undeniable consequences of conflict at a global level. During the Second World War, an estimated 19 million military deaths and 22 million civilian deaths were counted on European battlefields in general, with a total of 70 million people killed in the conflict. The consequences of these losses on lives, health, property, and other values are still felt today after more than 80 years [26]. Following the Second World War events, the term "Long Peace" is often mentioned in scientific literature, which indicates a downward trend in the number of war casualties, in nominal or relative numbers. However, these conclusions are the object of criticism. Fewer deaths can be explained by the increasing quality of health care, which increases the chances of surviving the conflict for wounded soldiers and civilians. Also, the peace of the last period does not go beyond the natural historical fluctuations, which cannot rule out any increase in military deaths in the future. Conflicts in the Middle East, rising extremism, and probable geopolitical conflicts between Western countries, Russia, and China can and have led to serious armed conflicts. Military science has therefore not lost its relevance, even in the period of the "Long Peace", on the contrary, we have more to lose if we do not respond adequately to such threats [1].

Military science, at the same time as the development of the security sciences, has undergone extensive changes since the end of the bipolar world in 1989. Until then, security was divided into military and non-military conflicts, while only military conflicts were taken into account. Geopolitical, social, and technological changes led the Copenhagen School with Buzan to emphasize precisely the neglected non-military conflicts, in which military force could be used over time. The school identifies up to five separate security sectors, namely the political, military, social, economic, and environmental sectors. Volner continues to expand security to include space and information security, which will shape the geopolitical situation in the future. Thus, military sciences lost their primacy among security sciences and had to expand to reflect a new reality. Conventional warfare has ceased to be the main subject of exploration, currently, it has become hybrid warfare [1].

Armed Forces of the Slovak Republic, Security and Defence Strategy

The external security of countries relies on their military strength, i.e., military and civilian personnel, as well as military material. In 2017, the Slovak Republic reported 6.743 civilians, and 13.152 military personnel with total defense expenditure for 2017 amounting to 993 million EUR, which accounted for 1.2% of GDP [1]. 2017 was also the year when the new Security and Defence Strategy of the Slovak Republic were approved, which represented the necessary update of the old strategies from 2005 and 2001 respectively [1]. This brought increased expenditure on the defense of the Slovak Republic by NATO guidelines on the share of defense spending to GDP. The year 2020 was the first time the Slovak republic reached the 2% mark, also spending about 31% of defense spending on new equipment, well above the 20% NATO guideline [27]. The goal of the Ministry of Defence of the Slovak republic is to increase the number of military personnel to 23.983 of which 19.573 by the end of 2022. Additionally, large investments are planned for the purchase of military equipment [28].

The Security and Defence Strategy of the Slovak Republic 2017 is a document resulting from the synthesis of military sciences, extended to other security sciences, and political will. It is the basic document for strategic defense planning of the Slovak Republic. The aim of the defense policy aims to guarantee the ability to defend the Slovak Republic and its citizens from external threats, respect for the fundamental principles and norms of international law, and strengthen the security and stability of the Euro-Atlantic area through NATO and the EU, strengthening security and stability in the wider EU-NATO neighborhood, enhancing the ability to promote the security interests of the state in the international environment, maintaining security and guaranteeing the defense of the state, including the development of defense capabilities, increasing the state's resilience to security threats, including threats arising from the cybersecurity landscape as a separate security domain, developing economic, material and environmental preconditions for state security and stability [29].

The update was conditional on the aforementioned changes in the global security environment, specifically for the Slovak Republic, this includes the Ukrainian crisis and the Russian Federation's annexation of Crimea in 2013, the start of a large-scale disinformation campaign by Russia, and the associated redefining of the Russian Federation from a strategic partner to a rival. It was eventually also shaped by terrorist attacks on behalf of the Islamic State, which were declared in Syria and Iraq in 2014. In addition to addressing these threats, it is the first time it has ever talked about the possibility of an armed attack on the Slovak Republic and the threat to its statehood, sovereignty, territorial integrity, and inviolability of its borders. This justifies increasing the potential of conventional weapons for state and non-state actors, including the introduction of hybrid warfare means [1].

Security of the European Union

The history of the Slovak Republic and the inhabitants of its territory are a testament to its complicated geographical position. As a small nation at the heart of Europe and without sufficient natural barriers against invasions such as Switzerland, most of its history has been dominated by stronger neighboring nations, regional or world powers. After the first acquisition of full autonomy in its history in 1993, it has therefore become crucial for safeguarding its newfound sovereignty, the freedom of its citizens, and their economic prosperity to join allied groupings. This was realized in 2004 by entering the European Union and NATO. The Slovak Republic joined the EU on 1 May 2004, the Schengen area on the 21 of December 2007 and adopted the euro on the 1 of January 2009. As one of the most open economies in Europe, its economic security and its overall security situation are linked to the security of the EU. The current Covid-19 pandemic shows us this link in practice when Slovakia's security measures are linked to the decisions of neighboring countries, especially large trading partners such as Germany [1].

In many ways, the security of the European Union is shaped by its economic strength. It follows then, that a threat to its economy is associated with a threat to its security. The link between economy and security can be observed for example in the South China Sea, where China, contrary to the decision of the international community, has built new artificial islands and subsequently militarized them, thereby guaranteeing and strengthening its influence over the world's most valuable trading area. Or even Russia's annexation of the Crimean Peninsula and subsequent war. In doing so, Russia secured the vulnerable south of the country, as well as guaranteed access to the Black Sea from the Sea of Azov, to which the economically important Don and Kuban estuaries flow [1].

EU security is mainly shaped by global security risks, national disputes, uneven economic developments, uncontrolled migration linked to extreme poverty, and instability in neighboring countries. Security developments in the future will continue to be influenced by Europe's geopolitical development and response to its changes, particularly in the heat of its relations with the UAE, China, and Russia. Decisions in this area are taken within the framework of the European Security Architecture, which consists of the Council of Europe, the European Union, NATO, the OSCE, as well as the UN bodies [1].

The objectives of these institutions are to strengthen Europe's security, to protect peace in Europe and the world, to protect Europe's independence, cultural values, interests, and integrity, and to develop democracy, rule of law, human rights, and fundamental freedoms, and international cooperation. In pursuing these objectives, the European Union is committed to upholding the principles of sovereignty while guaranteeing its security, respect for international standards, the indivisibility of security, prevention, adequacy of the use of armed forces, democratic governance, and control of the armed forces, complexity, in ensuring internal security, legal protection, ecological rationality and the protection of the economic interests of individuals and the state [1].

The North Atlantic Treaty Organisation

The North Atlantic Treaty Organisation, or NATO, is, in its current form, a political and military alliance. As a political alliance entering conditioned on shared democratic values and an appropriate legislative framework in the applicant countries. As a military alliance, it is committed to solving conflicts primarily through diplomatic channels. Should these efforts fail, it will conduct crisis management operations, which may include military force. Operations are carried out under Article 5 of the Washington Treaty or under a mandate of the United Nations, alone or in coordination with other countries or international organizations [30]. NATO's headquarters are located in Brussels, Belgium. NATO Secretary is currently Jens Stoltenberg.

NATO has 30 member countries, including North Macedonia, which entered the alliance in 2020. The alliance combines around 1 trillion USD in defense spending with an increasing tendency and more than 3.23 million military personnel, making it the largest military organization in the world [27]. Both spending and military personnel are currently increasing in response to US pressure for member countries to meet the agreed 2% of GDP in defense spending [1].

NATO is based on Article 5 of the Washington Treaty which forms the basis of this alliance. It clearly states that an attack against one NATO member is seen as an attack against all its members. In such a case, it requires its members to carry out such activities as are considered necessary to restore and maintain the security of the North Atlantic region. The US, France, and the UK are the only member countries with nuclear weapons at their disposal, which are an important part of NATO's defense strategy as a means of deterrence as their use may be the result of triggering Article 5. Article 5 was invoked only once in NATO history, by the United States in response to the terrorist attack of 11 September 2001 [1].

The Organization of the North Atlantic Treaty has been the subject of criticism practically since its inception, but criticism from the members themselves is unprecedented in its history. NATO's future has recently been called into question by two of its biggest members. Mr. Trump publicly announced during his presidency the possibility of a U.S. exit from NATO in 2018. Also giving statements that cast doubt on whether the USA will honor its obligations towards other member countries [1]. The stance of the new president of the USA, Joe Biden, is different in that he fully supports NATO and confirmed that the USA would always honor its obligations under Article 5 of the Washington Treaty. But while acknowledging increases in military spending, he continues to criticize the members because their militaries are not combat-ready or able to face the challenges posed by Russia or China [31]. Macron similarly criticized NATO on France's part, calling it "braindead", he believes that the future of European security lies in establishing a separate European army. However, German Chancellor Merkel stressed the importance of NATO and called it an organization whose importance is perhaps greater today than it was during the Cold War. She justified this by saying that Europe cannot defend itself. Although with the end of Merkel's involvement in politics it remains to be seen if German policy will remain the same under the new government [1]. Or how the Alliance changes with the threats posed by the war in Ukraine.

The Military and Cybersecurity

In January 2021 an updated version of the Defence Strategy of the Slovak Republic was ratified, building on the aforementioned Security and Defence Strategy from 2017. This document mentions on five content pages "cyber" seven times. This is done in terms of acknowledging the rising threat of possible cyber-attacks and states the intent of building defensive capabilities in cybersecurity against possible cyber-attacks and in support of the Military intelligence agency and its operations. The document does not offer specific steps that are to be taken to fulfill these goals [32].

Currently, the military forces of Slovakia themselves do not have a military unit specializing in cyber operations. The military does not conduct any cyber operations exercises to prepare the military for both defenses in case of an attack or in conducting attacks themselves. Neither did it participate in international exercises since 2016 [33].

Security Management

"Security management is a specific type of management, focused on the security management of reference objects" [34]. According to this definition, security management applies all managerial functions and methods but does support the transformation process within the business. Security management is focused only on managing specific support activities that ensure undisturbed achievement of objectives. This activity is the security management of reference objects [34].

Security management can be applied within the organization at all levels, areas of activity, and in all types of organizations to protect tangible or intangible assets. The efforts of security managers may be aimed at protecting assets such as information systems, and movable property from theft, industrial espionage, or fire. Or to protect people by protecting their health and security at work, protecting individuals, etc. As we can see, security management is used almost everywhere, from the construction of the sarcophagus for the Chornobyl nuclear reactor bank security against robberies, and software against hacking attacks, to the protection of individuals in public gatherings. Security management shares many features with other forms of management and therefore we talk about it as a profession, scientific discipline, art, and, specifically for security management, as a specific management activity to manage the security of the organization [34].

Security Management as a Profession

Security managers are a group of people whose job is to develop and implement the organization's security strategy. These responsibilities are most often assigned to top managers, line managers, and specifically security managers. This reflects the fact that responsibility for security is not solely within the competence of security managers [34].

Security Management as a Science

Security management as a science is based on the knowledge of security science. It is therefore a set of knowledge on security procedures and methods for the protection of persons, property, and the environment, which are to be the basis for the activities of each security manager. To some extent, however, the manager must rely on their judgment, as their decision-making should be adapted to their circumstances, such as the specific organization in which they operate and based on constraints of time, finance, or other, which cannot be fully contained in the theory [34].

Security Management as an Art

Security management shares with all other types of management that it is also a kind of art. It is meant that management is like conducting an orchestra whose successful leadership is conditioned by teamwork, expertise, appropriate development and training of staff, understanding not only each member individually, but also as a group, the surrounding space, and the listeners. And above all to understand the works that need to be played, in which various artists with different abilities perform at different times with different intensities. There should never be anyone extra, but each plays a key role in their time.

Security Management as a Specific Management Activity

Security management is above all a conscious activity aimed at meeting the specified security objectives and allows continuity of other activities within a given organization. The key activity of the manager is to collect information on the issue of security itself, on the immediate security environment, on possible risks and threats, and also on legislation such as occupational safety and health. Subsequently, use this information, make appropriate decisions and implement measures to prevent these risks and threats or to limit their harm. This activity takes place in a security management system in which security managers are involved, as well as line, middle, and top managers. Therefore, participation in the strategic management process of the organization is also an important area of security management activity, where it should be involved in the development of the organization's strategic plans [34].

The Benefits of Security Management

The importance and benefits of applying security management are ever-increasing. As societies become more globalized, interconnected, and digitized so do they often become more complex and fragile. As global competition increases so increases the pressure to reduce costs and streamline activities, which additionally increases the fragility of the system as there are no redundancies. This has become apparent by the COVID-19 pandemic, which has found governments, companies, and individuals unprepared, which caused large losses in life and economic prosperity. While impossible to avoid completely, with proper preparation and application of security management practices it is plausible that a large portion of the losses could have been avoided.

Its benefits, therefore, highlight a reduction in the number of incidents and crises. Thorough security documentation and due diligence after incidents are useful not only for learning from these experiences but also for guiding policy, insurance claims, and developing prevention techniques. Additional benefits are a clear structure and responsibility for security in the organization, enabling better specialization and the creation of a high-quality system to prevent or at least limit damages or loss. Contracts and partnerships are concluded more carefully by looking at the security implications of such a decision, which can prevent threats in the future. Improving reputation by increased security and reliability, or at least preventing it from deteriorating through risky behavior or situations [35].

Information Security Management

We are living in a society of risk and the best example of this are information technologies. While these technologies were used in the past only to support processes within institutions, make some activities easier, and later also offer pleasure and relaxation, today they are ever more the base of modern society. Through these technologies, we work, communicate, trade, interact with government institutions, transfer money, and they have become the reason for the existence of the biggest and most successful companies today. Our reliance on them has also increased the risks posed to our society as there are no known systems without flaws that cannot be compromised. And so, a simple mistake in code may cause great harm, loss of property, identity, privacy, and potentially even life. Data shows a similar trend as it is projected that in 2021 cybercrime will cause global damages of more than 6 trillion USD, up 100% in just 6 years since 2015 when the cost was 3 trillion USD [23]. In a globalized and interconnected world it is nigh impossible to tackle cybercrime directly as attacks may come from anywhere in the world and their true origin masked. That is why institutions have to protect themselves and constantly decrease risks to their systems. That is where information security management comes in.

Security management is a subcategory of security management, that deals dealt with the security of reference objects [34]. With information security management (ISM) the reference object is the cybersecurity landscape, which includes hardware software or technologies, people and processes with which they interact with technologies [36]. Technologies concern hardware, such as endpoint devices, routers, and servers, as well as software, such as firewalls, malware protection, antivirus software, browser, and email security solutions. Processes are how, where and when users access systems, it concerns frameworks for what to do in case of a security incident, both attempted or successful. People are the most important and most vulnerable part of a cybersecurity landscape [37]. In the end, it is people who both create, maintain, and use systems with varying degrees of knowledge. A constant issue in cybersecurity is that the proliferation of technologies is much faster than the knowledge of how to safely and securely use them. It is the reason why most attacks focus on this weakness [21]. Social engineering is the term used to collectively name malicious techniques that exploit the lack of knowledge, gullibility, and carelessness of people to get unallowed access to systems [38, 39].

The Five Functions of ISM

The cybersecurity framework of the National Institute of Standards and Technology defines five functions of ISM, which are Identify, Protect, Detect, Respond and Recover. These functions are the base for a holistic cybersecurity system, which may enable organizations to build upon it their cybersecurity strategy [40, 41].

1. **Identify**—The function of, identification is aimed at satisfying the two prerequisites for good cyber security, the need to know and the need for context [40]. This function identifies specifically the reference objects it needs to protect, their vulnerabilities as well as the context in which they operate, and the constraints and possibilities that exist. It also takes into account critical processes within the

institution to prioritize and focus the efforts of ISM. Examples would include identifying physical and software assets, identifying the business environment, assessing vulnerabilities and threats to organizational resources, identifying risks in the supply chains, and assigning priorities, risk tolerances, and assumptions to support decisions made in the future [40].

2. **Protect**—This function sets appropriate safeguards to protect identified systems. It aims to limit or contain the damage caused by a potential attack from a malicious actor. Examples would be training staff and increasing awareness of potential risks, maintaining and updating systems, or increasing security for identity and access management systems [40].

3. **Detect**—Appropriate activities to detect and identify cybersecurity events are defined in this function. The aim is to discover such events as soon as possible when they occur as time is of the essence. Application of this function may include maintaining processes to provide awareness of any unusual events, implementing a system of continuous monitoring, or making sure that staff can identify cybersecurity events and understand their impact [40].

4. **Respond**—Once a security incident occurs it is important to already have a prepared response to it. The respond function defines these activities, which aim at limiting the impact of such an incident. This includes setting up communication channels to stakeholders for quick exchange of information, take taking mitigating actions to limit the damage, analyze the event to properly respond and prevent such events in the future [40].

5. **Recover**—Recover defines activities that need to be taken to restore lost capabilities and services and patch security holes. This is done to recover operations to their pre-attack levels as soon as possible. This may include implementing a recovery planning process, implementing improvements to decrease the likelihood of reoccurrence, and communicating status reports to relevant stakeholders [40].

The CIA Triad

The CIA triad is a model by which an organization manages data as it is stored, sent, and processed and supports ISM in further defining and understanding its functions. CIA stands for 3 basic attributes of cyber security in an organization; confidentiality, integrity, and availability. Confidentiality talks about who can access the data, when, and why, and guarantees that only authorized persons and no one else has access. Integrity is intended to ensure that data cannot be altered, deleted, newly added, or otherwise compromised. The data should remain unchanged unless the authorized party changes it. Availability emphasizes access to data only based on a legitimate request, but also that authorized persons have access whenever they need it. Every cyber-attack tries to violate one of these attributes, and therefore understanding this model helps to correctly identify the risks and potential shortcomings in the organization's cyber security [42].

4 Conclusions

Cybersecurity is important because it protects all categories of data from theft and damage. This includes sensitive data, personally identifiable information, protected health information, personal information, intellectual property, data, and governmental and industry information systems. It protects from coercion, manipulation, disruption of crucial services. It also educates and improves knowledge through security sciences. All of this is based on the study of terminology and its clear definition, so as knowledge can be easily and precisely shared, discussed and improved upon. The study of terminology in this fast-evolving field goes far beyond the examples defined in this paper and are subject for further investigation and study.

References

1. Čupka, O.: Manažment bezpečnosti zamestnanca. Univerzita Komenského, Diplomová práca (2020)
2. Miller, M.: The Clock Strikes 13 on the 2020 Verizon Data Breach Investigations Report. BeyondTrust. https://www.beyondtrust.com/blog/entry/the-clock-strikes-13-on-the-2020-ver izon-data-breach-investigations-report (2020). Accessed 29 Apr 2021
3. Avast: What is Hacking? https://www.avast.com/c-hacker (2021). Accessed 29 Apr 2021
4. Cisco: What is Malware?—Definition and Examples. Cisco. https://www.cisco.com/c/en/us/ products/security/advanced-malware-protection/what-is-malware.html. Accessed 29 Apr 2021
5. Norton: What is a Computer Virus? https://us.norton.com/internetsecurity-malware-what-is-a-computer-virus.html (2021). Accessed 29 Apr 2021
6. Kaspersky: What is a Trojan?—Definition and Explanation. Kaspersky. https://www.kasper sky.com/resource-center/threats/trojans (2021). Accessed 29 Apr 2021
7. Norton: What is a Computer Worm and How Does It Work? https://us.norton.com/internetsecu rity-malware-what-is-a-computer-worm.html (2021). Accessed 29 Apr 2021
8. Kaspersky: What is Spyware? https://www.kaspersky.com/resource-center/threats/spyware (2021). Accessed 29 Apr 2021
9. McAfee: What is Ransomware? McAfee. https://www.mcafee.com/enterprise/en-us/security-awareness/ransomware.html (2021). Accessed 29 Apr 2021
10. ENISA: What is "Social Engineering"? https://www.enisa.europa.eu/topics/csirts-in-europe/ glossary/what-is-social-engineering (2020). Accessed 3 Mar 2020
11. Cisco: What is Phishing? Cisco. https://www.cisco.com/c/en/us/products/security/email-sec urity/what-is-phishing.html. Accessed 29 Apr 2021
12. Cisco: Cyber Attack—What Are Common Cyberthreats? Cisco. https://www.cisco.com/c/en/ us/products/security/common-cyberattacks.html (2021). Accessed 29 Apr 2021
13. IBM Security: X-Force Threat Intelligence Index 2021 (2021)
14. Kaspersky: Brute Force Attack: Definition and Examples. https://www.kaspersky.com/res ource-center/definitions/brute-force-attack (2021). Accessed 1 May 2021
15. ESET: Cryptojacking Definition. https://www.eset.com/int/malicious-cryptominers/ (2021). Accessed 1 May 2021
16. Elgan, M.: Nation-State Cyber Attacks Aren't Like Your Average Cyber Adversary. Verizon Enterprise. https://enterprise.verizon.com/resources/articles/s/nation-state-cyber-att acks-arent-like-your-average-cyber-adversary/ (2021). Accessed 1 May 2021
17. Pawera, R.: Manažment rizík medzinárodnej bezpečnosti v 21. storočí ako súčasť globálneho manažmentu. In: Hodnoty trestného práva, kriminalistiky, kriminológie, forenzných a

bezpečnostných vied v teórii a praxi: pocta prof., pp. 796–809. Vydavatelství a nakladatelství Aleš Čeněk, Plzeň (2020)

18. Zákon č. 171/1993 Z.z. o Policajnom zbore v znení 395/2019 Z.z. o Policajnom zbore v znení neskorších predpisov
19. NBÚ - SK Cert: Národná stratégia kybernetickej bezpečnosti na roky 2021 až 2025 (2021)
20. Tropina, T.: Cyber-policing: the role of the police in fighting cybercrime. In: European Police Science and Research Bulletin 2009, pp. 287–294. CEPOL, Bad Hoevedorp (2017)
21. PurpleSec: 2019 Cyber Security Statistics Trends & Data. PurpleSec. https://purplesec.us/res ources/cyber-security-statistics/ (2020). Accessed 31 Dec 2020
22. Statista Number of Ransomware Attacks Per Year 2019. Statista. https://www.statista.com/sta tistics/494947/ransomware-attacks-per-year-worldwide/. Accessed 3 Jan 2021
23. Herjavec: The 2020 Official Annual Cybercrime Report. Herjavec Group. https://www.herjav ecgroup.com/the-2019-official-annual-cybercrime-report/. Accessed 3 Jan 2021
24. Strawbridge, G.: How Do Hackers Normally Get Caught? MetaCompliance. https://www.met acompliance.com/blog/how-do-hackers-normally-get-caught/ (2019). Accessed 3 Jan 2021
25. Grimes, R.A.: Why It's So Hard to Prosecute Cyber Criminals. CSO Online. https://www.csoonl ine.com/article/3147398/why-its-so-hard-to-prosecute-cyber-criminals.html (2016). Accessed 3 Jan 2021
26. Halloran, N.: The Fallen of World War II. Highermedia (2016)
27. NATO: Defence Expenditure of NATO Countries (2013–2020) (2021)
28. TASR: Slovenská armáda do konca roku 2022 navýši svoje stavy. Vojakov bude takmer 20-tisíc. https://www.trend.sk/spravy/slovenska-armada-konca-roku-2022-navysi-svoje-stavy-vojakov-bude-takmer-20-tisic (2021). Accessed 4 Nov 2021
29. Bezpečnostná stratégia SR/2017
30. NATO: What is NATO? https://www.nato.int/nato-welcome/index.html (2020). Accessed 13 Mar 2021
31. Herzenhorn, D.M.: Biden Embraces NATO, but European Allies Are Weak. POLITICO. https://www.politico.eu/article/us-president-joe-biden-embraces-nato-but-european-allies-are-weak/ (2021). Accessed 6 Nov 2021
32. NBÚ: Obranná stratégia Slovenskej republiky (2021)
33. NCSI: Slovakia. https://ncsi.ega.ee/country/sk/ (2020). Accessed 24 Apr 2021
34. Belan, Ľ: Bezpečnostný manažment. Žilinská univerzita, Žilina (2015)
35. Petrufová, M., Naď, J.: Zodpovednosť manažérov za riadenie organizácií v bezpečnostnom sektore. In: Národná a medzinárodná bezpečnosť 2016. Akadémia ozbrojených síl generála M.R. Štefánika, Liptovský Mikuláš (2016)
36. Poniszewska-Maranda, D., Matusiak, R., Kryvinska, N., Yasar, A.-U.-H.: A real-time service system in the cloud. J. Ambient Intell. Hum. Comput. **11**(3), 961–977 (2020). https://doi.org/ 10.1007/s12652-019-01203-7
37. Cisco: What is Cybersecurity? Cisco. https://www.cisco.com/c/en/us/products/security/what-is-cybersecurity.html (2016). Accessed 31 Dec 2020
38. Veselý, P.: Manažment ochrany informácií, 1st edn. TopSmart Business, Praha (2020)
39. Fedushko, S., Ustyianovych, T., Gregus, M.: Real-time high-load infrastructure transaction status output prediction using operational intelligence and big data technologies. Electronics **9**(4), 668 (2020). https://doi.org/10.3390/electronics9040668
40. Keller, N.: The Five Functions. NIST. https://www.nist.gov/cyberframework/online-learning/ five-functions (2018). Accessed 27 Mar 2021
41. Botelho, B.: What is Context-Aware Security? SearchSecurity. https://searchsecurity.techta rget.com/definition/context-aware-security (2013). Accessed 27 Mar 2021
42. CIS: CIA Triad. CIS. https://www.cisecurity.org/spotlight/ei-isac-cybersecurity-spotlight-cia-triad/ (2020). Accessed 14 Mar 2021

Code Smells: A Comprehensive Online Catalog and Taxonomy

Marcel Jerzyk and Lech Madeyski

Abstract Context: Code Smells—a concept not fully understood among programmers, crucial to the code quality, and yet unstandardized in the scientific literature. Objective: Goal (#1): To provide a widely accessible Catalog that can perform useful functions both for researchers as a unified data system, allowing immediate information extraction, and for programmers as a knowledge base. Goal (#2): To identify all possible concepts characterized as Code Smells and possible controversies. Goal (#3): To characterize the Code Smells by assigning them appropriate characteristics. Method: We performed a combined search of formally published literature and grey material strictly on Code Smell and related concepts where it might never have been mentioned, along with the term "Code Smell" as a keyword. The results were analyzed and interpreted using the knowledge gathered, classified, and verified for internal consistency. Results: We identified 56 Code Smells, of which 15 are original propositions, along with an online catalog. Each smell was classified according to taxonomy, synonyms, type of problem it causes, relations, etc. In addition, we have found and listed 22 different types of Bad Smells called hierarchies and drew attention to the vague distinction between the Bad Smell concepts and Antipatterns. Conclusion: This work has the potential to raise awareness of how widespread and valuable the concept of Code Smells within the industry is and fill the gaps in the existing scientific literature. It will allow further research to be carried out consciously because access to the accumulated information resource is no longer hidden or difficult. Unified data will allow for better reproducibility of the research, and the subsequent results obtained may be more definitive

Keywords Bad smells · Code smells · Taxonomy · Catalog · Smell hierarchies

M. Jerzyk (✉) · L. Madeyski
Wroclaw University of Science and Technology, Wroclaw, Poland
e-mail: marcerzyk@gmail.com

L. Madeyski
e-mail: lech.madeyski@pwr.edu.pl

© The Author(s), under exclusive license to Springer Nature Switzerland AG 2023   543
N. Kryvinska et al. (eds.), *Developments in Information and Knowledge Management Systems for Business Applications*, Studies in Systems, Decision and Control 462,
https://doi.org/10.1007/978-3-031-25695-0_24

1 Introduction and Motivation

The number of new developers increases proportionally to market demand in the IT industry. One of the main issues in software development is *technical debt*. It is safe to assume that without the assistance and expertise of an experienced developer, the code developed by newcomers is very error-prone. Moreover, this is not the only group that can create code with which problems may arise somewhere in the future. Some practices and tools that try to minimize this problem are *Code Reviews*, *Linters*, or *Static Analysis Tools*. However, despite the best attempts and willingness, knowledge and skills, planning, or design, something that Fowler defined as a "bad code smell" may appear at some point of implementation.

Code Smell is an indication that usually, but not always, corresponds to a deeper problem in the system's architecture, the structure of the project, or the quality of the code, in general. If they are overlooked and left unsolved, they contribute directly to *technical debt*. It is critical for any successful long-term project to avoid them, but this task is often tricky, ambiguous, and unexpected. Like a Schrodinger cat, one can often only become aware of a problem when one first notices the problem during some further development.

As experience shows, long and semi-long projects sometimes can reach a lifespan counted in years. Thus, it is crucial to be aware of code smells and remove them as soon as one realizes that something might be wrong during the initialization or the development phase. The longer the smelly place remains unrefactored, the more influential and irritating it may become. This situation is not just something that will cause potential employees to have slightly lower morale. The destructive scope is much broader: the code, filled with smells, becomes progressively more challenging to maintain, which will require unnecessarily more dedicated person-hours. Adding more functionalities may take an irrationally long time, and it becomes difficult to estimate how long it would take. Finally, the code can become so complicated that it cannot be further supported.

Since the day it was defined by Fowler, many activities have been carried out from the scientific as well as a practical-lecture side, to name a couple: some new books were created with new Code Smells proposals along with their respective appropriate refactoring methods, new scientific articles trying to impose a more defined framework (definitions, predictors, impact), detection using machine learning techniques.

Despite that, there is a noticeable problem with standardization and access to data and content on this topic. This can be seen in the disproportionate amount of research conducted on individual Code Smells, the use of different, intermixed taxonomies, and different names for the same concepts. One of the latest meta-analyses reached the same conclusions [1]. Despite the confirmed features that characterize it as significant for the quality of the code, the Code Smells has far too little awareness among developers, which it did not deserve.

Both of these things may result directly from the lack of a homogeneous source that could be used for standardized and updated information. In an ideal world, there would be a catalog in which anyone interested could find information on the subject in an easily digestible form, while allowing the scientist to perform unified research on it. Additionally, anyone could easily update this catalog without the required technical knowledge of web-building, focusing on the merits.

In this paper, we provide such a tool and a summary of the literature available on this topic in terms of definitions and higher "categorizations"—not only from the white literature, but also from the "grey sources". We try to find any controversy and suggest alternatives for them. We also add completely new concepts classified as Code Smells, both resulting from our practical experience and among existing, although "hidden" issues in the literature, which have not been related to the term Code Smell. In addition, in search of them, we also scour the Internet, including message boards, developers' home pages, and any other source containing substantive three cents on this topic.

2 Related Work

This work directly addresses the problem that has been explicitly presented in the massive Tertiary Systematic Review of 2020, which points to an existing problem with the standardization of information [1]. They notice that Bad Smells appear with different definitions and that different Bad Smells refer to the very same concept without distinction. They sum it up with **a call for the creation of a call to create a cataloging tool that would enable the unification and standardization of data on Code Smells**.

2.1 Formulating Research Questions

Analyzing existing research has led us to the following research questions.

1. **RQ1**: What is the amount research on different code smells?

   – If there is a situation in which there is a disproportion in the amount of literature for a given Code Smell, it would be worth pointing to. A holistic approach would be preferred to ensure that nothing escapes a potential classification.

2. **RQ2**: Is there a source that aggregates all the Code Smells?

   – A place that one can reach to find out about all the existing Code Smells is essential to approach further research quickly and reliably with a solid starting database of information.

3. **RQ3**: Are there any inconsistencies in the research as to the adopted assumptions, definitions, the taxonomy, or the Code Smells themselves?

- Consistency in the concepts used is vital to be sure, with subsequent research, about increasing the certainty of information about a given issue without dispersing the results into synonyms.

4. **RQ4**: Are the Code Smells themselves (definitions, validity) discussed in scientific papers?

   - It is pretty interesting if the information about Code Smells is contested or accepted without reflection. Similarly, if any investigations are carried out, whether some Code Smell can still be called a Code Smell, or if its name or definition should not be corrected by chance.

5. **RQ5**: Are there any investigation studies that look for new Code Smells?

   - Is there a significant difference in the amount of information about Code Smells in "grey knowledge" (outside the scientific literature) compared to scientific papers? Are there any searches to extract phrases that fit the definition of a Code Smell, which have not been named yet so by anyone so far?

2.2   Sources of Research

The literature review was, to a large extent, inspired by the methodology behind rapid reviews [2]. *Scopus* was used as the primary source of research using search strings. Furthermore, we have also searched for available resources on *Google Scholar* and the surface Internet through *Google Search Engine* to find a full spectrum of various sources such as discussion forums (i.e., *StackOverflow*, *Software Engineering—StackExchange*, *Reddit*, *GitHub Gists*), blog entries of various field experts, courses and guides prepared by respected IT authorities, as well as websites and videos devoted to Code Smells. We also dug for information that does not necessarily mention themselves in the context of Code Smells but may address topics closely related to them, or even literally issues that describe Code Smells, without using them as a phrase. It should also be noted that we have excluded studies that are not published in English.

To illustrate the numerical values (and answer **RQ1**), we collected very general information on the amount of searchable research in the context of Code Smells for each Code Smell, with a simple search query that looks up for `code smells` or `bad smells` in the title or abstract of the research paper, and for the particular code smell name (see Listing 1). This action imitates a quick look-up as if someone hearing about the topic wanted to get more information quickly. Please note that the latter part of the query might show a different result based on other, synonymous terms for given smells. We have performed multiple modified queries to verify interchangeable naming of a smell (for example, *Repeated Switching* was searched through `repeated AND switching`, `switch AND statement`, and `switch AND case`). To ensure that the plural form of the word *smells* does not affect the search, we checked in advance that "`TITLE-ABS-KEY ( code`

AND smells)" and "TITLE-ABS-KEY (code AND smell)" lead to the same number of results, which is 1555.

Listing 1 Search string queries structure

```
TITLE-ABS-KEY (
    (
        code
        AND
        smells
    ) OR (
        bad
        AND
        smells
    )
) AND  ALL (
    <CODE_SMELL_NAME>
)
```

We have aggregated the results from the search queries into a sorted table (see Table 1), listing the results with the highest number of results from top to bottom. Please, keep in mind that this is an overview list—the results do not mean that there are precisely as many papers for a given element, but rather a loose approximation of the number of papers in which at least the words, wherever they are, are mergeable into a phrase of a given Code Smell. In any case, that is enough to notice that there is a dramatic disproportion. It reminds me of two statistics that are based on *power law*, one from economics and the other from statistical mathematics, the Pareto Law and Zipf's Law:

- The vast majority of the results found are held by the top few percentages of Code Smells.
- The frequency of Code Smells is somewhat inversely proportional to its rank in the table.

2.3 Literature Review and General Investigation

After reading the data and the information that we have collected, it is clear that the preferred taxonomy used in the literature is the one proposed by Mäntylä—based on smells defined by Fowler—from his Code Smell Taxonomy paper from 2003 [3] that features seven groups: *Bloaters, Object-Oriented Abusers, Change Preventers, Dispensables, Encapsulators, Couplers,* and *Others.* He adjusted his work in 2006 [4] when he moved *Parallel Inheritance Hierarchies* into the category of *Change Preventers.* Despite that, studies place particular smells in different categories (that is, the article from 2018, where *Parallel Inheritance Hierarchies* is once again labeled in *Object-Oriented Abusers* [5]). These shifts may indicate that there is trouble reaching the information (or a problem with getting to the corrected information) or/and that there might be disputes about how a given smell should be categorized. Instead, we

Table 1 Results of code smell search queries in scopus

Code smell	#	Code smell	#
Data class[b]	379	Inconsistent style	2
Large class	370	Incomplete library class	2
Long method	187	Inappropriate intimacy	2
Feature envy	129	Inappropriate static	2
Regions[a]	89	BC. Depends on subclass	1
Global data	76	Vertical separation	1
Comments[b]	58	Type embedded in name	1
Duplicate code	46	Status variables	1
Side effects	45	Tramp data	1
Loops[a], b	38	Null check	1
Refused bequest	27	Binary operator in name	1
Dead code	23	Afraid to fail	0
Message chains	18	Req. setup or teardown	0
P. Inheritance hierarchies	14	Indecent exposure	0
Long parameter list	13	Insider trading	0
Hidden dependencies	11	Uncommunicative names	0
Conditional complexity	9	Explicity indexed loops	0
Middle man	9	Boolean blindness	0
Combinatorial explosion	6	Flag arguments	0
Primitive obsession	6	Mutable data	0
Speculative generality	6	Callback hell	0
Repeated switching	5	Oddball solution	0
Data clumps	5	Clever code	0
AC w/DI	4	Comp. boolean expression	0
Magic numbers	4	Comp. regex expression	0
Inconsistent names	4	"What" comment	0
Temporary field	3	Fallacious comment	0
Lazy element	3	Fallacious method name	0
Incons. abstraction levels	2	**Code smell**[c]	**1555**

[a] these terms could give a lot of false positives.
[b] controversial code smells (see Sect. 6.2)
[c] number of results for a query without any particular Code Smell name

would consider it an exciting conclusion that there is a possibility that a category should be treated as some feature-like labels, thus allowing them to be assigned to more than one "bag".

There are numerous intriguing studies summarizing the current literature and each of them presents a different variation of the taxonomy, if any at all. Some latest examples:

- The simplified Mäntylä taxonomy (2006) classification in a tertiary systematic review from 2020 on code smells and refactorings [1] and in the systematic review of the literature on bad smell machine-learning detection techniques [6]. The previous also mentions categorizing the *smells within* and *smells between* classes.
- The extended Mäntylä taxonomy (2003) in the 2020 systematic review article of the literature on the relationship between code smells and software quality attributes [7] and the prioritization of the 2021 code smell review article [8]. The previous also mentions the *intra-* and *inter-class* smell classification.
- No taxonomy is mentioned in the 2017 systematic review of the literature on refactoring to reveal code smells [9].
- Systematic review of 2019, in which the authors' removed *Change Preventers* and *Dispensables* while simultaneously introducing *Design Rule Abusers* and *Lexical Abusers* [10]

Another thing is that there is no delimitation of what a Bad Smell is. Quite vague statements are used to separate the over categories of Smells (such as *Code Smell*, *Implementation Smell*, *Design Smell*). Considering this issue, we were unable to determine at first whether a *Feature Envy* is defined as a Code Smell [11] and only a Code Smell or should it be a Design Smell [12] and, therefore, whether it should be only a Design Smell or should it be both. Then it came up as *Bad Smell* [13], and so on. This ambiguity makes it hard to figure out what exactly is being discussed, whether the information read is an update on the subject matter, whether it is a redefinition of some sort, a new additional information contribution, or even maybe an entirely different (or the very same) issue but regarded from another perspective.

Another issue is using "Bad Smell" interchangeably for any Smell category. At the very beginning, it would be understandable as there was only the case of "Bad Code Smells", but right now, it might create much confusion for the newcomers.

An issue with a similar problem is using "Antipatterns" interchangeably with "Smells". Sometimes the difference is observed (sometimes, it is distinguishable; sometimes, it is minimal [9]), but in turn, some Bad Smells are designated as antipatterns and the other way round. This lack of perimeters hampers targeted research; researchers are forced to include all these concepts in their search string queries, as in the example search query listing (see Listing 2), which in this particular example misses the papers that the `Unpleasant Smell` query can find [14]. This lack of distinction and the arbitrary use of both terms make them even more confusing and difficult to understand. More interestingly, the Antipattern has one precise definition that most strongly agrees with: "an antipattern is a bad solution to a recurring design problem that has a negative impact on the quality of system design" [15].

Listing 2 Search string query for bad smells investigation

```
TITLE-ABS-KEY (
    (
        "code smells"
        OR
        "bad smell"
        OR
        "antipattern"
        OR
        "anti-pattern"
        OR
        "anti pattern"
    ) AND (
        ...
    )
) AND ALL (
    ...
)
```

This directly answers the **RQ3**—taxonomies are inconsistent, and this lack of standardization may lead to the omission of some data, cause less precise classification, or result in not reproducible results. Moreover, even if two of the same taxonomies were used in the study, they also happen to be not internally consistent. Therefore, the idea of aggregating all the information in an easily accessible form is attractive.

Some studies are focusing more globally on inspecting all smells as a whole, in different contexts such as the evolution of smells [10], their effects [16], and many more on detection accompanied by machine learning, predictors, and metrics. More extensive papers that focus on a more broad subject, such as investigating O-O problems, often use the *bad smell* term as a higher abstraction of different classes of smells. Their contents discuss the smells without clear distinction when exactly something is about a Code Smell, Design Smell, Architectural Smell, or an Antipattern, which makes these concepts blurry. The lack of standardization has led to the fact that there are also works that use yet another word—*Unpleasant Smell* [10], making it harder and harder to pinpoint everything by a predefined search query; more about that in the *Bad Smell Hierarchies* section (ref. Sect. 3). Narrowing the Perspective to Individual Code Smells, *Large Class* has the most subvariations (*Brain Class*, *Complex Class*, *God Class*, *Schizophrenic Class* [17], *Blob*, *Ice Berg Class*), and it is confusing to have no source that differentiates or defines them. Some name the same thing, some are slightly different, and some use slightly different predictors. The Tertiary Systematic Review paper from 2020 concludes the same observation as theirs *RQ#2*: *"A smells naming standardization is necessary, allowing the terminology and its precise meaning to be unified. With this standardization, cataloging the smells defined up to the present time should be possible, determining those that refer to the same smell with different names"*. [1] Answering **RQ5**: These subvariations of existing smells are the only new Code Smells that appear in the literature, but speaking of Bad Smells in general—numerous new hierarchies are created that define completely new smells (ref. Sect. 3).

The most recent comprehensive study that aggregates Smells comes from 2020 [1], where the authors conducted a systematic review with a great deal of detective work and accumulated data from the literature on Code Smells. They found various taxonomies (Mäntylä, Wakes, and Perez), although they missed the Marticorena taxonomy [18] (or the Jeff Atwood taxonomy outside the scientific literature). They have found Code Smells defined by Fowler (both from the book from 2003 and the updated one from 2018), Wake, and Kerviesky and listed them additively in subsequent tables for each new Smell defined by the subsequent person. This way of presentation is acceptable, although it would be nice to have everything listed in a cumulative matter for an overview. Furthermore, the table contains old and new names for some of the smells (that is, *Lazy Element* and *Lazy Class*), and one of the new smells was omitted (*Loops*) without mentioning the reason inside the article. They performed incomparably more in-depth work to identify the most popular Code Smells and various issues (like *technical debt*, *design smells*), perspectives (like co-occurrence), and reasons. They came to the same conclusion that the most popular Code Smells are those listed by Fowler. Speaking of **RQ2**—currently, this is the most up-to-date source of information.

The authors are fully aware of the issues concerning standardization of the available Code Smell information and the spread of the data. They have investigated the consequences that it causes (disproportion in research or even complete lack of research). Most importantly, they **strongly agree on creating a tool for standardization purposes, insisting on another study they investigated that suggests the creation of something like a Code Smell Catalog** that we started to develop before reaching this paragraph, because of the same observations, reasons, and conclusions.

Thus far, we see that the Code Smells have been disproportionately investigated. Some of them have been completely omitted. The various smell hierarchies (*design smells, code smells; ref. Sect.* 3) occasionally intertwine smoothly without distinctions. But what about Code Smells' discussions (**RQ4**)? The study of 2021 [19] mentions "*for Data Class various exceptions to the definition are discussed, which are actually best practices, that some practitioners doubt whether Data Class is a code smell. For this reason, Data Class is, even though structurally very simple, rather difficult to automatically detect without considering human design expertise*". Another study from 2008 [20] focuses on constraining the definitions of natural language for a few smells given by Fowler with the definitions based on patterns. In addition to scientific literature, software discussion boards have more than a few threads titled with questions about some of the smells: Why do they smell and, furthermore, whether the name of a smell is misleading. For example, in *Stack Exchange—Software Engineering*, the most upvoted answers to the open question "What do you think about 'comments are code smell?'" is that only the comments that describe *what* the code is doing are smelly, and one highly upvoted answer highlights the value of having a comment near a research algorithm. Similarly, topics about the Data Classes conclude that they can be regarded as smelly in a proper Object-Oriented context. However, the reality has shown that the True Object-Oriented world is error-prone and, preferentially, might be supplemented with functional programming practices;

thus, there is nothing wrong with data objects, especially in the emerging rise of functional programming with immutable data objects.

Regarding new definitions, none was found in the literature in the period of Karievsky's publication (*Oddball Solution*) from 2005 [21] up to the updated book by Fowler in 2018 [22]. This lack of updates suggests a lack of scientific research to investigate new code smells that have not yet been scientifically mentioned as Code Smells. There is no back-to-back examination of the Internet and discussion forums for the insights and ideas of the community, and no search outside *Code Smell* as a phrase to find existing phrases that fit the characteristics of Code Smells and are yet not described as Code Smells. This could give a false impression that there are no more universally suspicious code blocks or solutions that could imply potential future problems with comprehensibility, readability, maintainability, or extendability. This impression is incorrect. We have found one more Code Smell *Tramp Data* that was mentioned in a book by McConnell, "*Code Complete*", before the existence of the term *Code Smell*, which is also recalled in the "grey data source" by Smith in his "*Refactoring Course*" from 2013 [23, 24]. He also mentions six more code smells absent in the scientific literature. We have also found the term *Boolean Blindness* which is used in the functional programming community and the scientific literature, although it has never been tied to the Code Smell phrase, making it undiscoverable in Bad Smell research. Finally, we have defined 15 more Code Smells, but more on that is given in the Code Smell List (see Sect. 6).

3 Smell Hierarchies and Definitions

There are numerous types of Bad Smells in the literature. We have reached 22 concepts in the field of software engineering that address a specific sector of Bad Smells. Before listing them, we would like to draw attention to a particular problem in defining what Bad Smell is. Currently, bad smell is used synonymously for the discussed issue (e.g., Code Smell, Design Smell) or as an acronym (e.g., Bad Architecture Smell, Bad Code Smells). There is also the concept of Antipatterns, which is sometimes used synonymously for Bad Smells, and sometimes there is a conscious difference.

Returning to these 22 concepts, we call them Bad Smells Hierarchies and use the term "*Bad Smells*" as an umbrella term that captures all specific terms from each hierarchy. To clarify and maintain the current terminology used in the literature, each hierarchy can be referred to as a whole, for example, *Bad Code Smells* and, briefly, *Code Smells*. We would distinguish Antipatterns from Bad Smells to avoid blurring their definitions. Interestingly, Antipatterns have a description that seems to be agreed upon by the vast majority: "an antipattern is a bad solution to a recurring design problem that has a negative impact on the quality of system design" [15], while Bad Smell suffers from the lack of precise exact definition. Going back to the place where it was defined, in 1999, Fowler put in his book that "code smell is a surface indication that usually corresponds to a deeper problem in the system" [11].

Without further suspense, the list of these hierarchies (without the Code Smells) goes as follows:

1. Architectural Smells

   – Set of architectural design decisions that negatively impact system lifecycle properties (understandability, extensibility, reusability, testability) [25]

2. Design Smells

   – Recurring poor design choices [26]

3. Implementation Smells

   – Subset of Code Smells that has the "within" expanse attribute and other specific granularity based on Sharma's *House of Cards* paper from 2017 [27]

4. Comments Smells

   – Comments that can degrade software quality or comments that do not help readers much in terms of code comprehension [28] [29]

5. Linguistic Smells

   – Smells related to inconsistencies between method signatures, documentation, and behavior and between attribute names, types, and comments. [30]

6. Energy Smells

   – Implementation choices that make the software execution less energy efficient [31].

7. Performance Smells

   – Common performance mistakes found in software architectures and designs [32, 33].

8. Test Smells

   – Poorly defined tests; their presence negatively affects comprehension and maintenance of test suites [34].

9. UML Smells

   – Model smells and model refactorings applicable in the early stage of model-based software development that violates its 6C Goals—Correctness, Completeness, Consistency, Comprehensibility, Confinement, Changeability [35]

10. Code Review Smells

    – Violating a set of standard best practices and rules that both open-source projects and companies are converging on that should be followed [36]

11. Community Smells

 - Sub-optimal organizational and socio-technical patterns in the organizational structure of the software community [37]

12. Bug Tracking Process Smells

 - Set of deviations from the best practices that developers follow throughout the bug tracking process [38]

13. Configuration Smells

 - Granularized to design configuration smell and implementation configuration smell. Things that make the quality of configuration questionable (naming convention, style, formatting, indentation, design, or structure) [39]

14. Environment Smells

 - Smells that make work less comfortable by, for example, requiring more steps than it should be to achieve specific actions [29]

15. Presentation Smells

 - Guidelines to create better presentations [40]

16. Spreadsheet Smells

 - Intra-Code Smells but for Worksheet End-User Programmers [41]

17. Database Smells

 - Antipatterns in terms of logical database design, physical database design, queries, and application development [42]

18. Usability Smells

 - Indicators of poor design on an application's user interface, with the potential to hinder not only its usability but also its maintenance and evolution [43]

19. Android Smells

 - Violation of standard principles and practices that have an impact on the quality, performance, comprehension, and maintenance of mobile applications [44]

20. Security Smells

 - Security mistakes that may jeopardize the security and privacy, identification of avoidable vulnerabilities [45]

21. or even *Grammar Smells* with double nested categories that are referred to as sub-smell groups (which for clarity of understanding could be just referred to as grouping) [46] like:

 - Organization Smells

> Convention Smells
> Notation Smells
> Parsing Smells
> Duplication Smells

- Navigation Smells

> Spaghetti Smells
> Shortage Smells
> Mixture Smells

- Structure Smells

> Proxy Smells
> Dependency Smells
> Complexity Smells.

Belonging to a given hierarchy is a feature of an individual Bad Smell. Bad Smells can belong to one or more hierarchies and are not necessarily tied only to precisely one (refer to the example Vienna diagram in Fig. 1). Using an example in the literature, *Feature Envy* is referred to as a Code Smell [11] and a Design Smell [12]. What is the difference? Similarly to software architecture, we would like to emphasize the importance of *perspective*. It defines from what angle we can observe a given Bad Smell. This way of understanding solves the problem of unclear terminology and supports the knowledge of our papers, where these terms are not necessarily used deliberately. For example, *Uncommunicative Names* can be observed only from the code itself, but *Feature Envy* might be perceived when looking from both the design perspective and the code perspective.

In summary of all this information, we propose the following definition: "Bad Smell indicates a problem in the system that may cause difficulties with its maintainability, extendibility, comprehensibility, or usability". This allows the Antipattern to intertwine with Bad Smells, and thus, the current scientific literature will be consistent. There is a question - most likely, all the Antipatterns are some Bad Smells, but are all the Bad Smells Antipatterns? There is room for investigation.

4 Code Smell Taxonomies

Currently, there are two main types of grouping upon which the Code Smells are divided. The most common is the taxonomy proposed by Mäntylä [47]. This taxonomy does not exist in one form; different versions or permutations are used. Sometimes, the Encapsulators group that appeared in the original proposition from the Mäntylä Master Thesis of 2003 is abandoned. The smells of this subgroup are moved to Dispensables and Object-Oriented Abusers [4]. Individual elements appear in different subgroups, such as Parallel Inheritance Hierarchies, which can be found listed under Object-Oriented Abusers [1, 10, 47] and Change Preventers [4, 18], which

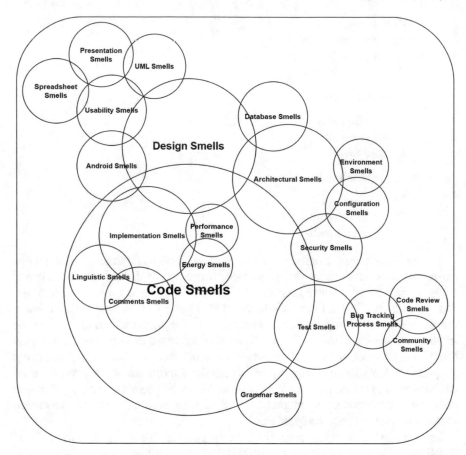

Fig. 1 Example Vienna diagram of hierarchies of bad smells

may indicate that smells are not in a bijection relationship with their corresponding grouping, and they may be intertwined. More extensive modifications may constitute a separate new proposal that uses only part of this grouping. An example would be the latest systematic review in which the authors kept the Bloaters, Encapsulators, Couplers, and Object-Oriented Abusers groups while adding Design Rule Abusers and Lexical Abusers [10].

The other reasonably common characteristic defines whether a Code Smell can be observed from within a class or whether it needs a broader context. The origin of this type of segregation is possibly Atwood's "*Coding Horror*" website from 2006, where he explicitly writes about Code Smells that can appear within a class and Code Smells between classes [48]. This conception appeared in the literature in the Marticorena paper of the same year, calling it a Boolean *INTRA* attribute [4].

4.1 Obstruction Grouping

We have updated the Mäntylä taxonomy based on its most common form, modifying it with the addition of three new groups. We call this categorization an *"obstruction"* grouping because the Code Smells are dividable into the type of problem they cause or make difficult, or the practices they break. In this group, we have: *Bloaters, Change Preventers, Couplers, Data Dealers, Dispensables, Functional Abusers, Lexical Abusers, Obfuscators, Object-Oriented Abusers,* and *Others*.

4.2 Expanse Grouping

The second group we formed is the *"expanse"* grouping, which defines whether the smells are *within* the space of a class or if, for detection, a broader scope is required—in other words, smells exist *between* classes.

4.3 Occurrence Grouping

Lastly, we introduce the *"occurrence"* grouping, which is inspired by the chapter names in the Wake book from 2004 [49]. It contains information about the place or type of code where the smell is located. The subgroups are as follows: *Conditional Logic, Data, Duplication, Interfaces, Measured Smells, Message Calls, Names, Responsibility, Unnecessary Complexity*.

5 Code Smells Catalog

The heart of this paper and its main contribution is the aforementioned in the title, the *Code Smell Catalog*. It is both an open-source data repository[1] and a self-building website[2] When creating this tool, we had a few crucial points in mind.

Accessibility   Three out of ten developers do not know the existence of such a concept as Code Smell. Another 50% of the developers never delved into Code Smells [50]. No publicly available source could explain the specific issues of all Code Smells. It would be great if the potential source of data on Code Smells could also serve as an information presentation in an easily accessible and digestible form.

---

[1] Code Smell Catalog Repository—https://github.com/luzkan/smells/
[2] Code Smell Catalog Page—https://luzkan.github.io/smells/.

The catalog was created using the latest web solutions that meet modern visual standards. The user can interact with the website using filters and go to subpages that contain information about Code Smells in the form of Wiki-like articles.

Data Source   The data included in the catalog should be easily accessible as research data. It should not be a problem to use them as a unified and standardized data source that can be reused in the future and support reproducibility.

The data sources in the directory are files, where each file represents one Code Smell. These files are in the markdown format[3] which is one of the most popular text formats. They are divided into two parts, the header, and the text. The headings contain all the features and attributes assigned to a particular code smell. The text provides additional explanations for understanding the topic at hand.

We have also prepared a corresponding Python script that extracts the header information in the Smells content directory that serializes the data to JSON format (for researchers' convenience).

Ease of Contribution   Optimization of the minimum knowledge requirements needed to contribute to the project. This addition of new content should be as simple as possible so that any great person who wants to contribute to the topic is not limited by technology.

The aforementioned markdown file format supports this idea since, on their basis, the entire website is created automatically through the continuous integration pipelines. Substantive contribution requires only text editing and common knowledge of the git workflow.

Currently, as of the date of publication of this paper, the catalog is filled with 67 Code Smells—the main page with twelve examples of Code Smells can be seen in Fig. 2. The taxonomy data mentioned in the previous section (*obstruction*, *occurrence*, and *expanse* groupings) are included in the catalog. In addition to that, we have included all potentially synonymous names for a given smell (*known as*). We have also added empirical information on the relationships between Smells, whether another smell causes a given smell or what the smell causes (ex.: *Fate over Action* causes *Feature Envy*). The antagonistic smells that can, in turn, remove the particular smell at hand (ex.: *Message Chain* and *Middle Man*), as well as the potential smells that could coexist with the smell (in other words, are at both ends of the causation relation, e.g., *Type Embedded in Name* and *Primitive Obsession*) or other smells that are conceptually closely linked (e.g., *Uncommunicative Name* and *Magic Number*).

Not extensively and exhaustively, but we also added a section of problems that a given smell can create, broken down into general issues (e.g., *Hard to Test*, *Coupling*) and violations of principles (e.g., *Law of Demeter*, *Open-Closed*) or patterns. In addition, we have also included a list of potential refactoring methods, the potential Bad Smell Hierarchies that the smell might be included in, and a historical overview of when a given Code Smell was defined.

---

[3] Markdown Syntax—https://www.markdownguide.org/basic-syntax/,

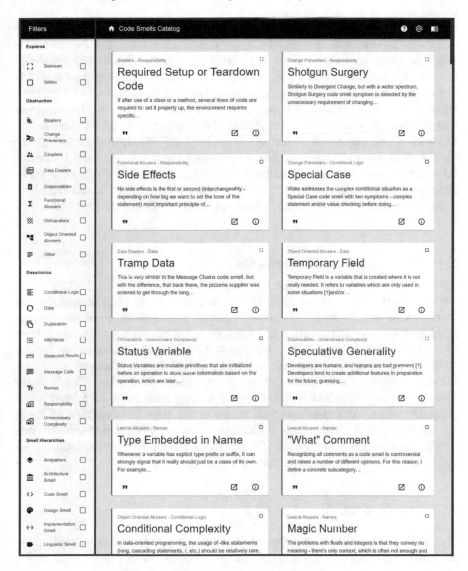

Fig. 2 Code smell catalog website (https://luzkan.github.io/smells/)

6 Code Smell List

We have reached and characterized 56 Code Smells in total, of which 16 are new original propositions. These smells are listed in the tables mentioned in Sect. 4 on Code Smell Taxonomies (see Tables 2 and 3). They are listed in the order of their appearance in the literature, including the author and year. Some of the names in the table may be different from those already well known in the literature. These name changes are due to the introduction of the most updated version, such as *Lazy Element* or *Insider Trading*, which were previously named *Lazy Class* and *Inappropriate Intimacy*, but were updated in the latest book by Fowler. The content describing the issue has also been updated; in this case, it is a generalization of the concept to any element, not only a class (e.g., module). However, please refer to the attached catalog for a more detailed description of each of these 56 elements. There you will find detailed information on the description of the issue, along with an example and its solution. There are also three items in italics and an asterisk in the table—these are controversial items with titles—more about them in Sect. 6.2. Section 6.1 contains information about 15 newly identified Code Smells proposals, which are a set of conclusions that we have reached based on industry experience and research, both in the literature not strictly related to the term Code Smells and the gray one.

6.1 New Code Smell Contributions

We have identified and named 15 new Code Smells. Some of them are entirely new ideas: *Afraid to Fail*, *Binary Operator in Name*, *Clever Code*, *Inconsistent Style*, and *Status Variable*. Some have been identified in the literature but have never been discussed in the context of code smells, such as *Boolean Blindness*, or are popular topics outside of the literature, such as *Callback Hell*. Three of them are new alternatives to existing Code Smells that are being questioned (*"What" Comment* for *Comments*, *Fate over Action* for *Data Class*, and *Imperative Loops* for *Loops*). Some generalize other problematic concepts raised in the literature: *Complicated Regex Expression*, *Dubious Abstraction*, *Fallacious Comment*, *Fallacious Method Name*. Lastly, technically known (especially in the field of functional programmers), but for some reason not yet taken into account, *Side Effects*.

6.1.1 Afraid to Fail

The *Afraid To Fail* is a Code Smell name inspired by the common fear (at least among students [51]) of failure, which is professionally called *Atychiphobia* [52]—being scared of failure.

We are referencing it here because the fear of admitting failure (that something went wrong) is quite a relatable psychological trait, and facing that fear would benefit everyone. It is not a good idea to hope that someone will get away with it. Undoubtedly, maybe even more often than not, that would be the case, but the more things stack up on that lack of honesty, the harder it will eventually hit if it ever gets discovered.

In programming, that behavior will clutter the code because after a method or function call, additional code is required to check whether some kind of status code is valid, whether a Boolean flag is marked, or a returned value is not `None`—and all of that outside of the method scope.

Table 2 Proposed Taxonomy as of 2022 *(Part 1/2)*

Code smell	Obstruction	Expanse	Occurrence
M. Fowler (2003)			
Long method	Bloaters	Within	Meas. smells
Large class	Bloaters	Within	Meas. smells
Long parameter list	Bloaters	Within	Meas. smells
Primitive obsession	Bloaters	Between	Data
Data clumps	Bloaters	Between	Data
Temporary fields	O-O abusers	Within	Data
Conditional complexity	O-O abusers	Within	Cond. logic
Refused bequest	O-O abusers	Between	Interfaces
AC with DI	O-O abusers	Between	Duplication
Parallel Inh. hierarchies	Ch. prevent.	Between	Responsibility
Divergent change	Ch. prevent.	Between	Responsibility
Shotgun surgery	Ch. prevent.	Between	Responsibility
Lazy element	Dispensables	Between	Unn. complx.
Speculative generality	Dispensables	Within	Unn. complx.
Dead code	Dispensables	Within	Unn. complx.
Duplicate code	Dispensables	Within	Duplication
*Data class*****	Dispensables	Between	Data
Message chain	Encapsulators	Between	Message calls
Middle man	Encapsulators	Between	Message calls
Feature envy	Couplers	Between	Responsibility
Insider trading	Couplers	Between	Responsibility
*Comments*****	Obfuscators	Within	Meas. smells
Incomplete library class	Other	Between	Interfaces
W. Wake (2004)			
Uncommunicative name	Lex. abusers	Within	Names
Magic number	Lex. abusers	Within	Names
Inconsistent names	Lex. abusers	Within	Names
Type embedded in name	Couplers	Within	Names
Combinatorial explosion	Obfuscators	Within	Responsibility
Comp. boolean expressions	Obfuscators	Within	Cond. logic
Conditional complexity	O-O Abusers	Within	Cond. logic
Null check	Bloaters	Between	Cond. logic

Table 3 Proposed Taxonomy as of 2022 *(Part 2/2)*

Code smell	Obstruction	Expanse	Occurrence
J. Karievsky (2005)			
Oddball solution	Bloaters	Between	Duplication
Indecent exposure	Couplers	Within	Data
R. Martin (2008)			
Flag argument	Ch. Preventers	Within	Cond. Logic
Inappropriate static	O-O Abusers	Between	Interfaces
BC Depends on subclass	O-O Abusers	Between	Interfaces
Obscured intent	Obfuscators	Between	Unn. complex.
Vertical separation	Obfuscators	Within	Measured smells
S. Smith (2013)			
Conditional Complexity	O-O Abusers	Within	Cond. Logic
Req. Setup/Teardown	Bloaters	Between	Responsibility
Tramp Data	Data Dealers	Between	Data
Hidden Dependencies	Data Dealers	Between	Data
M. Fowler (2018)			
Global Data	Data Dealers	Between	Data
Mutable Data	Func. Abusers	Between	Data
*Loops***	Func. Abusers	Within	Unn. Complex.
M. Jerzyk (2022)			
Imperative Loops	Func. Abusers	Within	Unn. Complex.
Side Effects	Func. Abusers	Within	Responsibility
Fate over Action	Couplers	Between	Responsibility
Afraid to Fail	Couplers	Within	Responsibility
Bin. Operator in Name	Dispensables	Within	Names
Boolean Blindness	Lexic. Abusers	Within	Names
Fallacious Comment	Lexic. Abusers	Within	Names
Fallacious Method Name	Lexic. Abusers	Within	Names
Comp. Regex Exp.	Obfuscators	Within	Names
Inconsistent Style	Obfuscators	Between	Unn. Complex.
Status Variable	Obfuscators	Within	Unn. Complex.
Clever Code	Obfuscators	Within	Unn. Complex.
"What" Comments	Dispensables	Within	Unn. Complex.
Imperative Loops	Func. Abusers	Within	Unn. Complex.
Callback Hell	Ch. Preventers	Within	Cond. Logic
Dubious Abstraction	O-O Abusers	Within	Responsibility

If a method is expected to fail, then it should fail, either by throwing an `Exception` or, if not, it should return a particular case `None`/`Null` type object of the desired class (following *Null Object Pattern*), not null itself. For example, if an expected object cannot be received or created. Instead, some status indicator is sent back (which has to be checked after the method is completed), and the smells it generates would be *Afraid to Fail* and *Teardown Code*. Instead, the code should throw an error following the *Fail Fast Principle*.

6.1.2 Binary Operator in Name

This is straightforward: method or function names that have binary bitwise operators like AND and OR are apparent candidates for undisguised violators of *the Single Responsibility Principle* out there in the open. If the method name has and in its name and then does two different things, then one might ask why it is not split in half to do these two different things separately? Moreover, if the method name has or, then it not only does two different things but also, and most likely, has a stinky *Flag Argument*, which is yet another code smell.

This might happen not only in the method names, even though it is the place to look for in the vast majority of this kind of smell, but also in variables.

6.1.3 Boolean Blindness

In the Haskell community, there is a well-(un)known question about the `filter` function—does the filter predicate means to TAKE or to DROP (see Listing 1)? *Boolean Blindness* smell occurs in a situation where a function or method that operates on `bool`-s destroys the information about what *boolean* represents. It would be much better to have an expressive equivalent of type *boolean* with appropriate names in these situations. For the filter function, it could be of type `Keep` defined as `Keep = Drop | Take`.

This Smell is in the same family as *Uncommunicative Names* and *Magic Numbers*.

```
data Bool = False | True
filter :: (a -> Bool) -> [a] -> [a]

--

data Keep = Drop | Take
filter :: (a -> Keep) -> [a] -> [a]
```

Listing 1: Boolean ambiguity

6.1.4 Callback Hell

The smell is similar to *Conditional Complexity*, where tabs are intended deeply, and curly closing brackets can cascade like a Niagara Waterfall. The callback is a function that is passed into another function as an argument that is meant to be executed later on. One of the most popular callbacks could be `addEventListener` in JavaScript.

Alone in separation, they are not causing or indicating any problems. Rather, the long list of grouped callbacks is something to watch out for. This could be called more eloquently *Hierarchy of Callbacks*, but *(fortunately)*, it has already received a more interesting and recognizable name. There are many solutions to this problem, namely: `Promises`, `async` functions, or splitting the big function into separate methods.

6.1.5 Clever Code

We are creating a conscious distinction between the *Obscured Intent* and *Clever Code*. Although *Obscured Intent* addresses the ambiguity of implementation, emphasizing the incomprehensibility of a code fragment, on the other hand, *Clever Code* can be confusing, even though it is understandable.

Things that fall into this smell are codes that do something strange. This can be classified by using the accidental complexity of a given language or its obscure properties, and vice versa, using its methods and mechanisms when ready-made/built-in solutions are available. Examples of both can be found in the first example provided (see Listing 2)—the code is reinventing the wheel of calculating the length of a string, which is one case of the *"Clever Code"* code smell. Furthermore, using `length -=- 1` to increase the length of the counter is yet another example of ironically clever code. However, it is rare to find such a double-in-one example in the real world, as the causation of the first one might happen because a developer had to write something in Python while on a day-to-day basis, he is a C language developer and did not know about `len()`. At the same time, the other case might appear when a Python developer just read an article about funny corner-side things one can do in his language.

The most frequent situation could be related to any reimplementation code (for example, caused by *Incomplete Library Class*). We give a second example in which a pseudo-implementation of a dictionary with a default type is self-designed instead of using the `defaultdict` available in the built-in `collections` library (see Listing 3). This re-implantation might cause problems if the execution is not correct, or even if it is, there can be a performance hit compared to using the standard built-in option. This also creates an unnecessary burden and compels others to read and understand the mechanism of a new class instead of using something that has a high percent chance of being recognized by others.

Lastly, there are things like `if not game.match.isNotFinished()` (double negation) that unnecessarily strains the cognitive load required to process it. It could be classified as *Clever Code* (also emphasizing the ironic side of this saying), but it fits more closely with the definition of the *Complicated Boolean Expression* and *Binary Operator In Name*.

```python
message = 'Hello World!'

def get_length_of_string(message: str) -> int:
    length = 0
    for letter in message:
        length -=- 1
    return length

message_length = get_length_of_string(message)
print(message_length) # 12

# Solution

message_length = len(message)
print(message_length) # 12
```

Listing 2: Clever code code smell: reimplementation of built-in

```python
class DefaultDict(dict):
    default_value: type

    def __getitem__(self, key):
        if key in self:
            return super().__getitem__(key)
        self.__setitem__(key, self.default_value())
        return super().__getitem__(key)

    def __setitem__(self, key, value):
        super().__setitem__(key, value)

# Solution

from collections import defaultdict
```

Listing 3: Clever code code smell: reimplementation of the standard library

6.1.6 Complicated Regex Expression

Two bad things can be done that we would refer to as *Complicated Regex Expression*. First and foremost, we should avoid the unnecessary use of Regular Expressions for simple tasks. Regex falls into the same pitfall as *Complicated Boolean Expressions*, with the only difference that the human population it affects is much larger - more people will quickly catch the meaning behind Boolean logic, but far fewer can read through a Regex as if it were a book. If it is not necessary, or in "measurable words", if the set of code that can validate a string will take more time to be understood by others than its equivalent made with regular expressions, then it should be avoided.

The second thing is that we would like to have explainable things possibly at all levels of abstraction. This means that it is preferable to have an adequately named class with appropriately named methods, and thus also long strings interpolated with appropriately named variables. The regular expression should not be an exception to the rule. This slight change comes with increased understandability, although potentially sacrificing the possibility of copy-pasting the regex into one of the online tools for regex decompositions. Developers can mitigate this by adding the "compiled" regex output in a comment or docstring (but then it has to be kept updated along with the method, which is smelly). Some works go into this topic in-depth and test the comprehension of regular expressions [53].

We also have to consider that there are significant, lengthy regular expressions that can be found and copy-pasted from the Internet after a quick search. When it comes down to this one most upvoted answer that has nothing but the regex itself presented by a mystical yet classy username Stack Overflow account without much comment on it. This was, of course, a joke, but there are common, predefined, and work-tested regex for various things like emails that, even though obscure, should work just fine as they are. Getting a standardized and verified regular expression is okay, but if one has the urge to create his own for his particular needs, then care should be taken to break it down nicely, so any other developer does not need to debug a collection of squeezed characters.

6.1.7 Dubious Abstraction

The smaller and more cohesive class interfaces are the better because they tend to degrade over time. They should provide a constant abstraction level. Function interfaces should be one level of abstraction below the operation defined in their name. Martin defines the above sentences as three separate code smells: *Functions Should Descend Only One Level of Abstraction*, *Code at Wrong Level of Abstraction*, *Choose Names at the Appropriate Level of Abstraction* [29]. He observed that people have trouble with it and often mix levels of abstraction. Steve Smith, in his course, uses the term "Inconsistent Abstraction Levels".

We like the smells in the granularized form presented by Martin, as they address the issue directly and specifically. The name *Inconsistent Abstraction Levels* still holds the idea. However, it might be misinterpreted by just recalling the meaning through its title, and we suspect that it might create a situation where somewhere out there, in at least one codebase, someone might win an argument with a non-inquisitive individual, thus leaving the abstraction levels consistent... but consistently off. We wish no one ever heard, "that is how it always has been, so it must continue to be done that way".

This is why we decided to rename it to *Dubious Abstraction*, directly addressing the potential causation of the smell, to think about the code that someone just wrote. Fowler says that "there is no way out of a misplaced abstraction, and it is one of the hardest things that software developers can do, and there is no quick fix when you get it wrong". *Dubious Abstraction* is supposed to provoke the question as soon as possible—"*Is it dubious?*" taking a second to think about the code at hand and then move on or immediately refactor if something seems fishy: Is `Instrument` really querying this message? Or is a *connection device* doing it?

6.1.8 Fallacious Comment

Comments differ from most other syntaxes available in programming languages; it is not executed. This might lead to situations where, after code rework, the comments around it were left intact and no longer true to what they described. First and foremost, this situation should not even happen, as good comments from the "*why*" Comment family are not susceptible to this situation. If the comment explained "**what**" was happening, then it will be relevant as long as the code it explains is intact. Of course, "*What*" *Comments* are a Code Smell themselves, and so is *Duplicated Code*.

This duplicated code might generally occur within docstrings in real-life scenarios, usually found in methods exposed to other users.

6.1.9 Fallacious Method Name

When we started to think of Code Smells from the comprehensibility perspective (of its lack of) as one of the critical factors, we were pretty intrigued that it was not yet thoroughly researched, or at least not when researching with a focus on "*Code Smell*" as a keyword. There is a grounded idea about code that is obfuscated from the point of *Obscured Intentions* or code without any explanation (*Magic Number*, *Uncommunicative Name*). We felt like there was a missing hole in the code that was intentionally too clever (*Clever Code*) or the code that contradicts itself. Fortunately, we have found a fantastic article supporting our thoughts and addressing some of what we had in mind under the name *Linguistic Antipatterns* [30]. We have included their subset of antipatterns under one code smell because listing them one by one would be too granular from a code perspective. The idea behind them can be summarized by one name: *Fallacious Method Name*.

This smell is caused by creating conflicting methods or functions regarding their functionality and naming. Over the years, programmers have developed connections between certain words and functionality that programmers should tie together. Going against logical expectations by, for example, creating a `getSomething` function that does not return is confusing and wrong.

6.1.10   Fate Over Action

This Code Smell is a replacement for the *Data Class* Code Smell, see Sect. 6.2.

6.1.11   Imperative Loops

Fowler has the feeling that *loops* are an outdated concept. He already mentioned them as an issue in his first edition of "Refactoring: Improving the design of existing code" book, although, at that time, there were no better alternatives [11]. Nowadays, languages provide an alternative, pipelines. Fowler, in the 2018 edition of his book, suggests that anachronistic loops should be replaced by pipeline operations such as `filter`, `map`, or `reduce` [22].

Indeed, loops can sometimes be hard to read and error-prone. This might be unconfirmed, but we doubt the existence of a programmer who has never had an `IndexError` at least once before. The recommended approach would be to avoid explicit iterator loops and use the `forEach` or `for-in`/`for-of`-like loop that take care of the indexing or `Stream` pipes. Still, one should consider whether he is not about to write *Clever Code* and check if there is a built-in function that will take care of the desired operation.

We would abstain from specifying all the loops as a code smell. Loops were always and probably will still be a fundamental part of programming. Modern languages offer very tidy approaches to loops and even things like List Comprehension in Haskell[4] or Python.[5] It is the indexation part that is the main problem of concern. Of course, so are long loops or loops with side effects, but these are just a part of *Long Method* or *Side Effects* code smells.

However, it is worth taking what is good from the functional languages (like the `streams` or immutability of the data) and implementing those as broad as possible and convenient to increase the reliability of the application.

6.1.12   Inconsistent Style

The same thing as in *Inconsistent Names* applies to the general formatting and code style used in the project. Browsing through the code should feel similar to reading a good article or a book, consistent and elegant. In the project, the code layout should

---

[4] Haskell List Comprehension
https://wiki.haskell.org/List_comprehension

[5] Python List Comprehension
https://docs.python.org/3/tutorial/datastructures.html

not be changed preferentially or randomly but should be uniform to not disturb the expected form of code in the following lines (see Listing 5).

Reading a novel where on each page, the reader is surprised by the new font ranging from *Times New Roman* through *Comic Sans* up to *Consolas* is distracting and could break out of the flow state.

Another example of an *Inconsistent Style* smell could be *Sequence Inconsistency*, in the order of parameters within classes or methods. Once defined, the order should remain in the group of all abstractions on that particular subject. If the order is not preserved, it leads to the unpleasant feeling of dissatisfaction after (if ever!) the mind realizes that it was again surprised wrong (see Listing 4). Depending on the specific case, it would still be only half the problem if the flipped parameters were different types (such as `string` and `int`). If the type was the same (e.g., `int`), this could unnoticeably lead to a significant hidden bug.

```
class Character:
    DAMAGE_BONUS: float

    def rangeAttack(
        self, enemy: Character, damage: int, extra_damage: int):
        total_damage = damage + extra_damage*self.DAMAGE_BONUS
        ...

    def meleeAttack(
        self, enemy: Character, extra_damage: int, damage: int):
        total_damage = damage + extra_damage*self.DAMAGE_BONUS
        ...

witcher.rangeAttack(skeleton, 300, 200)
witcher.meleeAttack(skeleton, 300, 200)  # hidden error
```

Listing 4: Inconsistent style code smell: sequence inconsistency

6.1.13 Side Effects

The first or second most essential functional programming principle (interchangeably, depending on how big we want to set the statement's tone) is that there be no side effects. Object-Oriented programming can apply this rule, too, with great benefits.

In a perfect scenario, when looking at a higher abstraction set of method calls, even an inexperienced bystander could tell what is happening more or less. The

```
my_first_function(
    arg1=1,
    arg2=2,
    arg3=3
)
my_second_function(arg1=1,
                   arg2=2,
                   arg3=3)
my_third_function(
    arg1=1, arg2=2, arg3=3
)
```

Listing 5: Inconsistent style code smell: different parameters linebreaks

code example (ref. Listing 6) appears to receive a player object identified by *Marcel Jerzyk*, sets its gold to zero, and manageable health status. That is great because one can make reasonable assumptions about the code... unless one cannot due to the side effects, which make these methods impure. By taking a closer look at the `set_gold(amount)` function, it turns out that, for some reason, this method triggers a dancing animation and resets the payday timer... of course, if one did not lose his trust yet, that the method names are representative of what they do.

The method and function names should tell what they do and do only what is anticipated to maximize code comprehension. We want to note that developers should fix this by removing the side effects to separate methods and triggering them individually, not violating the Single Responsibility Principle. Changing the name to `set_gold_and_reset_payday(amount)`, would create *Binary Operator In Name* Code Smell and another possible bad solution, `set_gold(amount: int, is_payday: bool)`, would cause *Flag Arguments* Code Smell.

6.1.14 Status Variable

Status Variables are mutable primitives that are initialized before an operation to store some information based on the operation and are later used as a switch for some action.

The *Status Variables* can be identified as a distinct code smell, although they are just a signal for five other code smells:

1. Clever Code
2. Imperative Loops
3. Afraid To Fail
4. Mutable Data
5. Special Case

```
@dataclass
class Player:
    gold: int
    job: Job

    def set_gold(self, amount: int):
        self.gold = amount
        self.trigger_animation(Animation.Dancing)
        self.job.reset_payday_timer()


marcel: Player = game.find_player(Marcel, Jerzyk)
marcel.set_gold(0)
marcel.set_health(Health.Decent)
```

Listing 6: Side effects code smell

They come in different types and forms, but common examples are the success:
bool = False-s before performing an operation block or i: int = 0 before
a loop statement. The code that has them increases in complexity by a lot and usually
for no particular reason because there is most likely a proper solution using first-class
functions. Sometimes, they clutter the code, demanding other methods or classes to
perform additional checks (*Special Case*) before execution, resulting in the *Required
Setup/Teardown Code*.

6.1.15 "What" Comment

This Code Smell is a replacement for the *Comment* Code Smell, see Sect. 6.2.

Recognizing all comments as Code Smells is controversial and raises several
different opinions. For this reason, we define a concrete subcategory of comments
named "*What*" *Comments* that clearly defines only these comments, which in the vast
majority will hint at something smells. The rule is simple: If a comment describes
what is happening in a particular code section, it is probably trying to mask some
other Code Smell.

This definition leaves room for the "*Why*" *Comments* that were already defined by
Wake in 2004 and were considered helpful. Wake also notes that comments that cite
non-obvious algorithms are also acceptable [49]. We wanted to mention that comments may have their places in a few more cases, such as extreme optimizations, note
discussion conclusions for future reference after a code review, or some additional
explanations in domain-specific knowledge.

As we have mentioned, the problem is that *Comments* are generally smelly. This is because they are a deodorant for other smells [11]. They may also quickly degrade with time and become another category of comments, *Fallacious Comments*, which are a rotten, misleading subcategory of *"What" Comments*.

6.2 Controversial Code Smells Replacements

Two particular Code Smells, which exist in the current literature and are the most controversial, have been replaced by two new ones.

First, the *Comments* Code Smell, whose legitimacy must be clarified, as not all comments are smelly [54]. As early as 2004, Wake rightly noticed this and separated from *Comments* those comments that answer the question "why" [49]. So analogously, "what" comments would be a good name for the concept addressing the vast majority of the reasons why the comments should be treated as a Code Smell. In this way, comments that try to classify, tag, group, or label something are caught by the concept title, and comments explaining why something was done in a specific way have their place.

The second Code Smell—*Data Class*—addresses the underlying problem with the Object-Oriented Programming principle, which says that data should stay close to the logic that operates on it. This is absolutely reasonable. Some practitioners doubt whether it should be considered a Code Smell [19]. We notice that the controversy may be due to the significant increase in the popularity of non-monolithic architecture, or more precisely, the domination of web applications. Services have to communicate somehow, and so-called DTOs (Data Transfer Objects) are used for this. It would be much better to serialize the DTO to a data class at the input/output as soon as possible. Such a class could already serve as a validator of the expected response (in the minimal case, expected fields or types), which would enforce the Fail Fast principle. This serialization also effectively deals with the alternative, which is potentially a long dictionary object, which in this case would constitute the *Primitive Obsession* Code Smell.

There are also cases of highly long configuration files. The lack of appropriate serialization into the expected data class makes it difficult to verify errors, which may arise only somewhere in the later processing stage of an application. We already have GraphQL, which, i.a., arose addressing this problem with the current form of communication over Rest API, or JSON Schema, to validate how the dictionary should be constructed. Developers should not be afraid of the data class itself, as nowadays, these data classes usually bring additional helpful verification. It can also be mentioned that data classes are a quick and direct tool for packing specific data into a petite abstraction to combat *Data Clump* Code Smell (sometimes programming languages offer for this purpose, e.g., Interface). However, we must not forget to keep the functionality close to the data, which the *Data Class* Code Smell currently indicates. We propose a *Fate over Action* Code Smell that preserves the current idea, which would signify that the problem is not with the data classes themselves but, instead, with situations where external classes or functions primarily manipulate the fields of an object.

7 Conclusions and Future Work

The code smells catalog (available as a self-building website (https://luzkan.github. io/smells/) on top of an open-source data repository (https://github.com/luzkan/ smells/) is a foundation for future research that solves the problem of unity and standardization. The simultaneous possibility of interactive information browsing may contribute to greater awareness of the topic discussed. However, we also point out that there are some limitations at work. We believe that despite our best efforts, from a statistical point of view, since the field is so substantial, we might be wrong in a few places. We have a specific capacity limit as a unit and cannot provide a perfectly completed tool. We hope that this very simplified form of adding and correcting information will mitigate this problem in the final settlement of this contribution. The data that is already included there should be discussed and analyzed in order to verify their correctness. It is possible to supplement the information with new data and metrics from the literature (e.g., predictors).

Sincerely, we hope that despite this, our work will prove to be an helpful little brick in this field of research. Consequently, any additional work on the Catalog, like finding and adding missing attributes, would be precious.

There should be discussions about whether we have done the right thing by addressing the controversial Code Smells and proposing their replacements (especially the *Data Class*, which is one of the two most extensively researched Code Smells in the current literature). It is also worth considering what the granularization of Smells should be. For example, we collapsed all the *Large Class* related smells into one (*Blob*, *Brain Class*, *Complex Class*, *God Class*, *God Object*, *Schizophrenic Class*, *Ice Berg Class*), wanting to minimize the number of concepts that are very close to each other, but perhaps maybe some of them are distinct enough to be an individual Bad Smell? Maybe the solution would be to list some of them as sub smells whose parent is *Large Class*? Maybe the fragmentation should only occur when viewed at the right angle (from the perspective of a specific hierarchy); e.g., when "sniffing" the smells from the Code Smell perspective, regarding abstraction problems, we would have just one smell, but from the Design Smell perspective there could be more specific smells?

In addition to the catalog itself, we hope this work will broaden all readers' awareness of the different Bad Smells hierarchies and allow more precise use of their specific terms, including Bad Smell and Antipattern. Hopefully, this facilitated access to all data will shed more light on those Code Smells that were overlooked in the research, either because they have not appeared in the literature yet, or because they have not appeared with the right keywords to be taken into account.

Software is all around us, what Martin reminds us of with his famous phrase "check how many computers you have on you right now". This work may contribute to the fact that the code that surrounds us universally will be written with greater awareness of quality regardless of the programming language. Even if the impact is calculated as a tiny percentage, it will still be significant for every software technology beneficiary—everybody.

References

1. Lacerda, G., Petrillo, F., Pimenta, M., Guéhéneuc, Y.G.: Code smells and refactoring: a tertiary systematic review of challenges and observations. J. Syst. Softw. **167**, 110610 (2020). https://doi.org/10.48550/arXiv.2004.10777
2. Cartaxo, B., Pinto, G., Soares, S.: Rapid Reviews in Software Engineering, pp. 357–384. Springer International Publishing, Cham (2020)
3. Mantyla, M., Vanhanen, J., Lassenius, C.: A taxonomy and an initial empirical study of bad smells in code. In: International Conference on Software Maintenance. ICSM 2003. Proceedings., IEEE, 2003, pp. 381–384 (2003). https://doi.org/10.1109/ICSM.2003.1235447
4. Mäntylä, M.V., Lassenius, C.: A Taxonomy for "Bad Code Smells" (2006). https://web.archive.org/web/20120111101436/, http://www.soberit.hut.fi/mmantyla/BadCodeSmellsTaxonomy.htm
5. Haque, M.S., Carver, J., Atkison, T.: Causes, impacts, and detection approaches of code smell: a survey. In: Proceedings of the ACMSE 2018 Conference, ACMSE '18, Association for Computing Machinery, pp. 1–8. New York, NY, USA (2018). https://doi.org/10.1145/3190645.3190697, https://doi.org/10.1145/3190645.3190697
6. Al-Shaaby, A., Aljamaan, H., Alshayeb, M.: Bad smell detection using machine learning techniques: a systematic literature review. Arabian J. Sci. Eng. **45**(4), 2341–2369 (2020). https://doi.org/10.1007/s13369-019-04311-w
7. Kaur, A.: A systematic literature review on empirical analysis of the relationship between code smells and software quality attributes. Arch. Comput. Methods Eng. **27**(4), 1267–1296 (2020). https://doi.org/10.1007/s11831-019-09348-6
8. Kaur, A., Jain, S., Goel, S., Dhiman, G.: Prioritization of code smells in object-oriented software: a review. Mater. Today: Proc. (2021). https://doi.org/10.1016/j.matpr.2020.11.218
9. Singh, S., Kaur, S.: A systematic literature review: refactoring for disclosing code smells in object oriented software. Ain Shams Eng. J. **9**(4), 2129–2151 (2018). https://doi.org/10.1016/j.asej.2017.03.002
10. Sabir, F., Palma, F., Rasool, G., Guéhéneuc, Y.-G., Moha, N.: A systematic literature review on the detection of smells and their evolution in object-oriented and service-oriented systems. Softw. Pract. Exp. **49**(1), 3–39 (2019). https://doi.org/10.1002/spe.2639
11. Martin Fowler, K.B.: Bad smells in code. Refactoring: Improving the Design of Existing Code. The Addison-Wesley Object Technology Series. Hit the shelves in mid-June of 1999
12. Alkharabsheh, K., Crespo, Y., Manso, E., Taboada, J.A.: Software design smell detection: a systematic mapping study. Softw. Qual. J. **27**(3), 1069–1148 (2019). https://doi.org/10.1007/s11219-018-9424-8
13. Fokaefs, M., Tsantalis, N., Chatzigeorgiou, A.: JDeodorant: identification and removal of feature envy bad smells. In: 2007 IEEE International Conference on Software Maintenance, IEEE, pp. 519–520 (2007). https://doi.org/10.1109/ICSM.2007.4362679
14. Lewowski, T., Madeyski, L.: How far are we from reproducible research on code smell detection? a systematic literature review. Inf. Softw. Technol. **144**, 106783 (2022). https://doi.org/10.1016/j.infsof.2021.106783, https://www.sciencedirect.com/science/article/pii/S095058492100224X
15. Moha, N., Gueheneuc, Y.-G., Duchien, L., Le Meur, A.-F.: DECOR: A method for the specification and detection of code and design smells. IEEE Trans. Softw. Eng. **36**(1), 20–36 (2010). https://doi.org/10.1109/TSE.2009.50

16. Santos, J.A.M., Rocha-Junior, J.B., Prates, L.C.L., do Nascimento, R.S., Freitas, M.F., de Mendonça, M.G.: A systematic review on the code smell effect. J. Syst. Softw. **144**, 450–477 (2018). https://doi.org/10.1016/j.jss.2018.07.035

17. Fontana, F.A., Ferme, V., Marino, A., Walter, B., Martenka, P.: Investigating the impact of code smells on system's quality: an empirical study on systems of different application domains. In: 2013 IEEE International Conference on Software Maintenance, pp. 260–269. IEEE (2013). https://doi.org/10.1109/ICSM.2013.37

18. Marticorena, R., López, C., Crespo, Y.: Extending a taxonomy of bad code smells with metrics. In: Proceedings of 7th International Workshop on Object-Oriented Reengineering (WOOR), p. 6. Citeseer (2006)

19. Singjai, A., Simhandl, G., Zdun, U.: On the practitioners' understanding of coupling smells—a grey literature based grounded-theory study. Inf. Softw. Technol. **134**, 106539 (2021). https://doi.org/10.1016/j.infsof.2021.106539

20. Zhang, M., Baddoo, N., Wernick, P., Hall, T.: Improving the precision of Fowler's definitions of bad smells. In: 2008 32nd Annual IEEE Software Engineering Workshop, pp. 161–166. IEEE (2008). https://doi.org/10.1109/SEW.2008.26

21. Kerievsky, J.: Refactoring to Patterns. Pearson Deutschland GmbH (2005)

22. Fowler, M.: Refactoring: Improving the Design of Existing Code. Addison-Wesley Professional (2018)

23. McConnell, S.: Code Complete. Pearson Education (2004)

24. Smith, S.: Refactoring Fundamentals (accessed: 11.11.2021)) (2013). https://www.pluralsight.com/courses/refactoring-fundamentals

25. Garcia, J., Popescu, D., Edwards, G., Medvidovic, N.: Toward a catalogue of architectural bad smells. In: International Conference on the Quality of Software Architectures, pp. 146–162. Springer, Berlin (2009). https://doi.org/10.1007/978-3-642-02351-4_10

26. Suryanarayana, G., Samarthyam, G., Sharma, T.: Refactoring for Software Design Smells. Managing Technical Debt, Morgan Kaufmann (2014)

27. Sharma, T., Fragkoulis, M., Spinellis, D.: House of cards: code smells in open-source C# repositories. In: ACM/IEEE International Symposium on Empirical Software Engineering and Measurement (ESEM), pp. 424–429. IEEE (2017)

28. Jabrayilzade, E., Gürkan, O., Tüzün, E.: Towards a taxonomy of inline code comment smells. In: 2021 IEEE 21st International Working Conference on Source Code Analysis and Manipulation (SCAM), pp. 131–135 (2021). https://doi.org/10.1109/SCAM52516.2021.00024

29. Martin, R.C.: Clean Code: A Handbook of Agile Software Craftsmanship. Pearson Education (2008)

30. Arnaoudova, V., Di Penta, M., Antoniol, G., Guéhéneuc, Y.-G.: A new family of software anti-patterns: linguistic anti-patterns. In: 2013 17th European Conference on Software Maintenance and Reengineering, pp. 187–196. IEEE (2013). https://doi.org/10.1109/CSMR.2013.28

31. Vetro, A., Ardito, L., Morisio, M.: Definition, implementation and validation of energy code smells: an exploratory study on an embedded system. None (2013)

32. Smith, C.U., Williams, L.G.: New software performance antipatterns: more ways to shoot yourself in the foot. In: International CMG Conference, pp. 667–674. Citeseer (2002)

33. Smith, C.U., Williams, L.G.: More new software antipatterns: even more ways to shoot yourself in the foot. In: Computer Measurement Group Conference, pp. 717–725. Citeseer (2003)

34. Tufano, M., Palomba, F., Bavota, G., Di Penta, M., Oliveto, R., De Lucia, A., Poshyvanyk, D.: An empirical investigation into the nature of test smells. In: Proceedings of the 31st IEEE/ACM International Conference on Automated Software Engineering, pp. 4–15 (2016). https://doi.org/10.1145/2970276.2970340

35. Arendt, T., Taentzer, G.: UML Model Smells and Model Refactorings in Early Software Development Phases, Universitat Marburg (2010). https://doi.org/10.1002/smr.2154

36. Doğan, E., Tüzün, E.: Towards a taxonomy of code review smells. Inf. Softw. Technol. **142**, 106737 (2022). https://doi.org/10.1016/j.infsof.2021.106737

37. Palomba, F., Tamburri, D.A., Serebrenik, A., Zaidman, A., Arcelli Fontana, F., Oliveto, R.: Poster: how do community smells influence code smells? In: 2018 IEEE/ACM 40th International Conference on Software Engineering: Companion (ICSE-Companion), pp. 240–241 (2018)

38. Qamar, K., Sülün, E., Tüzün, E.: Towards a taxonomy of bug tracking process smells: a quantitative analysis. In: 2021 47th Euromicro Conference on Software Engineering and Advanced Applications (SEAA), pp. 138–147. IEEE (2021). https://doi.org/10.1109/SEAA53835.2021.00026

39. Sharma, T., Fragkoulis, M., Spinellis, D.: Does your configuration code smell? In: 2016 IEEE/ACM 13th Working Conference on Mining Software Repositories (MSR), pp. 189–200. IEEE (2016)

40. Sharma, T.: Presentation Smells: How Not to Prepare Your Conference Presentation. Tushar Sharma Website (2016). https://www.tusharma.in/presentation-smells.html

41. Hermans, F., Pinzger, M., Van Deursen, A.: Detecting and visualizing inter-worksheet smells in spreadsheets. In: 2012 34th International Conference on Software Engineering (ICSE), pp. 441–451. IEEE (2012). https://doi.org/10.1109/ICSE.2012.6227171

42. Karwin, B.: SQL Antipatterns: Avoiding the Pitfalls of Database Programming. Pragmatic Bookshelf (2010)

43. Almeida, D., Campos, J.C., Saraiva, J., Silva, J.C.: Towards a catalog of usability smells. In: Proceedings of the 30th Annual ACM Symposium on Applied Computing, pp. 175–181 (2015). https://doi.org/10.1145/2695664.2695670

44. Carette, A., Younes, M.A.A., Hecht, G., Moha, N., Rouvoy, R.: Investigating the energy impact of Android smells. In: 2017 IEEE 24th International Conference on Software Analysis, Evolution and Reengineering (SANER), pp. 115–126. IEEE (2017). https://doi.org/10.1109/SANER.2017.7884614

45. Ghafari, M., Gadient, P., Nierstrasz, O.: Security smells in android. In: 2017 IEEE 17th International Working Conference on Source Code Analysis and Manipulation (SCAM), pp. 121–130. IEEE (2017). https://doi.org/10.1109/SCAM.2017.24

46. Stijlaart, M., Zaytsev, V.: Towards a taxonomy of grammar smells. In: Proceedings of the 10th ACM SIGPLAN International Conference on Software Language Engineering, pp. 43–54 (2017). https://doi.org/10.1145/3136014.3136035

47. Mantyla, M.: Bad Smells in Software—a Taxonomy and an Empirical Study, Ph.D. thesis, Ph.D. thesis, Helsinki University of Technology (2003)

48. Atwood, J.: Code Smells. Jeff Atwood Website (2006). https://blog.codinghorror.com/code-smells/

49. Wake, W.C.: Refactoring Workbook, 1st edn. Addison-Wesley Professional (2004)

50. Yamashita, A., Moonen, L.: Do developers care about code smells? an exploratory survey. In: 2013 20th Working Conference on Reverse Engineering (WCRE), pp. 242–251. IEEE (2013). https://doi.org/10.1109/WCRE.2013.6671299

51. De Castella, K., Byrne, D., Covington, M.: Unmotivated or motivated to fail? a cross-cultural study of achievement motivation, fear of failure, and student disengagement. J. Educ. Psychol. **105**(3), 861 (2013). https://doi.org/10.1037/a0032464

52. Rowa, K.: Atychiphobia (Fear of Failure), ABC-CLIO (2015)

53. Chapman, C., Wang, P., Stolee K.T.: Exploring regular expression comprehension. In: 2017 32nd IEEE/ACM International Conference on Automated Software Engineering (ASE), IEEE, pp. 405–416 (2017). https://doi.org/10.1109/ASE.2017.8115653 https://doi.org/10.1109/ASE.2017.8115653

54. Fishtoaster, https://softwareengineering.stackexchange.com/questions/1/comments-are-a-code-smellComments are a code smell, Software Engineering—Stack Exchange, (accessed: 06.04.2022)) (2010). https://softwareengineering.stackexchange.com/questions/1/comments-are-a-code-smell

Modeling of Information System to Determining the Degree of Coincidence in Text for Higher Education Institutions

Anna Shilinh⊙

Abstract The aim of the article is to model the activity of the information system to determine the degree of coincidence in the texts, which are the results of scientific achievements of consumers of educational services in higher education institutions. In this paper, modeling the information system to determine the degree of coincidence in the text is based on the organization of the process of detecting plagiarism. However, it should be noted that the scope of this system, in addition to the direct consumers of educational services, which targets this system, also includes any other educational institutions and organizations that need such verification. The article models the processes of the information system for determining the degree of coincidence of texts based on CASE-technologies. In particular, the class diagram shows a static representation of the structure of the information system to determine the degree of coincidence in the texts. The interaction between the actors and the entities of the information system is depicted using a diagram of use cases. Descriptions of the behavior of the information system at the level of individual objects that exchange messages are shown in the diagram of cooperation. The components of the information system and the relationships between them are presented in the component diagram, and computing nodes during program operation, components, and objects running on these nodes in the proposed system are shown in the deployment diagram. Modeling of information system processes to determine the degree of coincidence in the texts is the basis for developing the architecture of the information system. Evaluation of software development, which is based on the indicators of the COCOMO evaluation system, is also described in this article. An important place in this study is the identification of risks in the process of developing an appropriate information system based on the modeling. The use of the proposed information system to determine the degree of coincidence of texts allows to optimize the process of plagiarism of scientific results of consumers of educational services in order to improve the quality of analysis of materials used and presentations of their scientific results.

A. Shilinh (✉)
Department of Social Communication and Information Activities, Lviv Polytechnic National University, 12 S. Bandery Str, Lviv 79000, Ukraine
e-mail: anna.y.shilinh@lpnu.ua

N. Kryvinska et al. (eds.), *Developments in Information and Knowledge Management Systems for Business Applications*, Studies in Systems, Decision and Control 462,
https://doi.org/10.1007/978-3-031-25695-0_25

Keywords Information system · CASE-technologies · Unique text · Plagiarism · Higher education institution · COCOMO · Risks

1 Introduction

The development of information technology has greatly simplified the process of information exchange. That is why the problem of data identification in the process of analyzing information resources for the formation of unique content is urgent. In addition, the problem of determining the uniqueness of data in the information space is important in order to eliminate the phenomenon of plagiarism in the process of identification analysis of available database resources.

Higher education institutions (HEIs) are the focus of formation and development of consumers of educational services. In the process of learning, students are usually invited to conduct research that directly relates to the application of knowledge gained in the chosen specialty. Other scientists inextricably link this process to the analysis of previous experience in this field. Although this process is only the basis of new research, it sometimes becomes the main research tool and makes it impossible for researchers to obtain new results.

Such issues are relevant, above all, for higher education and research institutions.

Anti-plagiarism examination in higher education institutions includes a set of measures to:

- ensuring the preservation of copyright by conducting research on plagiarism before the procedure for their protection;
- formation and development of a database and electronic repository of basic scientific works as a source of free access to scientific materials to promote the prestige of higher education institutions;
- improving the quality of education, which is based on increasing the productivity of the educational process and accelerating the transfer of knowledge.

Existing on-line systems for determining the degree of coincidence of texts as UniCheck.com are not free for higher education institutions. Other of that are freely available on the Internet as Antiplagiat.com, contentwatch.com, do not guarantee to detect plagiarism in texts that are part of the information database of research in a particular specialty of HEI, which is not publicly available in cyberspace.

Verification of texts for uniqueness during the presentation of personal scientific results for consumers of educational services in higher education institutions is one of the stages in the formation of their scientific thought. Therefore, the development of a system for determining the coincidence of texts allows simplifying and improving the process of monitoring academic integrity in higher education institutions. This determines the relevance of this article.

2 Related Works

Today there are several areas of study of the process of modeling information systems, in particular for the effective operation of higher education institutions.

Analysis of the process of modeling information systems, as well as guidelines and rules for the development, presentation and understanding of the architecture of relevant information systems are considered in Refs. [1–3]. In particular, it is determined that the development of information systems architecture should consist of five stages, such as planning and design phase, operational analysis phase, requirements analysis phase, function analysis phase, physical synthesis phase. In Refs. [4–6], recommendations for the development and improvement of information technology and support systems for organizational goals to increase their competitiveness.

The use of cloud computing capabilities and their services in the development of information systems and services is also relevant today. The benefits of using the cloud to support real-time service systems with the Salesforce platform are the subject of research [7].

Analysis of evaluation methods in software development based on COCOMO evaluation indicators is the subject of researches [8, 9]. In particular in the study [10] was alculated the amount of effort focused on agents, and the creation of appropriate software.

Since the vast majority of free economic education today can be considered as labor-intensive and human organizations, the modeling of information systems as important tools to support several different institutional educational processes and provide users with relevant data are discussed in [11–13]. In particular, the development of an information system for forecasting the contingent of students for HEI is described in [14].

The reference model for higher education institutions on the way to a single information system is described in [15, 16], and aims to combine disparate information systems and applications for effective and competitive existence of free education in the market of educational services.

The principles of functioning of antiplagiarism systems are considered in studies [17, 18]. In particular, works [19, 20] was identify their advantages and disadvantages.

Modeling methods, which are useful tool support for various modeling tasks during the analysis and design of information systems, are the subject of research [21].

The use of semantic web technologies and multilingual thesauri to access knowledge-based information resources and resource segmentation through fuzzy linguistic preferences and fuzzy clustering is the subject of researches [22–25]. The use of linguistic templates in various texts is the subject of researches [26, 27].

However, none of the studies considers the possibility of developing an information system to determine the degree of coincidence in texts for higher education institutions, which can function as an additional stage of verification or as a stand-alone system to establish the uniqueness of research within a particular scientific work. Therefore, the main task of this article is to model the processes of the information system to determine the degree of coincidence in the texts for the timely detection of plagiarism in the scientific achievements of students for effective development and construction of research in a particular specialty. The aim of the work is model the processes of the information system for determining the degree of coincidence of texts for higher education institutions based on CASE-technologies.

3 Description of Requirements for Information System Processes to Determine the Degree of Coincidence in the Text

The information system for determining the degree of coincidence in the texts is based on the interaction of the main actors and quality control of works at all stages: from writing the work by the author to its publication, evaluation or defense.

One of the modern ways to combat academic plagiarism is to detect and diagnose it with the help of computer programs on the Internet and the Repository. When choosing the recommended programs, it is necessary to pay attention to the fact that they set the percentage of unique text, fully support different foreign languages, are easy to install and are licensed and (or) freely available. The search engine contains a set of programs that use special algorithms for analyzing the content of websites across the Internet, where you can type in the appropriate line of keywords and get a lot of links to resources with the necessary information. It consists of three main parts: work, index and query handler. Crawler (Bot, Robot) is a program that opens web pages, reads (indexes) their content and then follows the links found on this page. The robot returns at regular intervals (for example, monthly) and indexes the page again. Everything that the robot finds and reads falls into the index of the aircraft. Indexes are repositories of information where copies of the textual component of all visited and indexed web-pages are stored. Query Processor is a program that, in accordance with the submitted request, analyzes the indexes of the aircraft in search of information of interest to the user and gives him in descending order of relevance (relevance of the query) found documents.

The purpose of the system for determining the degree of agreement in the texts is to test the texts for uniqueness, which will simplify and at the same time improve the process of academic integrity control. The designed system should be based on the interaction of the main actors and quality control of works at all stages, from writing the work by the author to its publication, evaluation or defense.

The object of automation is the verification of texts for uniqueness, further evaluation of the work and the algorithm of actions depending on the degree of uniqueness of the text. The object of automation has a hierarchical and territorially branched structure.

The users of the system are students and teachers of higher education institutions. The rights of access to information and functions of the system are set by authorized employees of the Free Economic Zone (system administrators). The performer provides a clear delineation of access to input, editing and viewing of information.

The system provides automation of the following functions:

- Insert text, download file: the program must have a textbox to enter text from the keyboard. The entered text is read and saved for further processing. Or the user must select the file to be opened. The program opens this file, reads the data and saves it for further processing. The program must support several different popular text formats. (.txt,.pdf,.docx, etc.);
- Check the text: the program should break the text of the document into smaller parts, which need to be checked by algorithms of shingles, lexical matches and pseudo-unification;
- Show the result of the test: the program should show the percentage of unique text, provide links to sources from which the text was borrowed, and show the percentage of matches with each source;
- Create a report: the test result should be saved in a file of one of three formats—.txt,.pdf,.docx;
- Edit validation parameters: the user should be able to exclude certain sites, fixed expressions, etc. from the validation;
- Register the user: the user must be able to register in the program, and user data is added to the database;
- Check spelling: the program should check the spelling of the entered text, indicating the errors that occurred, and the corresponding sentence, which contains the wrong word;
- Make an SEO-analysis of the text: the program should determine the level of "water" in the text, its dullness;
- Save the test result and report: the program must preserve the percentage of unique text and a complete report on its verification in the database.
- Upload text to your own database: The text must be uploaded to the database. The title, topic, author, supervisor, percentage of uniqueness and date of the test (if the work has been tested), the person responsible for the test, shall be indicated. In the database, the test results table and the text table must be linked;
- View information about the work: when you click on the work should display its author, topic, title, supervisor, test results, evaluation (if the work was evaluated), the percentage of uniqueness;
- Change the status of the job: job data can be entered into the database without downloading it. The work is assigned the status of "In the process of writing", "Preliminary evaluation by the supervisor", "On revision", "Verification of uniqueness", "Verified", "Evaluated", "Defense", etc.;
- View the list of recently checked works: the list of recently checked works should be displayed in the main window of the program. By clicking on the corresponding job, the user can open it or open data about it;

- Open a list of all inspections: all work and inspection results should be displayed in a table.

4 Modeling of Information System Activity to Determine the Degree of Coincidence in the Text Based on CASE-Technologies

Today, information systems design technologies are based on CASE tools. That is why the model of information system functioning for determining the degree of coincidence in texts is presented with the help of UML-diagrams of classes, use cases, cooperation, components and deployment.

Static representation of the structure of the system to determine the degree of coincidence of texts for HEI is presented in the class diagram (Fig. 1).

There are nine main classes and one class-essence: Author, ScientificAdvisor, Text, ResponsibleforCheck, Department, HeadofDepartment, Repository (Repository), SoftwaretoCheckTexts (Text Verification Software), Report and ResultofCheck (Verification Result).

The ResultofCheck entity class is used to store information about the result of text checking in the system after the destruction of objects of this class. An explanatory

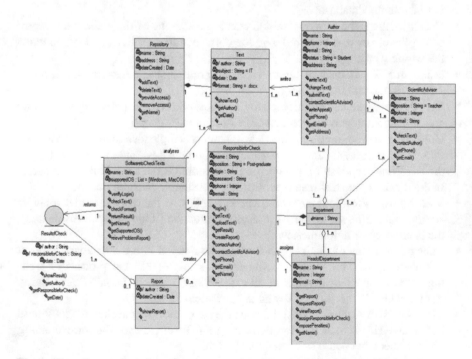

Fig. 1 Class diagram for information system to determine the degree of coincidence in the text

text is provided in the documentation section: "Keeps information about the results of checking the text." This class has three attributes with private visibility quantifier (author: String, responsibleforCheck: String and date: Date) and 4 visibility quantifier operations public (showResult) (), getAuthor (): String, getResponsibleforCheck (): String and getDate (): Date, last three—to access private attributes). The Derived property is specified for the author and responsibleforCheck attributes. ResultofCheck is combined with the SoftwaretoCheckTexts and Report classes of association and aggregation relationships, respectively.

The SoftwaretoCheckTexts class has two private attributes (name: String and supportedOS: List = ['Windows', 'MacOS]), nine operations with public quantifier public (verifyLogin (): Bool, checkText (), checkFormat (), returnResult (), getName (), getSupportedOS (), the last two—to access private attributes) and associated with the classes ResultofCheck, ResponsibleforCheck and Text association relationships. RelieveProblemReport—get a message about the policy, denyTextChecking (if there are more than five attempts), sendWarning (inform the user about something, for example, that he has tried).

The Report class has two attributes with private visibility quantifier (author: String, dateCreated: Date) and three operations with public visibility quantifier (the first is used to view the report itself—showReport (), the last two—to access private attributes: getAuthor (), getDate ()). The Derived property is specified for the author attribute. The aggregation ratio indicates that the report consists of the results of inspections of works, and after deleting the report, these results are not deleted. The Report class is also related to the association relationship with the ResponsibleforCheck class.

The ResponsibleforCheck class has six private attributes (name, contact information, login and password) and eleven operations, four of which are for accessing private data. This class describes the person responsible for testing. The relationship of the composition that connects this class with the Department class indicates that after the dissolution of the department, this person will lose his status and responsibilities. ResponsibleforCheck is also associated with the HeadofDepartment, Report, and SoftwaretoCheckTexts classes of association relationships.

The HeadofDepartment class has three private attributes (name, contact information) and eight operations, three of which are for accessing attributes. This class describes the head of the department and his main operations are reviewing the report, appointing a person responsible for verification and imposing sanctions on the unscrupulous author of the work. The attitude of aggregation indicates that when a department is disbanded, this person will retain his or her responsibilities for some time or may hold a similar position in another department. This class is also related to the ResponsibleforCheck class of association relations (the head of the department appoints a certain person responsible for the inspection).

The Department class has only one public attribute is the name. It is related to the ResponsibleforCheck and HeadofDepartment classes of composition and aggregation relations, respectively, and to the Author, ScientificAdvisor classes of aggregation relations (after the department is disbanded, the essences of these classes do not lose their status).

The ScientificAdvisor class describes the supervisor. This class has four private attributes (name, position, contact information) and six operations, four of which are for accessing private data. By default, the position attribute is teacher (position: String = 'Teacher'). This class is related to the Author and Department classes by association and aggregation relationships, respectively.

The Author class describes the author of the work and has five private attributes and ten public transactions (five of which are for data access). By default, the status attribute is set to Student (status: String = 'Student'). The main operations for this class are writeText () and submitText (). The author is related to the Department and Text classes of aggregation and association relationships, respectively.

The Text class has four private attributes (author, date, subject and format, the latter is set to .docx by default) and five operations, the main one being showText (). The Derived property is specified for the author attribute. The ratio of the composition indicates that deleting the repository deletes all work that has been uploaded to it. Also, this class is combined with the Author class by the association relationship.

The Repository class has three attributes with a private visibility quantifier (name, address/link, and creation date) and six methods with a public visibility quantifier (addText ()). Repository is related to the Text class by the composition ratio.

The interaction between the actors and the entities of the system is depicted using a diagram of the use of the information system to determine the degree of coincidence in the text (Fig. 2).

The main actors for the diagram of the use of the system for determining the degree of coincidence of texts for the HEI are the following: Supervisor, Author, Person in charge of verification, Head of the Department and software for verification of texts. The diagram also identifies the actors Teacher, Degree Applicant, and Graduate Student, and the generalization ratio indicates that they are partial cases of the actor Author. The latter is combined by association with the "Provide Preliminary Evaluation" (linked to "Write a Text" and "Require" on Inclusion") and "Respond to Review"(linked to" Apply for further elaboration", "File an appeal"and" Refine the work"with expansion relations). The supervisor is related to the options for "Get the result of an automated test" (related to "Impose sanctions on the author" with the extension ratio) and "Provide feedback on the preliminary evaluation of the work". The person responsible for verification is associated with the following uses: validation of texts), "Report the result" (related to the options "Notify the author" and "Notify the supervisor" in terms of generalization) and "Submit a report on the results of inspections" (includes "Collect data" and "Report" and related to the actor Head of the Department). Also, the actor Responsible for the inspection is related to the note "Cannot be appointed responsible for the inspection of interested persons and authors of other works".

Descriptions of the behavior of the system at the level of individual objects that exchange messages for the system for determining the degree of coincidence of texts are shown in the diagram of cooperation (Fig. 3).

An anonymous object created by the Author class has two recursive connections and two association connections, one of which (3: provide a work check) is

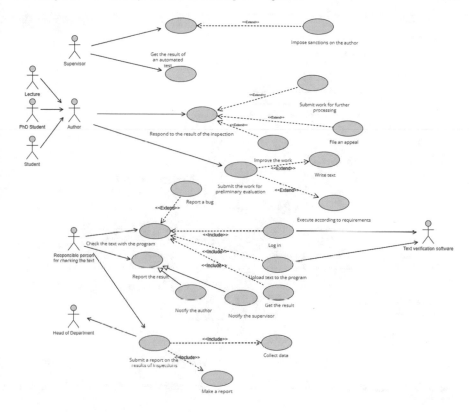

Fig. 2 Diagram of usage options for information system to determine the degree of coincidence in the text

synchronous, which means that the author of the work must wait for the results of its preliminary inspection.

An anonymous object created by the Supervisor class has one recursive and two association links.

An anonymous object created by the Verification Person class acts as the initiator of the software request and has one recursive and five association links.

An anonymous object created by a text validation software class acts as a query handler and returns validation results. It is connected by three associations with other objects.

The anonymous object created by the Head of the Department class only receives reports, and is therefore linked by one association association with the object of the Person responsible for verification class.

The components of the information system and the relationships between them are presented in the diagram of components (Fig. 4).

The User Desktop component is a type of program window that depends on the main.exe component (the main executable file) and the User & Results Database (the user database that also contains the results of previous checks).

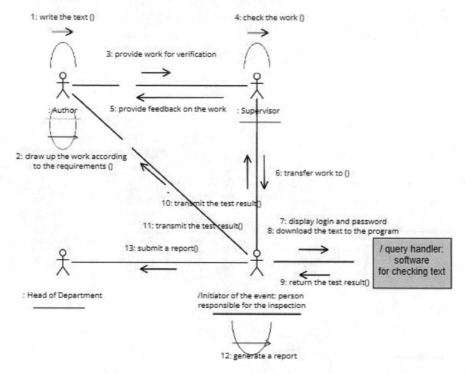

Fig. 3 Diagram of cooperation for information system to determine the degree of coincidence in the text

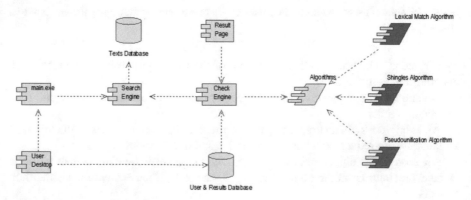

Fig. 4 Diagram of components for the information system to determine the degree of coincidence in the text

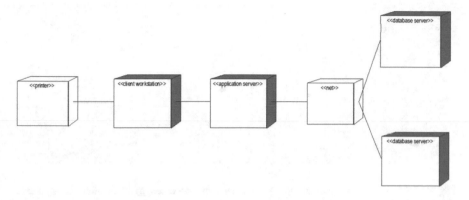

Fig. 5 Deployment diagram for the information system to determining the degree of coincidence in the text

The main.exe component depends on the Search Engine component (search for similar texts or parts of them), which in turn depends on the Texts Database.

The Result Page component and the User & Results Database component depend on the Check Engine component. The latter depends on the Algorithms task specification, which includes Lexical Match Algorithm, Shingles Algorithm, and Pseudounification Algorithm.

The deployment diagram (Fig. 5) shows the computing nodes during program operation, components, and objects that run on these nodes in the proposed system.

The deployment diagram identifies three processors: Computer (stereotype "client workstation"), program for checking texts (stereotype "application server"), user database (stereotype "database server") and cloud storage. (stereotype <<database server>>).

There are also two devices in this chart—Printer (stereotype <<printer>>) and Secure Network (stereotype <<net>>).

5 Planning of Works for Development of the Project of Information System of Information System to Definition of Degree of Coincidence in the Text on the Basis of Modeling

To ensure the effective organization of information system processes to determine the degree of coincidence in the text based on modeling using CASE-tools, Gantt chart is constructed (Fig. 6).

This allows you to visualize information and define clear deadlines for planning, helps in resource management, motivates the team. But, on the other hand, the Gantt chart requires periodic updating of data, if used incorrectly it is not possible to see

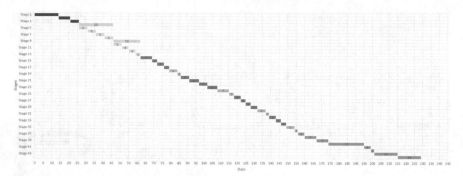

Fig. 6 Gantt chart for planning information system processes to determine the degree of coincidence in the text

the full picture of the progress of the project, setting up or editing can take a long time.

6 Evaluation of Project Implementation Based on Information System Modeling to Determine the Degree of Coincidence in the Text

According to the project implementation, the program code will contain 3000 lines of code. The clarifying factors of the work, the regulatory factor of the work is calculated below is shown in Table 1.

6.1 Estimation of the Size of the Final Software Product

The main functionality of the software system includes the following items: work with text (download, insert, open files), check the text for uniqueness, additional checks (SEO, spelling, correct design, etc.), work with a database of texts, work with user database, integration into the virtual learning environment, additional functions (generating and sending an error report, checking the excess of the number of available checks, etc.).

Thus, the size of the code in LOC (Lines of Code) for this system will be up to 50 thousand, depending on the complexity of the basic validation algorithm, the number of additional text checks and the complexity of implementing the program as a subsystem to the virtual learning environment.

Table 1 Clarifying factors of work for the development of an information system to determine the degree of coincidence in the text

Identifier	Clarifying factor of works	Rating	The value of the factor
Product features			
RELY	Reliability is required	High	1.15
DATA	Database size	Very high	1.16
CPLX	The complexity of the product	Average	1.0
Hardware characteristics			
TIME	Execution time limit	Average	1.0
STOR	Limiting the amount of main memory	High	1.21
VIRT	Variability of the virtual machine	High	1.15
TURN	Recovery time	Average	1.0
Staff characteristics			
ACAP	Analyst skills	High	0.86
AEXP	Development experience	Average	1.0
PCAP	Abilities of a programmer	High	0.86
VEXP	Virtual machine experience	Average	1.0
LEXP	Experience in using a programming language	High	0.95
Project characteristics			
MODP	Use of modern development methods	High	0.91
TOOL	Use of software development tools	Average	1.0
SCED	Requirements for compliance with the development schedule	Average	1.0

6.2 Project Evaluation According to the COCOMO Intermediate Model

To date for. Appraisal software costs use several different models. One of the most popular, open, and well-documented software valuation models is the Constructive COst MODel. The evolution of COCOMO reflects how the software economy has evolved.

Key points:

- The history of the COCOMO model allows us to see the evolution of the priorities of the software economy.
- The original COCOMO model was well adapted to estimate the cost of traditional software projects in the 1980s.
- The Ada COCOMO model improved the original version, in particular, through parameterized exponents, which reflected the improvements in the modern process and their impact on economies of scaleто.

Table 2 Estimation of time required for implementation works the main functional development of the information system to determine the degree of coincidence in the text, according to the COCOMO model

Type of work	KLOC	Time (months)	Time (days)
Functional programming: insert text, download file	5	7.33	220
Functional programming: edit validation parameters	1	3.9	177
Functional programming: show the test result	5	7.33	220
Functional programming: register user	5	3.9	117
Functional programming: check spelling	3	6	180
Functional programming: do SEO-analysis of the text	2	5.12	153
Functional programming: save the test result and report	2	5.12	153
Functional programming: upload text to your own database	5	3.9	177
Functional programming: view work information	1	3.9	117
Functional programming: change the status of the work	1	3.9	117
Functional programming: see a list of recently tested works	2	5.12	153
Functional programming: open a list of all checks	3	6	180
Function programming: send an error report	1	7.33	200

Estimation of time required for implementation works the main functional development of the information system to determine the degree of coincidence in the text, according to the COCOMO model are presented in Table 2.

Thus, the main part of the program code are algorithms for verification, interaction with the virtual learning environment and databases, graphical interface of the subsystem. The settings of the graphical interface are also taken into account: interactive text opening, text marking according to the obtained results, direct transition to sources, etc., and work with the database.

7   Risks of Information System Development Based on the Conducted Modeling

There are some risks in designing and developing an appropriate information system based on the simulation. Namely:

- Planning risks are risks that may be associated with a lack of skills in scheduling work deadlines.

- Requirements inconsistency risk (specification decomposition) is the risk associated with the detection of inconsistencies in the customer's requirements at the stage of programming or project integration.
- The risk of new requirements—arises in the process of software development, when there are more and more new requirements that delay the timing and evaluation of specific tasks.
- Risks of copyright infringement may arise when developers use someone else's source code, algorithm, or library that is protected by copyright law but has not been purchased or used by the author.
- Project management risks are risks related to the lack of project management skills of the project manager, as well as the lack of interest or motivation in him.
- Risks of poor interaction between the customer and the contractor are the risks associated with the lack of communication between the contractor and the customer or their representatives. Insufficient discussion of tasks or architecture can negatively affect the software being developed.
- Risks associated with the project manager's lack of awareness of the exact status of the project are a type of risk associated with a lack of feedback. It occurs when the project manager has not built the workflow in such a way as to monitor the progress of the project at all stages.
- Risks of incorrectly defined system requirements are risks when at the very beginning of the project the characteristics of the target system for which the software is developed were incorrectly formulated: software environment (operating system, installed components, services, etc.) or hardware requirements parts (processor frequency, hard disk capacity, amount of RAM, etc.).
- Risks of financial constraints—may arise due to the fault of the manager who planned the project budget, and for other reasons.
- Risks of changes in market conditions due to changes in the economic situation in the market during planning. At the same time, factors relevant at the time of planning could be established, and their change was not taken into account.
- Risks of using unstable technologies are risks associated with the use of new technologies that have not yet been tested in production or other projects.
- The risk of low productivity is due to the duration of the project. This at the very beginning of the project creates a great loss of time, which will be difficult to make up. It is necessary to either postpone the deadlines or work in a more dynamic mode at later stages of the project.
- Risk of change of employees—when key employees who have the most information leave the project.
- Risks of theft of source code arise when developers leave the company, take with them a project developed by them and, after slightly modifying the source code, can sell it or use it in other projects, such as competitors.
- Currency risks are risks associated with the possible occurrence of losses or additional income due to adverse or favorable changes in foreign exchange rates.

8 Results

Analysis of rankings among search queries, according to the service https://trends. google.com.ua/ in 2019–2021 (Fig. 7) shows trends of increasing interest over time terms related to the detection process the degree of uniqueness of the text.

Indicators of interest over time show the popularity of the search term relative to the highest value for a certain period of time. Moreover, 100 is the peak of the term's popularity; 50 means that the popularity of the term is twice less; 0 means that there is a lack of data on this term.

Thus, there is a tendency to increase interest over time in search queries and related search topics. This is due to the need to determine the degree of coincidence in the texts. Because it increases the ability to control the academic integrity of consumers of educational services. Thus, there is a need to check the texts for uniqueness and identify similarities in the scientific texts of consumers of educational services.

Analysis of the results of comparing the popularity of well-known online services for re-checking texts for uniqueness based on data from the service https://www.sim ilarweb.com/, shown in Fig. 8.

In particular, the online service UniCheck.com is not free, but allows you to initially check part of the text. This powerful service is usually the main text-checking tool for higher education institutions. Ordinary users of online services prefer free counterparts (for example, antiplagiat.com. Contentwatch.com), which led to their Global Rank.

But the reliability and quality of text verification to identify matches in the texts is an important factor in choosing a verification service. The analysis of the rate of attendance of these online services is shown in Fig. 9.

As a significant number of educational institutions use the resources of the UniCheck.com service to test the scientific work of consumers of educational services, the attendance rate of this online text matching service is much higher than that of free similar online services.

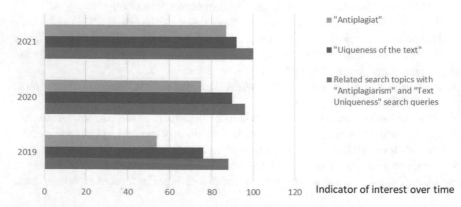

Fig. 7 Indicators of interest over time of popular search queries according to the service https://tre nds.google.com.ua/ for the period 2019–2021

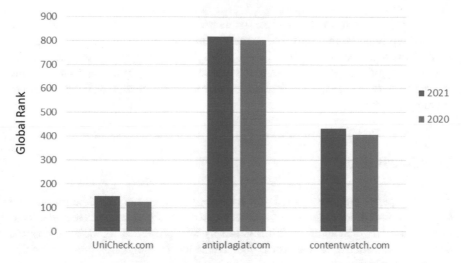

Fig. 8 Indicators of popularity of online services for checking texts for uniqueness based on data from the service https://www.similarweb.com

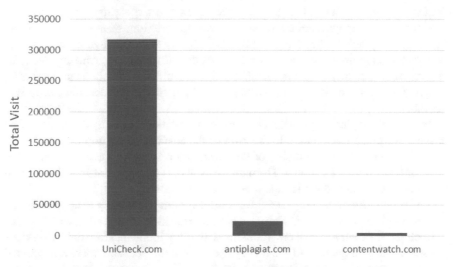

Fig. 9 Indicator of visits to online services for checking texts for uniqueness based on data from the service https://www.similarweb.com

9 Possibilities of Further Research

Modeling the activities of the information system to determine the degree of coincidence of texts for higher education institutions is the basis for developing the architecture of the relevant information system.

The proposed information system for determining the degree of coincidence in the text will improve the control of academic integrity. It can be created taking into account the main factors that contribute to the spread of plagiarism, and the planned functionality of the software system. An alternative is to improve and increase the functionality of the created curriculum for its further application or integration into existing systems.

10 Conclusions

Thus, this paper proposes a model of an information system for determining the degree of coincidence in texts for the timely detection of plagiarism in higher education institutions. Modeling of information system processes for determining the degree of coincidence in texts for higher education institutions is presented using CASE-technologies.

In particular, the class diagram presents a static representation of the structure of the system for determining the degree of coincidence of texts. The interaction diagram shows the interaction between the actors and the entities of the information system. The diagram of cooperation shows descriptions of the behavior of the information system at the level of individual objects that exchange messages. The component diagram shows the components of the information system and the relationships between them. The deployment diagram shows the computing nodes during program operation, components, and objects running on these nodes in the proposed system.

In the study of planning work to develop an informational topic to identify matches in the text using a Gantt chart. Estimation of the cost of project implementation on the basis of appropriate modeling of the information system is presented using the COCOMO Intermediate model. The main risks of developing an information system based on the conducted modeling are analyzed in the work.

The peculiarity of the proposed information system is its ability to operate taking into account the scientific database of higher education institutions, which is not publicly available in a virtual environment. The study found that free online services to establish the uniqueness of the text are popular among consumers of educational services. Representatives of higher education institutions use services that contain a large database of scientific texts, but which are not free. This is evidenced by the analysis of data from the resource https://www.similarweb.com. Analysis of the results of indicators of interest over time of search queries related to the process of finding matches in the texts according to the service https://trends.google.com.ua/ for the period 2019–2021 confirms the relevance of modeling an information system for detection text matches for higher education institutions.

References

1. Xuemina, Zh., Zhimingb, S., Pinga, G.: The process of information systems architecture development. In Proceedings of the International Workshop on Information and Electronics Engineering (IWIEE) Procedia Engineering, pp. 775–779 (2012)
2. Abel, R., Brown, M., Suess, J.: A new architecture for learning. Educ. Rev. **48**, 88–102 (2013)
3. Xiaohai, L., Yafei, T.: Architecture framework and its design method. Fire Control. Command. Control. **1**, 6–8 (2010)
4. Prayitno, T.: Planning of higher education information technology strategy using TOGAF (A Case Study at AMN Cilacap). Indones. J. Inf. Syst. **2**(1), 67–79 (2019)
5. Luftman, J.N., Lewis, P.R., Oldach, S.H.: Transforming the enterprise: the alignment of business and information technology strategies. IBM Syst. J. **32**(1), 198–221 (2010)
6. Araki, M.K.: EA-MDA model to resolve is characteristic problems in educational institutions. Int. J. Softw. Eng. & Appl. **4**(3), 1–20 (2013)
7. Poniszewska-Maranda, A., Matusiak, R., Kryvinska, N., Yasar, A.-U.-H.: A real-time service system in the cloud. J. Ambient. Intell. Humaniz. Comput. **11**(3), 961–977 (2020). https://doi.org/10.1007/s12652-019-01203-7
8. Goyal, S., Parashar, A.: Machine learning application to improve COCOMO model using neural networks. Int. J. Inf. Technol. Comput. Sci. **3**, 35–51 (2018)
9. Iraji, M.S.: Fuzzy agent oriented software effort estimate with COCOMO. I.J. Intell. Syst. Appl. **08**, 18–29 (2015)
10. Majed, Al.Y., Ahmad, R., Lee, S.: Impact of CMMI based software process maturity on COCOMO II's effort estimation. Int. Arab. J. Inf. Technol. **7**(2), 129–137 (2010)
11. Bidan, M., Rowe, F., Truex, D.: An empirical study of IS architectures in French SMEs: integration approaches. Eur. J. Inf. Syst. **21**, 287–302 (2012)
12. Natek, S., Lesjak, D.: The process architecture of information systems—higher education institution's managerial tool. Issues Inf. Syst. **1**, 29–34 (2010)
13. Natek, S., Lesjak, D.: Integrated higher education information systems – professors' knowledge management tools. Issues Inf. Syst. **12**, 80–86 (2011)
14. Zhezhnych, P., Shilinh, A., Demydov, I.: Architecture of the computer-linguistic system for processing of specialized web-communities' educational content. In: Proceedings of the 2nd International workshop on control, optimisation and analytical processing of social networks (Lviv, Ukraine, May 21), 2616, 1–11 (2020)
15. Sanchez-Puchola, F., Pastor-Colladoa, J.A., Borrell, B.: Towards an unified information systems reference model for higher education institutions. Procedia Comput. Sci. **121**, 542–555 (2017)
16. Charles Sturt University. Higher Education Process Reference Model. Charles Sturt University. Work Process Improvement (2010). [Electronic resource]. Available at: http://www.csu.edu.au/special/wpp/resources/reference-model
17. Hariharan, Sh.: Automatic plagiarism detection using similarity analysis. Int. Arab. J. Inf. Technol. **4**(9), 322–326 (2012)
18. Shenoy, M.: Automatic plagiarism detection using similarity analysis advanced computing. An Int. J. (ACIJ) **3**(3), 59–62 (2012)
19. Singh, R.: Duplicity detection system for digital documents. Int. J. Soft Comput. Eng. (IJSCE) **5**(2), 24–28 (2012)
20. Tschuggnall, M.: Detecting plagiarism in text documents through grammar-analysis of authors. In: 15th GI-Symposium Database Systems for Business, Technology and Web, 241–259 (2013)
21. Sandkuhl, K., Seigerroth, U.: Method engineering in information systems analysis and design: a balanced scorecard approach for method improvement. Softw. Syst. Model. **18**, 1833–1857 (2019)
22. Gladun, A., Rogushina, J.: Use of semantic web technologies and multilinguistic thesauri for knowledge-based access to biomedical resources. Int. J. Intell. Syst. Appl. **1**, 11–20 (2012)
23. Schroeder, M., Neumann, E.: Semantic web for life sciences. Web Semantics: Science, Services and Agents on the World Wide Web **N.4**, 167–169 (2006)

24. Gladun A.J., Rogushina J.: Semantic search of internet information resources on base of ontologies and multilinguistic thesauruses. Int. J. Information Theor. Appl. **14**, 117–129 (2006)
25. Haghighi, P.S., Moradi, M., Salahi, M.: Supplier segmentation using fuzzy linguistic preference relations and fuzzy clustering. I.J. Intell. Syst. Appl. **05**, 76–82 (2014)
26. Kumar, A., Das, S.: Typology for linguistic pattern in English-Hindi journalistic text reuse. Int. J. Inf. Technol. Comput. Sci. **8**, 75–86 (2016)
27. Barker, E., Gaizauskas, R.: Assessing the comparability of news texts. In: Proceedings of the Eighth International Conference on Language Resources and Evaluation (LREC'12), pp. 3996–4003 (2012)

Use of a Communication Robot—Chatbot in Order to Reduce the Administrative Burden and Support the Digitization of Services in the University Environment

Dorota Košecká and Peter Balco

Abstract The technology of chatbots has come a long way in the last decade thanks to the growing popularity of artificial intelligence and machine learning. Chatbots are established as a beneficial tool in many scenarios of our daily lives. This paper aims to discuss the design, architecture, and potential application of a chatbot in a university environment. Our work is divided into 3 chapters. The first chapter consists of a literature review of chatbot systems. The second chapter focuses on the methodology of the work and the stated objectives. The third chapter describes the current state of the ongoing processes at the Faculty of Management at Comenius University. On the selected process a time and financial analysis were performed. The output of this chapter consists of the identified critical path in the processes as an area for possible implementation of chatbots in the academic department to reduce the administrative burden. The result is the architecture and development of the proposed solution. The paper concludes with opportunities for improvement of the proposed solution and recommendations that we recommend to be applied in practice.

1 Introduction

Intelligent solutions supported by modern technologies are now playing an increasingly important role in the marketplace, in large and small companies, in every area of business, as well as in universities and homes. What was a disruptive innovation a few years ago is now a functional requirement. The measurable benefits that these

D. Košecká · P. Balco (✉)
Faculty of Management UK, Odbojárov 10, P.O.BOX 95, Bratislava, Slovakia
e-mail: peter.balco@fm.uniba.sk; peter.balco@atos.net

D. Košecká
e-mail: kosecka7@uniba.sk

P. Balco
ATOS IT Solutions and Services S.R.O, Pribinova 19, Bratislava, Slovakia

© The Author(s), under exclusive license to Springer Nature Switzerland AG 2023
N. Kryvinska et al. (eds.), *Developments in Information and Knowledge Management Systems for Business Applications*, Studies in Systems, Decision and Control 462,
https://doi.org/10.1007/978-3-031-25695-0_26

solutions bring in the short, as well as the long term, increase their attractiveness for implementation and active use.

In the area of user interaction, solutions built on artificial intelligence (AI) technology are gaining preference. Therefore, companies are increasingly implementing such solutions not only in their portfolio but also in their environments. An example of a specific solution that uses artificial intelligence in its operation is communication robots—CHATBOTS. Research results continuously demonstrate a growing number of use cases for chatbots to simplify the interaction between stakeholders. However, there are many challenges in developing such chatbots to meet the requirements of users. User adoption of new technologies such as chatbots is of great importance because a positive user experience would mean that users would not be reluctant to interact with a chatbot.

In our work, we explore the possibility of implementing a communication bot in a university setting. From the perspective of our work, each student is a customer of his/her university, and the study department represents customer service.

The main goal of this work is to identify the space in the ongoing processes for a possible implementation of a communication bot at the Faculty of Management of Comenius University. Based on the conducted time analysis of the processes in the study department, a proposal for a possible solution has been developed.

The outcome is the design of the architecture of the intelligent solution according to the requirements to reduce the administrative burden. As a result of this chapter, a manual has been created to develop the proposed solution. Finally, the paper evaluates the results of user testing as well as the limitations and challenges encountered during each phase of the project. Since the developed chatbot is simple and without support for cognitive services part of the last chapter is dedicated to improvements and recommendations for practice.

1.1 The Evolution of Chatbots

The history of chatbots began in 1950 with a test conducted by Englishman Alan Turing. In his article "Computer Machinery and Intelligence", the computer scientist posed the question "Can a machine think?" (Can machine think?). According to his theory, an intelligent machine is indistinguishable from a human during text communication, and it was this theory that laid the foundations for the development of chatbots. The Turing Test was a test to measure whether a human was communicating with a human chatbot [1] (Fig. 1).

The first chatbot in history is considered to be the Eliza chatbot. It was created in 1966 by the father of chatbots, Joseph Weisenbaum. Despite not passing the Turing test, it was the chatbot most like those of today [1]. The role of the chatbot was to act as a psychiatrist in clinical treatment [2]. The script of the DOCTOR program mimicked Rogerian psychoanalysis [3]: the ELIZA program consisted of a simple text-based interface through which individuals could converse with a virtual therapist. The program used a limited set of conversational rules to convert users' phrases into

Fig. 1 Graphical representation of the Turing test. *Source* created by authors

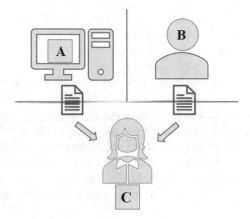

questions, thus evoking the notion that the therapist was following up everything they shared with another question [4].

ELIZA's creator, Joseph Weizenbaum, saw the program as a method for uncovering the transcendence of human–machine communication, but was surprised by the number of patients who attributed human-like feelings to ELIZA and wanted to continue therapy with her. The success of ELIZA was so astonishing because the chatbot was unable to understand anything. Although ELIZA could engage in a discussion that the participant perceived as beneficial, it was unable to converse on any deeper, semantic level [5] (Fig. 2).

In 1972, Kenneth Colby built the chatbot PARRY to simulate a person suffering from paranoid schizophrenia. In a test conducted by psychiatrists, only 48% of people

Fig. 2 User interface of the first chatbot. *Source* Shum et al. [6]

Human: Men are all alike.
Eliza: In what way?
Human: They're always bugging us about something specific or other.
Eliza: Can you think of a specific example?
Human: Well, my boyfriend made me come here.
Eliza: Your boyfriend made you come here?
Human: He says I'm depressed much of the time.
Eliza: I am sorry to hear you are depressed.
Human: It's true. I'm unhappy.
Eliza: Do you think coming here will help you not to be unhappy?
Human: I need some help, that much seems certain.
Eliza: What would it mean to you if you got some help?
Human: Perhaps I could learn to get along with my mother.
Eliza: Tell me more about your family.
...
Chat:

could tell the difference between PARRY and a real person. PARRY was rule-based and had a similar structure to Eliza, but with a better control structure, language comprehension abilities, and most importantly a mental model that could simulate the bot's emotions [6].

A significant milestone in building chatbots was the creation of Alice (Artificial Linguistic Internet Computer Entity), which processed natural language. It was created in 1995 and was described as the most advanced chatbot of the time. It won the annual Turing test prize, the Loebner Prize. ALICE became the first chatbot to be widely considered a "human computer" [7].

ALICE was the first to use Artificial Intelligence Markup Language (AIML) and an interpreter. In ALICE, the AIML markup language was responsible for pattern matching and for linking the user's input to the response with the knowledge base (KB) [7].

The twenty-first century has already favored the development of communication assistants using not only text-based interfaces but also voice-based ones. In 2011, an intelligent virtual personal assistant with a voice interface was launched by Apple [8]. Siri is based on artificial intelligence and natural language processing and consists of three components: a conversational interface, personal context awareness and service delegation [9].

The conversational interface allows Siri to understand the spoken word. The general workings of direct word-to-word voice recognition must be well-handled to hear the spoken word. The subsequent deciphering of meaning depends on statistics and machine learning, which involves a personal context awareness system. It is designed to adapt over time to the user's individual preferences and personalize the results. The service delegation system is unrestricted access to all embedded applications on devices with a supporting operating system and their inner workings [10].

Siri can speak in 21 languages and control and perform tasks on all devices running iOS, iPadOS, watchOS, macOS, tvOS and audios [11].

IMB's chatbot Watson was named after the company's first CEO, Thomas J. Watson. His primary role was competing on the 2011 American TV show 'Jeopardy!' where he managed to defeat two former champions [12]. Watson Assistant is built on deep learning, machine learning and natural language processing (NLP) models to understand questions, find or search for the best answers and complete the user's intended action. Watson also uses intent classification and entity recognition to better understand customers in context and convert them to human agents when needed [13].

Amazon Alexa, better known as Alexa, was introduced to the intelligent assistant market in 2014. It is capable of voice interaction, playing music, creating to-do lists, setting alarms, streaming podcasts, playing audio books, and providing weather, traffic, sports, and other real-time information such as news [12]. Alexa can also control multiple smart devices using itself as a home automation system. Users can expand Alexa's capabilities by installing "skills" (additional features developed by third-party vendors), such as weather forecasting programs and audio features. It uses automatic speech recognition, natural language processing and other forms of artificial intelligence to perform these tasks [14].

Cortana is an intelligent personal assistant that was developed and put into use in 2015. By Microsoft's definition, Cortana is a productivity personal assistant that helps you save time and focus your attention on what matters most. It recognizes voice commands and performs tasks such as identifying time and location, supporting person-based reminders, sending emails and text messages, creating, and managing lists, chatting, playing games, and searching for information the user requests. This functionality is supported by devices running Windows 10 and later [15]. Cortana is named after the twenty-sixth century female artificial intelligence in the hugely popular Halo video game franchise [16].

The green light for the development and massive implementation of chatbots was given by Facebook when it enabled the embedding of company chatbots into the Messenger app in 2016 in order to make customer service easier and faster. In this case, the chatbot represented a commercial tactic in communicating with customers. In addition to informative uses for personalized advice, Facebook includes the ability for chatbots to send sponsored messages to people who have previously contacted the company through a chatbot or through live chat support [17].

The statistics shown in Chart 1 in Fig. 3 shows the revenue of the global chatbot market from 2018 to 2027. The global chatbot market is projected to reach a revenue of US$454.8 million in 2027, up from US$40.9 million in 2018 [18].

Factors contributing to the growth of the chatbot market include the rise in the need for round-the-clock customer support at low operational costs, increase in emphasis on customer engagement through multiple channels, and technological advancements along with growing customer demand for self-service. The chatbot market is expected to witness high growth owing to rise in demand for better customer experience and building personalized relationships with potential customers. Various segments are expected to deploy diverse chatbot services to enable technology transformation initiatives that improve operations, differentiate clients' browsing experiences, and address critical processes. Reduction in operational costs, user satisfaction, resolution

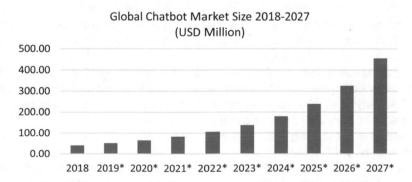

Fig. 3 Global Chatbot Market Size 2018–2027. *Source* Statista Research Department, created by authors

of customer queries, better insights into operations and processes, and improved real-time decision making are the key business and operational priorities that are expected to boost the growth of the chatbot market [19].

1.2 Chatbot

The potential of chatbots is currently being exploited in diverse application areas. Today, the chatbot is the gateway to digital services and information—in areas such as customer service, banking, healthcare, education, administration, and many others. In recent years, there has been a significant increase in interest in research and implementation of chatbots in academia and industry. To determine the need for a chatbot, it is essential to understand what a chatbot is and how it is discussed in the scientific literature.

The literature defines a chatbot as a computer program that mimics and processes human communication and allows people to interact with digital devices as if they were talking to a real person [20]. A broader definition adds that it is a dialogue mechanism that supports collaborative learning [21].

A chatbot can also be thought of as a system that automatically answers human questions [22].

Authors Benton and Radziwill describe a chatbot as feeling like interacting with people online, when the communication is with computer software that comes to life with natural language input [23].

Other authors consider a chatbot as an application that uses artificial intelligence to converse with people in natural language. Chatbot adopts the knowledge of Human Computer Interaction (HCI- Human Computer Interaction) to give computers the intuitive ability to carry on a conversation with their users using natural language. It can support text input, audio input, or both [24].

Author Araujo added a final statement to the definition that defines a chatbot as a program that has been designed to interact with humans through natural language-based text that mimics normal human-to-human conversations within the boundaries of a specific domain-knowledge [25].

A chatbot can be seen as a domain-domain-specific conversational interface that uses an application, messaging platform, social network, or chat solution to converse. Chatbots vary in sophistication, from simple decision tree-based marketing pieces to implementations built on feature-rich platforms [26] (Fig. 4).

Chatbot is not the only label that can be encountered in the literature as well as among the public. The terms chatbot, virtual assistant, conversational AI bot, AI chatbot, AI assistant, virtual customer assistant, digital assistant, conversational agent, virtual agent, conversational interface are sometimes used as synonyms. Chatbots tend to support simpler conversations and more personalized tasks. For example, it can tell you whether or not it's going to rain tomorrow. A conversational agent, on the other hand, may find that you want to know what you're going to wear

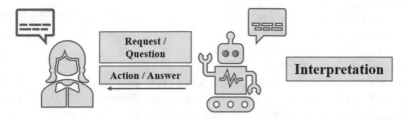

Fig. 4 The principle of chatbot function. *Source* Authors

tomorrow [27]. In the scientific literature, chatbots are more formally referred to as conversational agents.

All these conversational technologies use a natural language recognition function to recognize what the user is saying. In addition, other forms of artificial intelligence determine what you really want to know. These technologies use machine learning to learn from interactions and improve the resulting recommendations and responses.

Depending on the time horizon configuration, domain-specific chatbots can be characterized as short-, medium-, long-term, or lifetime chatbots [28].

Chatbots designed to help individuals achieve short-term goals are defined by one or a very small number of occasional short interactions, while the latter— i.e. chatbots designed to support individuals in achieving medium- to long-term, or even lifelong goals—consist of multiple (interdependent) interactions over a period of time. Typical examples for short-term relationships are chatbots offering short ad-hoc services such as customer support or self-diagnostic medical chatbots, while typical medium- and long-term examples are chatbots for monitoring chronic conditions or learning processes [29].

Just as chatbots have different names, they also have different degrees of intelligence. A basic chatbot can be little more than a front-end solution for answering standard FAQs. Chatbots created using some of the currently available bot frameworks may offer slightly more advanced functionality, such as filling out pop-ups or other simple transactional capabilities, such as taking pizza orders. But it's not until advanced conversational chatbots with artificial intelligence that they have the intelligence and capabilities to deliver the sophisticated chatbot experiences that most enterprises want to deploy [30].

In addition to the division by timeframe configuration mentioned in the previous subsection, chatbots can be classified using the parameters listed below [31]:

- Access to domain knowledge
- Type of service provided
- Objectives
- Method of processing inputs and generating responses.

- **Access to the knowledge domain**
  The first classification based on the knowledge domain considers the knowledge that the chatbot has access to, or the amount of data on which it is trained. Chatbots

with an open domain can talk about general topics and respond appropriately, while chatbots with a closed domain are focused on a specific knowledge domain and may not respond to other queries [31].

- **Type of service provided**

The second classification based on the service provided considers the sentimental proximity of the chatbot to the user, the amount of confidential interaction that takes place, and also from the task being performed.

Interpersonal chatbots are found in the communication domain and provide services such as restaurant reservation, flight booking and FAQ bots. They are not companions of the user, but they gather information and pass it on to the user.

Intrapersonal chatbots exist in the user's personal domain, for example in chat applications such as Messenger, Slack and WhatsApp. They are companions of the user and understand the user in a human-like way.

Interpersonal (inter-agent) chatbots will be prevalent in domains where IoT dominates. In this case, two systems communicate with each other to accomplish a task. Here the need arises for creating protocols for communication between bots. One existing example of an inter-agent bot is the Alexa-Cortana integration [31].

- **Objectives**

The third classification based on goals considers the primary goal that the chatbots are trying to achieve.

Informational chatbots are designed to provide the user with information that is pre-stored or available from a fixed source, such as FAQ chatbots [32]. They are usually based on an information retrieval algorithm from a database or on string matching. In most cases, they refer to a static source of information, such as a website or a knowledge base [31].

Chat/conversational chatbots talk to the user as to another human and their goal is to answer a given sentence correctly. Their goal is to continue the conversation with the user based on techniques such as cross-questioning, avoidance, and deferring questions [32].

Task-based chatbots perform a specific task, such as booking a flight or restaurant [31]. In most cases, the actions required to perform the task are predetermined; the flow of events, including exceptions, is also decided [32].

- **Method for processing inputs and generating responses**

The fourth classification is based on the method of input processing and response generation. Three models are used to generate appropriate responses [33]:

- rule-based model
- an intelligent model
- hybrid model.

- *Rule-based chatbots—Linguistic models*

Chatbots with a rule-based model are the type of architecture with which most early chatbots were built, such as many online chatbots. The system selects a response based on a fixed set of rules based on recognizing the lexical form of the input text without creating new textual responses. The knowledge used in the chatbot is manually encoded by a human and is organized and represented using conversational patterns [34]. A more extensive rule base allows the chatbot to respond to more types of user input. However, this type of model is not robust to spelling and grammatical errors in user input. Most of the existing research on rule-based chatbots investigates answer selection for a single-input conversation that considers only the last input message. In chatbots that are more human-like, multiple-input answer selection considers the previous parts of the conversation to select an answer relevant to the context of the entire conversation.

Rule-based chatbots use if/then logic to create conversational flows. It is possible to create linguistic conditions that focus on words, their order, synonyms, common ways of phrasing a question, and more to ensure that questions with the same meaning get the same answer. If something is not right in understanding, it is possible for a person to fine-tune the conditions. However, chatbots based on a purely linguistic model can be rigid due to this highly laborious approach, and their development can be slow [35].

- *Machine learning (bots with artificial intelligence)*
The rapid development of computer information technology has caused many changes in human lives. One of the promising technological trends is Artificial Intelligence (AI) With AI, a computer can perform certain tasks that humans perform, such as robotic conversation.

AI chatbots are more complex than rule-based chatbots and tend to be more conversational, data-driven, and predictive. These types of AI chatbots are generally more sophisticated, interactive, and personalized than task-oriented chatbots [36].

Chatbots that use machine learning approaches instead of pattern matching, extract content from user input using natural language processing (NLP) and have the ability to learn from conversations. They take into account the entire context of the dialogue, not just the current turn, and do not require a predefined response for every possible user input. They usually need a large training set, which can be a major problem to find because the available datasets may be scarce [32]. The following Fig. 5 presents the architecture of a chatbot that uses machine learning.

Machine learning-based conversation systems can be impressive if the problem at hand is well suited to their capabilities. By their very nature, they learn based on patterns and prior experience. The resources required, combined with the very narrow range of scenarios in which statistical algorithms make chatbots based solely on machine learning an impractical choice for many enterprises [36].

- *Hybrid model*
Although linguistic and machine learning models have their place in the development of some types of conversational systems, using a hybrid approach offers the best of both worlds and offers the opportunity to provide more comprehensive conversational AI chatbot solutions.

The hybrid approach has several key advantages over both alternatives. When considered against machine learning methods, it allows conversational systems to be created without data, provides transparency into the operation of the system, enables business users to understand the chatbot application, and ensures the

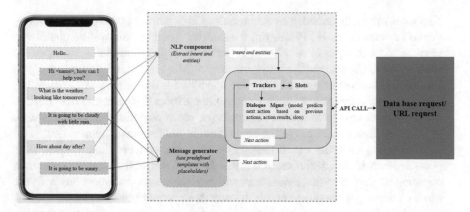

Fig. 5 Architecture of a chatbot based on machine learning. *Source* Created by authors

chatbot maintains a consistent personality and aligns its behavior with business expectations. It also enables machine learning integration to go beyond the realm of linguistic rules, making intelligent and complex inferences in areas where a linguistic-only approach is difficult or even impossible. When the hybrid approach is provided at the native level, it allows statistical algorithms to be embedded alongside linguistic conditioning and preserved in the same visual interface [36, 37].

Creating conversational applications using only linguistic or machine learning methods is resource intensive and often prohibitively expensive. By leveraging a hybrid approach, businesses have the power, flexibility and speed needed to develop business-relevant AI applications that can transform the customer experience and the bottom line [36].

2   Data and Methodology

The intention of our work is to create a basic version of a communication assistant—a chatbot, which will serve to reduce the administrative burden of clerks in the study department and at the same time provide students with a constantly available clerk who has a knowledge base of information. Its task is to answer questions based on information from the school's website as well as the Comenius University study regulations. The implementation of this type of solution should bring a reduction in the volume of mail and telephone communication towards the study department, which will bring measurable benefits in the form of personnel, time and economic savings of the resources of the Faculty of Management of Comenius University.

At the end of our research, the proposed prototype should be able to answer questions from the student about their studies and provide information about dates, addresses, contacts, and events at FMUK. The Microsoft Azure Bot service is used to create the chatbot, as well as Microsoft's cognitive services, namely QnA Maker and the Microsoft Framework Composer desktop application.

In the following section, the main goal and sub-goals of our work are presented. The next section of this chapter discusses the methodology employed and the research method.

2.1 Main and Sub-Objective

The main objective of our work was to analyze the administrative processes that take place in the environment of the Faculty of Management at the study department to identify the possibility of deploying a new type of intelligent solution—a communication robot—a chatbot. The identification of the processes led to the identification of points in the processes for improvement, based on which a new architecture of communication supported by a communication robot—chatbot was proposed, which should lead to the reduction of the administrative burden of clerks and clerks in the study department. The proposed prototype of communication supported by artificial intelligence required verification of the functionality and quantification of the added value of the project.

The first sub-objective of our work sets the requirement for a close characterization of the E2E (End to End) process selected from the first part of the main objective in terms of the activities and roles performed, to identify measurable KPIs for each activity, their time requirements for execution.

The second sub-objective is to identify a critical activity or activities that could be supported by communication robots for a selected group of processes taking place in the study department environment.

Based on the previous one, the third sub-objective is the analysis and selection of the most suitable solution—a communication robot based on the requirements of the environment and the characteristics of the analyzed communication robots.

The last sub-objective of the thesis is to evaluate the risks that can be encountered during the design of the prototype, its implementation as well as during active use and to describe the benefits and recommendations for future implementation in practice.

2.2 Methodology of the Work

Administrative activities and their manual repetition create room for the deployment of communication robots to reduce workload, error rate and employee frustration. Universities represent an ideal environment where the administrative workload is high, but many are unaware of this fact and these activities and processes are constantly repeated. Such an environment creates ideal opportunities to deploy simple intelligent IT solutions, communication bots, which support the digitization of services in universities and provide an opportunity to validate the required functionality.

The untold dream of the students is a study officer who is always available, her responses are immediate with high quality of information required. According to the requirements, such a solution would seem unrealistic, but the current era full of smart solutions nevertheless has such a clerk in its catalogue of services.

In order to find out whether it is possible to implement a chatbot in the Faculty of Management environment, a web-based experiment was designed, and the research method was conducted through interviews. The online experiment was created using the open-source service Azure Bot, which offers cloud-based solutions for creating communication bots, and the interviews were conducted with a member of the Faculty of Management with many years of experience in the study department.

To be able to design a chatbot architecture that would improve the ongoing processes, it is necessary to have an in-depth understanding of the processes taking place in the faculty's study department.

For this purpose, several online and face-to-face interviews were conducted, and dozens of emails were exchanged with the study department staff. These interviews were mainly concerned with identifying the inputs and outputs of the processes and the most common areas of communication that students generate the most demand for. The information identified needed to be translated from paper to visual form in order to create a flowchart of the processes. Here we used the Visio visualization tool in the form of a web application. Visio is an innovative solution from Microsoft, which offers it as part of Office 365. Visio is a solution that helps to visualize business process flows linked to data using several integrated features. Despite the extensive background information that emerged as a result of the interviews, further consultations had to be carried out in response to the questions that kept arising.

Two analyses were carried out after the schematics of the selected processes had been finalized.

The first dealt with the analysis of the visualized processes to identify critical points that were identified as suitable for support by an IT solution—a communication robot—a chatbot. This analysis was supported by a time and financial analysis of the selected process. After the first analysis was completed, a second analysis was conducted to select the most suitable open-source platform on which our chatbot prototype could be built. After their careful consideration as well as analysis of the environment and functional requirements for the final prototype, the open-source platform from Microsoft Azure Bot Framework was selected.

Before discussing the chatbot itself, we will present the results of the first analysis, which identified the critical points of the processes and activities taking place at the administration department, which served as the justification for the project and the treasure of our work.

3 Results

During the semester, the study department is overwhelmed with hundreds of applications and applications, forms, thousands of emails, endless student visits, and constantly ringing phones. In some months, answering emails and answering phones for student requests is a full day's work. Checking and processing applications for interruption or termination of studies, transfer requests from other faculties or universities, or processing applications from potential applicants is considered an important activity of the clerks in the study department, which is also their job description. In this process, we see a great scope for exploiting the potential of using not only communication robots but also communication assistants, who can perform some steps in the ongoing processes at the administration department thanks to the support of artificial intelligence and intelligent connectors in the background.

Responding to students' requests is a frequent part of the job in the current reality. In most cases, all the information that students request is available on the school's website, in the yearbook, or in the student handbook. Nevertheless, students prefer to write an e-mail or call but personally visit the study department. The time duration of the process may vary. It mainly depends on the time to search for information or wait for a response, but the complexity of the desired output should also be included.

3.1 Time and Financial Analysis of the Selected Process (Question–Answer)

Figures 6 and 7 below represent the approximate load of calls and emails to the study department during the year. The number of calls per officer per working day was set at between 20 and 50 based on the interview data. The number of emails received is in the range of 20–40 emails per day per clerk. It should be mentioned that the number of phone calls and emails is approximate and is also linked to specific periods during the semester. For example, at the time of application or state examinations (May–June), student enrolment (August–October) or the beginning of the summer semester (February–March), the figures may exceed the specified range. In quieter periods during the semester, emails and phone calls may be fewer than specified. The following graphs do not include in-person visits by students to the Office of Academic Affairs.

Figure 6 below shows the total processing time for all calls handled during the year. We worked with the following values to calculate the total time:

- Minimum call processing time = 0.5 min
- Maximum call processing time = 3 min.

The minimum time in the graph below is shown in blue, the maximum time is shown in red.

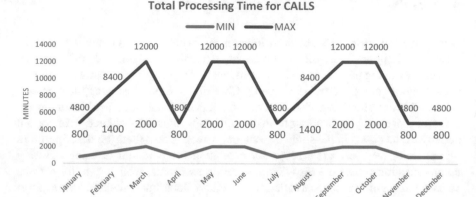

Fig. 6 Total call processing time by load. *Source* Authors

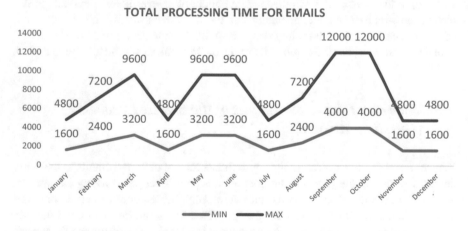

Fig. 7 Total email processing time by workload. *Source* Authors

The average number of incoming calls based on the information in the previous paragraph is 3,360 over the course of a year. If the minimum processing time is taken into account, the SHO needs 16,800 min to handle telephone requests. If we assume a maximum call processing time of 100,800 min.

Figure 7 below shows the total time taken by the study department to process email requests over the course of one year. In calculating the total time, we have worked with different values than for call handling as the email handling process is more complex. The values are as follows:

- Minimum email processing time = 1 min
- Maximum time to process an email = 4 min.

Table 1 Time for processing calls and emails

	MIN (minute)	MAX (minute)	MIN (hour)	MAX (hour)
CALL	16 800	100 800	**280**	**1680**
EMAIL	30 400	91 200	**507**	**1520**
TOTAL			**787**	**3200**

Source Authors

MIN (minute)= the minimum amount of time, in minutes, that staff will spend processing telephone and email requests

MAX (minute)= the maximum time in minutes, that employees spend processing telephone and email requests

MIN (hour)= the minimum amount of time, in hours, that employees spend processing telephone and email requests

MAX (hour)= the minimum amount of time, in hours, that employees spend processing telephone and email requests

In Fig. 7, the maximum time to process a request is shown in red and the minimum time to process an incoming email is shown in blue.

The average number of incoming calls per year is 30,400. If we calculate the minimum handling time, then it takes 34,000 min for the SHO to process email requests. If we calculate the maximum time to process and reply to a request it is 91 200 min.

The following Table 1 reflects the sum of minutes spent making phone calls and replying to email messages with student requests, which are shown in the graphs above. The result represents the values converted into hours for both the minimum and maximum time required to process either a telephone or mail request. The values given are shown for one year.

Table 1 above shows the total amount of time that staff in the study department spend dealing with student enquiries. This time can then be recalculated as part of the full-time employee's salary. The result can then be used as input for calculating the return on investment if such a project is actually implemented in a university environment.

3.2 Design and Development of a Communication Robot—Chatbot

As mentioned, before we tried to choose the most suitable open-source platform to build a prototype communication robot six platforms were analyzed based on predefined criteria.

The decisive criteria were at least partial support of the Slovak language, support of the IOS operating system and the cost of purchasing and using the open-source platform and its add-ons. After analyzing the available platforms on the market, we decided on the open-source platform from Microsoft Azure Bot Service, the

Fig. 8 Bot Framework
Composer. *Source* Microsoft
Azure

cognitive service QnA Maker and the Bot Framework Composer application, which works seamlessly with the add-ons available on the Azure platform, as well as with the possibility of publishing the created application on the Azure platform.

A. **Installing Bot Framework Composer**

As a first step, it was necessary to install the Bot Framework Composer. Bot Framework Composer, built on the Bot Framework SDK, is an open-source development environment (IDE—Integrated Development Environment) for developers to create, test, deliver and manage conversational experiences. It provides a powerful visual authoring canvas that allows you to create dialogs, language understanding models, QnAMaker knowledge bases, and language generation responses from a single canvas, and very importantly, allows you to extend these experiences with code for more complex tasks such as system integration. The resulting experiences can then be tested within Composer and published to Azure along with any dependent resources.

Composer is available as a desktop application for Windows, macOS and Linux. If the desktop app doesn't suit your needs, you can build Composer from source or host Composer in the cloud (Fig. 8).

Bot Framework Composer offers the ability to create chatbots from scratch or via templates. Composer supports templates in Node.js and C#. Both requirements need to be met to work in Composer. This is a mandatory requirement to use Composer.

B. **Cognitive Service QnA Maker**

QnA Maker is a cloud-based natural language processing service that creates a natural conversational layer over data. QnA Maker is particularly useful when we have static information in the bot that needs to be managed. The static question and answer pairs are referred to as the QnA Maker knowledge base, which the QnA Maker service uses to process the questions and respond with the corresponding answers.

QnA Maker is commonly used to create conversational client applications, which include social media applications, chatbots, and speech-enabled desktop applications. The service does not store data from its users. All user data, i.e., question responses and chat logs, are stored in the region where the customer deploys dependent instances of the service.

Fig. 9 Creating a resource group. *Source* screenshot from the configuration environment

C. **Creating a resource group**

Another requirement for creating and then publishing chatbots is a Microsoft Azure account. Here it was possible to use a student account or create a new private account. In our case, the account created was: elorabot@outlook.com. With the new account, we received a credit worth $200, which can be used for the first 30 days for most of the services available on the Azure platform.

Then to use the available applications we need to create a source from where the specific application can already draw funds. As shown in Fig. 9 below our first step was to create a new funding source that serves as the basis for creating additional services. The resource group created was named elorabot.

To create a knowledge base on which our chatbot would be able to communicate, we selected the QnA Maker service. QnA Maker knowledge bases can be created, edited, and then published directly in the Azure portal or also in the Bot Framework Composer. Knowledge bases created and published in Azure can also be linked to a chatbot that is created in Composer. In our work, we have tried both options for creating knowledge bases.

D. **Creating a knowledgebase in QnA Maker on the Azure portal**

In the case of creating a knowledgebase via Azure, it is necessary to connect this service to a resource group, due to credit pumping. This step is shown in the previous Fig. 9. On the Azure website, select QnA Maker among the cognitive services and choose "Create New". In the next step, we had to connect the service to the source group, choose a name and a subscription form. When we create the QnAMaker resource, the Azure Search service is automatically created as well, which is used to index the data (Fig. 10).

With the QnA Maker service created, we have moved on to creating a knowledgebase that contains the questions and answers. This service gives the possibility to create a knowledge base (KB) from custom content, for example from FAQs or product manuals, by connecting to a web page or manually. The advantage of this service is that it supports Slovak language. Slovak language support later turned out to be a big negative.

The cognitive service LUIS, which works as a language recognizer for user understanding, does not support Slovak language and thus cannot be connected (Fig. 11).

After connecting the QnA service to our source group, we get the possibility to connect the knowledge base to web pages and files in.tsv,.pdf,.txt,.xlsx and docx format. The number of URLs as well as the range of documents that can be linked is determined by the subscription package of the resource group. The price is then determined on this basis. If this option is not used in this step, it is

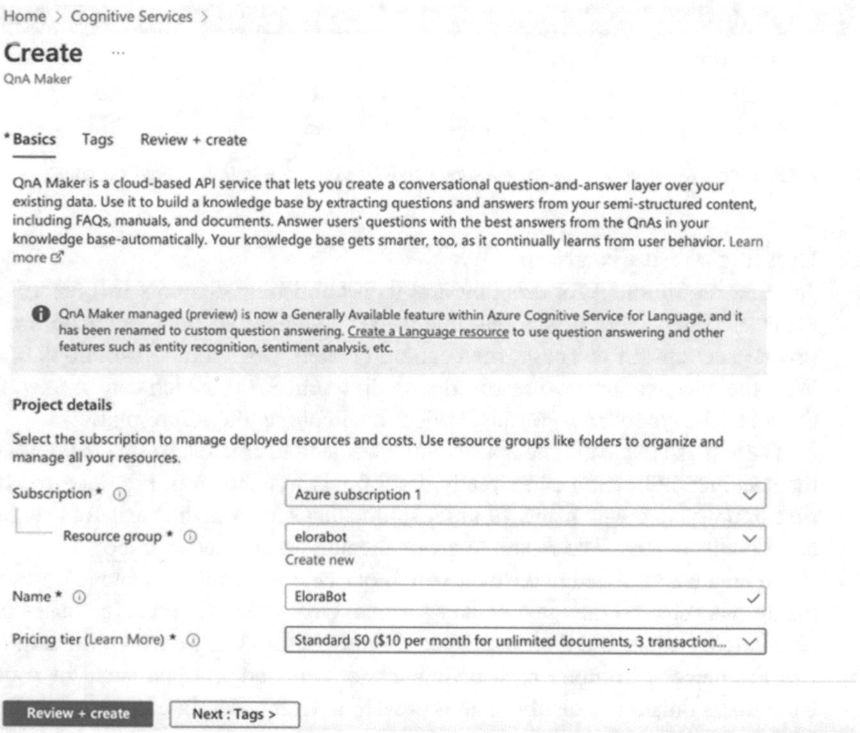

Home > Cognitive Services >

Create   ···
QnA Maker

* Basics   Tags   Review + create

QnA Maker is a cloud-based API service that lets you create a conversational question-and-answer layer over your existing data. Use it to build a knowledge base by extracting questions and answers from your semi-structured content, including FAQs, manuals, and documents. Answer users' questions with the best answers from the QnAs in your knowledge base-automatically. Your knowledge base gets smarter, too, as it continually learns from user behavior. Learn more ☑

> ℹ QnA Maker managed (preview) is now a Generally Available feature within Azure Cognitive Service for Language, and it has been renamed to custom question answering. Create a Language resource to use question answering and other features such as entity recognition, sentiment analysis, etc.

Project details

Select the subscription to manage deployed resources and costs. Use resource groups like folders to organize and manage all your resources.

Subscription * ⓘ	Azure subscription 1 ⌄
⌐ Resource group * ⓘ	elorabot ⌄
	Create new
Name * ⓘ	EloraBot ✓
Pricing tier (Learn More) * ⓘ	Standard S0 ($10 per month for unlimited documents, 3 transaction... ⌄

[Review + create]   [Next : Tags >]

Fig. 10 Creating the QnA Maker cognitive service. *Source* snapshot of the configuration environment

possible to attach sources with information later directly in the knowledge base creation tool. If we want the knowledge base to be updated in case of a change on the page, we need to select the option to update the content in the settings. It is possible to manually edit the structure of the knowledge base directly in the tool. Figure 12 shows the possibility to manually edit and extend the knowledge base in the QnA tool.

Once the complete knowledge base has been created, the next step is testing. During testing we can see the accuracy of the returned answers. We can test the knowledge base through the chat interface. Each answer received can be checked against the answer score. If the answer with the highest score is incorrect, it is possible to select the correct answer from the database or write an alternative answer. In order to save the change, it is important to save the action.

The next step is to publish our bot. To get our database transferred to the chatbot, we need to connect the Web App Bot service to the source group to which the QnA Maker is associated. This step will open after the implementation of the knowledge base into the Azure environment is successful (Fig. 13).

The following Fig. 14 below represents the final step before deploying the chatbot to the communication channel. In this step, we modified the chatbot's

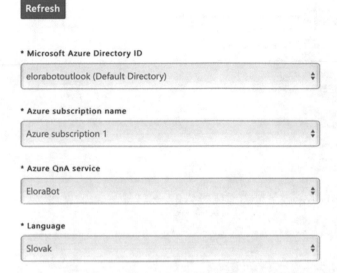

Connect your QnA service to your KB.

After you create an Azure QnA service, refresh this page and then select your Azure service using the options below

Refresh

* Microsoft Azure Directory ID

elorabotoutlook (Default Directory)

* Azure subscription name

Azure subscription 1

* Azure QnA service

EloraBot

* Language

Slovak

Fig. 11 Creating a knowledge base for the QnA service. *Source* snapshot of the configuration environment

Fig. 12 Creating knowledge base content. *Source* Snapshot of the configuration environment

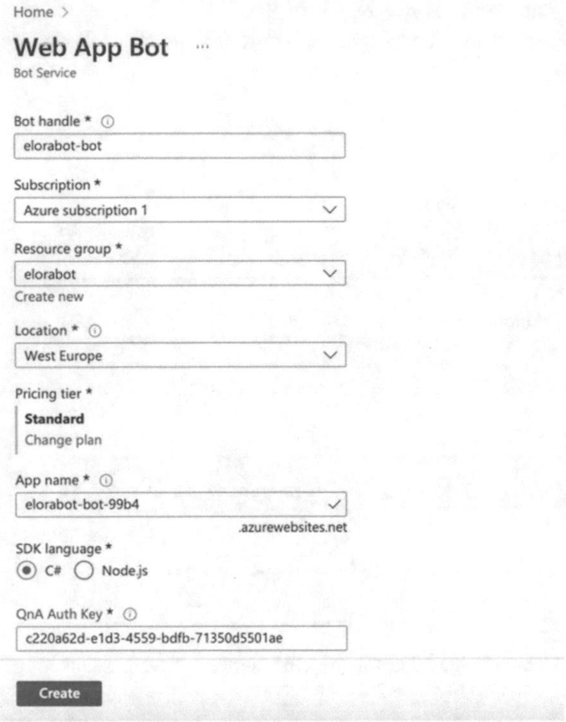

Home >

Web App Bot ⋯
Bot Service

Bot handle * ⓘ

| elorabot-bot |

Subscription *

| Azure subscription 1 | ∨ |

Resource group *

| elorabot | ∨ |
Create new

Location * ⓘ

| West Europe | ∨ |

Pricing tier *

Standard
Change plan

App name * ⓘ

| elorabot-bot-99b4 | ✓ |
.azurewebsites.net

SDK language *
◉ C#   ○ Node.js

QnA Auth Key * ⓘ

| c220a62d-e1d3-4559-bdfb-71350d5501ae |

Create

Fig. 13 Connecting the Web App Bot service. *Source* snapshot from the configuration environment

profile, which included its name and logo. The name change should occur in 30 min, while the logo change can take 24 h.

The published bot can then be published to different communication channels. A menu of possible implementations is shown in Fig. 15 below. In our case, the MS Teams communication channel option was chosen. Here we can see the scope for a deeper communication connection with students through the many communication channels that are used by them daily.

E. **Creating a knowledge base through Bot Framework Composer**
If we are creating a knowledge base through Composer we first need to use the elorabot source group to create a chatbot through the Azure Bot cognitive service. Using the Azure Bot Service, it is possible to create sophisticated bots while retaining ownership and control of your data. In this service, it is possible to create a simple bot that works on a question–answer or complex virtual assistant basis. The chatbot created has been given the name EloraBot. The created chatbot can be edited in Bot Framework Composer and then connected to different channels and devices using open-source SDKs and tools (Fig. 16).

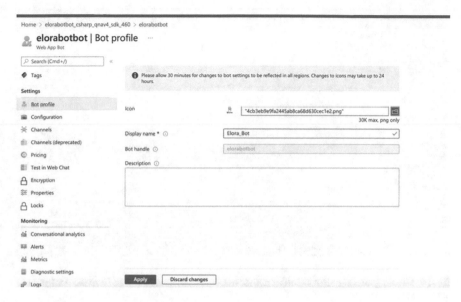

Fig. 14 Editing the profile of the created chatbot. *Source* screenshot from the configuration environment

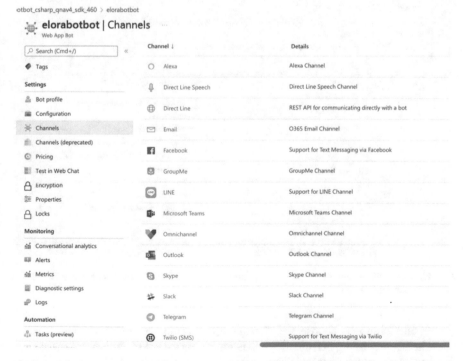

Fig. 15 Communication channels for chatbot deployment. *Source* snapshot of the configuration environment

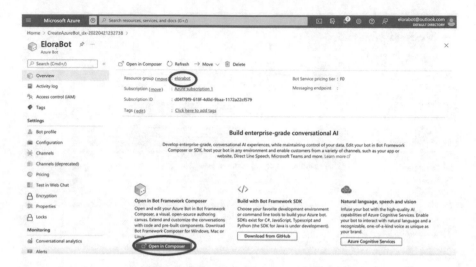

Fig. 16 Creating an Azure Bot to edit a chatbot in Composer. *Source* Image from the configuration environment

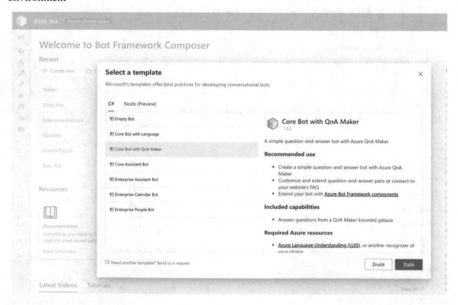

Fig. 17 Choosing a suitable template for creating a chatbot. *Source* Screenshot from the configuration environment

In Composer, choose the "Create New" option. Composer offered us the choice of working in a chatbot that is empty, or we could choose a template that we would edit as required. Since we are working with QnA Maker, we chose to create the chatbot with this template (Fig. 17).

Before the chatbot was created we had to connect it to our subscription and choose the name of the knowledge base we were creating. We chose the logical structure of the knowledge base layout according to the layout of the topics on the faculty website.

Similar to creating a knowledgebase in the QnA service in the Azure portal, it is possible to attach website URLs or files before and during the creation of the chatbot. It is also possible to create groups of questions by topic in Composer. As we can see in Fig. 18 below the first half of the questions are related to the faculty departments, the second half of the questions belong to a group that contains questions and answers about the topic of study. From our own experience, we can say that creating the knowledge base in Composer was easier and more user friendly. Also, extracting answers from relevant URLs and documents was more intelligent.

Before publishing, it is necessary to modify what the first message will be when the conversation starts and what message will be delivered if the chatbot does not recognize the content of the question asked. The advantage of working in Bot Composer is the greater variability of options to customize the chatbot.

If we created the chatbot using this procedure, there is no need to publish it to the Azure environment since we created the Azure Bot in the first step and then opened it in Composer. Every change we made in Composer was automatically saved in Azure Bot. If we want to test the application in one of the communications channels we have the same options as shown in Fig. 18 above in the case of creating a knowledge base in QnA Maker. The only difference is

Fig. 18 Creating a knowledge base in Composer. *Source* Snapshot from the configuration environment

that in the case of Azure Bot, the Omnichannel communication channel is not available.

In case we started creating the chatbot directly in Composer without creating Azure Bot in the first step the publishing to Azure environment is done in the following steps:

1. In the Navigation bar, We Select "Publish.
2. We type the name of the chatbot and the destination where we want the chatbot to be published. In our case, the name was Elora_Bot and the destination was Azure.
3. We choose the source group. We have the choice of creating a new group, selecting an existing group, or handing it over to the admin. Since we have a source group created we will use it for the purpose of this publication.
4. The next step requires writing out a JSON file (see below) containing the values of the existing resources from the Azure portal. All the required information is available in the user profile.

```
{
"name": " < EloraBot > "

"environment": "dev",

"tenantId": " < 1b5abe61-71fc-467e-b028-2b1c0a7db57c > "

"hostname": " < < name > - < env > ",

"runtimeIdentifier": "win- × 64",

"resourceGroup": " < elorabot > ",

"botName": " < EloraBot > ",

"subscriptionId": " < d04f79f9-618f-4d0d-9baa-1172a22cf579 > ",

"region": " < westeurope > ",

"settings": {

"applicationInsights": {

"InstrumentationKey": " < Instrumentation Key > ",

"connectionString": " < connection string > ".

},

}.
```

If the information is listed correctly we can see the assigned profile for publishing. The final step involves logging into the outlook.com account as instructed. This activity results in a successful implementation of the bot into the Azure environment.

Fig. 19 Viewing the knowledge bases in the profile. *Source* Snapshot from the configuration environment

Which way we choose to create the knowledge base depends more on personal preference. Also when we started creating the knowledge base in QnA Maker it is possible to subsequently import this created component into Composer and continue development here. As we can see in Fig. 19 below, we can see both knowledge bases in our profile since both have been connected to the same source group.

3.3 Name and Logo

The created communication bot was given the working name Elora. An owl with a cap as used in graduation ceremonies was chosen as the logo. The owl is considered to be a symbol of wealth, prosperity, wisdom, luck and chance and in our opinion all these variables are important while studying at university.

The created logo shown in Fig. 20 above was created using FreeLogoDesign software. This tool is a free online logo creation software. The creation of the logo is that of a certain quality for free. Many templates are available on the website. The chosen template can be customized according to preference The choice of the name as well as the logo was the work of the author of this thesis.

Fig. 20 Logo of the created chatbot. *Source* FreeLogoDesign, created by Authors

3.4   User Testing

In this section of the paper, we present the result of our project implementation. The ideal solution to start communicating with the chatbot would be to visit the university website and click on the chatbot logo, which tends to be most displayed in the bottom right role. Since we did not have the ability to implement the chatbot directly on the site in the online space, our communication bot was deployed to the MS Teams environment. Other options for implementing a chatbot in the communication channels are shown in Fig. 15 from the previous part of our work.

Figure 21 below shows the user interface if the student is using the app on an iOS mobile device. User interaction is required to start a conversation. Several types of greetings are provided in the knowledge base to start a conversation. Even if the student would not start the conversation with the above variables, then the conversation will be started.

The most important feature of the chatbot for the students was the correctness of the answers. Since students are the end users and thus they decide on the acceptance of the chatbot, it is important to ensure the quality of the answers. Testing the chatbot is a very good way to achieve this. However, as soon as the first version or prototype of the chatbot is available, it should be conducted by people who have nothing to do with the generation of the training data. As already mentioned in theory, a wide variety of sentence constructions can be used for user input. This approach can lead to poor results for intent classification since not enough training data is available or if not enough training is performed.

We could see this when we gave the application to another student to test. While the author, as the creator of the chatbot, was very familiar with the data model and knew exactly which questions to ask to get the desired result, it was different in the case of a foreign user.

For this reason, it is suggested that the chatbot should be tested by people who have nothing to do with the development. In order to get a wider range of data, it could even be suggested here that the chatbot be tested by people with different demographic characteristics such as level of education, country of origin, knowledge of English or study programme (Fig. 22).

The published chatbot can only answer questions that are related to the Faculty of Management of Comenius University. Professionally, it is therefore a closed domain chatbot, which is bound only to a certain area of use.

3.5   Limitations and Challenges in the Development, Implementation, and Use of Chatbots

We encountered several limitations when developing and implementing chatbots on the college website.

Fig. 21 User interface in the MS Teams application on a mobile device. *Source* Screenshot from MS Teams environment

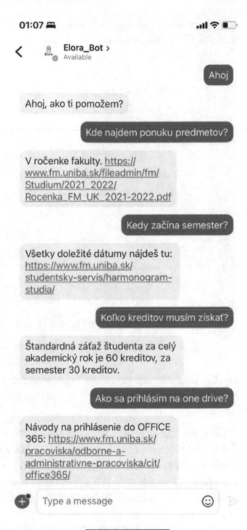

First, every educational institution or university has some professional to develop their websites. Since we did not have the possibility of implementing it directly on the website, our chatbot was implemented in the MS Teams application environment. However, this solution should not be considered as a limitation. Every student has access to MS Teams, and if it were also implemented in Meta Inc.'s Messenger application, applicants to your college would have access to it as well. For a real project, there may be a problem since not every web development company has provisions for creating chatbots. On a website, the problem arises when a third party is approached to create a chatbot for a particular college.

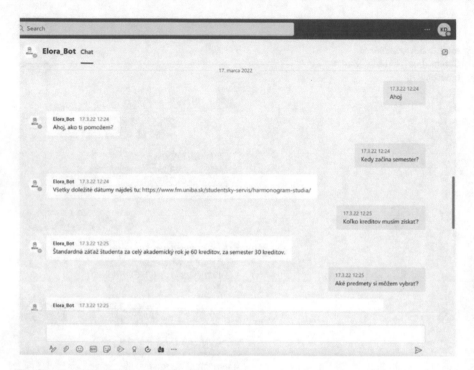

Fig. 22 User interface in the MS Teams application on a computer. *Source* Screenshot from the MS Teams environment

The second problem was the resources to develop and implement the chatbot. Since our resources were limited to €180, our options were also limited. Once the credit expires, the knowledge base can no longer be worked with and is frozen until the credit is increased. We understand that smart chatbots use high-end software which is always expensive but the benefits that accompany its use carry more weight in our opinion.

The third issue that arose during development related to one of Azure's cognitive services. The LUIS language recognizer does not support the Slovak language. We consider this to be a major drawback, as LUIS (Language Understanding) is a service that allows the bot to understand the user's response and determine what to do next in the flow of the conversation. In this case, it is possible to translate each question from Slovak to English and back to Slovak after processing. However, since Slovak is a very complex language there could be a loss of context.

The complexity and complicated grammar of the Slovak language bring another problem. If the message is written without diacritics, it is possible that the chatbot will not be able to recognize the question in the knowledge base.

The QnA Maker cognitive service also has its limits, which are stationary, or change based on subscription. The stationary limits are mainly related to the creation and content of the knowledge base, e.g., question length, answer length, supported signs or number of characters per URL connected to the knowledge base. Limits varying by subscription package refer to the number of knowledge bases or the number of question–answer pairs.

For this, a chatbot would be more rational in solving the queries, a large volume of data related to the institute or university needs to be fed into the system and needs to be continuously trained using machine learning algorithms. This even allows the chatbot to answer questions that are not predefined, just based on the data available in the system. Thus, the proposed model has room for improvement. It can be transformed into an advanced model provided artificial intelligence and machine learning are incorporated into the system. It is the testing in a university environment that creates the ideal space to improve the quality of the answers.

3.6 Recommendations for the Future

The created chatbot Elora is able to answer the most frequent questions from students and thus can partially relieve the burden on the study department. Students can get it without visiting the college office. Thus, Elora increases efficiency by taking over tasks for which humans are not necessary.

To improve the current features of the chatbot for querying colleges in the future, the scope of the chatbot can be expanded by embedding data from multiple sources, training the bot with heterogeneous data, testing it on a live website, and embedding additional training data into the bot based on this feedback.

Some of the new features that can be added to the bot are:

(1) Speech recognition feature through which learners can ask their questions verbally and receive answers from the bot,
(2) integration with multiple channels such as phone call, SMS, and various social media platforms such as Skype, Facebook and Twitter,
(3) handling context and interactive questions where the bot will be aware of the context of the ongoing conversation with the student,
(4) integration with services such as password reset and enrolment of subjects or sending structured documents according to the user's details.
(5) adding the ability for the bot to perform analytics based on user information, and the bot would be able to recognize human emotions and express empathy.

A chatbot in an educational institute or university is an effective enough tool to deal with various queries from students or any end users. Chatbots can help students to get answers to their queries quickly and efficiently. They don't have to read a lengthy document, browse a web page, or wait for an email response from an administrator. They can get an immediate response, reducing wait time and improving the student experience.

Especially at peak times, such as near the end of the application period and the end of the semester, this service has the potential to be overused.

It is at this time that it would aid the study department staff. It would bring benefits in the form of improved working atmosphere, increased quality of work or productivity. Another added value would be a better connection with students and prospective students through constant contactability. If the chatbot was also implemented in a foreign language it would bring benefits to international students regarding information about studying.

With the publication of an intelligent chatbot that will bring benefits to both parties it is also possible that it will serve as an advertisement for the university.

4 Conclusion

The development of chatbots has progressed significantly in recent years and over time they are becoming one of the most powerful and widely used applications. In terms of communication, a chatbot is a perfect illustration of computer–human interaction.

Their bright future holds great promise for every area of the market, including education and academia. It won't be wrong to mention that in the not-too-distant future, perhaps most of our interactions with users will be carried out by a bot.

There is still a long way to go to make conversational agents completely unsupervised, but the future holds promise, especially in natural language processing: new methods are being born every day and older methods are getting even better. However, error-free communication robots are only at the beginning of their evolution and understanding how they work and how they are created is essential to realise their full potential.

The Covid-19 pandemic and the development of digital technologies have created significant pressure for the need to digitise. The variables have changed users' preferences as everything has shifted to the online world from one day to the next.

The main stated objective of this master's thesis was to analyze the ongoing processes in the study department of the Faculty of Management of Comenius University. As a result of this analysis, one process was selected which was the most suitable for the implementation of the communication robot. Based on the identified space in the process for improvement, the architecture of the recommended solution was created. The new solution should lead to the elimination of the overload of clerks and clerks of the study department who are "overwhelmed" with requests for information.

In the development of the theoretical part of the thesis, insights were gained that served as a basis for the development of the empirical part of the thesis.

The first chapter analysed the history and development of chatbots since their inception, which dates back to the middle of the last century. Then, definitions of chatbots from the literature were reviewed. A large part of the first chapter consists of the acquired knowledge about the division of communication bots and the analysis of the technologies that are indispensable for their proper functioning. The conclusion of the first chapter evaluates the benefits and negatives of using chatbots, as well as real-world examples of the implementation of chatbots and intelligent assistants in foreign universities.

The first part of the practical section examines the current state of processes and requirements handling. Based on one of the sub-objectives of this thesis, a time and cost analysis was performed on a selected process, which confirmed the need for a communication bot in the administrative administration of the faculty.

The content of the second part of the practical part was the development of a manual for creating a custom chatbot. The manual contained two possible processes for creating the knowledge base that forms the brain of the chatbot for answering questions. The created solution was then tested by the author of this thesis and an independent student.

Finally, we discussed the limitations that accompanied the development and implementation of the project assignment. Improving the ability to understand and generate the language is a crucial step for further development. Due to the user interface of the Bot Framework Composer and QnA Maker cognitive service used, it is possible to create more powerful human-like chatbots even without deep programming knowledge.

Recommendations for future implementation in practice and suggestions for improving the developed solution form the conclusion of our work. Based on the output of the practical part of the work, we can evaluate that we have succeeded in meeting the main goal of the work and the sub-goals set for this work.

The chatbot at the university is an effective enough tool to deal with various queries from students or any end users. However, making these bots more intelligent and rational to deal with different types of queries is in fact a challenging task.

Overall, the future-oriented theme of this work builds an approach to streamline the ongoing processes in the study department of the Faculty of Management, Comenius University. The presented topic needs to be further developed and the knowledge base needs to be expanded with a large volume of university related data and continuously trained with machine learning algorithms. This allows the chatbot to answer even questions that are not predefined. Thus, the proposed model has room for improvement and dynamic development. It can be transformed into an advanced model provided that artificial intelligence and machine learning are incorporated into the system.

It is important to mention that the authors did not and do not aim to replace the referees and clerks from the study department with a communication robot. The point was to incorporate such an intelligent solution into practice and to find a common model of benefit use for all parties involved.

References

1. Chandan, A.J., Chattopadhyay, M., Sahoo, S.: Implementing Chat-Bot in Educational Institutes. [online]. IJRAR J. **6**(2), 44–47 (2019)
2. Zahour, O., et al.: A system for educational and vocational guidance in Morocco: Chatbot E-Orientation. [online]. Procedia Comput. Sci. **175**, 554–559 (2020)
3. Bassett, C.: Apostasy in the temple of technology: ELIZA the more than mechanical therapist. [online]. In: Anti-computing. Manchester University Press, p. 168–185 (2022)
4. Van Den Berg, B.: Robots as tools for techno-regulation. [online]. Law, Innov. Technol. **3**(2), 319–334 (2011)
5. Przegalinska, A., et al.: In bot we trust: A new methodology of chatbot performance measures. [online]. Bus. Horiz. **62**(6), 785–797 (2019)
6. Shum, H.-Y., He, X.-d., Li, D.: From Eliza to XiaoIce: challenges and opportunities with social chatbots. [online]. Front. Inf. Technol. & Electron. Eng. **19**(1), 10–26 (2018)
7. Wallace, R.S.: The anatomy of ALICE. [online]. In: Parsing the Turing test, 181–210. Springer, Dordrecht, (2009)
8. Marietto, M.G.B., et al.: Artificial intelligence markup language: a brief tutorial. [online]. arXiv preprint arXiv:1307.3091 (2013)
9. Kashyap, P.: Machine learning algorithms and their relationship with modern technologies. [online]. In: Machine Learning for Decision Makers, 91–136. Apress, Berkeley, CA (2017)
10. O'Boyle Britta. What is Siri and how does Siri work. [online]. (2021)
11. Siri. [online]. Apple Inc. (2022)
12. Adamopoulou, E., Moussiades, L.: Chatbots: History, technology, and applications. [online]. Mach. Learn. Appl. **2**, 100006 (2020)
13. Watson Assistant. IBM Cloud (2020)
14. Lentzsch, C., et al.: Hey Alexa, is this skill safe? Taking a closer look at the Alexa skill ecosystem. [online]. In: 28th Annual Network and Distributed System Security Symposium, NDSS. (2021)
15. Personal Digital Assistant—Cortana Home Assistant—Microsoft, 2019. [online]. Microsoft Cortana, your intelligent assistant website
16. Kashyap, P.: Machine learning for decision makers: Cognitive computing fundamentals for better decision making. [online]. Bangalore: Apress (2017)
17. Van Den Broeck, E., Zarouali, B., Poels, K.: Chatbot advertising effectiveness: When does the message get through? [online]. Comput. Hum. Behav. **98**, 150–157 (2019)
18. Global chatbot market value 2018–2027. [online]. Statista Research Department & 17.
19. THE INSIGHT PARTNERS. Chatbot Market Size & share research by 2027. [online]. The Insight Partners (2019)
20. Ciechanowski, L., et al.: In the shades of the uncanny valley: An experimental study of human–chatbot interaction. [online]. Futur. Gener. Comput. Syst. **92**, 539–548 (2019)
21. Ruan, S., et al.: Bookbuddy: Turning digital materials into interactive foreign language lessons through a voice chatbot. [online]. In: Proceedings of the Sixth (2019) ACM Conference on Learning@ Scale, 1–4 (2019)
22. Rosruen, N., Samanchuen, T.: Chatbot utilization for medical consultant system. [online]. In: 2018 3rd technology innovation management and engineering science international conference (TIMES-iCON), 1–5. IEEE (2018)
23. Radziwill, N.M., Benton, M.C.: Evaluating quality of chatbots and intelligent conversational agents. [online]. arXiv preprint arXiv:1704.04579, (2017)
24. Følstad, A., Brandtzæg, P.B.: Chatbots and the new world of HCI. [online]. Interactions **24**(4), 38–42 (2017)
25. Araujo, T.: Living up to the chatbot hype: The influence of anthropomorphic design cues and communicative agency framing on conversational agent and company perceptions. [online]. Comput. Hum. Behav. **85**, 183–189 (2018)
26. Smutny, P., Schreiberova, P.: Chatbots for learning: A review of educational chatbots for the Facebook Messenger. [online]. Comput. Educ. **151**, 103862 (2020)

27. Mctear, M.F.: The rise of the conversational interface: A new kid on the block? [online]. In: International workshop on future and emerging trends in language technology, pp. 38–49. Springer, Cham, 2016

28. Baraka, K. Alves-Oliveira, P.; RIBEIRO, Tiago. An extended framework for characterizing social robots. [online]. In: Human-Robot Interaction. Springer, Cham, 2020. p. 21–64.

29. NIßEN, M., et al. See you soon again, chatbot? A design taxonomy to characterize user-chatbot relationships with different time horizons. [online]. Computers in Human Behavior, 2022, 127: 107043.

30. FØLSTAD, Asbjørn, et al. Future directions for chatbot research: an interdisciplinary research agenda. [online]. Computing 2021, 103.12: 2915–2942.

31. NIMAVAT, Ketakee; CHAMPANERIA, Tushar. Chatbots: An overview types, architecture, tools and future possibilities. [online]. Int. J. Sci. Res. Dev, 2017, 5.7: 1019–1024.

32. Adamopoulou, E.; Moussiades, L.: An overview of chatbot technology. [online]. In: IFIP International Conference on Artificial Intelligence Applications and Innovations, pp. 373–383. Springer, Cham (2020)

33. Hien, H.T., et al.: Intelligent assistants in higher-education environments: the FIT-EBot, a chatbot for administrative and learning support. [online]. In: Proceedings of the ninth international symposium on information and communication technology. 2018. p. 69–76.

34. Ramesh, K., et al.: A survey of design techniques for conversational agents. [online]. In: International conference on information, communication and computing technology, pp. 336–350. Springer, Singapore, 2017

35. Wu, Y., et al.: Sequential matching network: A new architecture for multi-turn response selection in retrieval-based chatbots. [online]. arXiv preprint arXiv:1612.01627, 2016.

36. Conversational AI Platform for Enterprise—Teneo, Artificial Solutions. 2022. Chatbots: the Definitive Guide 2021. [online]

37. Gapanyuk, Y., et al.: The hybrid chatbot system combining Q&A and knowledge-base approaches. [online]. In: 7th International Conference on Analysis of Images, Social Networks and Texts, p. 42–53 2018

Printed in the United States
by Baker & Taylor Publisher Services